Heat Transfer Modeling

George Sidebotham

Heat Transfer Modeling

An Inductive Approach

 Springer

George Sidebotham
Mechanical Engineering Department
The Cooper Union for the Advancement
    of Science and Art
New York, NY, USA

ISBN 978-3-319-36046-1   ISBN 978-3-319-14514-3   (eBook)
DOI 10.1007/978-3-319-14514-3

Springer Cham Heidelberg New York Dordrecht London

Springer International Publishing AG Switzerland is part of Springer Science+Business Media
(www.springer.com)

Preface

There are many excellent textbooks on Heat Transfer. Why another?

In the course of teaching Heat Transfer to junior level Chemical and Mechanical Engineering students over several years and developing reams of notes, it became difficult for me to conduct classes in parallel with a traditional textbook. The order of topics and general approach I wanted to emphasize in my courses evolved to a point that deviated too much from any textbook I could find. So I used a sabbatical leave to write my notes up in the form of a textbook.

Unlike traditional texts, I wanted to follow the adage that "Heat Transfer takes time" and consider transient problems first, with steady-state applications being treated as a limiting case. I also wanted to introduce each topic with readily recognizable applications and build quantitative methods to analyze them as needed. This perhaps controversial "just in time" approach means that the order of development deviates from the accepted norm to some degree. The goal is that by showing a few specific examples in some detail and then drawing general conclusions, the reader will be in a position to develop his or her own approach to situations that arise in his or her practice. This approach is considered to be inductive, in contrast to a deductive approach in which general principles are taught first. Heat transfer is an invisible, massless, abstract phenomenon and it may help some students to visualize its nature with a tangible example first. I credit my mother, a religious educator, for teaching me the inductive method.

Only a few problem statements are given in each chapter, some of which are analyzed in detail as part of the development of the analysis methodology (with a suggestion given for the reader to attempt a solution before proceeding). These are called workshops, not problems, and form the basis for how I teach my courses. I use a part of class to lecture and a large part of class time for students to work together on problems that use those concepts right away. My hope is that instructors that adopt this textbook will write their own problem statements or draw from the many resources available to them. Better yet, I have found that students learn a lot and often enjoy making up their own problem statements.

A simple, recognizable situation, namely a cooling mug of coffee, is introduced early with an experimental solution and revisited repeatedly. A new modeling technique is introduced with each iteration, and virtually all traditional first level heat transfer concepts are thereby developed. The patient reader acquires a toolbox that can be applied to a broad range of applications. There naturally arises, however, a certain redundancy, and some topics are treated in multiple places, but with a different twist each time. This characteristic of an inductive method is consistent with the truth that "learning is an iterative process".

All of those variations are centered on the science, the knowing and/or predicting how quickly a mug of coffee cools. However, it is not engineering in the sense that, other than informing me how long I'll need to wait (the science), some human intervention is made to address a need (the engineering). I find it curious that, in this modern high tech world, when you buy a cup of coffee, it is usually too hot to drink when you get it. So you have to wait. Or add a few ice cubes to cool (and dilute) it. Furthermore, if you get distracted, the coffee becomes tepid.

Then it was brought to my attention that a pair of visionary young entrepreneurs (Dave and Dave) had engineered a solution and had brought "Coffee Joulies™" to market. So, with their blessing, I've added sections to appropriate chapters that demonstrate the principles involved. This example provides a good application that involves phase change, something that is generally not included prominently in traditional heat transfer texts. A testimonial of mine could be "*I like Coffee Joulies so much, I wrote a book about them*!"Another common situation is the opposite, namely if I put a warm beverage in a refrigerator, how long do I wait before it's cold? This problem, and an engineering solution called the Cooper Cooling process invented by Greg Loibl, a Cooper Union masters student of mine, is treated in workshops.

While addressing key traditional fundamental concepts in heat transfer practice, the emphasis is on simplified modeling philosophies, which carry into all walks of life. If it is determined that simplified analytical solutions are considered to be insufficient to address a specific need, basic numerical methods are preferred to complex analytical solutions. The goal is to develop the art of using theory as a design tool. Theory for theory's sake can obfuscate sound engineering judgment.

The practicing engineer must navigate a diverse array of sometimes inconsistent and/or conflicting nomenclature and definitions, so a good foundation in Engineering Thermodynamics is considered a prerequisite to this textbook. In the so-called *thermal* applications, terms like energy, heat, work, and temperature have specific meanings. Heat transfer is defined as the transport of energy across a system boundary driven by a spatial temperature gradient. Heat is a transported form of energy and is always associated with a boundary that has a surface area. Other forms of energy (internal, kinetic, and potential) are "storable" and are associated with a region of space that has a volume. Temperature is not a form of energy and is difficult to succinctly define, yet everybody knows what it is (or thinks they do), some sort of measure of hotness or coldness. Work is another transported energy form and is associated with a force acting through distance, or in the case of "electrical work" of electrons moving across the boundary against an electrical potential difference.

The word "heat" in common usage frequently means something different than that formal technical definition. My wife often says, "There's no heat in this house." I know what she means, and I know to respond, "I'll turn up the thermostat. In the meantime, put on a sweater." An hour later, the house is toasty. But in my head, I might say "You're right. There is technically no heat in this house. But the air temperature is sufficiently low relative to your body temperature that the rate at which you are losing heat is uncomfortable to you. Turning up the thermostat will send a signal for a fuel valve to open and a spark to ignite a flame in the boiler, where chemical energy will be converted to thermal energy. Then heat will be transferred from the hot gases to the water in the boiler, which will create steam that will flow into the radiators (which should be called convector/radiators) throughout the house. Then heat will be transferred to the air in the room, raising its internal energy, and thereby the air temperature to the new set point." I guess it's easier to just say "There's no heat in this house."

Modeling involves the creation of a simplified view of a complex system. In general, mathematical modeling (of anything, not just heat) involves two main steps: model development and its subsequent solution. They are not always purely sequential, and feedback from one to the other and an iterative solution is usually required for a good engineering solution. Modeling becomes a design tool when that solution leads to a decision that addresses the original problem statement, such as "what material and what dimensions should that heat sink be to keep the motherboard from overheating?"

Model development involves a derivation of governing equations from the application of fundamental laws of nature to a defined system, a specific limited portion of the universe. The art of this stage of modeling is to choose clear system boundaries and to make simplifying assumptions that ease the subsequent solution, while still capturing the "controlling physics." Frequently, complex systems are controlled by a simple balance between two or three "things." A determination of the extent to which that model captures reality is context-specific and a major part of engineering decision-making. Creation of the model is not the end, it is the start. The input is a word problem statement, and the output is a mathematical problem statement.

The next step is to solve the mathematical problem. This step can be trivial, easy, hard, or impossible, and there may be multiple approaches. It is good design to invoke a feedback response, a return to the model development with an eye toward either making additional simplifications or relaxing others, and ultimately arrive at a solution that is deemed adequate to address the original word statement. It is common to lose sight of the original problem statement once a detailed analysis is initiated.

The approach of this book is to start with a word problem statement (often with specific parameter values) and to develop *A* modeling strategy in detail. Not *THE* modeling strategy. A patient and resourceful reader will develop alternative models to the same initial problem statement, solve those, and then compare (qualitatively and quantitatively) to the examples given in this text.

Once they are implemented (and generally not before), models can be tested and then defined as being simple, simplistic, or complex. The testing criterion depends on the problem at hand and may involve comparison with experiment, parametric study (observe the effect of various parameters), or testing assumptions after the fact. Generally, a simple model is most desirable. It does the job with the least effort (and captures what is most important). A simplistic model misses something important. However, a legitimate, perhaps counterintuitive modeling strategy is to start as simple as possible, then determine whether the model is simplistic. If so, a more detailed model, perhaps guided by the first model, can be developed and then tested.

However, there appears to be a natural tendency to approach things from the opposite way, namely to make a very complex model right at the start. That first instinct can do more harm than good. How complex should a model be? Only as complex as it needs to be, no more. In plumbing, there appears to be a natural tendency to tighten fittings as much as you can, thinking that's better. But you can damage fittings that way and MAKE them leak. So how tight should they be? Only as tight as they need to be (perhaps a bit more). In other words, a complex model may seem to be better than a simple one. However, dealing with the complexity can be costly in time, effort, and money. Worse, the inherent beautiful simplicity of the underlying controlling aspects of problem that often (usually?) exist may be missed.

The general philosophical approach taken in this text is to start with as simple a model as possible (usually a "1-Node Model"), learn about the big picture from it, and then decide whether a more complicated model is needed to do the job. This strategy can be used in most (if not all) aspects of life. Numerical methods involve discretizing time to make a prediction of the future and/or discretizing space to create multinode models. Mathematically, differential equations are converted to algebraic equations. This textbook therefore is intended to be an introduction to numerical methods in general and a start toward computational fluid dynamics (CFD), an important and growing discipline.

The intention in this text is to favor an inductive over a deductive approach, in which the study of a specific case leads to a generalization that can be applied to other situations. Most engineering textbooks use a deductive approach in which general principles are developed and then applied to specific cases. Neither approach is inherently better than the other. They are just different tools.

All of the problems presented in the main body are solved using EXCEL spreadsheets, without writing VBA code. All of the problems could also be conducted by writing code in other programs. A few of the problems encountered really should NOT be solved using spreadsheets. A general rule is that if a nested loop is fundamentally required for a problem, program code is probably a better approach, otherwise whatever tool is most comfortable. It is really good practice developing programming skills to reproduce some of the solutions presented as part of the main text.

This textbook, targeted to upper level engineering students (but accessible to all life-long learners), is intended to be inexpensive to the student budget and a nurturing challenge to their intellect. It is dedicated to the Cooper Union students that I have had the honor and privilege of serving over the years. I feel blessed to have been given a chance to influence, in some incremental way, hundreds of young women and men on their way to becoming enlightened and loving human beings (through the intense development of engineering skills).

NOTE TO STUDENTS

This book is written for you. My hope is that many of you will gain confidence and apply the techniques developed here to situations that you define. You can have fun and challenge yourself at the same time.

NOTE TO INSTRUCTORS

My challenge to you is to draw from your experience and apply it using principles developed in this text. If you have taught Heat Transfer many times before, maybe you will see some things in a new light. The only missing technical element is the case of thermal radiation between opaque surfaces, which I never really can get to in a first course. I do emphasize radiation concepts throughout, however. My advice is to create in-class and homework-based group activities that use numerical methods early, even before students may have reached those topics in a straight reading of the text.

NOTE TO LIFE-LONG LEARNERS

If you've already taken a course in Heat Transfer and especially if your practice involves heat transfer, I hope that this textbook will help you hone your skills and begin to simplify your models and clarify your thought.

New York, NY, USA   George Sidebotham

Contents

Part III   Steady-State Conduction

Part IV Open Systems: Convection and Heat Exchangers

Part I

Modes of Heat Transfer

Thermal Circuits

1

This introductory chapter develops the concept of a thermal circuit and analytical and numerical methods used to analyze simple circuits. Subsequent chapters will apply and build on them using heat transfer examples from common experience. Table 1.1 introduces key elements, namely, resistors, capacitors, batteries, and current sources. Resistive/capacitive (RC) network diagrams are used as modeling tools to visualize the flow of heat, an abstract massless quantity. Hence, the term "thermal circuit." The focus of this book is to develop heat transfer modeling as a design tool, rather than as a pure science. The principles apply to other areas in engineering (and life) practice. Specific tools are developed in this textbook that pertain to heat transfer phenomena, most of which can be captured in thermal circuits.

1.1   RC Circuit Driven by a Temperature Source

The discharge of a capacitor through a resistor is modeled by the circuit of Fig. 1.1. This circuit is a model for the heating or cooling of an object placed suddenly into an environment at a different temperature, like a hot mug of coffee placed on a table, or a warm beverage placed in a refrigerator.

An electrical capacitor stores electrical charge (Q) on two plates. The capacitance (C) relates stored charge to voltage between the plates ($Q = C(V - V_{ref})$, where V_{ref} is a reference voltage, typically ground). A thermal capacitor stores thermal energy. For systems that do not experience phase change or chemical (or nuclear) reactions, the thermal capacitance relates the stored energy to the temperature ($E = C(T - T_{ref})$), where the stored energy is set to zero at some reference temperature, T_{ref}. An electrical resistor provides a pathway for charge to flow, driven by a voltage difference across it. The larger the resistance value, the larger the voltage difference required to move a given quantity of charge. Thermal resistors provide a pathway for heat to flow, heat being a specific form of energy that crosses a system boundary, driven by a temperature difference.

© Springer International Publishing Switzerland 2015

G. Sidebotham, *Heat Transfer Modeling: An Inductive Approach*,

DOI 10.1007/978-3-319-14514-3_1

Table 1.1 Circuit components and associated units in electrical and thermal applications

Component	Symbol	Electrical SI unit	Thermal SI unit
Resistor		Ohm	$\dfrac{^\circ\text{C}}{\text{Watt}}$
Capacitor		Farad	$\dfrac{\text{Joule}}{^\circ\text{C}}$
Battery		Volt	$^\circ$C
Current source		Ampere	Watts

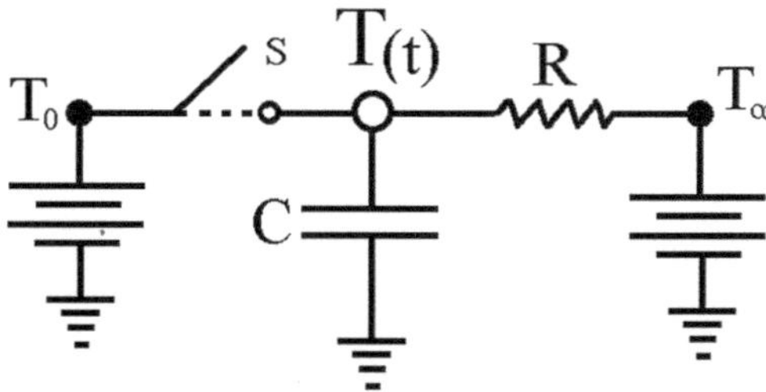

Fig. 1.1 RC circuit initiated by a temperature source (battery)

A major focus of this book is to develop estimation methods for thermal capacitance and resistance values based on the three modes of heat transfer (conduction, convection, and radiation). Heat is driven from one temperature to a lower temperature through a medium which impedes the flow of heat, modeled as a thermal resistance (R) to the flow. The ability of a material to store internal energy is determined by its thermal capacitance, C. The higher the capacitance of a material, the lower the change in its temperature when a quantity of heat is added or removed from it.

Temperature sources (batteries) are locations where the temperature is fixed, regardless of the rate at which energy is extracted in the form of heat. The components of Fig. 1.1 are two temperature sources (T_0 and T_∞), a switch (S) a resistor (R) and a capacitor (C). The normally open switch is momentarily closed (dashed lines), long enough to charge the capacitor (C) to the battery temperature (T_0), and then released. In response, heat (analogous to electrical current) flows from the capacitor through the resistor (R) to a second temperature source (T_∞), thereby reducing the thermal energy stored in the capacitor (analogous to stored electrical charge). Eventually, a final equilibrium state is reached at which point the capacitor is fully discharged. Theory is developed next that describes the response of the circuit given the fundamental laws of nature, key assumptions, and input parameter values.

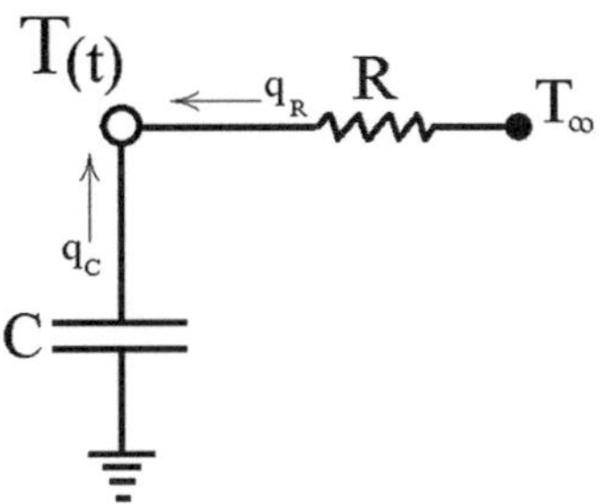

Fig. 1.2 Isolation of the main node of Fig. 1.1, after activation by the switch. The *arrows* show the direction assumed for current flow used in Kirchhoff's current law

In this text, a large filled "node" is generally used to indicate a temperature value that is fixed by a battery of known temperature (which may change with time in a prescribed way). The battery symbol will generally not be shown. For transient problems, an open "node" is one to which thermal capacitance is assigned, while small filled symbols (none in this case) indicate nodal temperatures that can be determined based on neighboring capacitive-carrying and fixed nodal temperatures.

The fundamental law of nature governing this case is the conservation of charge for the electrical case (the net flux of current into a node is zero: Kirchhoff's current law) and the First Law of Thermodynamics for the thermal case (developed formally in the next chapter). Figure 1.2 shows the circuit after the starting point, with current (q, with units of energy per unit time in the thermal case) assumed to flow into the node from both the capacitor and the resistor.

Mathematically, this charge balance, or Kirchhoff's current law, is:

$$q_R + q_C = 0$$

The current through the resistor is related to the voltage (T) by Ohm's law, namely:

$$q_R \equiv \frac{(T_\infty - T)}{R}$$

The current is driven by the voltage difference across the resistor. The larger the resistance, the lower the current. The (perhaps) familiar equation $V = IR$ is Ohm's Law, written in the form here of $I = V/R$, which is really an unfortunate misrepresentation of the relationship between voltage and current across a resistor. It should be $\Delta V = IR$, where ΔV is the voltage difference across the resistor, not the absolute value of the voltage at one specific point.

To establish the relationship between capacitor voltage and charge, consider an event of time duration Δt, during which time a quantity of charge, Q_{exit}, flows out. Conservation of charge stated in words is:

$$\text{Initial Charge} = \text{Final Charge} + \text{Charge that Exited}$$

$$Q_{initial} = Q_{final} + Q_{exit}$$

Rearranging and dividing by the time elapsed:

$$\frac{Q_{final} - Q_{initial}}{\Delta t} = \frac{\Delta Q}{\Delta t} = \frac{-Q_{exit}}{\Delta t}$$

In the limit of short duration events ($\Delta t \rightarrow 0$), the left hand side becomes a first derivative, and the right hand side represents the rate at which charge flows out, the electrical current, q_c:

$$\frac{dQ}{dt} \equiv -q_c$$

The total charge stored in the capacitor is related to the temperature by:

$$Q \equiv -C(T - T_{ref})$$

Taking the derivative and assuming the capacitance does not change with time:

$$\frac{dQ}{dt} \equiv -C\frac{dT}{dt}$$

The conservation of charge applied to a thermal capacitor is therefore:

$$C\frac{dT}{dt} = -q_c$$

Care has been taken in defining the signs. For the capacitor, a positive q_c represents current leaving the capacitor (and entering the node) which results in a decrease in temperature, accounted for by the negative sign. For heat to flow into the resistor, the temperature difference is expressed as that of the neighbor (T_∞) subtracted from the capacity-carrying nodal value (T). If that nodal temperature is higher than its neighbor, this term will have a negative value.

Inserting these component laws into the charge balance yields the governing first-order ordinary differential equation (ODE) for the circuit:

$$-C\frac{dT}{dt} + \frac{(T_\infty - T)}{R} = 0$$

with an initial condition $T(0) = T_0$.

This governing equation (and its initial condition) is the result of the model, yet it is not the solution to the problem. Some calculus (and the assumptions that R, C, and T_∞ are constant) is needed to solve it.

It is not necessary, but can be helpful, to define a dimensionless variable for temperature that starts at a value of 1 and approaches a value of 0 at equilibrium (when $T = T_{eq}$).

$$\theta \equiv \frac{T - T_{eq}}{T_0 - T_{eq}}$$

Formally, equilibrium, or steady-state, is achieved when nothing changes with time, accomplished by setting the time derivative to zero:

$$\frac{dT}{dt}\bigg|_{eq} = \frac{(T_\infty - T_{eq})}{RC} = 0$$

Since R and C are finite, the equilibrium temperature is therefore equal to the source T_∞. For more complicated circuits, it may be more difficult to solve for an equilibrium value.

A dimensionless time can be defined by normalizing the time by a characteristic time constant (τ^*), whose value will be determined shortly:

$$\hat{t} \equiv \frac{t}{t^*}$$

Inserting these variables into the governing ODE and rearranging:

$$-\frac{C}{t^*} \frac{d\big[T_{eq} + (T_0 - T_{eq})\theta\big]}{d\hat{t}} + \frac{(T_\infty - \big[T_{eq} + (T_0 - T_{eq})\theta\big])}{R} = 0$$

Since T_{eq} is constant, its time derivative is zero, and $T_{eq} = T_0$ so the governing differential equation becomes:

$$\frac{d\theta}{d\hat{t}} + \left(\frac{t^*}{RC}\right)\theta = 0$$

It is now natural to define the time constant as

$$t^* \equiv RC$$

The values of R and C have been assumed to be constant (technically, their product is constant). There are situations where R and/or C may vary with time, and therefore this model would not apply.

The governing equation becomes devoid of all parameters:

$$\frac{d\theta}{d\hat{t}} + \theta = 0$$

with initial condition $\theta(0) = 1$.

In a first-order system (i.e., governed by a linear first-order differential equation), the rate of change of some quantity is proportional to the negative of itself.

The solution to this linear first-order differential equation, plotted in Fig. 1.3, is

$$\theta = e^{-\hat{t}}$$

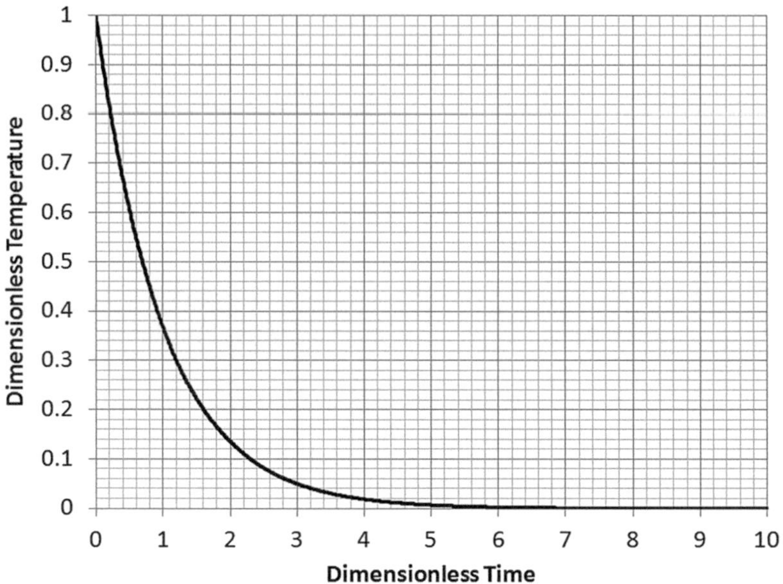

Fig. 1.3 Solution of the RC circuit in dimensionless variables

Or, in terms of dimensional variables:

$$\frac{T - T_\infty}{T_0 - T_\infty} = e^{\frac{-t}{RC}}$$

Alternatively, this equation can be expressed in terms of the time by taking the natural logarithm of both sides and rearranging:

$$t = -RC \ln\left(\frac{T - T_\infty}{T_0 - T_\infty}\right)$$

Example

Q: How much time is needed for the temperature to drop from an initial value of 100 °C to a final value of 70 °C, given that the ambient (T_∞) is at 25 °C and the time constant RC has a value of 1.613 h?

A: $time = -RC \ln\left[\frac{T - T_\infty}{T_0 - T_\infty}\right] = -1.613 \ln\left[\frac{70 - 25}{100 - 25}\right] = 0.819\,h$

In this case, the dimensionless temperature has a value of 0.6, and, referring to Fig. 1.3, the corresponding dimensionless time is approximately ½.

To generalize, in one time constant, the dimensionless temperature drops to within $1/e = 0.368$ of its initial value, or it is about 2/3 of the way to equilibrium. After two time constants, it is close to 10 % of its equilibrium value. A value of 5 time constants is a generally accepted value to indicate that equilibrium has been reached, for most engineering purposes.

1.2   RC Circuit Driven by a Current Source

Figure 1.4 shows an RC circuit driven by a current (heat) source. This circuit is a model for an object, initially at equilibrium with its environment, suddenly exposed to a radiant heat source, like parking a car in the sun.

In contrast to a voltage source (battery) which fixes the voltage regardless of the current flow in (or out), a current source fixes the current (heat transfer rate) regardless of the voltage (temperature). Initially, with the switch open, the voltage (T) is at the fixed value T_∞ (fixed by a battery, not shown, but implied by the filled symbol). At some initial time, the switch is held closed, and current from the source, $\dot{Q} = q_s$ as indicated, flows steadily into the node. Current sources in thermal systems can include chemical reactions, certain forms of radiation (i.e., solar radiation) or dissipation of electrical energy.

Kirchhoff's current law for this circuit is:

$$q_S + q_R + q_C = 0$$

Introducing the component expressions:

$$\dot{Q} + \frac{T_\infty - T}{R} - C\frac{dT}{dt} = 0$$

with an initial condition $T(0) = T_\infty$.

As before, a dimensionless temperature can be defined that starts at an initial value of unity, and is driven toward an equilibrium value:

$$\theta \equiv \frac{T - T_{eq}}{T_0 - T_{eq}}$$

The equilibrium value is obtained by setting the time derivative equal to zero:

$$C\frac{dT}{dt}\bigg|_{eq} = \frac{\left(T_\infty - T_{eq}\right)}{R} + \dot{Q} = 0$$

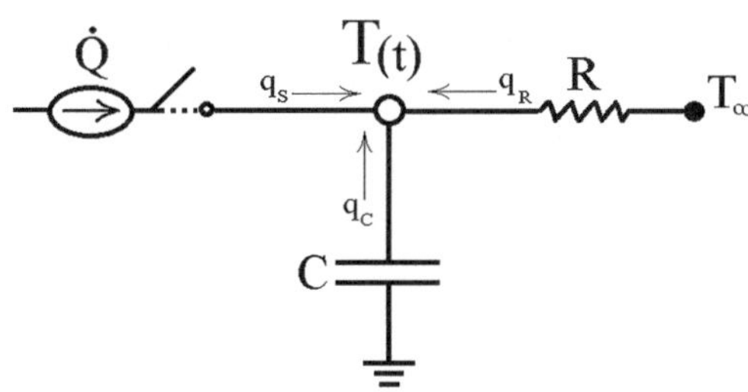

Fig. 1.4 RC circuit driven be a current source

Solving for the equilibrium temperature:

$$T_{eq} = T_\infty + R\dot{Q}$$

R and $\dot{Q}$ are inherently positive quantities, and therefore the equilibrium temperature is above ambient. A dimensionless time can be defined by normalizing the time by a characteristic time (t^*), whose value will be determined shortly:

$$\hat{t} \equiv \frac{t}{t^*}$$

Inserting these variables into the governing equation and rearranging:

$$\frac{C(T_0 - T_{eq})}{t^*}\frac{d\theta}{d\hat{t}} = \left[\dot{Q} + \frac{T_\infty - T_{eq}}{R}\right] - \frac{(T_0 - T_{eq})\theta}{R} = 0$$

The term in brackets is identically zero from the definition of the equilibrium temperature, and therefore, upon further rearrangement:

$$\frac{d\theta}{d\hat{t}} + \left(\frac{t^*}{RC}\right)\theta = 0$$

The same time constant as before naturally emerges, namely:

$$t^* \equiv RC$$

The governing equation becomes:

$$\frac{d\theta}{d\hat{t}} + \theta = 0$$

with initial condition $\theta(0) = 1$.

This governing equation is exactly the same as for the voltage source. The only difference is in the value of the equilibrium temperature. The underlying physics of a voltage source and current source are very different, however.

1.3   RC Circuit Driven by Multiple Sources

Figure 1.5 shows an RC circuit in which a node (at T) is driven by a current source and two separate voltage sources (at T_1 and T_2, with resistance R_1 and R_2). Kirchhoff's current law for this circuit is:

$$q_S + q_{R1} + q_{R2} + q_C = 0$$

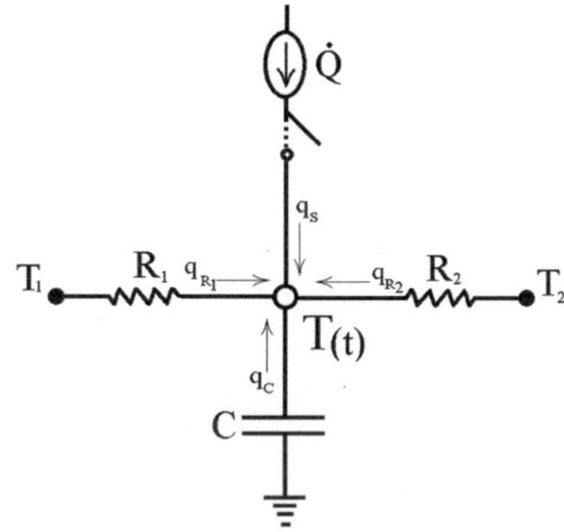

Fig. 1.5 RC circuit driven by multiple sources

Introducing the component expressions:

$$\dot{Q} + \frac{T_1 - T}{R_1} + \frac{T_2 - T}{R_2} - C\frac{dT}{dt} = 0$$

with an initial condition $T(0) = T_0$, where T_0 can be taken to be any initial temperature. If, as suggested by Fig. 1.5, the switch is initially open for a sufficiently long time prior to closing, the initial temperature will be between T_1 and T_2. Precisely where depends on the relative values of R_1 and R_2, as will be clear in the next section.

A general form of Kirchhoff's current law for a node "i" in thermal contact with multiple neighbors "j" and a current source is:

$$C_i \frac{dT_i}{dt} = \dot{Q}_i + \sum_j \frac{T_j - T_i}{R_{ij}}$$

where R_{ij} is the thermal resistance between node "i" and "j". As before, a dimensionless temperature can be defined that starts at an initial value of unity, and is driven toward an equilibrium value:

$$\theta \equiv \frac{T - T_{eq}}{T_0 - T_{eq}}$$

The equilibrium value is obtained by setting the time derivative equal to zero:

$$C\frac{dT}{dt}\bigg|_{eq} = \dot{Q} + \frac{\left(T_1 - T_{eq}\right)}{R_1} + \frac{\left(T_2 - T_{eq}\right)}{R_2} = 0$$

Solving for the equilibrium temperature:

$$T_{eq} = \frac{\dot{Q} + \dfrac{T_1}{R_1} + \dfrac{T_2}{R_2}}{\dfrac{1}{R_1} + \dfrac{1}{R_2}}$$

A general form of the equilibrium temperature for a node "i" in thermal contact with multiple neighbors "j" and a current source is:

$$T_{i,eq} = \frac{\dot{Q}_i + \sum\limits_{j} \dfrac{T_j}{R_{ij}}}{\sum\limits_{j} \dfrac{1}{R_{ij}}}$$

A dimensionless time can be defined by normalizing the time by a characteristic time (τ), whose value will be determined shortly:

$$\hat{t} \equiv \frac{t}{t^*}$$

Inserting these variables into the governing equation and rearranging:

$$\frac{C(T_0 - T_{eq})}{t^*} \frac{d\theta}{d\hat{t}} = \left[\dot{Q} + \frac{T_1 - T_{eq}}{R_1} + \frac{T_2 - T_{eq}}{R_2}\right] - \left(\frac{1}{R_1} + \frac{1}{R_2}\right)(T_0 - T_{eq})\theta = 0$$

The term in brackets is identically zero from the definition of the equilibrium temperature, and therefore, upon further rearrangement:

$$\frac{d\theta}{d\hat{t}} + \frac{t^*}{C}\left(\frac{1}{R_1} + \frac{1}{R_2}\right)\theta = 0$$

It now makes sense to define the time constant as:

$$t^* \equiv \frac{C}{\left(\dfrac{1}{R_1} + \dfrac{1}{R_2}\right)}$$

Or defining an equivalent resistance as:

$$\frac{1}{R} = \frac{1}{R_1} + \frac{1}{R_2}$$

Then

$$t^* \equiv RC$$

Notice that the time constant does not involve the current source. The governing equation becomes:

$$\frac{d\theta}{d\hat{t}} + \theta = 0$$

with initial condition $\theta(0) = 1$.

This governing equation is exactly the same as for the temperature source. The only difference is in the value of the equilibrium temperature, and an equivalent resistance rather than a single resistance.

1.4   Resistors in Series

In heat transfer, adding layers of insulation adds thermal resistance in series. For example, putting on a sweater introduces an additional barrier to heat flow from the torso and arms to the environment. Figure 1.6 shows a simple circuit, called a voltage divider, in which two resistors are wired in series, separated by two temperature sources (batteries).

An equivalent circuit with a single resistor (R) between the two batteries is also shown. The relationship between R and the individual resistors R_1 and R_2 is sought.

There is only one channel for current to flow from T_1 to T_2 (if T_1 is greater than T_2, otherwise the direction is reversed). Conservation of charge dictates that the current flowing through both resistors must be the same.

For the equivalent circuit, the current through the equivalent resistor (R) must equal the current entering R_1, consistent with the directions of the arrows. That is, for resistors added in series:

$$q_{12} = q_{R1} = q_{R2}$$

where q_{12} represents the current from T_1 to T_2. Inserting the component laws for R_1 and R_2 separately, making the signs consistent with Fig. 1.6:

$$q_{R1} = q_{12} = \frac{T_1 - T}{R_1}$$

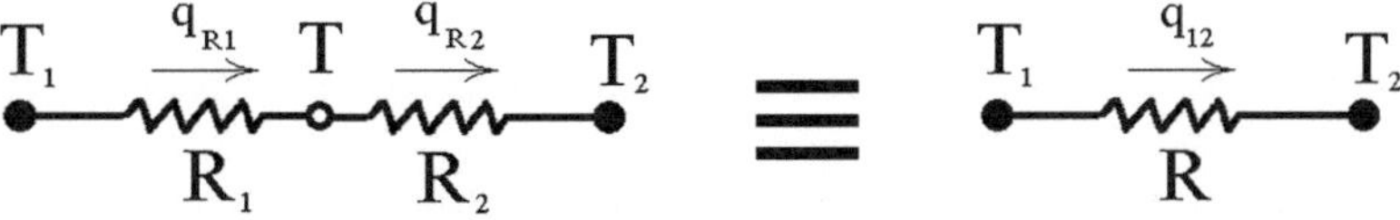

Fig. 1.6 A channel with two resistors in series, and the equivalent circuit

$$q_{R2} = q_{12} = \frac{T - T_2}{R_2}$$

Expressing these equations in terms of the intermediate temperature (T) between the two resistors:

$$T = T_1 - R_1 q_{12}$$

$$T = T_2 + R_2 q_{12}$$

Subtracting these two equations:

$$0 = T_1 - T_2 - (R_1 + R_2)q_{12}$$

Rearranging for the heat transfer rate:

$$q_{12} = \frac{T_1 - T_2}{R_1 + R_2}$$

For the equivalent circuit:

$$q_{12} = \frac{T_1 - T_2}{R}$$

The equivalent resistance is therefore:

$$R = R_1 + R_2$$

The equivalent resistance of resistors in series is the sum of the resistances. The currents are equal through the resistors. The intermediate temperature can be expressed in a number of ways. Rearranging for the intermediate temperature three different (but equivalent) ways:

$$T = \frac{\dfrac{T_1}{R_1} + \dfrac{T_2}{R_2}}{\dfrac{1}{R_1} + \dfrac{1}{R_2}}$$

$$T = T_1 - \left(\frac{R_1}{R_1 + R_2}\right)(T_1 - T_2)$$

$$T = T_2 + \left(\frac{R_2}{R_1 + R_2}\right)(T_1 - T_2)$$

Adding resistors in series impedes current flow. The voltage in between the two resistors is intermediate. If the resistors are equal in value, the intermediate temperature will be halfway between T_1 and T_2. If R_1 is greater than R_2, a larger voltage drop will be required across R_1 than across R_2, and the intermediate temperature will be closer to T_2 than to T_1.

Example   Let $T_1 = 100\,°C$ and $T_2 = 25\,°C$ and let $R_1 = 10\,°C/W$ and $R_2 = 30\,°C/W$. Calculate (a) the equivalent resistance, (b) the current, and (c) the intermediate temperature for the series resistance circuit of Fig. 1.6.

(a) The equivalent resistance is $R = R_1 + R_2 = (10 + 30)\,°C/W = 40\,°C/W$. Notice that the equivalent resistance is larger than either of the two resistors.
(b) The current is

$$q_{12} = \frac{T_1 - T_2}{R_1 + R_2} = \frac{100 - 25}{10 + 30} = 1.875\,W$$

(c) The intermediate temperature is

$$T = T_1 - \left(\frac{R_1}{R_1 + R_2}\right)(T_1 - T_2) = 100 - \left(\frac{10}{10 + 30}\right)(100 - 25) = 81.25°C$$

1.5   Resistors in Parallel

In winter, heat loss through an exterior wall takes places across its windows and the remainder of the wall simultaneously. The rate of heat loss through the windows is different than that through the solid wall, even though the air temperatures are the same inside and outside. The total heat loss across the wall is the sum of these two "thermal channels," and these channels act in parallel.

Figure 1.7 shows two resistors wired in parallel between two batteries, and an equivalent circuit with a single resistor, whose relationship to R_1 and R_2 is sought. Unlike the series circuit, there are two distinct channels through which current can flow. In a heat transfer application, heat lost through a wall of a building takes place through two parallel channels, through the windows and through the remainder of the walls.

A nodal balance at either side of the circuit states that the total current entering (or leaving) the circuit is the sum of the currents flowing through each resistor, and the voltage drop is the same across both resistors:

$$q_{12} = q_{R1} + q_{R2}$$

Inserting component laws:

$$\frac{T_1 - T_2}{R} = \frac{T_1 - T_2}{R_1} + \frac{T_1 - T_2}{R_2}$$

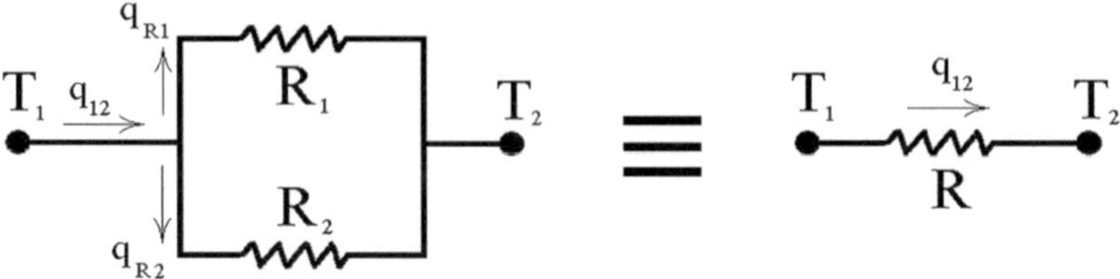

Fig. 1.7 RC channel with two resistors in parallel, and the equivalent circuit

Eliminating the common voltage drop:

$$\frac{1}{R} = \frac{1}{R_1} + \frac{1}{R_2}$$

The equivalent resistance can be expressed as:

$$R = \frac{1}{\frac{1}{R_1} + \frac{1}{R_2}}$$

or

$$R = \frac{R_1 R_2}{R_1 + R_2}$$

The equivalent resistance is less than the smaller of the two resistances. In heat transfer, heat flows through a wall in a house through parallel channels, i.e., across solid walls and across windows. The inside and outside temperatures are the same for windows and solid walls, and the total heat transfer rate is the sum of the heat through these parallel channels.

Example Let $T_1 = 100\,^{\circ}\text{C}$ and $T_2 = 25\,^{\circ}\text{C}$ and let $R_1 = 10\,^{\circ}\text{C/W}$ and $R_2 = 30\,^{\circ}\text{C/W}$. Calculate (a) the equivalent resistance and (b) the current for the parallel resistance circuit of Fig. 1.7. Compare the results to the series circuit that had the same batteries and resistors.

(a) The equivalent resistance is

$$R = \frac{R_1 R_2}{R_1 + R_2} = \frac{(10)(30)}{10 + 30} = 7.5\,^{\circ}\text{C/W}$$

The resistance is less than the smaller of the two, while the resistance of the series circuit was $40\,\Omega$, greater than the larger of the two.

(b) The current is

$$q_{12} = \frac{T_1 - T_2}{R} = \frac{(100 - 25)^{\circ}\text{C}}{7.5^{\circ}\text{C/W}} = 10\,\text{W}$$

This current is much larger than the 1.875 W for the series circuit.

1.6   Iterative Solution Methods

"I'm thinking of a number from 1 to 100"

50

Higher

75

Lower

65

Lower

62

Correct. You read my mind!

This dialogue is an example of an iterative method. There is a correct solution, but not an obvious way to get to it directly. So a feedback loop is used in which a guess is made, followed by feedback, a subsequent guess using an algorithm, and eventually, a fairly rapid convergence to the correct solution is found. In this case, the algorithm is "guess a value that is midway between the latest high and low values" (This is the so-called regula-falsi method in which the correct result is initially bracketed, and then methodically arrived at by closing in on it).

There are other algorithms that may or may not work. For example. . .
"I'm thinking of a number from 1 to 100"
1. Higher. 2. Higher. 3. Higher. 4. Higher. . . (etc etc. . .) . . .61. Higher. 62. Correct! You read my mind!

This algorithm, start at 1 and count up, found the same answer, not as quickly, but so what, if the only objective is the right answer. A brute force algorithm like this is perfectly legitimate in many instances.

Here is an algorithm that is not likely to work.
"I'm thinking of a number from 1 to 100"
40. Higher. 40 Higher. 40. Higher. Etc. . .

The algorithm here is "always guess the same number." It would only work if the first guess happened to equal the correct answer.
Here is an algorithm in which the feedback is useless.

"I'm thinking of a number from 1 to 100"
50
Wrong!
25
Wrong!
. . ..

The guesser will eventually find 62, but not because of the feedback.

Iterative procedures are quite common in heat transfer applications. There are formal methods that will be introduced for specific types of problems, and iterative procedures can by "invented" as needed to solve problems for which a direct solution is not possible. You'll see.

There is a fundamental difference between an iterative and predictive method. In an iterative method, guesses are made, based on feedback, and hopefully the algorithm works in the sense that the desired final solution is obtained. The "trajectory" of the guessed solutions is really not important. It is the final result that matters. In a predictive method, each step in time is a prediction of what the actual temperature is. The prediction of the temperature after a finite time step is not a "guess," it is a simulation based on the model. Some important predictive numerical methods are introduced next.

1.7   Numerical Solution 1: Forward Explicit Formulation (Euler Method)

A few techniques for solving transient problems numerically are introduced in this introductory chapter, using the basic first-order system to demonstrate the pros and cons of each method, namely,

$$\frac{dT}{dt} = \frac{(T_\infty - T)}{RC}$$

$$Initial\,Condition: \quad T(0) = T_0$$

Note that the values of R and C could be variables.

The derivative in the governing equation can be expressed as a forward finite difference approximation by taking a finite step in time (Δt), using a superscript to represent time as p (present) or p + 1 (future). There is no past in this method.

$$\frac{T^{(p+1)} - T^{(p)}}{\Delta t^{(p)}} = \frac{T_\infty^{(p)} - T^{(p)}}{R^{(p)} C^{(p)}}$$

The right hand side represents the rate of heat transfer (divided by C) at the present time. Rearranging and solving for the future temperature in terms of the present:

$$T^{(p+1)} = T^{(p)} + \frac{\Delta t^{(p)}}{R^{(p)} C^{(p)}} \left(T_\infty^{(p)} - T^{(p)} \right)$$

with initial condition $T^{(0)} = T_0$. In effect, this method extrapolates the slope forward into time by a finite step (the numerical parameter Δt), without any adjustments during that step. Notice that in the limit of Δt going to zero, the Euler method becomes the definition of the first derivative.

The unit of the RC combination is time, and is a characteristic time constant of the problem. Engineering decisions can often be made knowing no more than that.

This Euler method is called an "explicit" method because the future predicted value is expressed explicitly in terms of present values, which are known, starting with the initial condition. Such will also be true for more complicated models (with more than one unknown nodal temperature). It is easy, and somewhat intuitive, to implement, and is very powerful. It can run into problems as systems become more complex, but a legitimate strategy is to implement the method, and if it fails to produce acceptable results, it can dictate an appropriate "Plan B."

"If you know the present, you can tell the future." Of course, you only ever know so much of the present.

1.7.1   Variable Step Size Explicit Methods

The step size (Δt) is a numerical parameter. It is relatively easy to incorporate logic that calculates a new value for the step size at every step. Three variations of this method are:

1.7.1.1 Fixed Step

A fixed step size method is one where the step size is chosen at the start of the simulation and used throughout:

$$\Delta t^{(p)} = \Delta t$$

This method seems to be the most obvious and is recommended as the methods are being learned, yet it can have its drawbacks in practice. The change in the independent variable is held constant in this method.

1.7.1.2 Fixed Change

A less obvious, but occasionally useful, method is to allow the dependent variable to change by a fixed amount:

$$T^{(p+1)} - T^{(p)} = \Delta T$$

This method works well if it is known that the temperature change will monotonically change from beginning to end, which is often the case in heat transfer. The step size at each step is given by:

$$\Delta t^{(p)} = \frac{R^{(p)} C^{(p)} \Delta T}{T_\infty^{(p)} - T^{(p)}}$$

1.7.1.3 Fixed Fractional Change

In this method, the step size is calculated that will result in a change in the dependent variable by a fixed fraction, f, of its deviation from its equilibrium value T_{eq}, in this case T_∞:

$$T^{(p+1)} - T^{(p)} = f\left(T^{(p)} - T_{eq}^{(p)}\right)$$

The step size for this method is then:

$$\Delta t^{(p)} = f R^{(p)} C^{(p)} \left|\frac{T_{eq}^{(p)} - T^{(p)}}{T_\infty^{(p)} - T^{(p)}}\right|$$

Absolute values are needed to guarantee a positive step size. For the RC circuit studied for now, R and C are constants, and the equilibrium value for T is T_∞, making the temperature ratio equal to unity. Therefore, this method will result in a fixed step size for this case, but not in general.

Also, this method works best, conceptually, for dependent variables that are inherently positive, like absolute temperature, as opposed to temperature expressed in Celsius. For example, a 1 °C change in a temperature of 100 °C is a 1 % change, whereas the same 1 °C change in a temperature of 10 °C is a 10 % change. What if the temperature were 0 °C? or -10 °C?

1.8   Numerical Solution 2: Backward Implicit formulation

Express the derivative in the governing equation as a backward finite difference approximation by taking a finite step in time (Δt), and introduce a superscript to represent time as p (present) or $p - 1$ (past). There is no future, yet...

$$\frac{T^{(p)} - T^{(p-1)}}{\Delta t} = \frac{T_\infty - T^{(p)}}{RC}$$

Rearranging and solving for the present temperature in terms of the past:

$$T^{(p)} = \frac{T_\infty + \left(\dfrac{RC}{\Delta t}\right) T^{(p-1)}}{1 + \left(\dfrac{RC}{\Delta t}\right)}$$

with initial condition $T^{(0)} = T_0$.

"The present is a result of the past." But we want to predict the future! Let the past become the present, and the present the future by updating "p" to "p + 1";

$$T^{(p+1)} = \frac{T_\infty + \left(\dfrac{RC}{\Delta t}\right) T^{(p)}}{1 + \left(\dfrac{RC}{\Delta t}\right)}$$

It is not clear why this method is called an implicit method in this "1-node" example because the future predicted value (p + 1) is expressed explicitly in terms of present values (p). That will not be true for more complicated models (with more than one unknown nodal temperature).

1.9   Numerical Solution 3: Central Implicit formulation (Crank–Nicolson Method)

This method is introduced here, but it is not recommended that it be used in heat transfer applications, because it violates the Second Law of Thermodynamics for large step sizes, whereas the backward method yields correct equilibrium values. There are other applications in numerical analysis for which the improved accuracy of this method are advantageous.

Express the derivative in the governing equation as a central finite difference approximation by taking a half finite step in time ($\Delta t/2$), and introduce a superscript to represent time as $p - 1/2$ (past) or $p + 1/2$ (future). There is no present, yet...

$$\frac{T^{(p+\frac{1}{2})} - T^{(p-\frac{1}{2})}}{\Delta t} = \frac{T_\infty - T^{(p)}}{RC}$$

There are three times depicted, past, present and future. To reduce that to two, approximate the present as the average of the past and future:

$$\frac{T^{(p+\frac{1}{2})} - T^{(p-\frac{1}{2})}}{\Delta t} = \frac{T_\infty - \dfrac{T^{(p+\frac{1}{2})} + T^{(p-\frac{1}{2})}}{2}}{RC}$$

Let the past become the present, and the future further into the future by updating "p" to "p + 1/2". Rearranging and solving for the future temperature in terms of the present:

$$\frac{T^{(p+1)} - T^{(p)}}{\Delta t} = \frac{T_\infty - \dfrac{T^{(p+1)} + T^{(p)}}{2}}{RC}$$

$$T^{(p+1)} = \frac{T_\infty + \left(\dfrac{RC}{\Delta t} - \dfrac{1}{2}\right) T^{(p)}}{\dfrac{1}{2} + \left(\dfrac{RC}{\Delta t}\right)}$$

with initial condition $T^{(0)} = T_0$.

It is not clear why this method is called an implicit method in this "1-node" example because the future predicted value is expressed explicitly in terms of future values. That will not be true for more complicated models (with more than one unknown nodal temperature).

Workshop 1.1: Numerical Solution of First-Order RC Circuit

Consider the model problem of an RC circuit shown, after activation of an initial state by the switch. Write an Excel workbook or program code that compares the exact solution (as lines, no markers) and the numerical predictions (as markers with straight lines between them) from the forward, backward, and central difference methods for different values of time step, Δt. Allow user input, but use default values from the following table:

Add a worksheet or code to implement the Euler method with a variable step size (fixed change).

Input parameter	Symbol	Value	Unit
Initial temperature	T_o	100	°C
Ambient temperature	T_∞	0	°C
Resistance value	R	10	°C/W
Capacitance	C	0.1	J/°C
Numerical time step	Δt	0.5	s

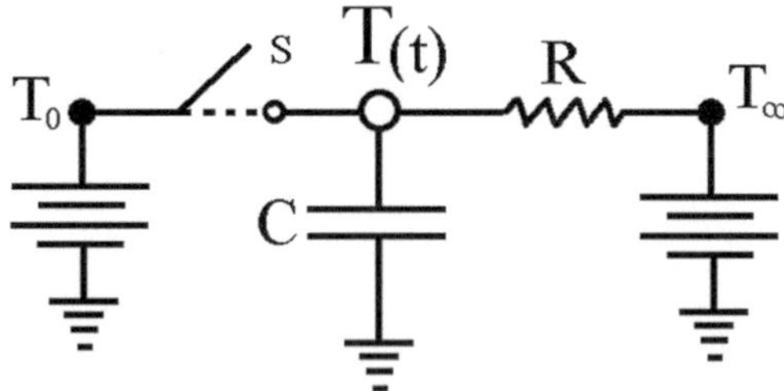

Note: A detailed solution follows. It is advised that this solution not be consulted prior to doing this workshop yourself. The intent is for as much "self-discovery" to occur as possible.

Numerical Accuracy

The three numerical methods are compared with the exact solution in Figs. 1.8, 1.9, and 1.10 for fixed numerical time steps of 1.0 s, 0.5 s, and 0.1 s, respectively. The numerical methods are applied to the RC model problem, with a time constant (RC) of 1 s. The exact solution is shown as a solid line. For the case where the step size of the numerical approximation equals the physical time constant of the system ($\Delta t = 1$ s here), the explicit method achieves equilibrium in a single step, and then stays there. The backward implicit method decays too slowly, and the central method is the most accurate. The accuracy of all methods increases as the step size decreases. For practical purposes, a step size of 10 % of the underlying time constant can be considered sufficiently small to simulate the model accurately for all methods.

Numerical Stability

Figure 1.11 shows the effect of time step on numerical stability for the Euler method (explicit forward difference method) for step sizes between one and two time constants. The method involves calculating the initial slope, and extrapolating forward at that slope for a full step. As shown previously, a step size of 1.0 will bring the temperature to ambient (0 °C here) in a single step. For a larger step, the

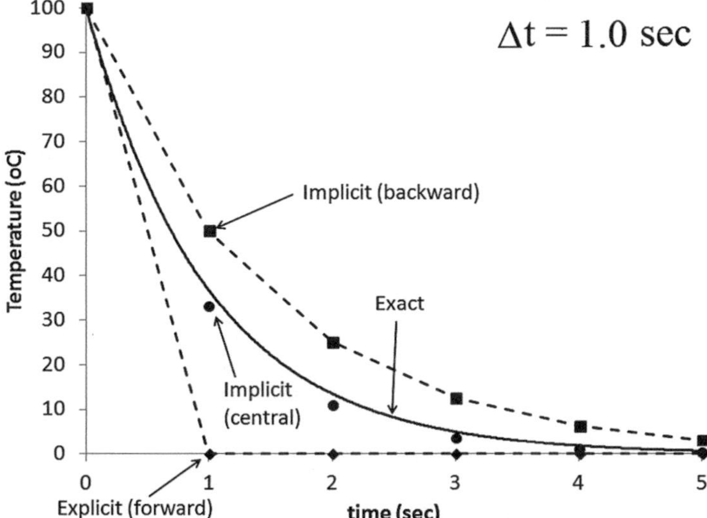

Fig. 1.8 Comparison of three numerical techniques on the transient model problem. The time constant (RC) has a value of 1 s for all three numerical techniques. The exact solution is shown as a *solid line*, and *straight dashed lines* between points are drawn for the explicit and backward implicit method

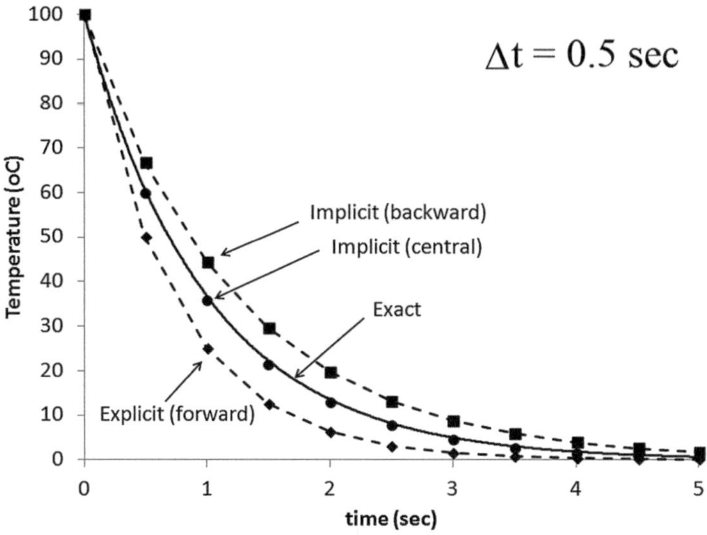

Fig. 1.9 Same as previous with a step size of 0.5 s

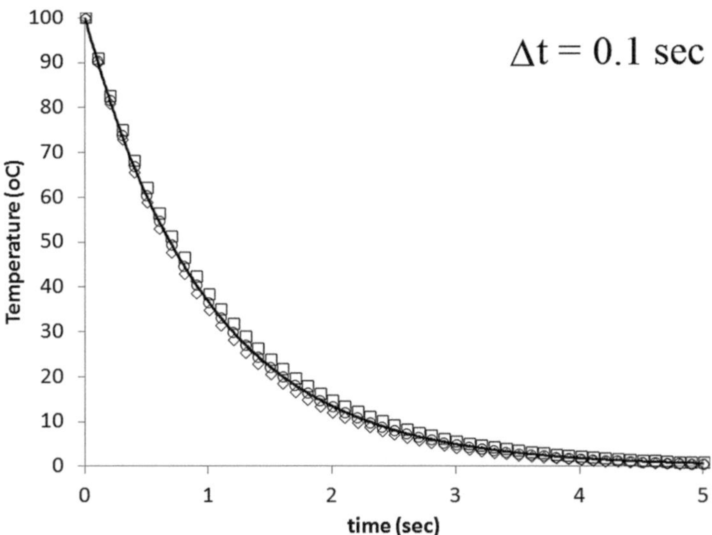

Fig. 1.10 Same as previous with a step size of 0.1 s

solution after one step will overshoot its equilibrium value. The next step will evaluate a positive slope, and will overshoot its equilibrium. The temperature will resemble a decaying periodic function, and eventually settle to the correct equilibrium value. This behavior is numerically stable (the function remains finite), but in

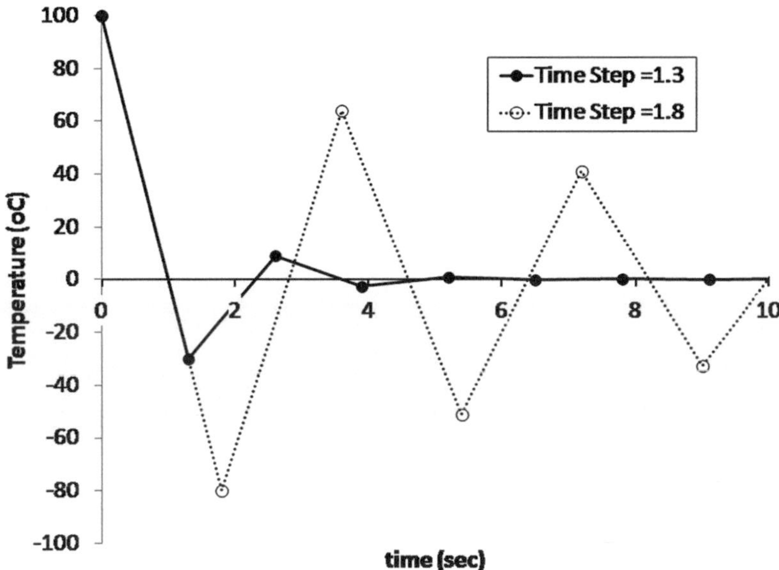

Fig. 1.11 Explicit forward difference method for step sizes between one and two time constants

a heat transfer context, each step in time violates the Second Law of Thermodynamics: a warm object cooling to a cold environment cannot become colder than that environment without some other agent (such as evaporation). Therefore, a step size of 1 time constant is considered to be the maximum step size allowable for the explicit Euler method.

Figure 1.12 shows the behavior of the Euler method for a step size of 2 time constants and a slightly larger value (2.01). For a step size of 2, the solution bounces back and forth between +100 and −100 °C, and will do so indefinitely. For a step size of 2.01, the solution grows in absolute value with each step, and will eventually grow to infinity (resulting in a computer error once the size limit of the program being used is reached). A step size of 2 time constants is therefore shown to be the maximum step size for numerical stability for the Euler method. This instability can be seen in the mathematics of the recursion relation for the Euler method, namely,

$$T^{(p+1)} = T^{(p)} + \frac{\Delta t}{RC}\left(T_\infty - T^{(p)}\right)$$

in which the absolute value of the future temperature increases with step size (Δt). The limit for the Euler method is:

$$\lim_{\Delta t=\infty} T^{(p+1)} = \infty$$

In general, a numerical method is said to be unstable if an inherently stable physical problem grows without bounds as successive steps are taken. It is possible, in general, that a problem is inherently unstable, but that is not the case here.

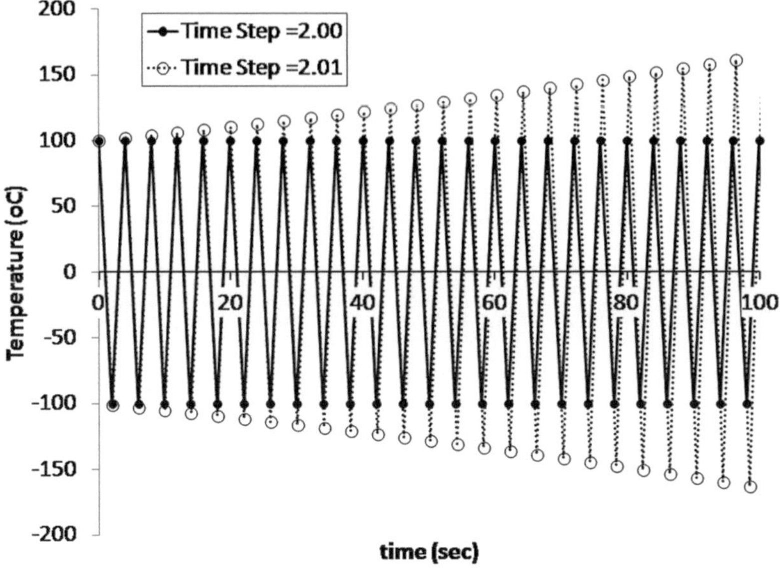

Fig. 1.12 Explicit forward difference method for step sizes of 2.00 and 2.01

Figure 1.13 shows the behavior of the backward and central implicit methods for a step size of 10 (which would be wildly unstable for the forward method). These methods are shown to be inherently stable, numerically, meaning that their solution will not increase indefinitely, but rather will approach equilibrium no matter how large a step size is taken. For the backward method, the recursion relation is

$$T^{(p+1)} = \frac{T_\infty + \left(\dfrac{RC}{\Delta t}\right) T^{(p)}}{1 + \left(\dfrac{RC}{\Delta t}\right)}$$

Its limit is

$$\lim_{\Delta t = \infty} T^{(p+1)} = T_\infty$$

This limit shows that for large step sizes, the solution approaches the equilibrium value in a single step. On the other hand, for the central method, the recursion relation is

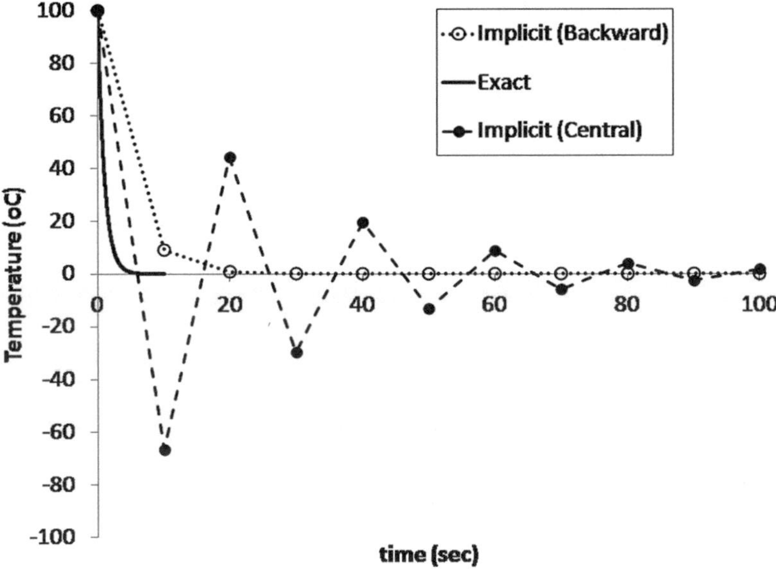

Fig. 1.13 Implicit methods (backward and central difference) with a step size of 10 time constants

$$T^{(p+1)} = \frac{T_\infty + \left(\dfrac{RC}{\Delta t} - \dfrac{1}{2}\right)T^{(p)}}{\dfrac{1}{2} + \left(\dfrac{RC}{\Delta t}\right)}$$

Its limit is

$$\lim_{\Delta t = \infty} T^{(p+1)} = 2T_\infty - T^{(p)}$$

There is no real physical meaning that can be attributed to this limit. For very large step sizes, the central method will eventually settle to equilibrium, but will do so as a decaying periodic function. Its behavior violates the Second Law, and in no way resembles the true behavior of a first-order system.

In future chapters, when multi-node systems are considered, it is possible that the inherent time constant of one or more nodes is much smaller than all the others, the so-called stiff numerical problem. In these cases, an implicit method, which is harder to implement than the explicit method, might be advantageous. While the central difference method might appear to have better accuracy for step sizes less than the inherent time constant of the system, the backward difference method is arguably preferred when the numerical step size is much larger than the inherent time constant of the most rapidly changing nodal temperature.

Workshop 1.2: Numerical Solution of RC Circuit with Time Varying Ambient Temperature

Consider the RC circuit shown (with R = 1 °C/W and C = 1 J/°C as defaults), after activation of an initial state by the switch ($T_0 = 0$ °C). In this circuit the ambient temperature varies with time in a prescribed way. Specifically,

$$T_\infty = T_{average} + A \sin\left(2\pi N t\right)$$

where $T_{average}(=50\,°C)$ is the average temperature, $A(=50\,°C)$ is the amplitude and $N(=1\,Hz)$ is the frequency.

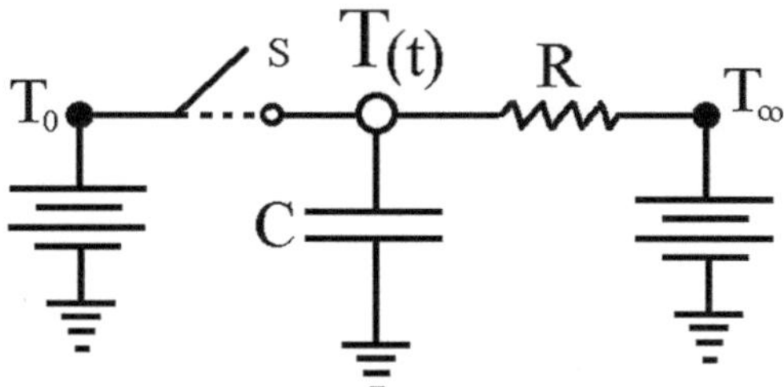

Write an Excel workbook or program code that implements the Euler method and plots T and T_∞ vs. time (on the same axes) over enough cycles to reach a "steady-state" (in which there is no cycle to cycle temperature variation). Complete the Results table in order to investigate the system amplitude and phase response to source frequency. An analytical solution to this problem can be developed for those so inclined.

Model Development

An energy balance applied to the temperature node is:

$$C\frac{dT}{dt} = \frac{T_\infty - T}{R}$$

The Euler method uses a forward finite difference approximation to the first derivative:

$$\frac{dT}{dt} \approx \frac{T^{(p+1)} - T^{(p)}}{\Delta t}$$

where Δt is the numerical step size, the superscript (p) is an index referring to the present time, and (p + 1) is one step into the future. Inserting and rearranging for the future temperature in terms of the present:

$$T^{(p+1)} = T^{(p)} + \frac{\Delta t}{C}\left(\frac{T_\infty^{(p)} - T^{(p)}}{R}\right)$$

For numerical accuracy, the step size should be much smaller (say 10 %) than the smaller of the inherent response time (RC), or the cycle time (1/N).

RESULTS (run cases with different step sizes to observe the effect, but don't report them)

N	Δt	Amplitude	Phase lag (deg)
0.01	0.1		
0.1	0.1		
1	0.1		
10	0.01		
100	0.001		

Lumped Capacity Systems and Overall Heat Transfer Coefficients

2

This chapter introduces two practical heat transfer concepts; lumped capacity and overall heat transfer coefficients. The term "lumped capacity system" means that a finite region of space is considered to be at some average temperature, even though there could be hot or cold spots in it. The example of a cooling mug of coffee will be used extensively in subsequent chapters, in which case an analysis of the average temperature will be conducted, even though the coffee is somewhat hotter than the mug, and there are cold spots of coffee near the mug wall. In numerical analysis, subdivisions of a physical object can be made, treating each subdivision as a lumped capacity element.

The "overall heat transfer coefficient, U" is a parameter that captures the basic idea of heat transfer, namely, energy transfer that is driven by a temperature difference. For example, "R value" is the metric used to rate the thermal performance of insulating materials, and it is the inverse of an overall heat transfer coefficient, U. Applied to a wall on a cold day, heat must flow from the inside air to the inner surface of an exterior wall, then through the wall, which has paint, sheet rock, studs, insulation, sheathing, vapor barrier, shingles, paint, and then from the outer surface of the wall to the cold air outside. This textbook builds theoretical tools for estimating all these effects. The R value is a measure of the overall heat transfer coefficient (actually, its inverse) and includes all the complicated details into a single metric.

The focus of this chapter is on the thermal capacity of systems, which depends on the material properties of density and specific heat. The next chapter will focus on the thermal resistances of systems, where thermal conductivity and emissivity are key material properties. The remaining chapters will build analysis tools for a widening range of applications.

In dealing with complicated situations that involve heat transfer, the approach will be to start with as global a picture as is reasonable, generally called a 1-node model, in which all the material in question is lumped into a single "node," and an overall heat transfer coefficient between this system and its environment estimated (which can be rather involved). This first step is the ultimate in the process of

© Springer International Publishing Switzerland 2015
G. Sidebotham, *Heat Transfer Modeling: An Inductive Approach*,
DOI 10.1007/978-3-319-14514-3_2

numerical analysis in which space is "discretized." Time is also discretized using one of the methods from Chap. 1 (or others that have been developed). If more detail is needed, the system in question can be broken into two nodes, and then an interaction between the two nodes occurs, and both of them may interact with the environment. Then it is logical to define three nodes, or then to "n" nodes, or into "n × m" nodes for a two-dimensional space, or into "n × m × p" nodes for 3D. Each node is treated as a lumped capacity system.

That's a road map of the next several chapters. Welcome aboard!

2.1   A Cooling Mug of Coffee

I, like many, *NEED* a cup of coffee as soon as possible when I wake up. It takes a few minutes to prepare the coffee maker and run it through its cycle. Then I pour hot coffee into a mug. It's *TOO HOT!* Grrrrrr... So maybe I drop a few ice cubes in it so I can get a good gulp to start my day. Diluting the coffee is not ideal, but it's worth it for that first cup. I wonder if I can predict how long it would take to just wait for it. A goal of this book is to satisfy that curiosity.

2.1.1   Problem Statement

A volume $V = 0.41$ L of coffee at 1 atm, 92 °C is quickly poured into a room temperature (21 °C) ceramic mug (shown schematically in Fig. 2.1) with a mass of $m_{mug} = 0.365$ kg, inner diameter $D = 0.083$ m, inner height $H = 0.098$ m, outer height $H_o = 0.109$ m, and nominal wall thickness $w = 0.0040$ m. The mug of coffee is placed on a table in a still room. Estimate the time it takes for the water to cool to 60 °C (above which it is too hot) and to 45 °C (defined as "tepid").

2.1.2   Experimental Solution

An experiment was conducted using a ceramic mug which closely approximates the mug defined in the problem statement. The actual mug tested has rounded edges, a handle, and other deviations from the schematic. The idealizations implied in the schematic diagram represent an early modeling step. Water was brought to a rolling boil on a stovetop and quickly poured into the mug. A handheld infrared thermometer was used to measure the free surface and the outside wall temperatures as functions of time. The raw data are plotted in Fig. 2.2 for the entire cooling event, the end of which, after some 4 h, is defined by the coffee reaching ambient. After an initial adjustment period, the top and side temperatures fall more or less together and asymptotically approach ambient. The event will be broken down into distinct stages in the following paragraphs. The raw data is reported in Appendix 1.

It is emphasized that this experiment was conducted with a handheld device that measures surface temperature only and was not in a controlled laboratory

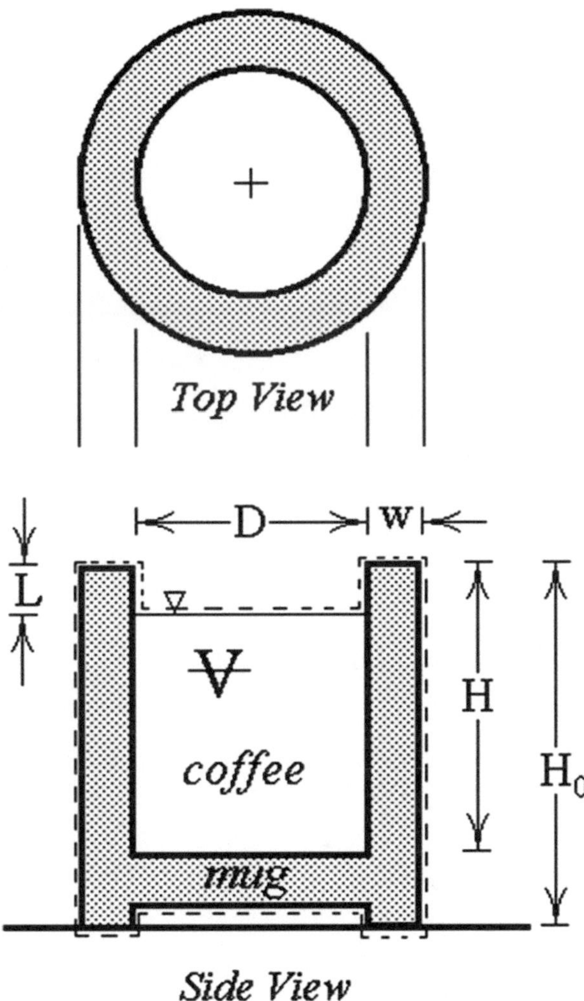

Fig. 2.1 Schematic of coffee mug with geometry defined

environment, nor with a detailed error analysis. There is much to be said, but not here, about the experimental uncertainty in this experiment that would raise into question many fine details of these temperature curves, especially during the initial period as the mug is heated. Readers of this text are encouraged to conduct their own experiments, and apply the theoretical methods developed in this text to those situations.

Figure 2.3 shows the first 10 min of the cooling event. At the start (immediately after pouring the coffee), the mug is at ambient temperature and the top surface is near the boiling point of water at atmospheric pressure. Note that the IR temperature sensor measures the surface temperature of the water, and strong evaporative effects of water near its boiling point likely produce temperatures noticeably below that of the interior. There is an initial phase, called the "mug heating phase" which

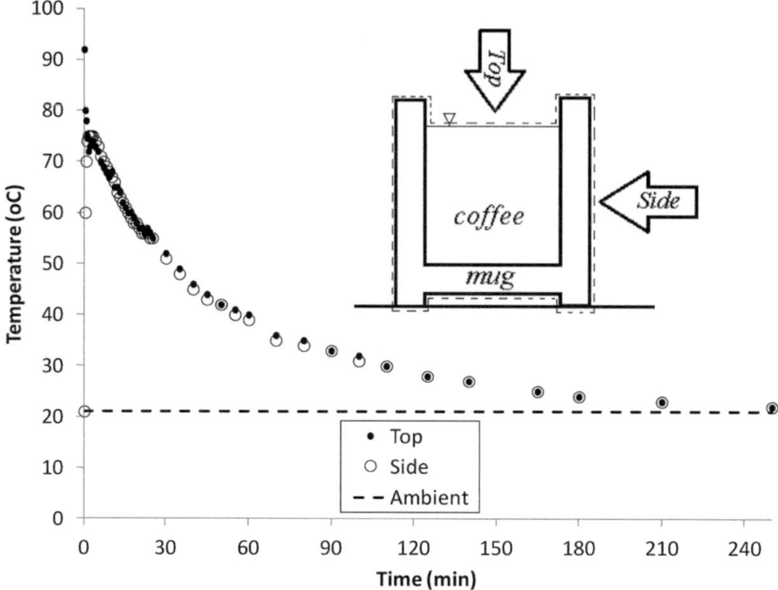

Fig. 2.2 Temperature of the free surface (*top*) and middle of the outer side wall (*side*) as a function of time for the entire cooling event

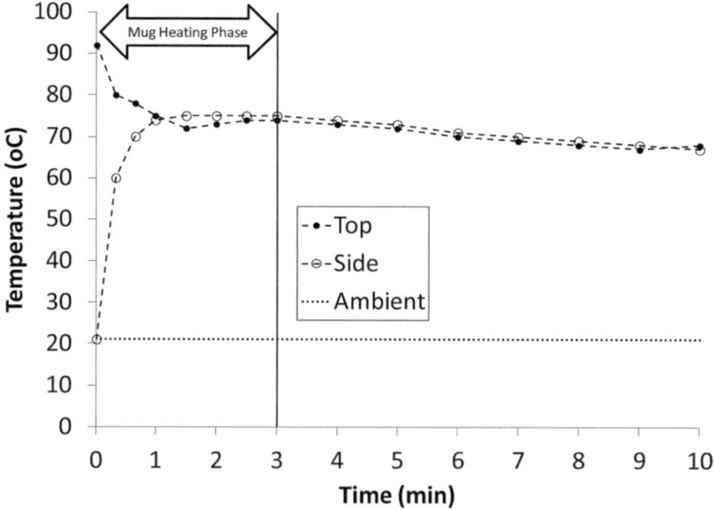

Fig. 2.3 First 10 min of the cooling event

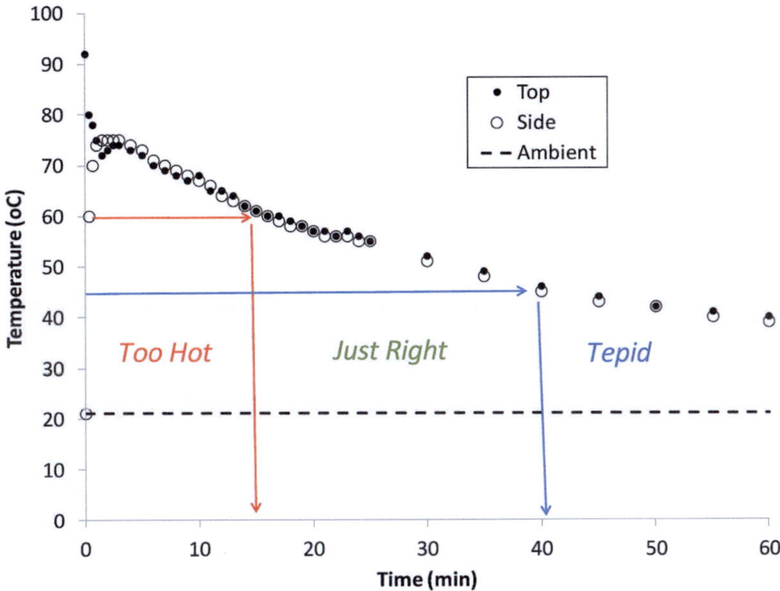

Fig. 2.4 Temperatures for the first hour of the cooling event

will be investigated in some detail in future chapters. This phase lasts about 3 min, during which time the side temperature rises to a peak, and the top (surface) temperature drops quickly to a slightly lower temperature than the side. At the end of the mug heating phase, the two temperatures are close to each other at approximately 72 °C.

Figure 2.4 shows the first hour of the event, during which time the coffee passes between practical drinking stages. For the first 16 min, the coffee is too hot to drink. The subjective definition of "too hot" at 60 °C was determined by testing with small sips (in a separate experiment). The temperature falls after the mug heating phase as intuition would suggest. During the subsequent cooling phase, the indicated temperature of the free surface is slightly lower than that of the sides of the mug most of the time. Between 16 and 41 min, the coffee is drinkable, but not tepid. The sampling rate was changed during this time, and by the end, the temperature indicated by the IR detector of the top surface is slightly higher than that of the side wall.

So we're done, right? After pouring the coffee, I need to wait about 15 min before I can start sipping, and then 25 min later (40 min after pouring), it'll be too late. But what if I put a lid on it, or was using a travel mug, or had a paper cup, or a bigger mug, or was really interested in a much larger (or smaller) scale type of practical problem? The main point of this example is to use a simple experiment to compare to theoretical predictions. The skill of making such a prediction from "first principles" is the main objective of this text, to which we now turn.

2.2   Lumped Capacity Model Development

The first step of a thermodynamic analysis (of which heat transfer is a subset) is generally to split the universe into two parts, system and surroundings. Doing so produces a control volume (the region of space that encompasses the system) and a control surface (the boundary between the system and its surroundings). A lumped capacity heat transfer model captures the global transient response of the average temperature of a system to its surroundings at a different temperature. Many practical problems fall into this category. Furthermore, the basic concepts developed here form a basis for more complicated analyses when more detail is required. It is easy, but wrong, to conclude that the lumped capacity approximation means that the temperature of an object is uniform. Proper application of heat transfer methods (next chapter) allow for significant spatial variation in temperature, while still maintaining a lumped capacity model.

2.2.1   Conservation of Energy

The First Law of Thermodynamics (conservation of energy) is the applicable fundamental law for this event. Choosing an imaginary boundary surrounding the mug of coffee, shown as dashed lines in the problem sketch, and assuming a closed system (one in which mass does not cross the boundary, meaning evaporation is neglected here), the First Law states that the energy (E) stored within the system can be changed (Δ) by either heat entering (Q_{in}) or work leaving (W_{out}) the system across the system boundary:

$$\Delta E = Q_{in} - W_{out}$$

Expressing the change in energy in terms of an initial and final state:

$$E_{final} - E_{initial} = Q_{in} - W_{out}$$

Heat is energy that crosses a defined system boundary due to a temperature difference, the focus of this text. Work is energy that crosses the boundary due to a force (mechanical or electrical) acting through a distance. The First Law applies to a specific process that links an initial with a final state. The stored energy (E) is a state property, or a snapshot. Its value at any time is independent of what went on before or after. On the other hand, heat and work are forms of energy that are associated with energy transfer across the system boundary. If a snapshot is taken, heat and work are not apparent. On the other hand, if a movie is taken, heat and work will be "seen" to cross the boundary. Heat and work are "process-dependent" variables.

   In this form, the elapsed time between the initial and final states does not come into play. In traditional thermodynamic analysis, the prime objective is to relate overall changes in energy content to the total heat and/or work exchanged. In traditional heat transfer analysis, where the details of the process as it occurs are often the objective, a rate basis is generally more advantageous. Dividing the energy balance by the elapsed time between states (Δt):

$$\frac{\Delta E}{\Delta t} = \frac{Q_{in}}{\Delta t} - \frac{W_{out}}{\Delta t}$$

Consideration of the process that takes place over a "short" time interval produces the illusion of a snapshot (rather than a process, with a before and after), in which the rate of change of energy is related to the rate of heat and work exchange. Taking the limit $\Delta t \rightarrow 0$ and defining a nomenclature $q = Q_{in}/\Delta t$ and $w = W_{out}/\Delta t$ which have units of energy per unit time (or J/s = Watt in SI units, or Btu/h in English units), the First Law for closed systems on a rate basis becomes the following first-order ordinary differential equation:

$$\frac{dE}{dt} = q_{in} - w_{out}$$

Note that thermodynamics texts tend to use $\dot{Q}$ for q and $\dot{W}$ for w.

2.2.2 Lumping the Stored Energy into a Single Node

In general, stored energy can be in the form of kinetic energy, potential energy, and/or internal energy (u). Each element of mass ($dm = \rho dV$) within the defined control volume has, in general, a different specific energy, the word "specific" meaning per unit mass. The total stored energy is the sum of the energy of all the elemental masses. Expressed as an integral:

$$E \equiv \iiint\limits_{CV} \rho \left(\frac{V^2}{2} + gH + u \right) dV$$

In this expression, the triple integral is evaluated over the control volume (CV) defined within the system boundary by differential (i.e., vanishingly small) volume elements (dV). In general, the density (ρ), velocity (V with a line through it to distinguish from volume), height (H) in a gravitational field (g) and internal energy (u) can all vary with position.

For the problem at hand, the kinetic energy associated with motion of the liquid and any changes in its potential energy as it rises and falls are assumed from the outset to be negligible in comparison to the changes in internal energy that arise during the event. There is motion of the liquid, even if it is not being stirred due to mild density variations, but it seems plausible that the energy associated with that motion is not important. Without a much more detailed analysis, the neglect of kinetic and potential energy cannot be further evaluated, but kept in the back of the mind. The heat transfer principles developed in this text can be applied to those situations, where kinetic and/or potential energy are important.

In general, the internal energy of a system consists of its nuclear, chemical, latent (i.e., phase change), and sensible (thermal, or that energy associated with its temperature) components. However, if nuclear or chemical reactions are absent and there is no phase change within the system boundary, the only energy form which changes is the sensible component which relates internal energy to the temperature through the specific heat (c_v, in J/kg/K), a thermodynamic function defined as:

$$c_v \equiv \left. \frac{\partial u}{\partial T} \right|_v$$

where the partial derivative is taken at constant specific volume and indicates that the internal energy can be expressed as a function of two other thermodynamic variables (i.e., specific volume and temperature), in general. Note that lower case "c" is used to denote specific heat (with units of J/kg/K), while upper case "C" is used to denote total capacitance (J/K). However, for solids and liquids that are not undergoing phase change, it is generally acceptable for engineering purposes to assume that the internal energy depends solely on temperature (not pressure, or volume, or any other thermodynamic variable), and therefore the partial derivative becomes a total derivative, and the subscript "v" is no longer required:

$$du \equiv cdT \quad \text{for liquids and solids}$$

Defining a reference temperature (T_{ref}, analogous to a reference height for potential energy in a gravitational field) at which point internal energy is defined to have a value of u_{ref} (which is often set to zero), the internal energy is obtained by integration from the reference temperature to the actual temperature (T):

$$\int_{u_{ref}}^{u} du = u - u_{ref} = \int_{T_{ref}}^{T} cdT \equiv \bar{c}(T - T_{ref})$$

The last expression defines formally this temperature averaged specific heat (which, in general, still varies with temperature).

In the present case, with the control volume consisting of two very different substances (mug and coffee), the spatial triple integral that expresses the total energy stored can be split into two:

$$E \equiv \iiint_{coffee} \rho \bar{c}(T - T_{ref})dV + \iiint_{mug} \rho \bar{c}(T - T_{ref})dV$$

There are two material properties (density and specific heat) that vary with position because the temperature varies with position, and these integrals cannot be further evaluated at this point. However, it is generally justifiable in engineering practice to also assume that density and specific heat are suitably chosen constants. Defining an average temperature for each subsystem (coffee and mug), the total energy content can be expressed as:

$$E = (\rho c V)_{coffee}\left(\overline{T}_{coffee} - T_{ref}\right) + (\rho c V)_{mug}\left(\overline{T}_{mug} - T_{ref}\right)$$

The combination $\rho c V$ for a given subsystem has units J/K and is given the shorthand nomenclature of a capital C, with an appropriate subscript. That is:

$$(\rho c V)_{mug} = C_{mug}$$

and similarly for the coffee. It is understood that the density and specific heat are assumed constant with respect to space, yet could still be functions of temperature (an effect which could be modeled if required).

Numerical values of specific heat can be estimated in a number of ways, but generally can be searched for on the Internet or in tables from textbooks to obtain reasonable approximations. Whenever possible, a careful measurement of an actual sample (Appendix 2) and/or a value obtained from a manufacturer's specification sheet for an actual material in an application is the best practice.

2.2.3 Thermodynamic Analysis of Mug Heating Phase

At the start of the event, the mug and coffee have very different temperatures. Empirically (Fig. 2.3), after a few minutes, they are at nearly the same temperature. A demonstration of why the mug temperature and coffee temperatures are so close to each other is delayed until more heat transfer tools have been developed. The thermodynamic tools are in place to predict the value of the temperature at the end of the mug heating phase. Recalling the basic First Law expression:

$$E_{final} - E_{initial} = Q_{in} - W_{out}$$

Having defined the system to be the coffee plus mug, and the process to start at an initial temperature where the mug is at ambient temperature and the coffee is at the temperature it is poured into the mug (near the boiling point of water):

$$\begin{aligned}(C_{mug} + C_{coffee})\left(T_{final} - T_{ref}\right) - \left[C_{mug}\left(T_{mug,initial} - T_{ref}\right) + C_{coffee}\left(T_{coffee,initial} - T_{ref}\right)\right] \\ = Q_{in} - W_{out}\end{aligned}$$

The coffee and mug are assumed to be at the same final temperature (which is the initial temperature of the main cooling of the coffee and mug together). All the terms with the reference temperature cancel, and the First Law can be rearranged to solve for the final temperature:

Table 2.1 Parameters related to thermal capacity (derived quantities in bold)

Property	Coffee	Mug
Density (kg/m^3)	1,000	1,700
Specific heat (J/kg/K)	4,200	800
Volume (cm^3)	410	**197**
Mass (kg) = Density × volume	**0.410**	0.365
Capacitance (J/K)	**1,722**	**292**

$$T_{final} = \frac{C_{mug}T_{mug,initial} + C_{coffee}T_{coffee,initial}}{C_{mug} + C_{coffee}} + \left[\frac{Q_{in} - W_{out}}{C_{mug} + C_{coffee}}\right]$$

The combined system exchanges heat with the environment (Q_{in}) but the primary heat exchange takes place internally between the coffee and mug, which does not appear directly here because the system boundary does not include it. It is plausible that the heat loss to the environment is negligible in the relatively short time (3 min) of the mug heating phase (compared to the 4 h cooling event). The work exchange is identically zero in this case. Assuming, therefore, an overall adiabatic process ($Q_{in} = 0$), the First Law reveals the final temperature of the mug heating phase to be a capacitance-weighted average temperature:

$$T_{final} = \frac{C_{mug}T_{mug,initial} + C_{coffee}T_{coffee,initial}}{C_{mug} + C_{coffee}}$$

Table 2.1 shows property values assumed for the heat capacitance. Coffee is assumed to have thermal properties of liquid water at nominal room conditions. The mug values are typical of a ceramic material. In Appendix 2, a rough measurement of the specific heat of the actual mug used is described. A majority, 85.5 % of the total capacitance is associated with the coffee. The heat capacity of the mug itself is therefore considered to be small but not negligible, and it is easy to account for this "thermal mass." Note that the mass of the mug itself is comparable to that of the coffee, yet its capacitance is small because its specific heat is less than a quarter of that of the coffee. Liquid water (and therefore coffee) has an exceptionally high specific heat value, one of the many unique features of water.

Inserting values for this case:

$$T_{final} = \frac{(292\,\text{J}/°\text{C})(21\,°\text{C}) + (1,722\,\text{J}/°\text{C})(92\,°\text{C})}{(292 + 1,722)\text{J}/°\text{C}} = 81.7\,°\text{C}$$

Consider the experimental value (from Fig. 2.3) of approximately 75 °C for the peak side wall temperature (the top surface temperature is a few degrees lower). Is this good agreement or bad? It could be argued both ways. There is uncertainty in the experimental temperature values, and these values are taken at two places on the defined control surface. The theoretical temperature calculated is an average temperature of the system. The temperature of the liquid in the interior is expected to be at a somewhat higher temperature (how much higher will be investigated in subsequent chapters from a theoretical, not experimental, point of view). This trend is consistent with the discrepancy of the measured and predicted temperatures. What theory and experiment agree on is the qualitative comparison, namely, that the final temperature of the mug/coffee combination is much closer to the initial coffee temperature than the initial mug temperature. The theory explains the observed result because the capacitance of the coffee is roughly six times that of mug.

By the way, my Mom likes really hot coffee, so when I make her a cup, I pour boiling water into a mug first to preheat it, then dump that water out (after about 3 min) before filling it with coffee. The final temperature for the mug heating phase (initial temperature for the mug/coffee cooling phase) is then much closer to the boiling point of water, the way she likes it.

2.2.4 1-Node Lumped Capacity Model

In conversation, it is no problem to imagine and speak of a cup of coffee as being at a certain temperature. When pressed, there would be little problem of acknowledging that perhaps the temperature of the outside of the mug may be somewhat different from the temperature of the inside liquid, for example. This trend might be very obvious if a styrofoam mug were used, or if a cardboard holder placed outside a paper cup to protect the hand were used. Furthermore, the temperature of the coffee near the walls of the mug or near the surface may be different than that in the center. So what is really meant by THE temperature of the coffee? A simplified mental image (or model) in which the entire object in question has been lumped into a single representative entity has already been intuitively made. That intuition is now turned into a formality.

The simplest model lumps the entire system (coffee plus mug) into a single "node" by defining a single average temperature to both triple integrals of the energy content, and assuming constant properties (ρ and c):

$$E = \left(C_{mug} + C_{coffee}\right)\left(\overline{T} - T_{ref}\right)$$

The rate of change of the stored energy is then simply:

$$\frac{dE}{dt} = C\frac{d\overline{T}}{dt}$$

where $C = C_{mug} + C_{coffee}$ is the total thermal capacitance of the combined coffee/ mug system. The temperature is best thought of as an average temperature of the system, not necessarily that the mug and coffee are at the same temperature. The bar over the temperature will be dropped from here on, keeping in mind that T means "average temperature."

Modeling, in this case, means expressing the almost infinite complexity implied by the imposing triple integral in a much simpler (but not simplistic) mathematical expression. As a result, clarity is gained at the expense of detail.

$$\frac{d}{dt}\iiint\limits_{CV} \rho\left(\frac{V^2}{2} + gH + u\right)dV \xrightarrow{\;\;\;Modeling\;\;\;} C\frac{d\overline{T}}{dt}$$

And we're halfway to an RC circuit.

2.2.5   Overall Heat Transfer Coefficient: Empirical Value

The other two terms of the First Law, namely, heat and work, are now addressed. Work represents mechanical energy transfer across the boundary, such as from expanding pistons, rotating shafts, electrical input and the like, which do not exist here, so the work is identically zero. Heat represents the flow of thermal energy across the system boundary, driven by a temperature difference. The rate of heat transfer between the coffee (at its average temperature value) and the ambient can be expressed in the form of Ohm's Law:

$$q_{in} = \frac{T_\infty - \overline{T}}{R}$$

This model states that the rate of heat transfer is proportional to the temperature difference driving it and inversely with a property of the system called the thermal resistance (R). The thermal resistance represents the barrier to heat flow between the system and surroundings that is NOT accounted for in the temperature difference driving the heat. The thermal resistance can vary with temperature, in general, although a suitably chosen average value is frequently justified.

Alternatively, the heat transfer rate from a system to its surroundings can be modeled as being proportional to the driving temperature difference and to the total area of the control surface, A. The proportionality factor (not necessarily a constant) is thus defined as an effective overall heat transfer coefficient, U. Note that there is a nomenclature problem in that U is used to represent both internal energy and overall heat transfer coefficient, which mean very different things. In heat transfer:

$$q_{in} = UA\left(T_\infty - \overline{T}\right)$$

The "R value" of building materials (given the symbol $\mathfrak{R}$ to distinguish it from the thermal resistance R) is the inverse of U, which is the thermal resistance times the system surface area:

$$\mathfrak{R} = 1/U = RA$$

Putting it all together, the energy balance becomes:

$$C\frac{d\overline{T}}{dt} = \frac{T_\infty - \overline{T}}{R}$$

which is the same differential equation obtained for the electrical circuit, with the same solution if, and it can be a big "if," R, C, and T_∞ are constant:

$$\frac{\overline{T} - T_\infty}{T_0 - T_\infty} = e^{\frac{-t}{\tau}}$$

where T_0 is the initial temperature and τ is a time constant, defined by:

$$\tau = \frac{C}{UA}$$

The response time of a lumped capacity system increases with its total capacitance (or thermal mass), and decreases with its surface area, and with the overall heat transfer coefficient, U. In the present example, estimates of the capacitance and surface area are possible. Therefore, by measuring the time constant from an analysis of the cooling curve, an empirically determined value of the overall heat transfer coefficient (U) can be obtained by comparing the prediction of this lumped capacity model for different values of U. The capacitance has already been calculated, and the total surface area of the control surface is calculated from the geometry of Fig. 2.1 as shown in Fig. 2.5. More than half of the surface area is found on the side wall, and the remainder is distributed between the bottom, the inner rim (i.e., the area above the liquid on the inside surface, which obviously depends strongly on how much liquid is poured in the mug), the top free surface, and the top edge (assumed to be squared off). The heat transfer process is quite different on all these surfaces, as will be explored in subsequent chapters, but the global view taken at this point does not distinguish that detail.

Figure 2.6 shows the measured free surface temperature (taken to be the best representation of the average coffee temperature available experimentally) and predictions of the cooling curve for three input values of the overall heat transfer coefficient (U). The initial temperature is taken to be that predicted for the end of the mug heating phase (81.7 °C). After the first 10 min, the data is reasonably banded between U values of 10 and 20 W/m^2/°C. A U value of 15 W/m^2/°C yields a reasonable agreement over the entire event, giving a time constant of $\tau = 45.1$ min from:

$$\tau = \frac{C}{UA} = \frac{(1{,}792 + 272)\dfrac{J}{°C}\left|\dfrac{min}{60\,s}\right|}{\left(15\dfrac{J}{s\,m^2°C}\right)(495.6\,cm^2)\left|\dfrac{m^2}{10{,}000\,cm^2}\right|} = 45.1\,min$$

Since the data points do not fall exactly on any line, the response is not truly first order which could be due to a variation in the parameters (R, C, or T_∞) during the

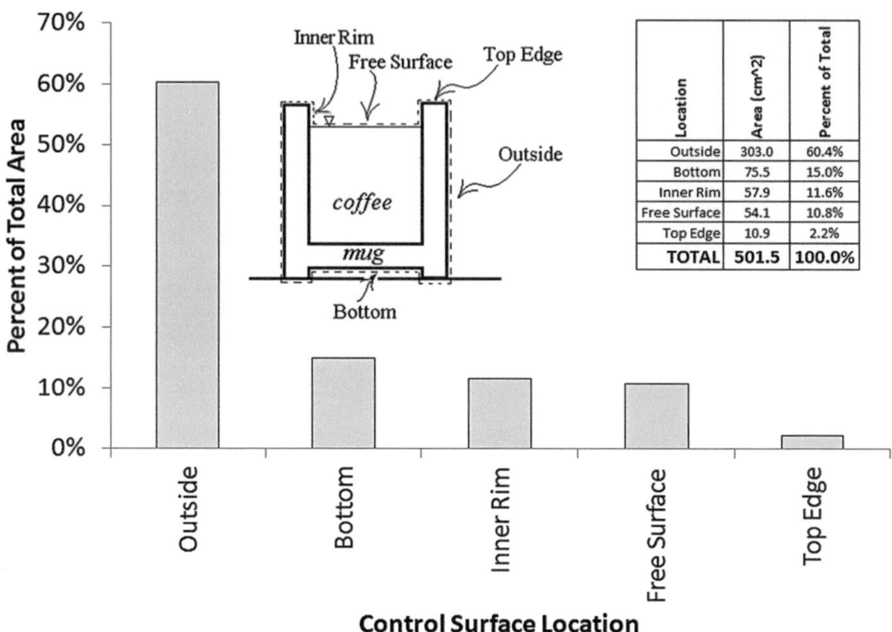

Location	Area (cm^2)	Percent of Total
Outside	303.0	60.4%
Bottom	75.5	15.0%
Inner Rim	57.9	11.6%
Free Surface	54.1	10.8%
Top Edge	10.9	2.2%
TOTAL	**501.5**	**100.0%**

Fig. 2.5 Surface area of system boundary broken into regions. For the bottom surface, the small extra area inside the outer rim is obtained assuming the trapped air gap height is the same as the mug thickness

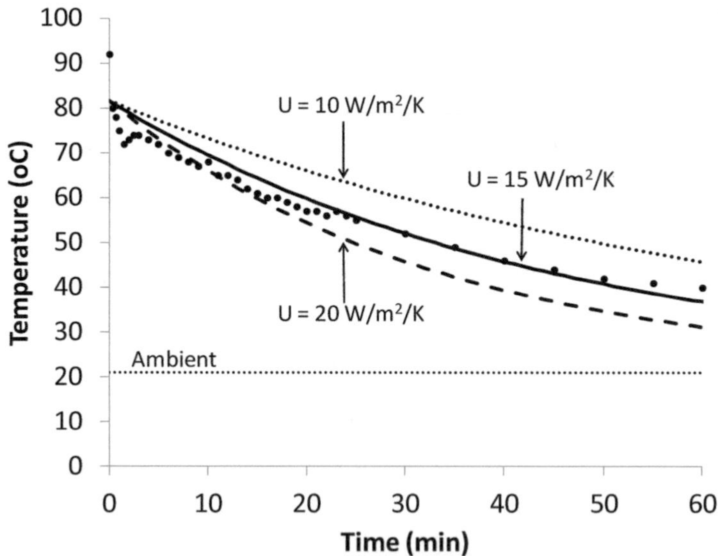

Fig. 2.6 Empirical determination of the overall heat transfer coefficient, U for the entire event. Experimental points for the free surface are shown as markers

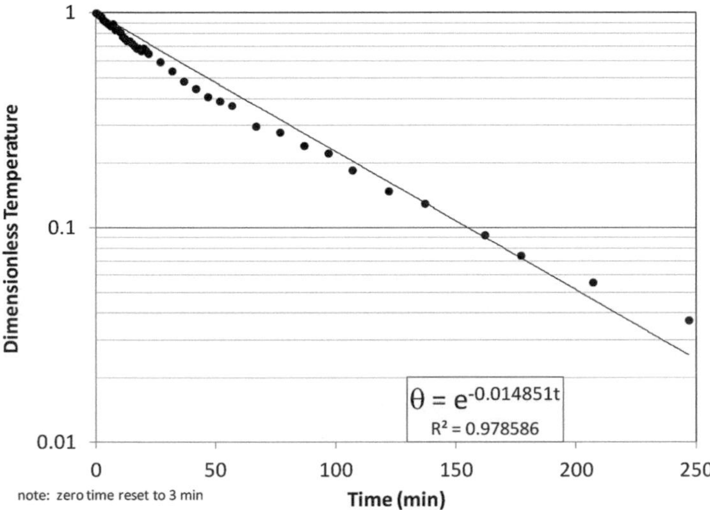

Fig. 2.7 Dimensionless temperature vs. time on semi-logy axes. Initial time in this plot is 3 min into the cooling event, beyond which time the coffee and mug temperatures are observed to fall together

event (R being the most likely candidate) or phenomena neglected (like evaporation) may be noticeable. Also, a constant reminder that the experimental data is of the free surface of the coffee, and the lumped model is of an average temperature of the system.

There are more sophisticated ways to analyze or test for first-order behavior than this admittedly brute force approach. However, those methods do not necessarily improve the basic understanding. Brute force is often acceptable, if not preferable, to getting side-tracked by unwarranted analysis. That is an engineering judgment call.

A more sophisticated way to present the same data is to plot the dimensionless temperature vs. time on semi-logy axes, as in Fig. 2.7. Here the initial time is reset to zero at 3 min into the cooling event because prior to that mug heating makes a lumped capacity model unreasonable, during which time the coffee and mug do not fall together; rather, the coffee is heating the mug. Investigation of those effects require separating the mug and coffee into separate subsystems, as is considered in a subsequent chapter. The line on the plot shows an exponential curve fit of the entire event, with the initial value forced to unity. For a pure first-order response, all data points would fall on this line. The experimental curve initially falls faster than the overall response, and slower near the end of the event, as thermal equilibrium is approached. The R^2 value of 0.98 suggests that despite these deviations, this event is reasonably characterized by a first-order event, with a time constant of the inverse of the number in the exponential of the curve fit:

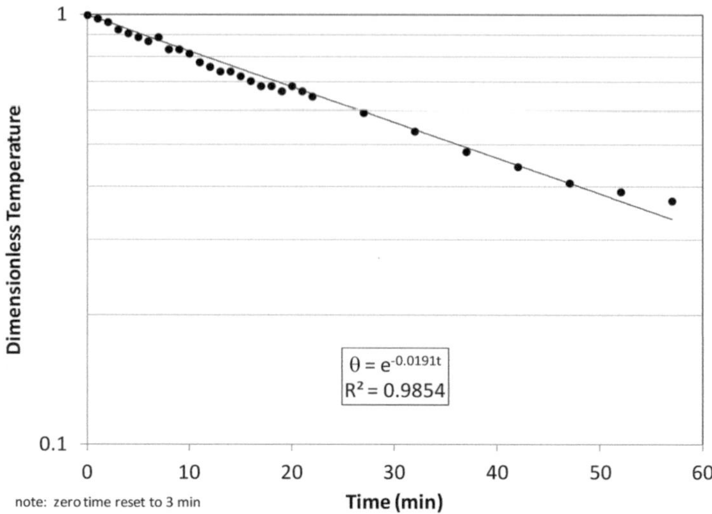

Fig. 2.8 Dimensionless temperature vs. time on semi-logy axes considering only the first hour of the event. Initial time in this plot is 3 min into the cooling event at which time the measured surface temperature is 74.0 °C, beyond which time the coffee and mug temperatures are observed to fall together

$$\tau_{FullEvent} = \frac{C}{UA} = \frac{1}{0.014851} = 67.3\,\text{min}$$

The time constant obtained from Fig. 2.7 is targeted toward a simulation of the entire cooling event, from hot coffee poured into the mug, until the mug and coffee reach thermal equilibrium with their environment. In practice, the objective is to understand when the coffee is within an acceptable range of temperature for drinking, not once it has become tepid. Figure 2.8 shows a semi-logy plot for the first hour of the event. The time constant during the first hour of the cooling event is shorter than the time constant that characterizes the entire event:

$$\tau_{1stHour} = \frac{C}{UA} = \frac{1}{0.0191} = 52.3\,\text{min}$$

The time constant for the first hour is 22 % shorter than that of the full event. That's a big difference, despite the fact that the "R^2" value was 98 % for the full event. That suggests the high R^2 values can give a rather misleading sense of confidence.

The final result sought from the engineering problem statement, namely, the time it takes for the coffee, once poured, to cool first to 60 °C, and then to 45 °C, thereby defining a window of opportunity. Expressing the first-order cooling curve in terms of the time:

$$t = -\tau \ln\left(\frac{T - T_\infty}{T_0 - T_\infty}\right)$$

The initial temperature (T_0) is taken to be the equilibrium temperature the mug and hot coffee come to during the mug heating phase (81.7 °C), and the time constant is the experimentally based value of 45.1 min. Therefore, the time to reach 60 °C is:

$$t_{60\,°C} = -(45.1\,\text{min})\ln\left(\frac{60.0 - 21.0}{74.0 - 21.0}\right) = 13.8\,\text{min} + 3\ \text{min}\ \textit{offset} = 16.8\,\text{min}$$

The time to reach 45 °C is:

$$t_{45\,°C} = -(45.1\,\text{min})\ln\left(\frac{45.0 - 21.0}{74.0 - 21.0}\right) = 35.7\,\text{min} + 3\,\text{min}\ \textit{offset} = 38.7\,\text{min}$$

Of course, this window of opportunity agrees with the direct data plots (Fig. 2.4) because the time constant was empirically determined, not developed from first principles. The next chapter introduces the main theoretical heat transfer tools that can be used to predict this behavior from first principles.

A word about units and absolute temperature is in order. Several quantities have temperature as one of the units, and those units can be specified as either °C or K because the quantity itself is based on a temperature difference. For example, the units of c_v are J/kg/K or J/kg/°C because c_v is the change in energy divided by the change in temperature. This author prefers writing these units in the "code" version J/kg/K rather than the J/kg·K version (which looks like K is in the numerator) or the J/(kg·K) version. As another example, the overall heat transfer coefficient U is based on the driving temperature difference, and therefore it is W/m^2/K or equivalently W/m^2/°C. Some quantities require absolute units, such as the universal gas constant whose units must be J/kmol/K and not J/kmol/°C.

2.3   A Melting Cup of Ice/Water

At times, I like a nice tall glass of iced coffee. Everyone knows that if you place a cup of iced coffee on a table top, the ice will eventually melt, diluting the coffee in the process. The question is, how long does that take? This problem involves heat transfer, where phase change occurs in the system. In addition, it is another scenario where an overall heat transfer coefficient can be measured.

2.3.1   Problem Statement

An ice/water slurry is prepared by filling a plastic cup with 0.295 kg of ice and adding liquid water to completely fill the cup. Solid (ice) and liquid are initially at the melting point (0 °C). The cup is placed on a table in a still room at 18 °C.

Determine the time it takes for half of the ice to melt, and the overall heat transfer coefficient, U, based on the exposed surface area of the cup plus slurry to the environment. The cup has an inner diameter $D = 0.074$ m, inner height $H = 0.144$ m, outer height $H_o = 0.166$ m, and nominal wall thickness $w = 0.0038$ m.

2.3.2　Experimental Solution

To prepare this experiment, ice from a conventional household freezer is placed in the cup, weighed, and allowed to sit at rest in a refrigerator for sufficient time that the cubes warm from the freezer temperature to the melting point, but not long enough to begin to melt. A lot could be said about how to determine that, either experimentally or theoretically. To make a long story short, the change in internal energy of solid ice from freezer temperature to melting point is much less than the change in internal energy of solid melting to liquid. As a result, when removed from a freezer space and placed in the cup, the ice will first warm up to near the melting point before it begins to melt. That can be proved, just not yet. The liquid water is prepared by pouring water from a larger container packed with ice and liquid water filling the void spaces and filled to the brim. In a rough experiment, the cup dimensions and initial masses of ice and water were those given in the problem statement. The cup was placed on an inverted plastic container (approximately 5 cm high) with trapped air inside to minimize (but not eliminate) the heat transfer through the bottom surface. The test was terminated when, visually, the ice was approximately half melted. Then the remaining liquid and solid were separated and the masses measured.

It is more difficult to monitor the progress of this event by measuring temperature than the cooling of hot coffee because the phase change takes place at a fixed temperature, namely, the melting point. A measurement of the temperature cannot distinguish a slurry that is mostly solid from one that is mostly liquid. It is possible to roughly measure the height of the bottom of the ice/liquid slurry as a preliminary measure. However, the shape of the ice likely changes with time, and therefore the packing ratio of solid to liquid volume changes as well.

On the other hand, it would seem that the masses of the liquid and solid could be measured as the event unfolds. However, that would require separating the phases, making the measurement of liquid and solid, and then putting them back into the cup. This procedure would introduce a significant loading error since the experiment called for not disturbing the event, and keeping it stationary. A loading error in measurements occurs when the act of taking a measurement changes the event that is to be measured. It is a closely related concept to the Heisenberg uncertainty principle in quantum mechanics. An alternative would be to repeat the experiment many times, and change only the length of time of the event. At the end of each experiment, the liquid would be separated from the solid by pouring it off.

In the experiment conducted, after 3.5 h (210 min), the remaining ice and liquid were separated and the mass of ice remaining was measured to be 0.125 kg (42.3 % of original).

2.3.3   Theoretical Model Prediction: Energy Analysis

While the percentage of liquid and solid changes continually, the temperature of the slurry does not change with time, as a first approximation. In practice, since ice has a lower density than liquid water, as ice melts, the slurry will separate into two parts; an ice/liquid slurry in good contact floating above a single phase liquid below. The liquid on the bottom portion of an ice/water slurry is not constrained to the melting point. In fact, for an undisturbed scenario like this, the liquid water will tend to stabilize at 4 °C, the temperature where the liquid water density is maximum. Assuming a constant system temperature makes theoretical modeling simpler than the cooling mug of coffee. Since the rate of heat transfer is driven by the temperature difference, it remains constant, unlike the hot coffee cooling where the heat transfer rate decreases as the coffee approaches ambient temperature.

Defining the cup and slurry (solid and liquid) as a closed system, an energy balance is:

$$\frac{dE}{dt} = \frac{d\left(E_{cup} + dE_{solid} + dE_{liquid}\right)}{dt} = \dot{Q}_{in} - \dot{W}_{out}$$

The work done by the system on the surroundings is identically zero. The heat transfer rate can be expressed as before with an overall heat transfer coefficient, $\dot{Q}_{in} = UA(T_{\infty} - T)$ where U is an overall heat transfer coefficient, A is the surface area of the system, T_{∞} is ambient temperature, and T is the average system temperature. In this model, the system will be assumed to remain at the melting point as long as some ice remains.

The energy (E) of each component in general consists of three components: internal, kinetic, and potential energy. There is little motion in this example, and the energy of each component is therefore the internal energy per unit mass, u. For the cup, which does not change phase, there is only sensible internal energy. That is:

$$E_{cup} = m_{cup}c_{cup}\left(T - T_{ref}\right)$$

Since nothing in this equation is changing with time, the first term of the energy equation will be zero when the time derivative is applied. The thermal capacitance of the cup is not involved in this constant temperature model.

The internal energy of the solid and liquid water have different specific internal energies, given by u_L for the liquid and u_S for the solid. Their difference is the heat of fusion (or heat of melting): $u_L - u_S = u_{LS} = u_{melt}$ which for water at 0 °C has a value of 330 kJ/kg. The internal energy of the system is thus:

$$E_{cup} = m_{cup}c_{cup}\left(T_{cup} - T_{ref}\right) + m_{liquid}u_L + m_{solid}u_S$$

The total mass of water remains constant: $m_{H_2O} = m_{liquid} + m_{solid}$

Therefore, the mass of liquid can be eliminated:

$$E_{cup} = m_{cup}c_{cup}\left(T_{cup} - T_{ref}\right) + (m_{H_2O} - m_{solid})u_{liquid} + m_{solid}u_s$$

Taking the time derivative:

$$\frac{dE}{dt} = -u_L\frac{dm_{solid}}{dt} + u_s\frac{dm_{solid}}{dt} = -u_{melt}\frac{dm_{solid}}{dt}$$

The energy balance becomes:

$$-u_{melt}\frac{dm_{solid}}{dt} = UA(T_\infty - T)$$

Every parameter can reasonably be modeled as being constant during this event, and therefore a simple integration, with the initial mass of ice being $m_{ICE,0}$ yields:

$$m_{solid} = m_{ICE,0} - \left[\frac{UA(T_\infty - T_{MP})}{u_{melt}}\right]t$$

Note that the system temperature has been set equal to the melting point of the slurry ($T_{MP} = 0\,°C$). The mass of ice decreases linearly with time, and it is an easy matter to calculate the time it takes to melt the ice to any value. In particular, the total melting time of the ice (t_{melt}) is determined by setting $m_{ice} = 0$ and solving for the time:

$$t_{melt} = \frac{m_{ICE,0}u_{melt}}{UA(T_\infty - T_{MP})}$$

The only difficult parameter to evaluate is the overall heat transfer coefficient.

2.3.4 Results

The results are reported in Fig. 2.9 (spreadsheet structure) and Fig. 2.10 (plot of theory and experiment). For the calculation (Fig. 2.9), the surface area was taken to be the sum of the side surface and the top surface. The bottom surface was assumed to be well insulated, as attempts to do so were made. The initial ice fraction (0.522) was determined using "goal seek" to set the known initial mass by changing the ice fraction. The worksheet was copied twice and values entered in the overall heat transfer coefficient to encompass the experimental result (Fig. 2.10).

INPUT PARAMETERS

Tmp	0	oC	Melting point
Tinf	18	oC	Ambient temperature
Di	0.074	m	Inner diameter of cup
Hinner	0.144	m	inner height of cup
Houter	0.166	m	outer height of cup
thick	0.00381	m	thickness of cup
rhoS	915	kg/m^3	ice density
fice	0.522		initial mass fraction of ice
umelt	330000	J/kg	heat of fusion

VARIED INPUT PARAMETER

U	5	W/m2/K	overall heat transfer coefficient

DERIVED PARAMETERS

Vtotal	6.18E-04	m^3	total cup volume
Atop	5.22E-03	m^2	surface area top
Asides	0.0424	m^2	surface area of outside
Atotal	0.0476	m^2	total surface area
mice,0	2.95E-01	kg	initial mass of ice
tmelt	22704	sec	melting time
tmelt	6.31	hrs	melting time

DATA PLOT

time(min)	mice (kg)
0	0.295
378.4	0

Fig. 2.9 Spreadsheet calculation of the melting cup of ice

The heat transfer coefficient is determined empirically to be a value between 5.0 and 5.5 $W/m^2/K$. This value is of the same order of magnitude, but lower (by a factor of roughly 3) than the cooling mug of coffee. The external convection environment in the two examples are comparable (object sitting in a still room). There are differences, however, such as the thermal conductivity of the material, the effect of temperature on heat transfer coefficient, and direction of the plume created that might account for the observed difference in overall heat transfer coefficient. These effects constitute much of the subject of ensuing chapters.

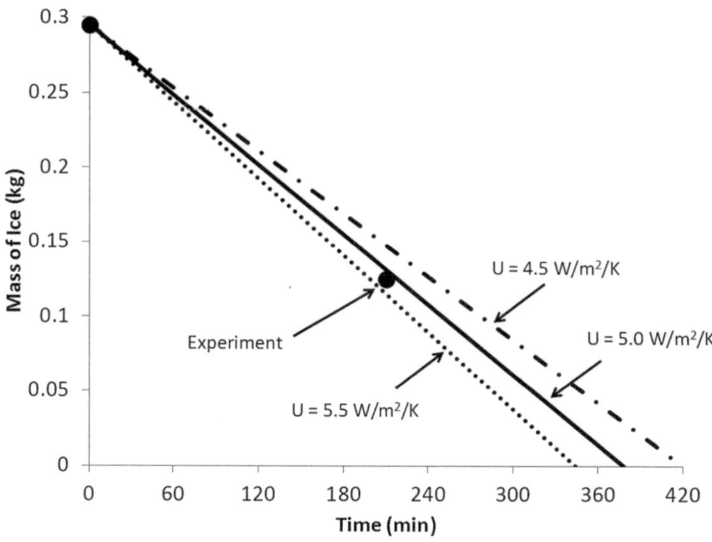

Fig. 2.10 Model predictions for several values of heat transfer coefficient and experimental results for the melting cup of ice

Appendix 1. Raw Data for Cooling Mug of Coffee

Time (min)	Top (°C)	Side (°C)
0	92	21
0.33	80	60
0.67	78	70
1	75	74
1.5	72	75
2	73	75
2.5	74	75
3	74	75
4	73	74
5	72	73
6	70	71
7	69	70
8	68	69
9	67	68
10	68	67
11	65	66
12	65	64
13	64	63
14	62	62

(continued)

(continued)

Time (min)	Top (°C)	Side (°C)
15	61	61
16	60	60
17	60	59
18	59	58
19	58	58
20	57	57
21	57	56
22	56	56
23	57	56
24	56	55
25	55	55
30	52	51
35	49	48
40	46	45
45	44	43
50	42	42
55	41	40
60	40	39
70	36	35
80	35	34
90	33	33
100	32	31
110	30	30
125	28	28
140	27	27
165	25	25
180	24	24
210	23	23
250	22	22

Appendix 2. Measurement of the Mug-Specific Heat

Problem Statement

A mug of mass $m_{mug} = 0.335$ kg is placed in a large pot of warm water at $T_{mug} = 41\ ^\circ C$ and maintained for several minutes. The mug is removed, placed on a countertop, and 500 mL of water at $T_{water} = 14.5\ ^\circ C$ is poured into it. After approximately 2 min, the temperature of water in the mug settles at $T_{final} = 17.5\ ^\circ C$. Determine the specific heat of the mug.

Analysis

The First Law of Thermodynamics is applied to a system that consists of the mug and water poured into it:

$$\Delta E = Q - W$$

The heat Q represents heat exchange between the mug/water and the environment during the 2 min settling period, and while not zero, is assumed to be negligible. There is no work exchange, and evaporation effects are neglected. Therefore, the change in energy is zero, or the initial energy equals the final energy:

$$E_{initial} = E_{final}$$

Considering the energy of the two components (water and mug) relative to any reference temperature, T_{ref}, and assuming constant specific heats:

$$m_{mug}c_{mug}\left(T_{mug} - T_{ref}\right) + m_{water}c_{water}\left(T_{water} - T_{ref}\right)$$
$$= m_{mug}c_{mug}\left(T_{final} - T_{ref}\right) + m_{water}c_{water}\left(T_{final} - T_{ref}\right)$$

The terms involving the reference temperature all cancel, and the energy balance can be solved for the specific heat of the mug:

$$c_{mug} = \left(\frac{m_{water}}{m_{mug}}\right)\left(\frac{T_{final} - T_{water}}{T_{mug} - T_{final}}\right)c_{water}$$

Inserting numerical values, using the density of water to be 1,000 kg/m^3 and the specific heat of liquid water to be 4,200 J/kg/K:

$$c_{mug} = \left(\frac{0.5}{0.335}\right)\left(\frac{17.5 - 14.5}{41.0 - 17.5}\right)4200\,\text{J/kg/K}$$

$$c_{mug} = 800\,\text{J/kg/K}$$

Workshop 2.1. Cooling Coffee Data

In a rough experiment, hot coffee was poured into a room temperature ceramic mug located on a table in a still room. A handheld infrared temperature sensor was used to measure the temperature of the free surface (top) of the coffee, and the outer side wall temperature (side) as a function of time. The data is contained in an excel file (CoffeeData.xlsx, listed also in Appendix 1) which has columns for time, top surface temperature, and side wall temperature.

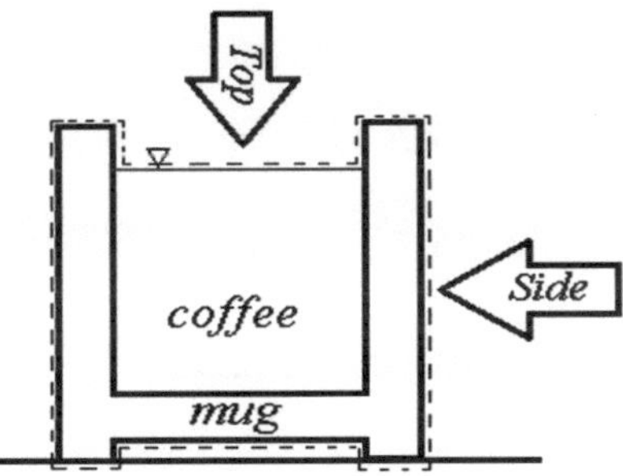

IN CLASS EXERCISE: (follow this general approach in documenting future workshops)

- Open a WORD file and save it as "WKSP_2_1_****.docx", where **** is the first four letters of your last name. Add front matter to the DOC file (i.e., name, date, title). You will be writing a short technical memo, whose target audience is your peer group (but someone not doing this exercise...). Write a few sentences of Introduction.
- Download "CoffeeData.xlsx", or enter raw data from Appendix 1. Rename it "CoffeeData_****.xlsx".
- Prepare a well-formatted plot of the top and side temperatures vs. time for the entire cooling event. Remove grid lines. Add an indication on the plot that indicates clearly the window of time for which the coffee is considered to be drinkable (define over 60 °C as too hot, and 45 °C as too cold). Copy the figure and paste special (as an "enhanced metafile") into the DOC file and adjust its size. Add a numbered figure caption. Write a few sentences to describe the figure.
- In excel, make a copy of the figure and format the x-axis to focus on the first 5 min of the cooling event. Format, copy, and paste special, and add a few sentences.
- In excel, format the x-axis to be on a logarithmic scale, showing the entire event. Format, copy, and paste special, and add a few sentences.
- Write a concluding paragraph, and submit the DOC file.

Workshop 2.2. Euler Method Simulation of Cooling Mug of Coffee

A mug of coffee, initially at $T_0 = 75$ °C, is placed on a table in a still room at $T_\infty = 21$ °C. After 20 min, the coffee temperature is measured to be 57 °C. The coffee continues to cool for a total of 250 min. The total capacitance of the coffee plus mug is 2,010 J/K, and the total surface area is 496 cm². Use a numerical method to determine the inherent time constant for this problem and overall heat transfer coefficient, and simulate the entire cooling event.

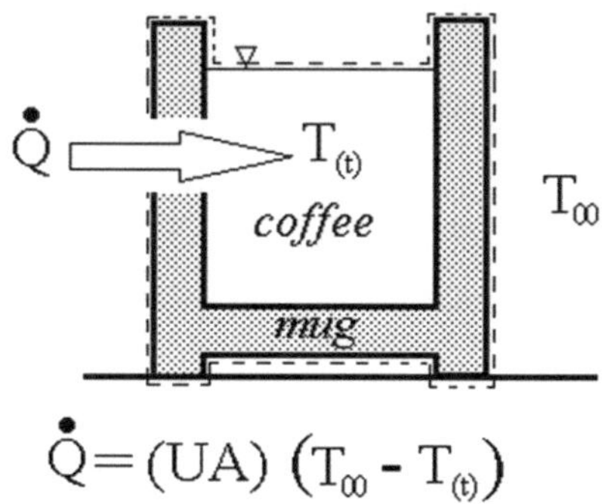

$$\dot{Q} = (UA)\,(T_\infty - T_{(t)})$$

Model Development

The First Law of Thermodynamics applied to the system defined by coffee plus mug (neglecting evaporation, making it a closed system with no mass flows across the boundary) is:

$$\frac{dE}{dt} = \dot{Q}_{in} - \dot{W}_s$$

There is no shaft work, no significant potential energy or kinetic energy changes (so E is internal energy), and no phase change (so the internal energy $E = C(T - T_{ref})$, where C is a total heat capacitance (in J/K), treated as a constant, and T_{ref} is a reference temperature). Assume that the rate of heat transfer is proportional to the instantaneous difference between ambient and coffee temperature (proportionality constant UA, where U is an effective overall heat transfer coefficient, NOT internal energy) and A is the total surface area of the system. The First Law becomes:

$$\frac{dC(T - T_{ref})}{dt} = C\frac{dT}{dt} = UA(T_\infty - T)$$

Upon rearrangement:

$$\frac{dT}{dt} = \frac{(T_\infty - T)}{(C/UA)} = \frac{(T_\infty - T)}{\tau}$$

The combination C/UA must have units of time for dimensional consistency, and that is defined as the time constant for this problem, $\tau = C/UA$. Note that the rate of change of temperature is proportional to the temperature difference driving the heat transfer, which decreases with time, whether it is a heating event (a cold beer) or a cooling event (hot coffee).

Numerical Solution

Forget that this first-order ODE with an initial condition has a simple closed-form (i.e., analytical) solution. The goal of the workshop is to develop a powerful numerical technique that will be applied to problems for which an analytical solution is impossible.

The Euler method "discretizes" time and expresses first derivatives as forward difference approximations with a small but finite step in time (Δt), namely,

$$\frac{dT}{dt} \approx \frac{T_{future} - T_{present}}{\Delta t} = \frac{T^{(p+1)} - T^{(p)}}{\Delta t}$$

The superscripts are not powers, they are indices, with (p) meaning present and (p + 1) meaning one step into the future. Inserting this approximation into the governing ODE:

$$\frac{T^{(p+1)} - T^{(p)}}{\Delta t} = \frac{\left(T_\infty - T^{(p)}\right)}{\tau}$$

Solving for the temperature one step into the future yields the Euler formula for this case:

$$T^{(p+1)} = T^{(p)} + \Delta t \left[\frac{\left(T_\infty - T^{(p)}\right)}{\tau}\right]$$

Notice that on the right hand side, the temperature has been given a superscript (p) because it is changing with each step. The ambient temperature and time constant are not changing, so they do not need superscripts. They are fixed input parameters. The time step is a numerical parameter that can be changed by the user. A simulation is considered accurate if the predicted results are insensitive to the choice of time step.

Guided Spreadsheet Implementation

- Open "CoffeeData.xlsx" and rename it "WKSP2_******.xlsx".
- On a blank worksheet renamed "SIMULATION", enter all fixed physical quantities in a block labeled "INPUT PARAMS" (i.e., T_0, T_∞, C, A). These values are known and will not change. Below that enter a block labeled "RESULTS" and enter a time constant in a cell to any value. It is a constant in the simulation, but its value is unknown and will be solved for (It is a type of "eigenvalue"). Add a cell to calculate $U = C/\tau/A$ in SI units (W/m^2/K).
- Below that, start a labeled block "NUMERICAL SIMULATION". Enter a numerical step size, Δt, with a value of 0.5 min (but you can change this later to explore its effect...).
- In a "header row," label three columns; for index (p), time (min), and temperature (°C).
- In the next row, enter 0 for "p" and time. Enter reference to the cell, where initial temperature is (T_0). This row represents the initial condition, and these values are known inputs.
- In the next row, enter the three appropriate formulas, each of which refers to the initial condition row, and to input or numerical parameters. Use dollar signs ($) to lock in rows and/or columns as appropriate for copying the formula down. The index (p) is incremented by 1, the time is incremented by Δt, and the temperature is incremented by the Euler formula for this problem.
- Copy this row down until the time elapsed equals the total simulation time (250 min, which requires 500 rows for this time step).
- Create a scatter plot (with straight lines, no markers) of temperature vs. time, and move it near the top of "SIMULATION" so you can monitor it.
- Add a series that enters the two data curves. Change its format to "markers with no lines."
- Manually change the value of the time constant to obtain a better fit (it will never be perfect, there are assumptions in the model that are never completed satisfied...).
- Now use "goal seek" (in Data, What if) by setting the temperature in the cell at 20 min to 57 °C by changing the time constant.
- Play around with the value of the numerical time step (Δt), then set it back to 0.5 min.
- Upload the file.

C'mon, admit it. This was fun!

Workshop 2.3. Floating Thermometer Response Time

All physical (as opposed to optical) temperature measuring devices are subject to both static and dynamic errors. A static error occurs when the sensor is in thermal equilibrium with its environment, but its temperature is different than the object or fluid it is intended to measure. For example, a temperature sensor placed in sunlight will be hotter than the fluid surrounding it. A dynamic error occurs when the temperature being measured changes with time, and the temperature sensor is not in thermal equilibrium. This workshop focuses on dynamic response, specifically, the determination of the inherent response of a particular thermometer to a sudden change in temperature. The response depends on the nature of the fluid the thermometer is placed in.

A measurement of the response time of a temperature sensor in a particular fluid environment can be obtained by the following general steps:

1. Precondition the temperature of the temperature sensor to a different temperature than the fluid.
2. Suddenly place the thermometer into the fluid.
3. Measure the indicated temperature as a function of time until the thermometer achieves equilibrium with the fluid.
4. Analyze the data for response time, and overall heat transfer coefficient.

In designing the experiment, the temperature of the fluid should not change substantially in the time it takes for equilibrium to be achieved, which can require patience. It is important to know the final equilibrium temperature to determine the true response time.

A response time experiment was conducted with a floating thermometer, an inexpensive liquid-in-glass sensor designed to measure the temperature of consumer aquariums. The construction consists of a glass cylinder of length 10.8 cm and outer diameter 1.1 cm (with rounded ends). Tiny balls made of lead are located at one end which causes the thermometer to float vertically in water, with approximately 10.0 cm submerged (0.8 cm above the surface). Above the lead balls is a reservoir of sensing liquid and a smaller cylinder in which the thermometer fluid rises and falls with temperature. A calibrated scale placed adjacent to this inner cylinder allows for measurement. For thermal modeling purposes, the mass of the whole system can be considered to be dominated by the lead.

Measurement of Response Time

The response was determined for two different fluid environments: ambient air and nominally room temperature water. In both cases, the thermometer was preconditioned to a lower temperature by allowing it to equilibrate with a cold water bath. The raw data is given in Fig. 2.11. Blank columns are left for the calculation of the dimensionless temperature and the response time, which can be calculated for each individual data point. Note that the initial temperature is the first data point. The fluid temperature (T_{inf}), reported at the top of each table, was determined after approximately five times longer than the last reported data point.

In addition to calculating the response time associated with each point, prepare plots of the raw data (temperature vs. time) and a plot of the dimensionless

FLUID: Water

T_{inf} (oC) =　16.6

time(sec)	Temp (0C)	$\theta = (T - T_{inf})/(T_{init} - T_{inf})$	Response Time (sec)
0	5		NA
6	9		
16	14		
27	17		
39	19		

FLUID: AMBIENT AIR

T_{inf} (oC) =　16.6

time(sec)	Temp (0C)	$\theta = (T - T_{inf})/(T_{init} - T_{inf})$	Response Time (sec)
0	6		NA
13	8		
23	9		
32	10		
43	11		
56	12		
72	13		
97	14		
129	15		
184	16		

Fig. 2.11 Spreadsheet templates and raw data for dynamic response of a floating thermometer

temperature vs. time on semi-logy axes. An exponential fit (with the x-intercept forced to a value of unity) can be added to each curve. The degree to which the system obeys a first-order response is determined by the consistency of the individual response times, or equivalently from the linearity of the exponential fit.

Measurement of Overall Heat Transfer Coefficient

The overall heat transfer coefficient (U) for the two cases can be determined from the definition of the response time, namely, $\tau = RC = C/(UA)$. An estimate of the total capacitance (C) can be made by assuming that the thermal properties (i.e., specific heat) is approximately that of the lead used to weigh it down. The mass of the thermometer can be determined by measuring the fraction of the length that is above water when floating, and using Archimedes' principle (the weight of an object equals the weight of the displaced fluid).

Heat Transfer Modes: Conduction, Convection, and Radiation

3

Heat transfer is energy transfer driven by a temperature difference. A first step in estimating the rate of heat transfer between, say, a hot mug of coffee and its environment is to identify the modes of heat transfer and the thermal channels of heat flow between two locations. There are different pathways, or thermal channels, that heat can take from hot coffee to ambient; directly from the coffee to the air above the free surface, through the side wall of the mug, or through the bottom of the mug and through the floor on which the mug rests.

Within each of these pathways, heat flow encounters discontinuities in material properties, from coffee, to mug, to air. The mathematical expression of the different underlying physics associated with these discontinuities gives rise to three modes of heat transfer: conduction, convection, and radiation. In contrast, there are two mechanisms of heat transfer (conduction and radiation) that refer to the physical phenomena that give rise to the energy transfer. The mode of heat transfer by convection is based on a conduction mechanism. A thermal channel is defined as a "pathway" for heat to flow from a hot fluid or surface to a cold one, expressible in terms of a series combination of heat transfer modes.

In this chapter, the practical application of a single-paned window is used to develop the engineering tools (and the thinking process behind them) used in heat transfer analysis. This example involves a single thermal channel. The basic method is applied more generally, and to the coffee/mug problem with its multiple channels, in subsequent chapters.

3.1   Example: Single-Paned Window

On the north side of a building, (so that there is no solar gain), a single-paned glass window (shown schematically in Fig. 3.1) of thickness $\Delta x = 0.012$ m, height $H = 2$ m, and width $W = 8$ m (into the page) is exposed on the inside to room air maintained at $T_H = 25\ ^\circ$C. The outside of the window is exposed to a crosswind of 10 miles per hour ($= 4.47$ m/s) and an outside air temperature of $T_C = 0\ ^\circ$C.

© Springer International Publishing Switzerland 2015
G. Sidebotham, *Heat Transfer Modeling: An Inductive Approach*,
DOI 10.1007/978-3-319-14514-3_3

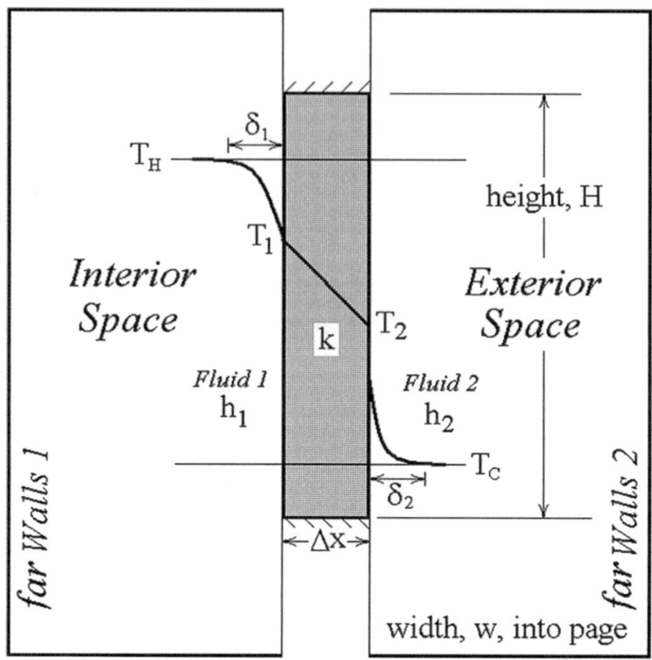

Fig. 3.1 Cross-section of single-paned window with representative temperature distribution. Both surfaces are exposed to a fluid (air, in this case) and are enclosed by far walls which exchange thermal radiation with the window

Estimate the rate of heat transfer through the window (in Watts), the temperature of the inside and outside surfaces of the glass (T_1 and T_2) and the daily energy cost associated with heat loss through the window, assuming the price of heating to be $p = \$20/GJ$.

3.2   Heat Transfer Modes and Thermal Channels

The three modes of heat transfer used in engineering practice are introduced in this section. A brief definition of the modes is followed by the development of a thermal channel and its representation as a resistance network, and finally the mathematical models used for the modes are defined.

Conduction: Heat transfer through a solid is called conduction, and is governed by Fourier's Law, defined shortly. In the window problem, conduction heat transfer takes place between the inner and outer surfaces of the glass of a known thickness Δx. In the mug of coffee example, conduction occurs across the mug wall.

Convection: Heat transfer between a fluid (gas or liquid) and a solid surface is called convection and is governed by Newton's Law of Cooling. There are two convective surfaces in the window problem, the inside surface, which is exposed to room air, and the outside surface, which is exposed to outside air (with a crosswind). In reality, the physical mechanism of convection is conduction across a thin layer of fluid, called a thermal boundary layer, shown with thicknesses δ_1 and δ_2 in Fig. 3.1. However, the thickness of these layers is unknown, and they are heavily influenced by the motion of the fluid.

Radiation: Heat transfer between two solid surfaces that "see" each other and thereby exchange electromagnetic energy, or radiation, is governed by the Stefan–Boltzmann Law. Like convection, radiation is associated with a specific surface, and there are two of them in the window problem. However, the physical mechanism of heat transfer is very different from that of convection. Accelerating charged particles (electrons and protons) in the solid emit photons (or electromagnetic waves) that carry energy with them. Simultaneously, photons that emanate from other surfaces that impinge on a surface can be absorbed. Therefore, a net exchange of energy occurs. Radiation is much more difficult to model than conduction and convection, in general. Good engineering judgment, from experience, dictates when it is essential to include it, and how to do so in an approximate way.

Mass Transfer: In addition to the three modes of heat transfer, energy flows associated with mass crossing the boundary can be important. There are two mechanisms for mass transfer effects: advection and diffusion. Advection (often called "convection" but with a very different meaning than the mode of convection between a fluid and solid surface) is the energy associated with bulk fluid flow across a boundary. The enthalpy is the natural thermodynamic property that accounts for both internal energy flowing in and the work required to do so (the "flow work"). For example, fluid flowing into a heat exchanger carries energy with it, and exits the heat exchanger at a different energy level. The other way energy crosses the boundary is by molecular diffusion (driven by concentration gradients). Evaporation effects fall into this category. These effects are developed in Chapter 13, and results will be discussed in relation to the cooling mug of coffee problem.

As will be seen, conduction and convection are naturally expressed in an Ohm's Law form, where the heat transfer rate (current) is proportional to a temperature difference (voltage drop), and inversely with a thermal resistance (which may be a variable). There are important practical cases where radiation can be modeled in Ohm's Law form as well and others where it can be modeled as a current source. These cases are developed in this chapter. Cases where these models do not capture the controlling physics are left to other textbooks. Mass transfer by advection can be modeled as a dependent current source (i.e., the current source depends on the enthalpy values entering and exiting, developed in Chapters 11 and 12). Mass transfer by diffusion can be modeled either as a current source, or in an Ohm's Law fashion, but with effective R values that are extremely temperature dependent.

A lumped capacity (or 1-node) thermal resistance model of the single-paned window is shown in Fig. 3.2. All the capacitance of the window ($C = \rho c V = \rho c \Delta x H w$) has been lumped into a single node, placed at the midway point of the glass, and given a temperature with a bar over it to indicate that it is an average window

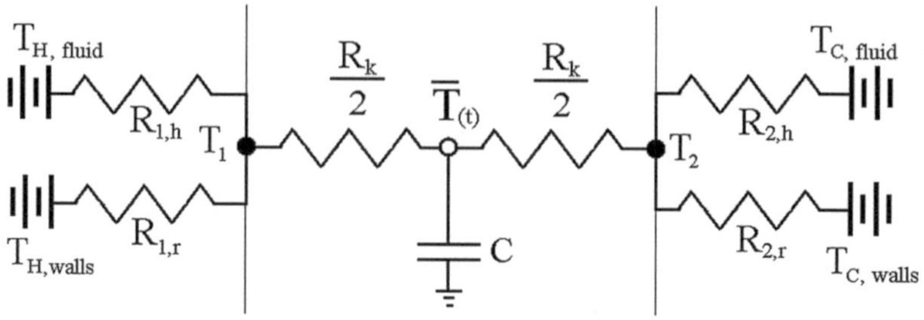

Fig. 3.2 Lumped capacity (i.e., 1-node) model of a single-paned window

temperature, and it varies with time, in general. This node is represented with an open circle, indicating that it carries capacitance and is governed by a differential equation, in the general case. Nodes are placed at the inner and outer surfaces of the window because these locations are discontinuous in space (air on one side, glass on the other), and there is a distinct change in heat transfer mode from one side to the other. These temperatures are unknown (prior to analysis), but can be related to the capacitive node through algebraic manipulation and are therefore represented by filled circles.

There are four fixed temperatures represented by batteries; fluid and far wall, on each side of the window. The fluid (air) in the immediate vicinity of the window exchanges heat by the mode of convection, and a thermal resistance with a subscript "h" is used (because "h" will be used in defining a convection coefficient). In addition, each surface exchanges heat with "walls" that can be far removed from the window. The average temperature of these walls is treated as fixed. The window surfaces exchange heat with these removed surfaces through the mode of radiation, and a thermal resistance with a subscript "r" is used for these modes. For example, the outer window surface "sees" anything that an observer inside the building looking outside would see; other buildings, the street, the sky. The inner window surface "sees" what someone would see looking in: ceiling, floor, walls, furniture, people, etc. In general, the fluid temperature in the vicinity of the window and the far wall temperatures could be different from each other. A wisely chosen average value is required if the floors, ceilings walls, and occupants are at widely different temperatures.

The mode of heat transfer between the inner and outer window surfaces and the midpoint of the window is conduction, and conductive thermal resistances ($R_k/2$) are placed between each surface node and the interior node. The total conductive resistance between the inner and outer surface (R_k) has been split into two equal resistance values.

The RC network of Fig. 3.2 has four thermal channels; from each of the four fixed ambient temperatures to the average temperature of the window itself. After setting up the general transient solution, this problem will be reduced to a single thermal channel from inside to outside by considering only the steady-state solution, and setting the fluid and radiation wall temperatures equal to each other. Then heat will flow from inside to outside through a single thermal channel.

A nodal energy balance applied to the window node is:

$$C\frac{d\overline{T}}{dt} = \frac{T_{H,fluid} - \overline{T}}{R_{1,h} + R_k/2} + \frac{T_{H,walls} - \overline{T}}{R_{1,r} + R_k/2} + \frac{T_{C,fluid} - \overline{T}}{R_{2,h} + R_k/2} + \frac{T_{C,walls} - \overline{T}}{R_{1,r} + R_k/2}$$

Each of the four heat transfer rate terms on the right side model the four thermal channels. If the five thermal resistances are treated as constants, this equation can be recast into a general first-order system with a steady-state temperature and an effective RC time constant. Alternatively, the energy balance can be solved numerically, and that would allow for models for which the thermal resistances depend on temperature.

The temperatures of the two window surfaces (T_1 and T_2) are not directly solved for in the governing equation. However, they are interior temperatures within thermal channels and, as derived in Chap. 1, they can be expressed in terms of their neighbors as:

$$T_1 = \frac{\frac{T_{H,fluid}}{R_{1,h}} + \frac{T_{H,walls}}{R_{1,r}} + \frac{\overline{T}}{R_k/2}}{\frac{1}{R_{1,h}} + \frac{1}{R_{1,r}} + \frac{1}{R_k/2}} \quad \text{and} \quad T_2 = \frac{\frac{T_{C,fluid}}{R_{2,h}} + \frac{T_{C,walls}}{R_{2,r}} + \frac{\overline{T}}{R_k/2}}{\frac{1}{R_{2,h}} + \frac{1}{R_{2,r}} + \frac{1}{R_k/2}}$$

However, the window problem at hand is really a quasi-steady state problem. The problem statement states that the inside and outside temperatures are fixed. Even if the inside or outside temperatures were changing with time, in practice, it is expected that they would do so gradually, and the window would maintain the quasi-steady temperature obtained by setting the time derivative to zero. Formally, it would have to be shown that the rate of heat flowing into the capacitor is small compared to the rate of heat that flows from the inside surface (T_1) to the outside surface (T_2). Here is a case that could certainly be tested by running a transient simulation as outlined, and then neglecting the capacitance effects. The problem statement would be modified to specify the manner in which either the hot or cold ambient temperatures varied with time.

At steady-state, then, heat does not flow into or out of the capacitor, and in representing the network, the capacitor can be removed. In this case, heat from the inside window surface to the outside surface takes place through two series resistors of $R_k/2$. The average temperature of the window (by a voltage divider) equals the geometric average of the inside and outside temperatures. The simplified thermal resistance network that carries no capacitance is shown in Fig. 3.3. In general, for problems that are considered to be at steady-state, RC networks need not show the inherent capacitances associated with material because there is no change in energy stored, as is assumed in this case.

A further simplification can be made if it is reasonable to assume that the fluid temperatures and the radiation wall temperatures are equal. That is, if

$$T_{H,fluid} = T_{H,walls} = T_H \quad \text{and} \quad T_{C,fluid} = T_{C,walls} = T_C$$

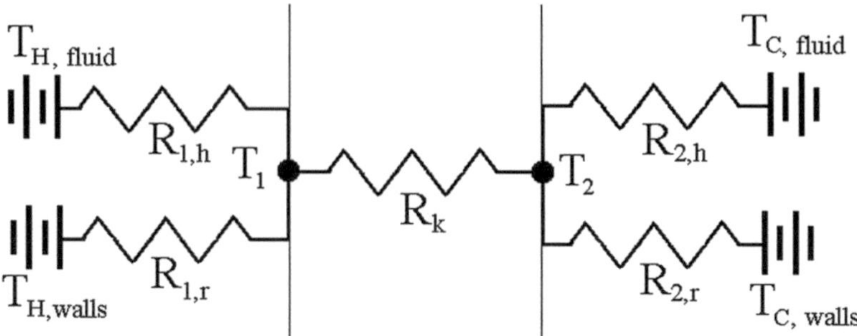

Fig. 3.3 Single-paned window resistance network at steady-state

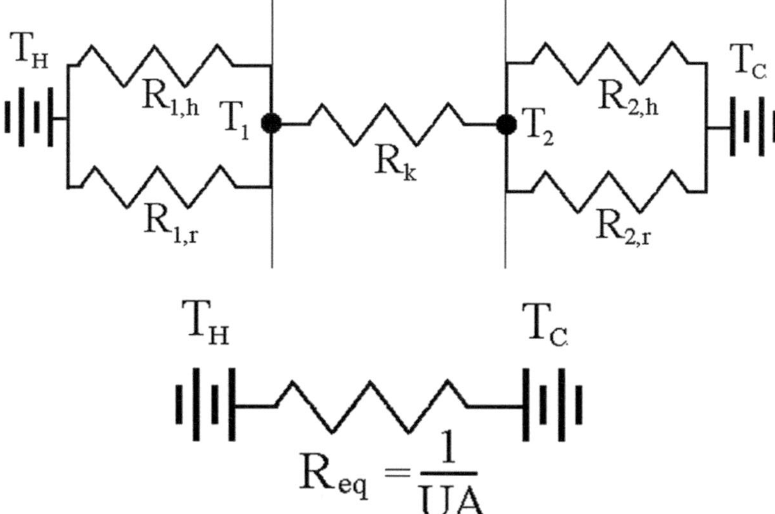

Fig. 3.4 Single-paned window resistance network for steady-state conditions, with equal fluid and wall temperatures, and shown in terms of a single equivalent resistance

then only a single battery on either side can be used, and the simplified resistance network is shown in Fig. 3.4. Notice now that radiation and convection act in parallel with each other, and in series with conduction (and the convection/radiation on the other side). Care must be taken in assuming that these two driving temperatures are the same. There are important practical cases where an object exchanges heat with a radiation partner at a very different temperature than the fluid in the immediate vicinity of the object.

This problem has been reduced to a model of heat flow through a single thermal channel from one temperature source to another. Within this channel, all three modes of heat transfer are involved, with distinct temperature drops along the way.

The heat transfer rate through the window, as sought from the original problem statement, is expressed in terms of the overall temperature drop and thermal resistances and, equivalently, in terms of an overall heat transfer coefficient (U):

$$q = \frac{T_H - T_C}{R_{eq}} = UA(T_H - T_C)$$

where the equivalent thermal resistance is:

$$R_{eq} = \left(\frac{1}{R_{1,h}} + \frac{1}{R_{1,r}}\right)^{-1} + R_k + \left(\frac{1}{R_{2,h}} + \frac{1}{R_{2,r}}\right)^{-1}$$

The basic laws that govern the modes of heat transfer, and methods for estimating resistance values, are developed in the following sections.

3.3 Fourier's Law of Conduction

Focusing first on the conductive heat transfer that takes place across the window itself between the inner and outer surfaces of the glass, Fig. 3.5 shows a schematic of a solid wall consisting of two surfaces (1 and 2) with solid material between them that are maintained at different average temperatures. It is emphasized that these surface temperatures are not generally equal to their adjacent fluid temperatures. Conduction through the wall is governing by Fourier's Law, which states that, at steady-state, the rate of heat transfer between two surfaces (1 and 2) with solid material between them is proportional to the area of the surfaces (A_{12}, which can be an average area if they are different), the temperature difference between them ($T_1 - T_2$), and inversely with the distance between them (Δx_{12}, which can be an average distance). The proportionality factor (not necessarily a constant) is a material property called the thermal conductivity (k_{12}), which has SI units W/m/K.

Fourier's Law of Conduction is expressed mathematically as:

$$q_{conduction,\,1\longrightarrow 2} = \frac{k_{12}A_{12}(T_1 - T_2)}{\Delta x_{12}} = \frac{T_1 - T_2}{R_{12}}$$

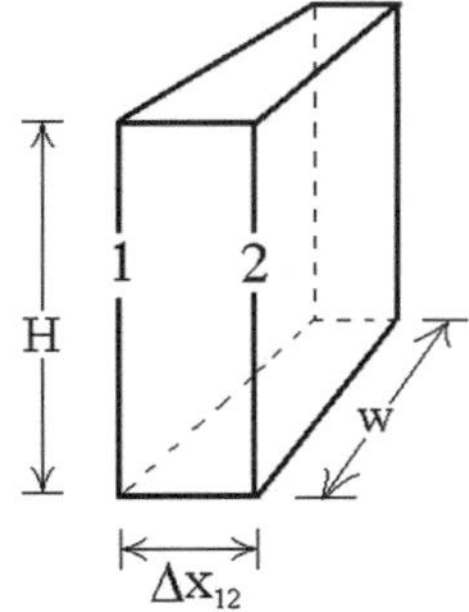

Fig. 3.5 Rectangular solid wall. The surface on the left is considered to be surface 1, and the surface on the right surface 2. All six other faces are considered to be "edge" faces

where a generic definition of a conductive resistance has been introduced as:

$$R_{12} = \frac{\Delta x_{12}}{k_{12} A_{12}}$$

In general, since the area of the two surfaces can be different, and the distance between them might vary somewhat, these expressions are approximate for a more general geometry than a plane wall.

Figure 3.6 shows the thermal conductivity of a variety of materials (solids, liquids, and gases) at nominal room temperature ($25\,°C$). The thermal conductivity of most materials is a mild function of temperature, but a suitably chosen average value can be used for most engineering applications.

For the window problem, the conductive resistance between in inner and outer surfaces is:

$$R_{12} = \frac{\Delta x_{12}}{k_{12} A_{12}} = \frac{0.012\,\text{m}}{(0.96\,\text{W/m/}°\text{C})(2\,\text{m} * 8\,\text{m})} * \left| \frac{1,000\,\text{W}}{\text{kW}} \right| = 0.781\,\frac{°\text{C}}{\text{kW}}$$

Alternatively, Fourier's Law can be expressed in terms of the heat flux crossing a given area at a location x ($A_{(x)}$, where the area is normal to and can vary with x) by postulating that the rate of heat transfer is proportional to the surface area, as before, and to the temperature gradient perpendicular to the area. The thermal conductivity is the proportionality factor. That is:

$$q_{conduction,x} = -kA_{(x)}\frac{\partial T}{\partial x}$$

The minus sign accounts for the direction of heat flow from high to low temperature, according to the Second Law of Thermodynamics. That is, if the temperature increases in the direction of x ($dT/dx > 0$), then heat will flow in the negative x direction.

The Appendix contains a derivation of the thermal resistance for a general geometry in which the area can be expressed as a function of a single dimension. The result for the general formula is:

$$R_{12} = \frac{1}{k_{12}} \int_{x_1}^{x_2} \frac{dx}{A_{(x)}}$$

This formula is applied to three well-defined geometries: rectangular, cylindrical, and spherical. The result $R_{12} = \Delta x_{12}/(k_{12} A_{12})$ is the exact result for a rectangular geometry, or a plane wall (Cartesian, where the area and distance are constant). Across a cylindrical shell (where the distance is constant but the area varies), the thermal resistance for a pipe of length L into the page, outer radius r_2 and inner radius r_1 is:

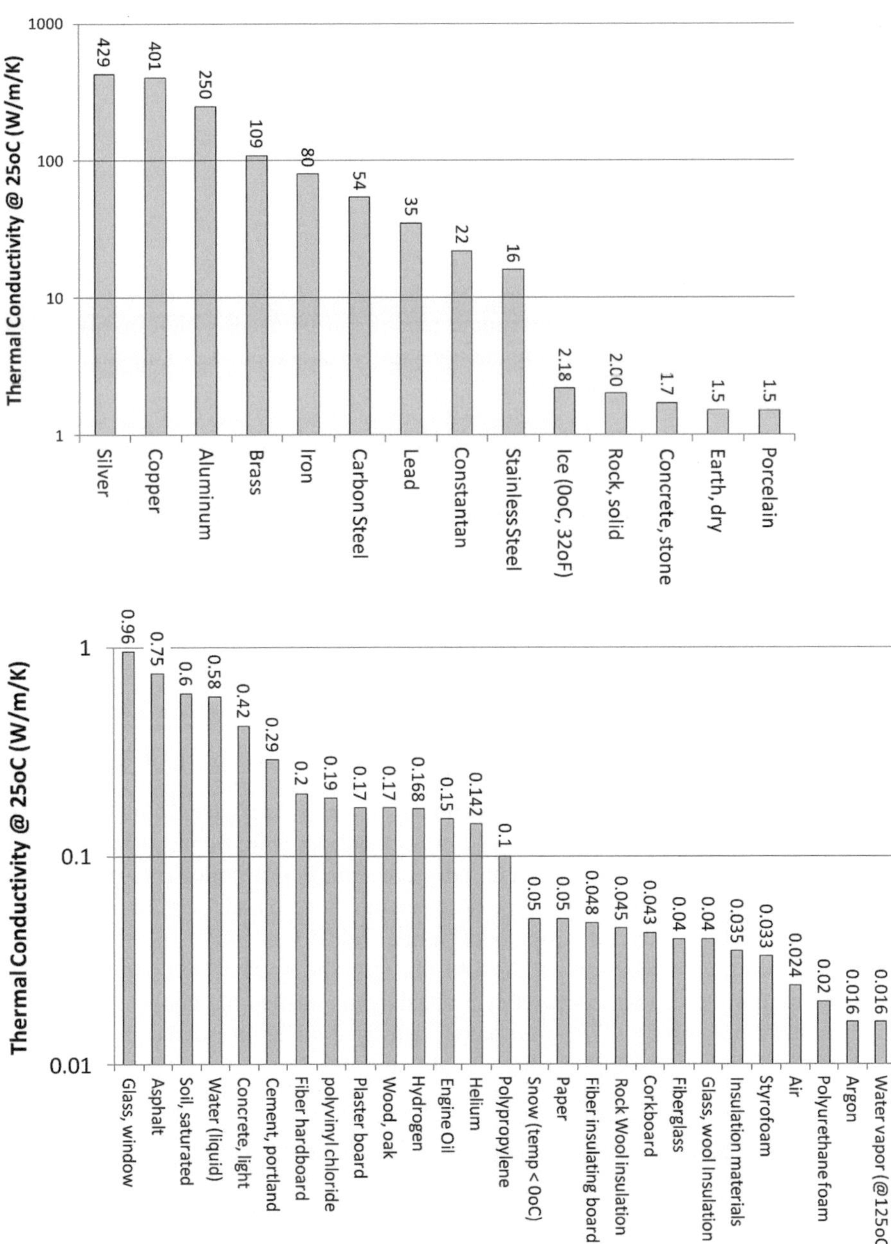

Fig.   3.6 Thermal   conductivity   of   various   materials.   Adapted   from   http://www.engineeringtoolbox.com/thermal-conductivity-d_429.html, accessed 1/30/2011

$$R_{12} = \frac{\ln\left(\frac{r_2}{r_1}\right)}{2\pi k_{12} L} \quad \textit{for a cylindrical shell}$$

For a spherical shell, with outer radius r_2 and inner radius r_1, the thermal resistance is:

$$R_{12} = \frac{1}{4\pi k_{12}}\left(\frac{1}{r_2} - \frac{1}{r_1}\right) \quad \textit{for a spherical shell}$$

For any other geometry, an approximate model can be used, which involves first defining two surfaces, calculating an average distance between them, and an average area. A general approach is:

1. Identify two surfaces of the solid in question between which a conductive resistance is to be estimated.
2. Calculate the area of these two surfaces and the geometric average (A_{12}).
3. Estimate an average distance between the two surfaces (Δx_{12}). This step can be difficult to conceptualize for complicated shapes.

The conductive resistance is $R_{12} = \frac{\Delta x_{12}}{k_{12} A_{12}}$

3.4   Newton's Law of Cooling

The mode of heat transfer between a solid surface and an adjacent fluid is termed convection. It is governed by Newton's Law of Cooling, which states that the rate of heat transfer between a surface and an adjacent fluid is proportional to the surface area and to the temperature difference between the surface and the fluid just outside a thermal boundary layer (a region of influence within which the temperature changes from the surface temperature to the free stream fluid temperature). The proportionality factor (not necessarily a constant) is termed the convection coefficient, with the symbol "h" (not enthalpy, a totally different quantity). Newton's Law of Cooling is expressed mathematically as:

$$q_{convection} = hA_{surface}\left(T_{surface} - T_{fluid}\right) = \frac{T_{surface} - T_{fluid}}{R_{convection}}$$

A thermal resistance associated with a convection boundary has been introduced, which is inversely proportional to the convection coefficient and the surface area. In some models, the convection coefficient can be a function of temperature (and thus variable) and the surface area could change with time (i.e., a moving piston changes the area of the cylinder wall exposed to the gases it contains).

The mechanism of convection is actually conduction. Heat flows by conduction across the thermal boundary layer, which is usually thin compared to the size of the

Table 3.1 Approximate convection coefficient ranges (SI units)

	Gases (W/m²/°C)	Liquids (W/m²/°C)
Natural convection		
A natural convective flow environment is one in which there is no imposed flow velocity on the surface. Gravity causes an induced buoyant flow, however	1–10	100–1,000
Forced convection		
A forced convective flow environment is one in which the surface experiences an identifiable imposed velocity. The wind chill effect is a common experience of forced convection	10–100	1,000–10,000

physical object (except in microscales), driven by a temperature difference. That is, heat is conducted from the fluid to the surface through the fluid in the thermal boundary layer of thickness δ.

$$R_{convection} = \frac{1}{hA_{surface}} \sim \frac{\delta}{k_{fluid}A_{surface}}$$

That is:

$$h \sim \frac{k_{fluid}}{\delta}$$

As might be intuitive, the thermal conductivity of the fluid is directly linked to the convection coefficient (a measure of the effectiveness of the heat transfer between the fluid and an adjacent solid). In addition, the thinner the thermal boundary layer, the higher the convection coefficient. This thickness is influenced by the fluid motion, as well as fluid properties (thermal conductivity, density, and specific heat) and will be explored more fully later.

Chapter 9 develops a procedure for estimating convection coefficients using dimensionless quantities (Nusselt number correlations) that characterize the flow and temperature fields. Until then, appropriate values (or formulas) for convection coefficients will be given in problem statements. Table 3.1 shows the order of magnitude range expected for convection coefficients in SI units. These rules of thumb can be used as a first guess, or as a reality check for a detailed calculation. The art of making appropriate choices for numerical values for convection coefficients and/or whether they are treated as variables (in space and/or tempera-ture) is developed with experience.

For the window problem as stated, the inside window surface is exposed to room air that is nominally still. That is, there is a natural convection with air as the fluid, and a convection coefficient value between 1 and 10 W/m²/°C is expected. The outside window is exposed to a crosswind. That is, there is a forced convection with air as the fluid, and a convection coefficient between 10 and 100 W/m²/°C is expected.

For the window problem considered here, the ballpark estimates for the window problem are:

$$h_1 = 5.0\frac{W}{m^2\,K} \text{ for the inside surface (typical of natural convection in air)}$$

$$h_2 = 15.0\frac{W}{m^2\,K} \text{ for the outside surface (typical of forced convection in air)}$$

The convective thermal resistances in the window problem are therefore:

$$R_{1,h} = \frac{1}{h_1 A} = \frac{1}{(5\,W/m^2/K) * (16\,m^2)} \left| \frac{1,000\,W}{kW} \right| = 12.5\frac{°C}{kW}$$

$$R_{2,h} = \frac{1}{h_1 A} = \frac{1}{(15\,W/m^2/K) * (16\,m^2)} \left| \frac{1,000\,W}{kW} \right| = 4.17\frac{°C}{kW}$$

These resistances are significantly larger than the conductive resistance (0.78 °C/ W), indicating that there is a larger barrier to heat flow at the two convective surfaces compared to the conductive resistance across the window itself. However, radiation acts in parallel, and it must be estimated before a final conclusion can be made on that score...

3.5   Stefan–Boltzmann and Kirchhoff's Radiation Laws

The mechanism of radiation heat transfer is fundamentally different than that of conduction. Conduction involves the direct interaction of adjacent molecules that vibrate at different energy levels (indicative of temperature), and energy is transferred from regions of higher to lower temperature. On the other hand, radiation heat transfer is the result of energy transport associated with electromagnetic waves, or photons, which requires no medium. Each photon carries energy with it in the amount $ℏν$ (where $ℏ$ is Planck's constant and $ν$ is the frequency) at the speed of light.

A black body is one that absorbs all photons incident upon it. The ability of an object to absorb photons is closely related to its ability to generate (emit) photons. The Stefan–Boltzmann Law states that the total power emitted from a black body with a surface area A and absolute temperature T is given by:

$$q_{BlackBody} = \sigma A T^4 \quad (\text{in Watts})$$

where $\sigma = 5.669(10)^{-8}\,W/m^2/K^4$ is the Stefan–Boltzmann constant. This equation must be in absolute (K or R) temperature units. This energy flux represents an integration over all wavelengths of light. This law does not depend on the material in any way, other than its blackness. However, no real bodies are perfectly black, and they emit less total radiation than a black body. For engineering purposes, the introduction of the so-called grey body defines a material property, the emissivity (ε), that represents the ratio of the actual emitted power to what a black body would emit:

$$\varepsilon = \frac{q_{GreyBody}}{q_{BlackBody}}$$

The emissivity is a crude, but very successful, modeling tool for estimating radiation heat transfer. There is an extremely complex relationship between wavelength and emissivity, influenced strongly by quantum mechanical effects. The molecules of a real body tend to emit radiation strongly in localized bands of frequency. Emissivities of representative material surfaces are shown in Fig. 3.7. Highly reflective materials tend to have low emissivities (bottom plot). Most other materials tend to have somewhat higher emissivities.

Not only do surfaces emit radiation, but they can also absorb incident radiation. The absorptivity (α) is defined as the fraction of incident light that is absorbed by the material. At a given wavelength of light, the absorptivity equals the emissivity (Kirchhoff's Law of Thermal Radiation).

However, the emissivity can be very different at different wavelengths. For example, glass has a relatively high emissivity (and absorptivity) for infrared electromagnetic radiation, but a low emissivity (and absorptivity) to visible light. Therefore, windows are transparent to visible light, but are opaque (with a high emissivity) to infrared, which is where terrestrial thermal radiation is centered.

To add further to the complexity, while a given surface is emitting radiation (from the vibration of its molecules), photons that strike it from other sources can either be transmitted (pass through, like visible light through a window), absorbed near the surface, or reflected. Keeping track of all these phenomena involves complicated geometric calculations, and a general treatment of radiation is beyond the intended scope of this first course in heat transfer.

Fortunately, useful engineering approximations to radiation heat transfer can often be made, so that radiation can often be included in a thermal resistance model. In such limiting cases, radiation heat transfer involves heat transfer from (or to) a surface. Radiation is like convection in that way. However, because the physical mechanism is entirely different, the net heat transfer rate from (or to) the surface is the algebraic sum of radiation and convection. That is, radiation and convection act in parallel, not series. Convection involves thermal contact with the fluid immediately adjacent to the surface. Radiation involves thermal contact with distant surfaces as electromagnetic waves travel between them. Two situations allow for simplified models: an object completely surrounded by a much larger object, and a radiation source imposed on an object.

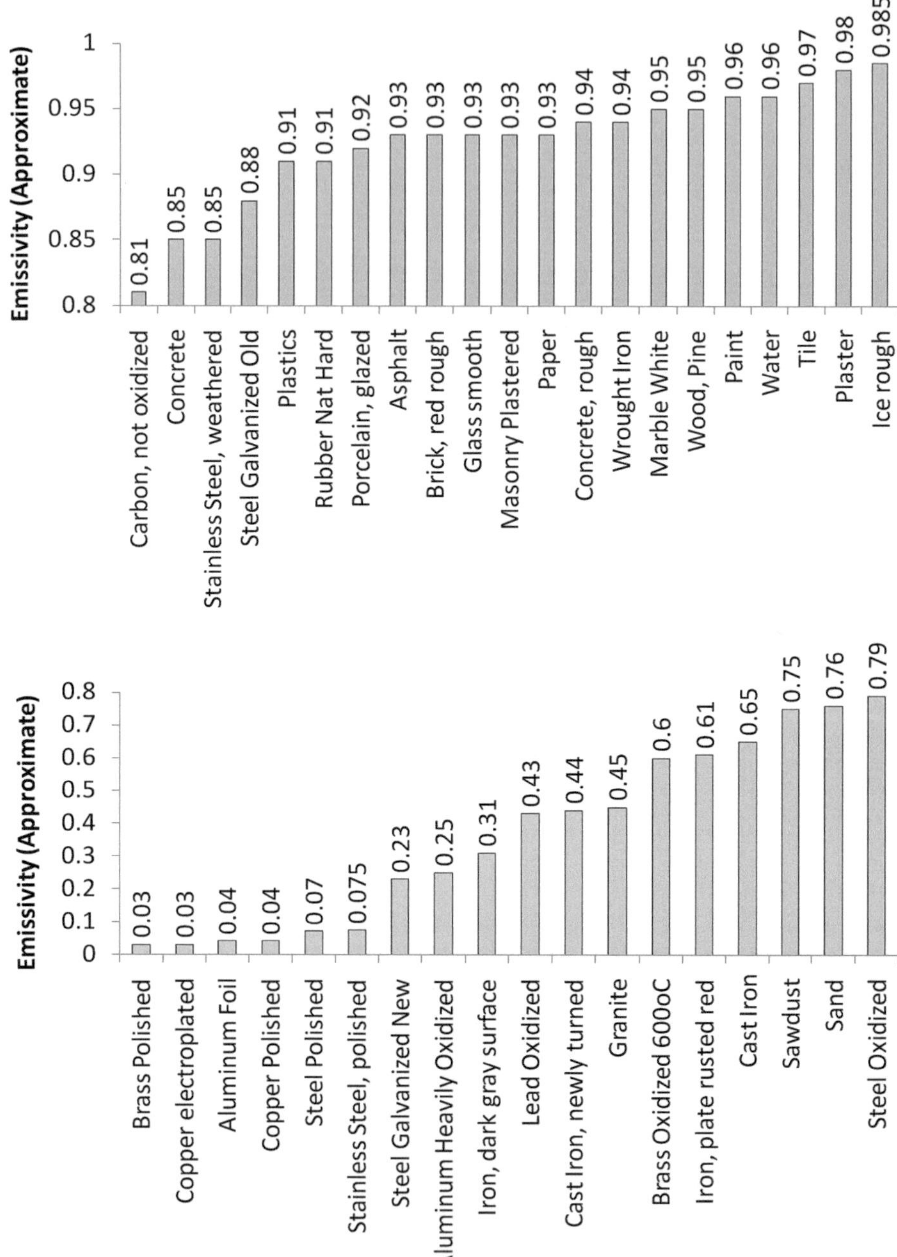

Fig. 3.7 Emissivity of various materials. The uncertainty (not shown) of all these values is quite high. Data adapted from http://www.engineeringtoolbox.com/emissivity-coefficients-d_447.html, accessed 2/5/2012

3.5.1 Case 1: Surrounded Object

If two objects, separated by a gas (or vacuum) that does not significantly absorb radiation, can "see" each other, they exchange heat via radiation. When a convex object (one that cannot see itself) with a surface temperature of T_1 is completely surrounded by a much larger object with a surface temperature T_2 (shown schematically in Fig. 3.8), the rate of heat transfer due to radiation between them can be approximated as:

$$q_{rad,1\longrightarrow 2} = \varepsilon_1 \sigma A_1 \left(T_1^4 - T_2^4\right)$$

where

ε_1 is the emissivity of object 1 (with a value between 0 and 1)
$\sigma = 5.669(10)^{-8}\,\mathrm{W/m^2/K^4}$ is the Stefan–Boltzmann constant
A_1 is the surface area of object 1

These temperatures must be in absolute units (K or R).

Since this relationship is not linear with respect to temperature, it cannot be used in this form for Ohm's Law analogy (i.e., Ohm's Law is NOT $V^4 = IR$). However, by expanding the temperature term, it can be rewritten in an alternate form that isolates on the temperature difference $(T_1 - T_2)$, namely:

$$q_{rad,1\longrightarrow 2} = \varepsilon_1 \sigma A_1 \left(T_1^2 + T_2^2\right)(T_1 + T_2)(T_1 - T_2) = h_r A_1 (T_1 - T_2)$$

where a radiation coefficient, h_r, has been "defined" as:

$$h_r = \varepsilon_1 \sigma \left(T_1^2 + T_2^2\right)(T_1 + T_2)$$

The heat transfer rate can now be expressed as a thermal resistance between the two objects:

$$q_{12} = \frac{T_1 - T_2}{(1/h_r A_1)} = \frac{T_1 - T_2}{R_{rad}}$$

Since the rate of heat transfer by radiation is now expressed in terms of a temperature difference, the temperatures of the nodes in a resistance network need not be in absolute units, as long as the radiation coefficient (h_r) is calculated using absolute temperature.

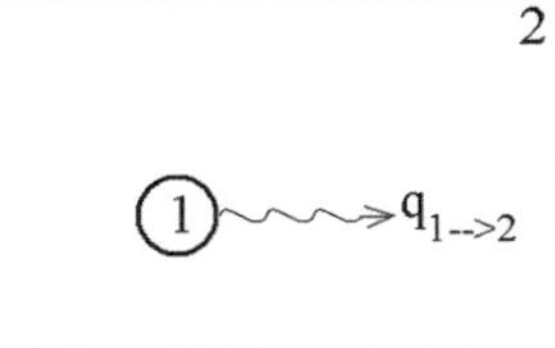

Fig. 3.8 Radiation exchange between a surrounded convex object and a surrounding cavity

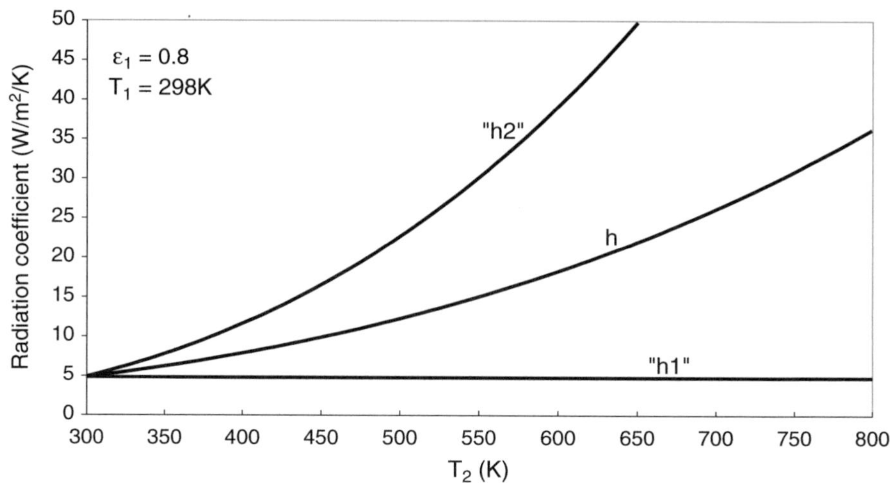

Fig. 3.9 Radiation coefficient for object 1 completely surrounded by object 2

The radiation coefficient depends strongly on the two surface temperatures, but only on the emissivity of the smaller, surrounded, object. If these surface temperatures are not known beforehand, the radiation coefficient cannot be evaluated initially, and an iterative solution technique is needed. However, an excellent first (possibly last) approximation can be made by "guessing" the surface temperatures. To help make an intelligent guess, notice that, for $T_2 > T_1$, the value of h_r can be bracketed by:

$$"h_{r1}" = \varepsilon_1\sigma 4T_1^3 < h_r < \varepsilon_1\sigma 4T_2^3 = "h_{r2}"$$

Figure 3.9 shows these radiation coefficients for a surface 1 maintained at room temperature for temperatures up to 800 K. For moderate temperature differences between object 1 and 2, the radiation coefficient can be reasonably assumed to be constant, on the order of 5–10 W/m²/K.

Since radiation and convection act in parallel, radiation is important when the radiation coefficient is of the same order or much larger than the convection coefficient. This situation occurs almost always when the convection environment is natural convection of a gas (in which case $h_{convection}$ is of the order of 1–10 W/m² °C). Radiation can be important in other convection environments, especially at high temperatures, such as in furnaces.

3.5.2 Case 2: Radiation Source

Solar radiation incident on the Earth is approximated by a different model; the sun does not completely surround the Earth. However, solar radiation can be treated as a current source (Fig. 3.10).

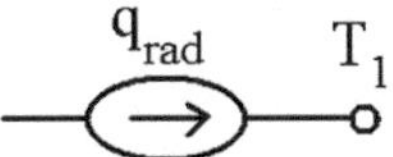

Fig. 3.10 Radiation modeled as a current source to a surface at temperature T_1

Fig. 3.11 Thermal resistance model of the Earth

A flat surface held perpendicular to the sun's rays outside the influence of the atmosphere receives a solar flux of 1,400 W/m^2 (the "solar constant"). This radiation is partially attenuated as it travels through the Earth's atmosphere. On a bright sunny day around noon, a flat plate held normal to the sun will receive at most approximately 1,000 W/m^2. Some of this radiation is reflected, some is absorbed, and the rest is transmitted. How much radiation is absorbed depends on the material. Black objects absorb a lot; reflective objects absorb a little.

Other radiation sources can be treated as a current source. For example, in a toaster oven, electrical energy is converted to thermal energy in a resistive element, creating a glowing hot wire coil. Thermal radiation emitted by the coils is absorbed by the surface of the toast which warms, dries, chars, and eventually burns.

3.5.2.1 Example: Estimate the Steady-State Temperature of the Earth

The Earth is bathed in a constant bombardment of photons from the sun, a portion of which is absorbed by the atmosphere, ground and oceans, the rest being reflected. This absorption of energy is modeled as a current source, labeled "Solar" in the resistance network shown in Fig. 3.11. Simultaneously, the Earth is modeled as an object completely surrounded by a larger "object," namely outer space. The spectral distributions of the absorbed solar and emitted terrestrial electromagnetic waves are very different (the solar flux peaks in the visible light spectrum, while the emitted flux peaks in the infrared).

At steady-state, the rate that energy is absorbed by the Earth (from the sun) equals that emitted by the Earth (to space):

$$q_{\text{solar absorbed}} = q_{\text{emitted}}$$

$$\alpha I_0 A_{shadow} = \varepsilon \sigma A_{surface}\left(T_E^4 - T_{space}^4\right)$$

where α is the net integrated absorptivity of the Earth (equal to approximately 0.9), I_0 is the solar constant (1,400 W/m^2), $A_{shadow} = \pi R_E^2$ is the shadow area the Earth ($R_E = 6.378$ km is the Earth's radius), $A_{surface} = 4\pi R_E^2$ is the surface area of the Earth, ε is the net integrated emissivity of the Earth (approximately equal to the absorptivity), T_E is the temperature of the Earth in K, and T_{space} is the temperature of outer space (0 K). In this example, the use of the resistance coefficient h_r is not

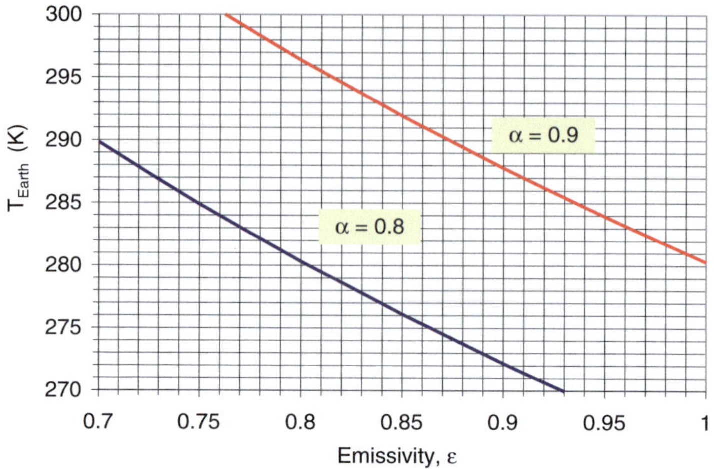

Fig. 3.12 Predicted Earth's temperature as a function of radiation parameters

really needed because outer space is near absolute zero, and the equation can be solved directly for the average temperature of the Earth:

$$T_E = \left[\left(\frac{\alpha}{\varepsilon}\right)\left(\frac{I_0}{4\sigma}\right)\right]^{0.25} = \left(\frac{\alpha}{\varepsilon}\right)^{0.25} 280.1\,\text{K}$$

Note that this temperature is independent of the size of the spherical planet, and a first approximation to other planets can easily be made using the appropriate solar constant based on the planets distance from the sun.

The temperature of the Earth from this model is plotted in Fig. 3.12 as a function of emissivity for two values of absorptivity (0.8 and 0.9). As emissivity increases, the temperature decreases. As absorptivity increases, the temperature increases. Notice that for $\alpha = \varepsilon$, the predicted temperature of the Earth is $280\,\text{K} = 7\,°\text{C} = 44\,°\text{F}$, which is about right. This simplified model captures the controlling physics of a rather complex problem!

Comment on Global Warming

This model yields insight to the much discussed and controversial greenhouse effect, in which radiation emitted from the Earth's surface is absorbed by gases in the atmosphere. Light emitted from the Earth's surface and absorbed by the atmosphere would result in a decrease in the Earth's effective emissivity, and result in a higher temperature. The extent to which an apparent warming trend can be attributed to the combustion of fossil fuels will likely remain controversial forever, as climate is a fundamentally nonlinear chaotic process. Nevertheless, anything that can be done to responsibly reduce our reliance on fossil fuels is a good idea.

3.5.3   Case 3: Radiation Between Infinite Plates

Consider two plates that are placed close together so that the gap between them is much less than the width and height, i.e., a double-paned window. There is thermal radiation exchange between the two inner surfaces that enclose the gap. However, these two surfaces that "see" each other do not fit the thermal radiation criterion of an object completely surrounded by another, case 1. However, the following derivation yields a simple formula for radiation between these surfaces that can be cast into an Ohm's Law form.

Photons are continually being emitted in random directions by both surfaces (1 and 2) at a rate that is governed by the Stefan–Boltzmann Law. In a short period of time (Δt), the photon energy that is emitted by surface 1 (the hotter surface, say) per unit area is:

$$n_1 = \varepsilon_1 \sigma T_1^4 \Delta t$$

The photon energy simultaneously emitted by surface 2 is:

$$n_2 = \varepsilon_2 \sigma T_2^4 \Delta t$$

For the case of infinitely large plates, all of the photons emitted by 1 will strike surface 2 and vice versa. That is, the fraction of photons emitted by surface 1 that strike surface 2 is unity. One of the three things will happen to emitted photons. A fraction (α_2) will be absorbed, a fraction will be transmitted (τ_2, i.e., the photon passes through) and a fraction will be reflected (ρ_2). Mathematically:

$$\alpha_2 + \tau_2 + \rho_2 = 1$$

Similarly, for photons that strike surface 1:

$$\alpha_1 + \tau_1 + \rho_1 = 1$$

There is no energy exchange between surface 1 and 2 associated with transmitted photons. However, photons emitted by 1 that are reflected by 2 will return and strike surface 1, their original source. In turn, some of these photons will be reflected back to surface 2. Eventually, all the photons emitted by surface 1 will be either absorbed by surface 2, absorbed by surface 1, or transmitted (in either direction). Any photons that are emitted by 1 and ultimately absorbed by 2 results in an energy exchange between surface (and vice versa), expressed as:

$$q_{1\,to\,2} = n_1(\alpha_2 + \alpha_2\rho_1\rho_2 + \alpha_2\rho_1\rho_2\rho_1\rho_2 + \alpha_2\rho_1\rho_2\rho_1\rho_2\rho_1\rho_2 + \cdots)$$

The first term represents the photons emitted by 1 that are absorbed by 2. The second term represents photons that reflect from 2, then reflect from 1 and are absorbed by 2 on the second pass. The third term represents those photons that reflect again. In summation form:

$$E_{1\,to\,2} = n_1\alpha_2 \sum_{i=0}^{\infty} (\rho_1\rho_2)^i = \text{energy absorbed by surface 2 of photons emitted by 1.}$$

Simultaneously, the number of photons emitted by 2 and ultimately absorbed by 1 is given by:

$$E_{2\,to\,1} = n_2\alpha_1 \sum_{i=0}^{\infty} (\rho_1\rho_2)^i = \text{energy absorbed by surface 1 of photons emitted by 2.}$$

Recognizing the binomial expansion (a Taylor series expansion), namely:

$$\sum_{i=0}^{\infty} (\rho_1\rho_2)^i = \frac{1}{1 - \rho_1\rho_2}$$

The net energy exchanged is:

$$E_{12} = E_{1\,to\,2} - E_{2\,to\,1} = \frac{(n_1\alpha_2 - n_2\alpha_1)}{1 - \rho_1\rho_2}$$

The rate of energy exchange (per unit area) is therefore:

$$q_{12} = \frac{E_{12}}{\Delta t} = \frac{\left(\varepsilon_1\alpha_2\sigma T_1^4 - \varepsilon_2\alpha_1\sigma T_2^4\right)}{1 - \rho_1\rho_2}$$

In formal radiation theory, Kirchhoff's Radiation Law dictates that at a given wavelength, the emissivity and absorptivity are equal ($\alpha_1 = \varepsilon_1$ and $\alpha_2 = \varepsilon_2$). Therefore:

$$q_{12} = \left[\frac{\varepsilon_1\varepsilon_2}{1 - \rho_1\rho_2}\right]\sigma\left(T_1^4 - T_2^4\right)$$

Note that if either surface has a low emissivity, the thermal radiation exchange between surfaces is small. Therefore, in practice, only one surface need be coated with a "low-e" coating to substantially reduce the radiation exchange across the gap of a double-paned window.

Opaque surfaces are those which do not allow electromagnetic energy to pass through them. That is:

$$\tau = 0 \quad \text{and} \quad \rho = 1 - \alpha = 1 - \varepsilon \quad \text{for opaque surfaces}$$

For window glass, the transmittance of visible light is very high, by design. However, the longer wavelength thermal radiation for nominal Earth's ambient temperatures is strongly absorbed or reflected, and very little thermal radiation at those wavelengths passes through. That is, window glass is generally opaque to thermal radiation.

$$q_{12} = \left[\frac{\varepsilon_1 \varepsilon_2}{1 - (1 - \varepsilon_1)(1 - \varepsilon_2)}\right]\sigma\left(T_1^4 - T_2^4\right)$$

$$= \left[\frac{\varepsilon_1 \varepsilon_2}{\varepsilon_1 + \varepsilon_2 - \varepsilon_1 \varepsilon_2}\right]\sigma\left(T_1^4 - T_2^4\right) \quad \text{for opaque surfaces}$$

Expressed in an Ohm's Law form (by expanding the temperature term twice):

$$q_{12} = h_{rad}(T_1 - T_2)$$

where

$$h_{rad} = \left[\frac{\varepsilon_1 \varepsilon_2}{1 - (1 - \varepsilon_1)(1 - \varepsilon_2)}\right]\sigma\left(T_1^2 + T_2^2\right)(T_1 + T_2)$$

Expanding the denominator:

$$h_{rad} = \left[\frac{\varepsilon_1 \varepsilon_2}{\varepsilon_1 + \varepsilon_2 - \varepsilon_1 \varepsilon_2}\right]\sigma\left(T_1^2 + T_2^2\right)(T_1 + T_2)$$

As before, the temperatures in the radiation coefficient are in absolute units (Kelvin), while in Ohm's Law form they can be expressed in Celsius. Note that the radiation coefficient is zero if either surface has a zero emissivity (smooth surfaces). A "low e coating" would only need to be applied to one inner surface to minimize the radiation effect. Also, the coefficient in brackets evaluates to unity if both emissivities are unity.

3.5.4   Radiation in the Single-Paned Window Problem

Both surfaces of the single-paned window can be treated as a completely surrounded object. The exposed surface on the inside "sees" three walls and the floor and ceiling of the room it is in.

Kirchhoff's Law can therefore be expressed as:

$$q_{H,walls \longrightarrow 1} = \varepsilon_1 \sigma A_1 \left(T_{H,walls}^2 + T_1^2\right)(T_{H,walls} + T_1)(T_{H,walls} - T_1)$$

Kirchhoff's Law can be put into an Ohm's Law form:

$$q_{H,walls \longrightarrow 1} = h_{r,1} A_1 (T_{H,walls} - T_1) = \frac{T_{H,walls} - T_1}{R_{r,1}}$$

where the radiative coefficient for surface 1 is:

$$h_{r,1} = \varepsilon_1\sigma\left(T^2_{H,walls} + T^2_1\right)\left(T_{H,walls} + T_1\right)$$

Similarly, for the outside window surface:

$$q_{2\longrightarrow C,walls} = h_{r,2}A_2\left(T_2 - T_{C,walls}\right) = \frac{T_2 - T_{C,walls}}{R_{r,2}}$$

where the radiative coefficient on the outside depends on the outside window surface temperature (T_2) and the outside walls ($T_{C,\ walls}$) that the outer window surface "sees":

$$h_{r,2} = \varepsilon_2\sigma\left(T^2_{C,walls} + T^2_2\right)\left(T_{C,walls} + T_2\right)$$

The radiative coefficient is a strong function of absolute temperature (and is inherently positive). In the window example, the surface temperatures (T_1 and T_2) are not known a priori. However, an iterative procedure can be set up to determine it. That is, make a first approximation by "guessing" T_1 and T_2, then evaluating for the heat transfer rates and then the surface temperatures. Then use this updated "guess" for T_1 in a second iteration and continue to iterate as needed. A good first approximation that will demonstrate the sensitivity of the radiative coefficient to temperature is to set T_1 equal to the inside wall temperatures, and T_2 to the outside wall temperatures. This guess will overestimate $h_{r,1}$ and underestimate $h_{r,2}$. A value for the emissivity of 0.9 will be used on both surfaces.

$$h_{r,1(FIRST\ APPROX)} = \varepsilon_1\sigma\left(T^2_{H,walls} + T^2_{H,walls}\right)\left(T_{H,walls} + T_{H,walls}\right) = 4\varepsilon_1\sigma T^3_{H,walls}$$

$$h_{r,1(FIRST\ APPROX)} = 4(0.9)\left(5.669(10)^{-8}\right)(25 + 273)^3$$

$$h_{r,1(FIRST\ APPROX)} = 5.40\,\text{W/m}^2/^\circ\text{C}$$

Notice that the temperature unit for radiative coefficient can be either $^\circ$C or K, since it was defined from Newton's Law of Cooling from a temperature difference, whereas K must be used to evaluate h_r. For the outside window surface:

$$h_{r,2(FIRST\ APPROX)} = \varepsilon_2\sigma\left(T^2_{C,walls} + T^2_{C,walls}\right)\left(T_{C,walls} + T_{C,walls}\right) = 4\varepsilon_2\sigma T^3_{C,walls}$$

$$h_{r,2(FIRST\ APPROX)} = 4(0.9)\left(5.669(10)^{-8}\right)(0 + 273)^3$$

$$h_{r,2(FIRST\ APPROX)} = 4.15\,\text{W/m}^2/^\circ\text{C}$$

Notice that both of these radiative coefficients are of the same order of magnitude as the convection coefficients. That means that radiation is important and should not be neglected in this problem. The corresponding radiative resistance values are:

$$R_{1,r} = \frac{1}{h_{r,1}A} = \frac{1}{(5.4\,\text{W/m}^2\,°\text{C})(16\,\text{m}^2)}\left|\frac{1,000\,\text{W}}{\text{kW}}\right| = 11.5\,\frac{°\text{C}}{\text{kW}}$$

$$R_{2,r} = \frac{1}{h_{r,2}A} = \frac{1}{(4.15\,\text{W/m}^2\,°\text{C})(16\,\text{m}^2)}\left|\frac{1,000\,\text{W}}{\text{kW}}\right| = 15.1\,\frac{°\text{C}}{\text{kW}}$$

This treatment of thermal radiation is a preliminary treatment and should be considered a good means to establish if radiation is important. If radiation is shown at this level of analysis to be dominant, a more thorough development of radiation theory might be warranted.

3.6 Single-Paned Window: Results

With all the pieces in place, the resistance network, and its reduction to a single equivalent resistance, is shown with resistance values in (°C/kW) in Fig. 3.13.

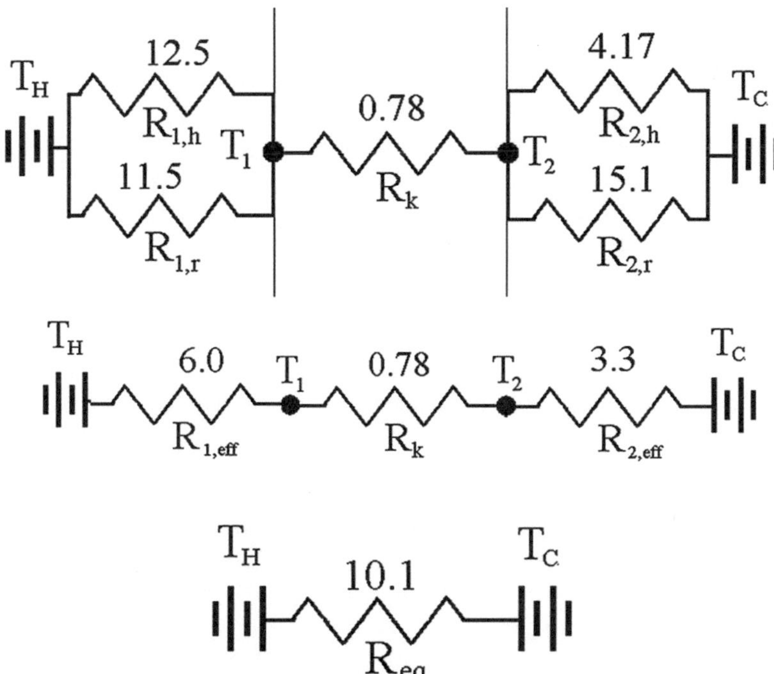

Fig. 3.13 Single-paned window resistance values (in °C/kW) and its reduction to a single equivalent resistance. This calculation is the first iteration, in which the surface temperatures were assumed in order to calculate the radiative resistances

The window surface temperatures can be solved for using voltage divider formulas. For example, the general formula derived in Chap. 1 for a node "i" with thermal neighbors "j":

$$T_i = \frac{\sum\limits_{j,\,neighbors} \dfrac{T_j}{R_{ij}}}{\sum\limits_{j,\,neighbors} \dfrac{1}{R_{ij}}}$$

Referring to the middle network of Fig. 3.13 and using a series resistance "jump" across the immediate neighbor to a known temperature:

$$T_1 = \frac{{}^{T_H}\!/_{R_{1,eff}} + {}^{T_C}\!/_{\left(R_k+R_{2,eff}\right)}}{{}^{1}\!/_{R_{1,eff}} + {}^{1}\!/_{\left(R_k+R_{2,eff}\right)}} = \frac{{}^{25}\!/_{6.0} + {}^{0}\!/_{(0.78+3.3)}}{{}^{1}\!/_{6.0} + {}^{1}\!/_{(0.78+3.3)}} = 10.1\,^{\circ}\mathrm{C}$$

$$T_2 = \frac{{}^{T_H}\!/_{\left(R_{1,eff}+R_k\right)} + {}^{T_C}\!/_{\left(R_{2,eff}\right)}}{{}^{1}\!/_{\left(R_{1,eff}+R_k\right)} + {}^{1}\!/_{\left(R_{2,eff}\right)}} = \frac{{}^{25}\!/_{(6.0+0.78)} + {}^{0}\!/_{(3.3)}}{{}^{1}\!/_{(6.0+0.78)} + {}^{1}\!/_{(3.3)}} = 8.18\,^{\circ}\mathrm{C}$$

Recall that the radiative resistances used a first approximation that the surface temperatures were equal to the adjacent fluid temperatures. A second approximation (iteration) uses the temperatures just calculated to give an improved guess at the surface temperatures, and therefore improved resistance estimates. This iterative procedure is demonstrated in Fig. 3.14, which is a screenshot of a spreadsheet designed for that purpose. The top block contains input parameters (which can be simply changed in the spreadsheet to run different scenarios), followed by calculation of the fixed thermal resistances in the model developed here. Finally, the iteration table shows the effect of improved "guesses" for this case study. Each row represents a different iteration (index "k"). The two temperatures are the "guesses" for the temperatures. The first iteration (k = 0) is an input set equal to the fluid temperatures (but any values could be manually placed). Then the radiative coefficients are calculated based on these guesses, followed by the radiative resistances, effective resistances (convection and radiation in parallel), and finally the net equivalent resistance. For the next iteration, the two temperatures are calculated based on the resistance values calculated from the previous iteration, and therefore are improved "guesses." It is apparent that only three iterations are needed to converge to within three significant figures. However, in light of the uncertainty in the conductive resistances (k value), and both convective resistances, it can be argued that the first approximation is sufficient for this case.

This general procedure for performing iterative calculations arises frequently in heat transfer applications, and will be returned too often, in different variations.

Finally, to answer the original question of the problem statement, the rate of heat transfer through this window (of area 16 m^2) is 2.436 kW. The total energy

INPUT PARAMETERS

w=	8	m, window width
H=	2	m, window height
Dx=	0.012	m, window thickness
TH =	25	oC, inside air and wall temperatures
TC=	0	oC, outside air and wall temperatures
k=	0.96	W/m/K, thermal conductivity of glass
h1=	5	W/m2/K, convection coefficient on inside
h2=	15	W/m2/K, convection coefficient on outside
e=	0.9	emissivity of glass

FIXED THERMAL RESISTANCES

Rk=	0.78125	oC/kW
Rh1=	12.5	oC/kW
Rh2=	4.166667	oC/kW

ITERATION TABLE

k	T1	T2	h1,r	h2,r	R1,r	R2,r	R1,eff	R2,eff	Req	q(kW)
0	25	0	5.40	4.15	11.57	15.05	6.01	3.26	10.05	2.487
1	10.05736	8.114674	5.01	4.34	12.48	14.40	6.25	3.23	10.26	2.437
2	9.779702	7.875645	5.00	4.34	12.50	14.42	6.25	3.23	10.26	2.436
3	9.776913	7.873849	5.00	4.34	12.50	14.42	6.25	3.23	10.26	2.436
4	9.776881	7.873826	5.00	4.34	12.50	14.42	6.25	3.23	10.26	2.436
5	9.776881	7.873826	5.00	4.34	12.50	14.42	6.25	3.23	10.26	2.436
6	9.776881	7.873826	5.00	4.34	12.50	14.42	6.25	3.23	10.26	2.436

Fig. 3.14 Worksheet designed for single-paned window with iteration procedure to calculate radiative coefficients. The values for the first iteration (k) differ from the calculations in the text because of round-off error in those hand calculations

transferred is the rate multiplied by the time, and the energy cost is the energy price multiplied by the total energy transferred:

$$Energy\ Cost = \left(\frac{\$20}{GJ}\right) * \left(\frac{2.436\,kJ}{s}\right) * \left|\frac{GJ}{10^9\,J}\right| * \left|\frac{10^3\,J}{kJ}\right| * \left|\frac{3,600\,s}{h}\right| * \left|\frac{24\,h}{day}\right|$$

$$= \frac{\$4.21}{day}$$

This energy cost estimate scales with the total surface area of window (a parallel heat transfer concept). It may not seem like a lot, but it can add up over the course of a year, and over a building that has many windows.

Engineering Significance: Notice that since the conductive resistance is much smaller than the effective resistances on both sides of the window, increasing the thickness of the window is not an effective means to reduce heating costs associated with windows. On the other hand, introducing a second pane of glass with an air gap (or other gas) between them has the effect of introducing more convective/radiative thermal resistances to the thermal channel. Indeed, double-paned windows very effectively reduce the heat transfer rate. Workshop 3.1 develops simplified models for analyzing double-paned windows.

Appendix. Conductive Thermal Resistances in Various Geometries

Consider the cylindrical shell of inner radius r_1, outer radius r_2, and length L (into the page) shown schematically in cross-section in Fig. 3.15. The inner surface is maintained at a temperature T_1 and the outer at T_2. Conceptually, the shell can be represented in rectangular coordinates by "cutting" the projection at a random place, and straightening it out to reveal the projection shown of a wall of constant thickness ($\Delta r = r_2 - r_1$), but with a cross-sectional area that varies from $2\pi r_1 L$ on one side to $2\pi r_2 L$ on the other. The coordinate "r" on the original view will be replaced by the coordinate "x" in the projection.

At steady-state, the heat transfer rate crossing the inner surface must equal the rate crossing the outer surface. Furthermore, the heat transfer rate at any surface in between must be equal. Invoking Fourier's Law, in which the conductive heat transfer rate in the x direction is proportional to the area normal (i.e., perpendicular) to the heat flow and to the temperature gradient (with the thermal conductivity, k, being the proportionality factor):

$$q_x = -kA_{(x)} \frac{dT}{dx} = constant \ @SS$$

The goal of the analysis is to express the conduction between the two surfaces in terms of an Ohm's Law form:

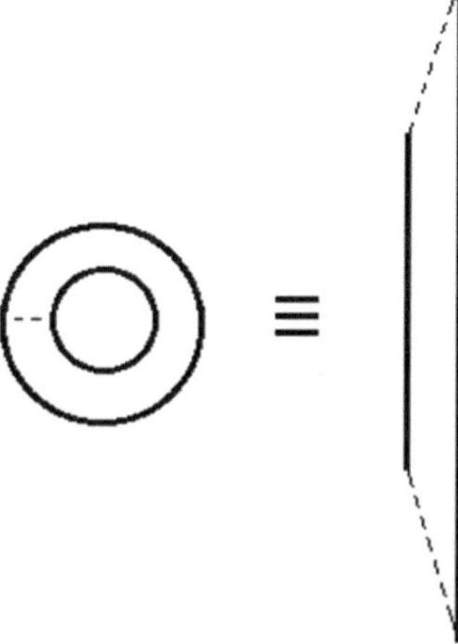

Fig. 3.15 Schematic of a cylindrical shell and its projection as a one-dimensional rectangular section

$$q_{12} = \frac{T_1 - T_2}{R_{12}}$$

In general, the thermal conductivity is a mild function of temperature. Separating the variables (T and x) in the Fourier's Law expression:

$$kdT = -q_x \frac{dx}{A_{(x)}}$$

Integrating from inner to outer surface (where the position goes from x_1 to x_2 and the temperature from T_1 to T_2, and noting that q_x is a constant):

$$\int_{T_1}^{T_2} kdT = -q_x \int_{x_1}^{x_2} \frac{dx}{A_{(x)}}$$

Defining an average thermal conductivity on the left hand side:

$$\int_{T_1}^{T_2} kdT \equiv \bar{k}(T_2 - T_1)$$

The heat transfer rate can therefore be expressed as:

$$q_x = q_{12} = \frac{-\bar{k}(T_2 - T_1)}{\int_{x_1}^{x_2} \frac{dx}{A_{(x)}}} \equiv \frac{T_1 - T_2}{R_{12}}$$

A general form for the conductive thermal resistance is therefore:

$$R_{12} = \frac{1}{\bar{k}} \int_{x_1}^{x_2} \frac{dx}{A_{(x)}}$$

For well-defined geometries (i.e., Cartesian, cylindrical, spherical coordinates), this integration can be carried out directly. Other geometries in which the appropriate area can be expressed in terms of a single coordinate are possible as well. For geometries that are not well defined, an approximation for conductive resistance between two surfaces follows as:

$$R_{12} \approx \frac{(average\ distance\ between\ surfaces)}{(\bar{k})(average\ area\ of\ surfaces)}$$

Applied to Cylindrical Coordinates

For the cylindrical shell:

$$R_{12} = \frac{1}{\bar{k}} \int_{x_1}^{x_2} \frac{dx}{A_{(x)}} = \frac{1}{\bar{k}} \int_{r_1}^{r_2} \frac{dr}{2\pi rL} = \frac{1}{2\pi\bar{k}L} \ln\left(\frac{r_2}{r_1}\right)$$

If the approximate resistance value is expressed in this geometry:

$$R_{12} \approx \frac{(average\ distance\ between\ surfaces)}{(\bar{k})(average\ area\ of\ surfaces)} = \frac{r_2 - r_1}{\bar{k}\left(\frac{2\pi r_1 L + 2\pi r_2 L}{2}\right)} = \frac{r_2 - r_1}{\pi \bar{k} L(r_1 + r_2)}$$

The ratio of the exact to approximate resistance values is:

$$\frac{R_{12,EXACT}}{R_{12,APPROX}} = \frac{\left(1 + {r_2}/{r_1}\right)\ln\left({r_2}/{r_1}\right)}{2\left({r_2}/{r_1} - 1\right)}$$

The latter form (in terms of the ratio of radii) is obtained by dividing numerator and denominator by r_1. A plot of this ratio is shown in Fig. 3.16. It is clear that the approximate value underestimates but approaches the exact value when the ratio of radii approaches unity, the thin shell approximation. However, the error is less than 10 % up to r_2/r_1 ratios of 3.

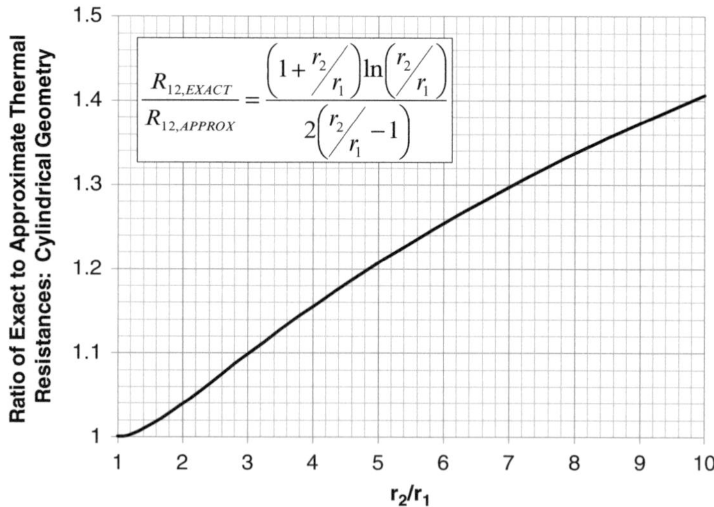

Fig. 3.16 Comparison of exact thermal resistance to approximate thermal resistance for a cylindrical shell

Exercises: Derive the exact conductive thermal resistance for

- A spherical shell with inner radius r_1 and outer radius r_2 (see how the floor conductive resistance is treated in Appendix 1 of Chapter 4...)
- A rectangular duct of aspect ratio (width to height) AR, length L, with inner width w_1, and thickness Δx
- A truncated cone with smaller area A_1, larger area A_2, and a distance L between them (with insulated sides)...

Workshop 3.1. Double-Paned Window Models

In an attempt to reduce the heat loss across a window, the single-paned window from the opening example of this chapter is replaced with a double-paned window in which a second pane of glass is added, with a gap between the plates (g). Estimate the rate of heat loss and the energy cost as a function of gap width, and compare to the single-paned window result. Compare the effects of typical window glass (with an emissivity of 0.93 on both surfaces) and the performance with one of the inner surfaces coated with a "low-e coating" (emissivity of, say 0.3). Also, compare the replacement of air with argon.

It is a good idea to stop reading and think about this problem first and draw your own thermal resistance network before going on.

Model Development

Heat transfer across the window is still a single thermal channel, but inclusion of a second pane with a gap introduces additional thermal resistances. There is radiation exchange between the two gaps (modeled as case 3: radiation between infinite plates), and there is a parallel heat exchange mechanism by conduction and/or convection. Two models for the latter are considered at this point: narrow gap and wide gap.

Narrow Gap Model

In this model, the air (or other gas, like Argon) that is trapped between the two panes of glass is assumed to be stationary. Therefore, it behaves like a solid and heat exchange between the inner surfaces takes place through a straight conduction mode. However, radiation between the two inner plate surfaces acts in parallel. A thermal resistance network (at steady-state) for this model is shown in Fig. 3.17. There are two parallel channels across the gap, a conductive resistance and a radiative resistance.

The gap resistance is given by:

$$R_{k,gap} = \frac{g}{k_{gas}A}$$

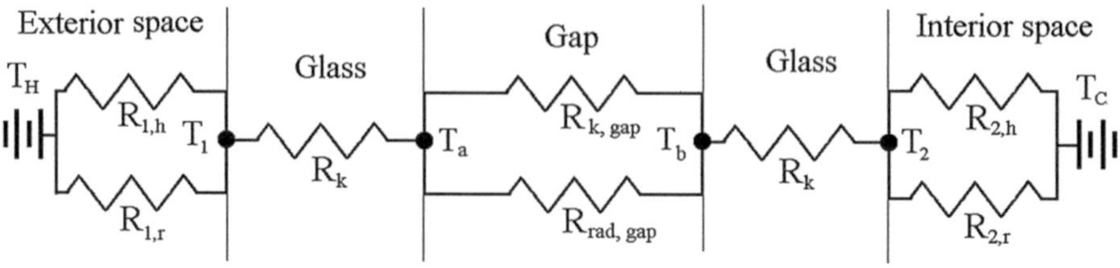

Fig. 3.17 Thermal resistance network for a double-paned window with the narrow gap model

The total heat transfer rate is:

$$q_{12} = \frac{T_H - T_C}{R_{eq}} = UA(T_H - T_C)$$

where the equivalent resistance is:

$$R_{eq} = \frac{1}{\frac{1}{R_{r,1}} + \frac{1}{R_{h,1}}} + R_k + \frac{1}{\frac{1}{R_{rad,gap}} + \frac{1}{R_{k,gap}}} + R_k + \frac{1}{\frac{1}{R_{r,2}} + \frac{1}{R_{h,2}}}$$

In terms of equivalent resistances where parallel channels exist:

$$R_{eq} = R_{H,equiv} + R_k + R_{gap,equiv} + R_k + R_{C,equiv}$$

The equivalent gap resistance can be expressed as:

$$R_{gap,eq} = \frac{R_{k,gap}}{\frac{R_{k,gap}}{R_{rad,gap}} + 1} = \frac{g}{k_{air}A}\left(\frac{1}{\frac{h_{rad,gap}g}{k_{air}} + 1}\right)$$

The equivalent gap resistance increases with the gap width. Conceptually, the gap acts like an insulating layer of material added, and since the thermal conductivity of air (0.023 W/m/K) is much lower than that of glass (0.96 W/m/K), this added layer is more effective than simply making the glass thicker. However, the radiation channel across the gap reduces the effective insulating property (notice the dimensionless parameter in the denominator). Given an effective radiation coefficient of $h_{rad,gap} = 6$ W/m^2/K, from experience, and $k_{air} = 0.023$ W/m/K, the value of the gap width where the conduction and radiative resistances are equal (setting the dimensionless parameter to unity) is $g = k_{air}/h_{rad,gap} = 0.0038$ m $= 3.8$ mm. That is, for gaps below this, radiation is dominant, and conduction dominates for wider gaps. Since manufactured double-paned windows set the gap spacing to be typically 10–20 mm, this comparison suggests, tentatively, that radiation effects are minor in comparison to straight conduction for actual windows. However, the assumption that air in the gap does not move must be questioned. The air in the gap does not remain stationary, and an additional convective-like mode of heat transfer occurs, and described in the following wide gap model.

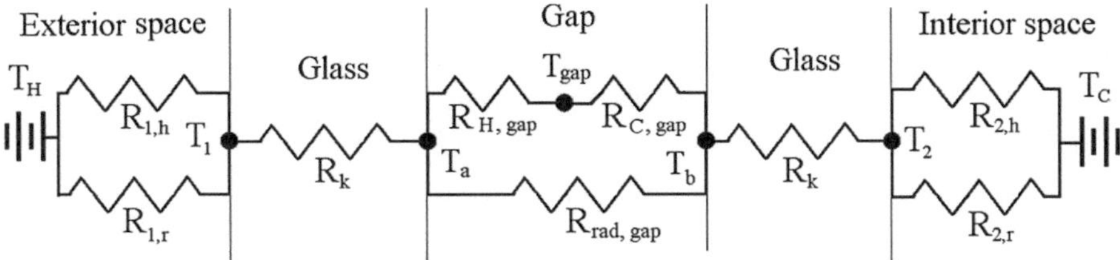

Fig. 3.18 Thermal resistance network for a double-paned window using the wide gap model

Wide Gap Model

Since the gap contains a fluid and there is a temperature difference between the two panes of glass, the fluid near the colder pane is denser than the fluid near the hotter plane. A buoyancy force will cause the hotter fluid to rise and the colder to sink, and a natural circulation pattern will develop in the gap. Fluid friction opposes the motion. In this introductory chapter, a conceptual model is used in which there is a boundary layer adjacent to each inner surface that is small compared to the gap width. Therefore, the bulk of the fluid in the middle of the gap is considered to be stagnant, and a classic natural convection mode of heat transfer governed by Newton's Law of Cooling exists between each glass surface and the bulk fluid in the gap, which is at an intermediate temperature to that of the two glass surfaces. A thermal resistance network for this model is shown in Fig. 3.18.

In the wide gap model, heat flows from the hotter inner surface (T_a) into the bulk of the gap (at T_{gap}) by a natural convection mode, and in series, from the bulk fluid to the colder inner surface (T_b). The equivalent resistance of this channel is:

$$R_{eq} = \frac{1}{\frac{1}{R_{r,1}} + \frac{1}{R_{h,1}}} + R_k + \frac{1}{\frac{1}{R_{rad,gap}} + \frac{1}{R_{H,gap} + R_{C,gap}}} + R_k + \frac{1}{\frac{1}{R_{r,2}} + \frac{1}{R_{h,2}}}$$

As with the narrow model, the equivalent resistance is:

$$R_{eq} = R_{H,equiv} + R_k + R_{gap,equiv} + R_k + R_{C,equiv}$$

The equivalent gap resistance can be expressed as:

$$R_{gap,eq} = \frac{\left(R_{H,gap} + R_{C,gap}\right)}{\left(\frac{R_{H,gap} + R_{C,gap}}{R_{rad,gap}}\right) + 1}$$

The gap resistance consists of two convective resistances in series, and a reducing factor due to radiation. Notice that the size of the gap does not affect the prediction of this model.

Conceptually, and in practice, the conductive resistance across the glass itself was shown to be negligible for the single-paned window, so that the heat transfer rate is controlled by two convective/radiative equivalent resistances in series that are of comparable magnitude. The second pane of glass introduces two additional convective resistances that are of the same order, and therefore, the heat transfer rate through the double-paned window should be approximately half that of the single-paned window. Adding a third pane of glass (a triple-paned window) introduces two more convective resistances, so the rate of heat transfer is approximately a third of that of a single-paned window. Notice the diminishing returns. A second pane cuts the original heat transfer rate by 50 %, a third pane cuts it by only an additional 17 %, a fourth pane would cut an additional 8 %, and so on.

Additional Workshop Ideas

Workshop 3.2. Heat Loss Through Walls

Compare the rate of heat loss per unit area (report results as "R-Value" in $°C\ m^2/W$) across an exterior wall constructed with 2×4 studs, spaced by 16 inches with that of 2×6 studs, spaced by 24 inches. Consider the inner surface to be finished with sheetrock, and the outer surface to be finished with ¼″ plywood, a vapor barrier, and siding. Consider the cases with and without insulation in the gaps. Note that the actual dimensions of a 2×4 are not 2×4 in.

Workshop 3.3. "Critical Radius of Insulation"

The outer surface of a 100 m long electrical wire of radius $r_{wire} = 0.005$ m is not to exceed $T_1 = 100\ °C$ (it doesn't matter what the wire material is, just what its outer temperature is). The wire is exposed to a fluid at $T_{inf} = 0\ °C$ with an effective convective/radiative coefficient of 15 $W/m^2/K$. Calculate the steady-state rate of heat loss (from which maximum amperage could be determined) for a bare wire. Next, a layer of insulation (with thermal conductivity $k_{ins} = 0.15$ W/m/K) is to be added. What insulation thickness (Δx) will maximize the heat transfer rate between the pipe and ambient (hence amperage)? What is the corresponding heat transfer rate and outer surface temperature of the insulation? Wait a minute… Isn't insulation supposed to REDUCE the heat transfer rate? What's up with that? Use resistance networks (and exact formula for conductive resistance across insulation layer).

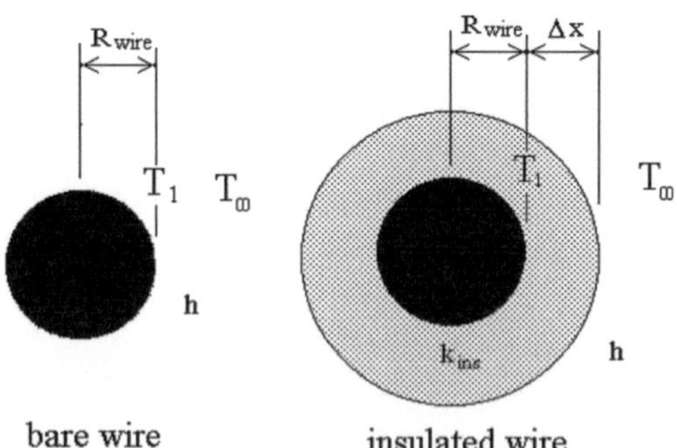

bare wire   insulated wire

Workshop 3.4. "Wearing Short Pants"

Compare the steady-state rate of heat loss from your legs for a hot summer day with long pants versus short pants. For both scenarios, draw a thermal resistance network (omit capacitances since steady-state values are specified.). The skin can be set at a fixed temperature of, say 32 °C (a value which can vary due to thermoregulatory effects). For the case of short pants, the skin is exposed to a fluid at one ambient temperature and as an enclosed surface in radiation thermal contact with walls that could be at a different temperature than the fluid. For example, in summer, the exterior wall of the room could be at a higher temperature than the room, and in winter, the wall could be cold. For the case of long pants, there is an air gap between the skin and the pants, which can be modeled as two natural convection resistances (skin to air, air to pants) and a radiation between infinite planes (skin to pants). Conduction occurs across the pants and radiation/convection on the outside. What if it were a sunny day? Compare black and white pants in that case (note that due to very different wavelengths, the absorptivity of solar radiation is higher for black pants than white pants, but the emissivity at the Earth's temperatures can be about the same and a value near unity).

Workshop 3.5. "Temperature of the Planets"

Estimate the average temperature of all the planets, assuming first that the effective absorptivity of solar insolation equals the effective thermal emissivity of thermal radiation. Obtain the average distance from the sun from a published or online source. Compare the published temperature values. Then use that data to determine the effective ratio of solar absorptivity to emissivity for each planet. For this problem, the solar flux (intensity times area) leaving the sun is the same exiting any spherical shell enclosing the sun. That is, $I(4\pi r^2) = Constant$, where "I" is the solar intensity ($=1{,}400$ W/m^2 for the Earth) and "r" is the distance from the sun center to the distance of the spherical shell ($=1.49(10)^{11}$ m for the Earth).

Part II

Transient Conduction

1-Node Transient Models

4

A basic philosophy of this book is to develop a modeling approach that considers "the big picture" first, and then more detail only if needed. Since heat transfer rates are driven by temperature differences, the first step in analyzing heat transfer problems is usually to solve for the temperature field, which depends on spatial position and time. Numerical analyses involve the discretization of space (breaking the physical region into discrete subelements) and time (taking finite steps forward in time).

This chapter considers "1-node" models, where the region of space in question is lumped into a single node. Time dependence often results in easily solved ordinary differential equations, or ones that can be easily solved numerically. By treating a defined system as a single object with average properties, that is, as a single node, a first approximation can often be made that is either sufficient to make design decisions, or directs attention to intelligently develop a more detailed approach. Furthermore, key functional relationships and associated insights are obtained with relatively little mathematical complexity.

The so-called exact solutions involve breaking the spatial region of influence into an infinite number of pieces (by making them infinitesimally small) and by stepping forward in time by an infinitesimal amount. There is certainly mathematical elegance and historical relevance in the exact solutions of these partial differential equations, but the philosophy implied in solving problems that way is the opposite of that adopted here. Furthermore, what is often lost in translation is that the "exact solution" applies to a problem that has gone through an extensive set of approximations to the physical problem to make it "fit" a model that has a closed-form mathematical solution. There is a certain false sense of security in how "exact" those models are to real physical problems.

The first section considers the cooling mug of coffee as a lumped element. Next, the heating or cooling of a general homogeneous object is investigated in a general way. Firstly, the straightforward case is considered in which the size of the node is constant, but its temperature changes. Secondly, several cases are considered in which the size of the node changes with time. This class of problems serves to

© Springer International Publishing Switzerland 2015
G. Sidebotham, *Heat Transfer Modeling: An Inductive Approach*,
DOI 10.1007/978-3-319-14514-3_4

introduce important concepts of thermal boundary layers and thermal lengths that will be useful in subsequent chapters, particularly in gaining a fundamental understanding of boundary layer phenomenon in fluids that form the basis for estimation of convection heat transfer coefficients.

The applications developed in this chapter correspond to cases that are developed in traditional textbooks, with exact solutions. The methods developed here can be extended to a much broader set of problems.

4.1   Lumped Capacity Model for Coffee/Mug Problem

In this section, the lumped capacity model is applied to the coffee/mug problem. All of the capacitance of the system (mug plus coffee) is lumped into a single capacitor. A thermal resistance network of multiple channels (top, side, and bottom) is developed to model parallel paths for heat to flow from the coffee interior to the ambient. A first model includes all convection and conduction modes (Model 1). Then effects of radiation on the outer surfaces are included (Model 2). Finally, the effect of evaporation at the coffee-free surface is included (Model 3). In all these models, the resistance values are treated as constants, suitably chosen to represent an average value during the cooling event. Use of constant resistances results in a first-order system, and the entire event is characterized by a single overall heat transfer coefficient (U) and a time constant (RC).

Engineering Problem Statement (Repeated)
A volume $V = 0.41$ L of coffee at 1 atm, 92 °C is quickly poured into a room temperature (21 °C) ceramic mug (shown schematically in Fig. 4.1) with a mass of $m_{mug} = 0.365$ kg, inner diameter $D = 0.083$ m, inner height $H = 0.098$ m, outer height $H_o = 0.109$ m, and nominal wall thickness $w = 0.0040$ m. The mug of coffee is placed on a table in a still room. Estimate the time it takes for the water to cool to 60 °C (above which it is too hot) and to 45 °C (defined as "tepid").

4.1.1   Model 1: Neglect Radiation and Evaporation

Figure 4.2 shows a resistance network that breaks down the total heat flux from the hot coffee (at T, representing an average temperature of the interior of the coffee) to ambient (at T_∞) into three parallel channels (top, side and bottom), each of which consists of several resistances in series. Each individual resistance represents a mode of heat transfer between one identifiable surface or fluid and another, with nodes to identify each surface. Appendix 1 details the calculation of the individual resistance values.

In this model, all the capacitance of the system (coffee plus mug) has been lumped into a single node, and therefore the smaller filled nodal temperatures can be related to the main capacity-carrying node, T, and the known ambient temperature through the methods developed in Chap. 1. To reiterate, open symbols are

Fig. 4.1 Schematic diagram
of coffee mug with geometry
defined

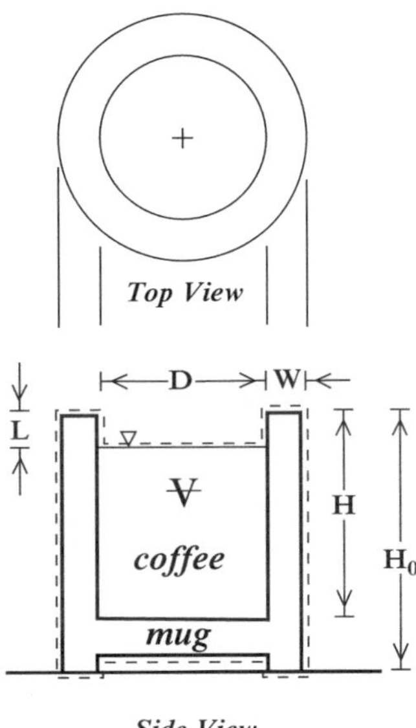

reserved for nodes for which capacitance has been attributed, and therefore their temperature is governed by a differential equation, not an algebraic relationship between neighbors. At times, big filled symbols (rather than a battery symbol) will be used to represent known temperatures. Small filled symbols represent unknown intermediate temperatures of nodes that have no capacitance assigned to them.

The results of this model will be compared to the experimental data from Chap. 2. It is noteworthy that the experimental method (an IR detector) gave direct measurements of the outer surface of the side wall (T_{os}) and of the free surface (T_s), not of the bulk average.

For the side channel, heat must first be transferred from the interior of the liquid to the inner surface of the mug side wall. The resistor R_{cs} (for coffee/side) represents that convection heat transfer mode. Next, the heat must cross the side wall through the mug, from the inner surface to the outer surface, and that conduction mode is represented by the resistor R_{ws} (wall/side). Finally, heat must flow from the outer wall to the ambient fluid, and that is a convective mode represented by R_{as} (ambient/side). These three thermal resistances act in series, and therefore the total resistance of the side channel is their algebraic sum. Once numerical values are estimated, judgments about their relative importance can be made.

For the bottom channel, heat must flow from the interior of the coffee to the inner surface of the bottom, or floor through a convective resistance (R_{cb} for coffee/bottom). Heat must then cross through the mug wall through a conductive resistance (R_{wb} for wall/bottom). Heat flows from the bottom of the mug across an air gap then into the floor through a conductive path (R_{fb} for floor/bottom).

Fig. 4.2 Lumped capacity thermal resistance model superimposed on a schematic of a mug of coffee (*top*) and then a stepwise breakdown to a single equivalent thermal resistance

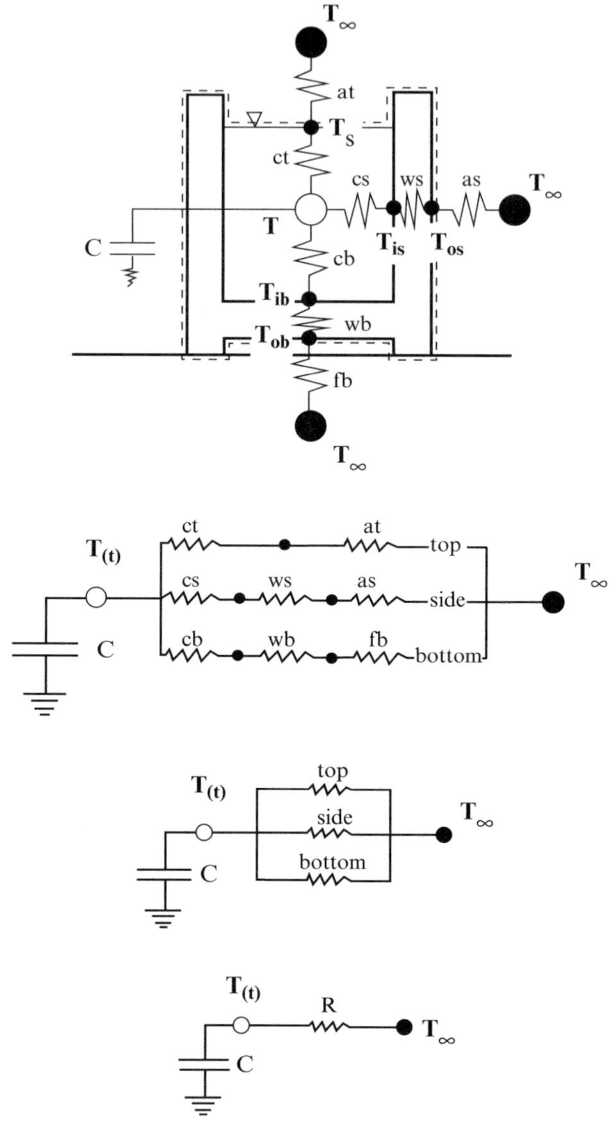

For the top channel, the free surface of the coffee does not rigidly fit into the modes defined in Chap. 3. Convection was defined as heat transfer between a fluid and a solid surface. However, the coffee-free surface is a boundary between two fluids (air and coffee). Heat flows from the interior to the surface and then from the surface to the ambient above it, and so two resistances in series that are like convection resistances are introduced for the top channel (R_{ct} for coffee/top and R_{at} for air/top).

INPUT QUANTITIES

D=	0.083	m
H=	0.098	m
H0=	0.106	m
V=	4.10E-04	m^3
w=	0.004	m
mmug=	0.365	kg
kmug=	1.5	W/m/K
cmug=	800	J/kg/K
rhocoffee=	1000	kg/m^3
ccoffee=	4200	J/kg/K
"h"floor =	4.7	W/m/K
hair/sides=	6.8	W/m2/K
hair/top=	6.8	W/m2/K
hcoffee/mug=	470	W/m2/K
hcoffee/surface=	300	W/m2/K
hevap=	1.00E-04	W/m2/K
hradiation =	1.00E-04	W/m2/K
Tcoffee, init=	92	oC
Tmug,initial=	21	oC
Tambient=	21	oC

RESULT

RC=	111.3	min	Theory
U =	6.0	W/m2/oC	

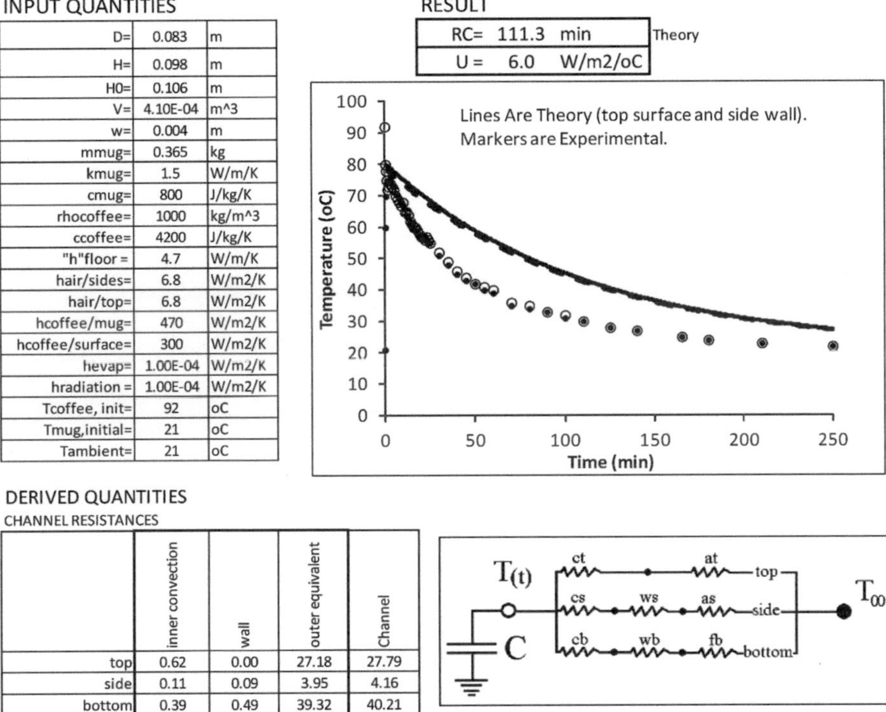

DERIVED QUANTITIES

CHANNEL RESISTANCES

	inner convection	wall	outer equivalent	Channel
top	0.62	0.00	27.18	27.79
side	0.11	0.09	3.95	4.16
bottom	0.39	0.49	39.32	40.21

Fig. 4.3 Spreadsheet implementation of extended lumped capacity Model 1: neglect radiation and evaporation

Figures 4.3 and 4.4 show results from a spreadsheet implementation of the model. The block of input quantities represents input values. Several of these inputs have been calculated in other spreadsheets (and future chapters) and manually input into this spreadsheet. The values chosen for the convection coefficients (6.8 W/m^2/°C for air/surface and 470 for coffee/surface locations) were determined in a separate spreadsheet (detailed in a future chapter) as a suitable average value for the entire cooling event.

All quantities in this block are fixed constants. More sophisticated models might allow for variable coefficients, but then a numerical simulation of the event would be required. On the other hand, the assumption of constant heat transfer coefficients (suitably chosen averages) allows for a closed-form solution to the lumped capacity model, with a net determined time constant and an overall effective heat transfer coefficient (RC and U, shown in the results block). Figure 4.4 applies to Model 3 discussed shortly (which includes radiation and evaporation effects). The effects of radiation and evaporation are "turned off" in Fig. 4.3 by setting the corresponding "h" values to a small number (a "zero" cannot be used, as the model divides by this value), resulting in effectively infinite resistance values.

DERIVED PARAMETERS

H-L	0.076	m	Inner exposed height
L	0.022	m	air column length
C_{mug}	292	J/oC	Mug Capacitance
C_{coffee}	1722	J/oC	Coffee Capacitance
$A_{top,bottom}$	0.0054	m^2	free surface area, inner mug bottom
$A_{outside}$	0.0372	m^2	outer surface, top rim, inner surface
$A_{middleside}$	0.0285	m^2	average area across mug side wall
A_{inside}	0.0198	m^2	inner submerged side wall
A_{total}	0.0501	m^2	Total exposed surface area
$T_{initial}$	81.7	oC	Initial temperate after mug heating

INDIVIDUAL RESISTANCE VALUES (oC/W)

Resistor	Value	mode	location A	location B
R_{cb}	0.393	convection	coffee	bottom inner surface
R_{cs}	0.108	convection	coffee	side inner surface
R_{ct}	0.616	convection	coffee	free surface
R_{ws}	0.094	conduction	side inner surface	side outer surface
R_{wb}	0.493	conduction	bottom inner surface	bottom outer surface
$R_{at,evap}$	18.5	mass transfer	free surface	ambient
$R_{at,rad}$	28.0	radiation	free surface	ambient
$R_{at,conv}$	27.2	convection	free surface	ambient
$R_{as,rad}$	4.1	radiation	side outer surface	ambient
$R_{as,conv}$	3.95	convection	side outer surface	ambient
R_{fb}	39.3	mixed	bottom outer surface	ambient (thru floor)

Fig. 4.4 Derived parameters used in Models 1, 2, and 3, with the following exceptions: for Model 1, the thermal resistances $R_{at,evap}$ and $R_{at,rad}$ evaluate to large (effectively infinite) values by setting the corresponding heat transfer coefficients to low values (they can't be zero, since the model divides through by them). In Model 2, the thermal resistance $R_{at,evap}$ is large

The plot of temperature vs. time in Fig. 4.3 shows that the predicted cooling is noticeably slower than the experimental result. Recall that an experimental value of U was banded between approximately 10 and 20 $W/m^2/K$ (the experimental result being not a pure first-order response). The predicted overall U value of 6.0 $W/m^2/K$ is below that range.

The table of channel resistances reveals more about the controlling routes for heat to flow from the coffee to ambient. The net channel resistance is much lower through the side wall than through the bottom channel, which is lower than the top channel. This effect is largely due to the surface area (recall the side surface area is more than half the total exposed area). Within the side channel, the resistance is dominated by the outer equivalent resistance. One result of this effect is that the outer surface temperature is much closer to the coffee temperature than ambient, as confirmed experimentally.

4.1.2 Model 2: Add Thermal Radiation

Radiation effects have not been included in the model of Fig. 4.2. The alternative model of Fig. 4.5 adds radiation effects. Here, the mug of coffee sits in a room whose walls and ceiling (the big outer box) may, in general, be at a different temperature ($T_{\infty w}$) than the air in the room. The exposed surfaces on the sides and top of the coffee/mug system are in direct contact with the air surrounding them (and modeled as convection), but in addition, these exposed surfaces "see" the walls of the room it is in. The physical mechanism of this radiation heat transfer is fundamentally different than the convection. Notice that now there are two "ambients," fluid and radiation wall.

Figure 4.6 shows this resistance network recast. At the top and side surfaces exposed to air, the channels split, or branch. Consider, for example, the side channel. In order for heat to flow from the coffee ultimately to ambient, it must first flow from the coffee to the submerged portion of the inner surface of the mug. From there, heat flows across the mug wall to its outer surface. The outer surface branches into two channels, one to the fluid immediately adjacent to it (at T_∞) and the other to the walls of the room (at $T_{\infty w}$). Notice that the temperature at the mug outer surface is represented by an open symbol. The temperature there cannot be related to two other known temperatures by a simple voltage divider calculation (recall Chap. 1). This network cannot be modeled as a simple RC network because of the branching.

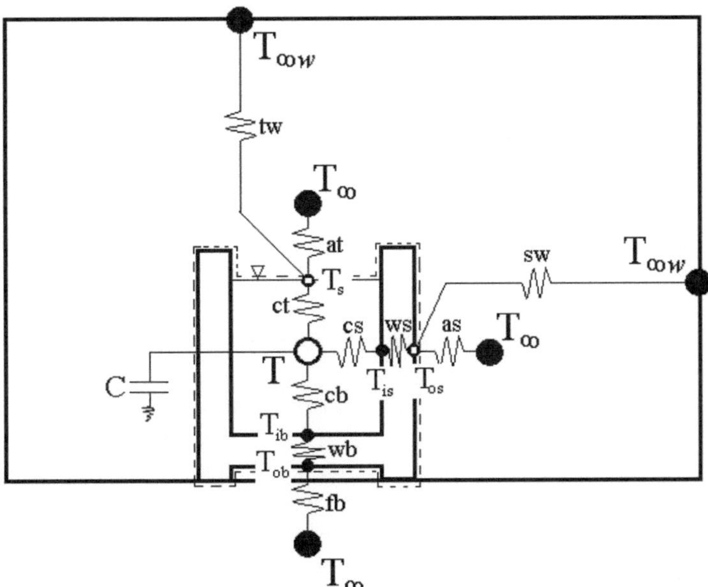

Fig. 4.5 Resistance network superimposed on a mug of coffee schematic that includes radiation effects

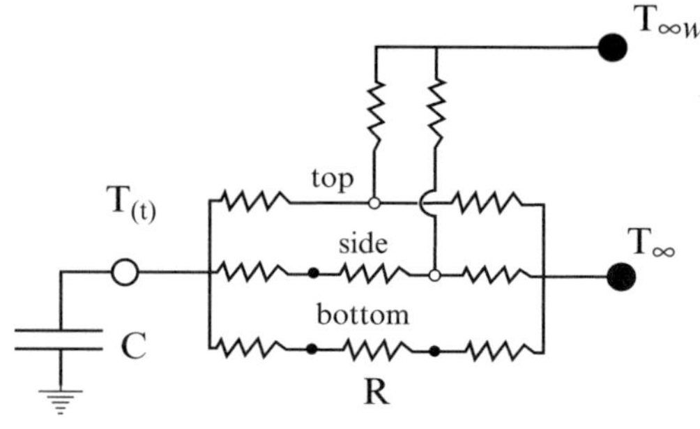

Fig. 4.6 Another thermal resistance network for a mug of coffee

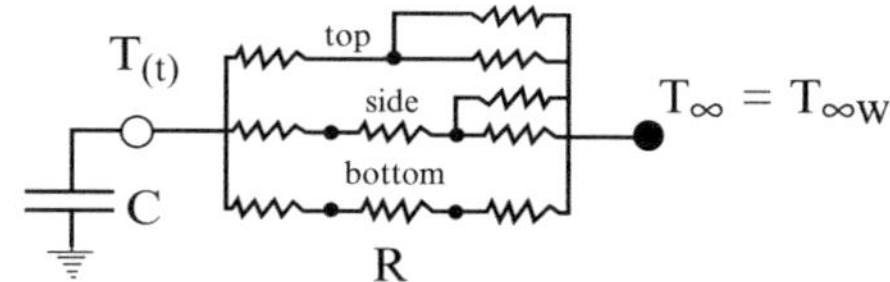

Fig. 4.7 Resistance network when the walls of the room are at the same temperature as the air within it

Figure 4.7 shows the resistance network for the case when the walls of the room and the air within it are assumed equal. The convective and radiative resistances act in parallel on the top and bottom surfaces, and the entire network can be reduced to a single R value, so that a simple RC circuit model can be used. When radiation and convection act in parallel (to the same ambient temperature), the net heat transfer coefficient is the sum of the two modes: $h_{equiv} = h_{convection} + h_{radiation}$.

Figure 4.8 shows the spreadsheet results when these radiation effects are included. The value chosen for the radiation coefficient (6.6 W/m^2/°C) was determined in a separate spreadsheet (detailed in a future chapter) as a suitable average value for the entire cooling event. The model now shows good agreement between theory and experiment, with perhaps a slightly slower initial cooling rate. The outer equivalent resistance value is approximately half of what it was for the model that neglected radiation.

4.1.3   Model 3: Add Evaporation Effects

Figure 4.9 shows the resistance network with evaporation effects included as a third parallel channel between the free surface of the coffee and ambient. Figure 4.10 shows the spreadsheet result when this effect is included. The energy exchange associated with the evaporation is modeled with an equivalent convection coefficient (h_{evap}). While this energy exchange is by a mass heat transfer mechanism, not a heat transfer mechanism, it can be expressed as being proportional to the temperature difference (with the coefficient having a strong temperature dependence). The selection of a specific constant value requires judgment, and a formal way of doing that is the subject of Chapter 13. In the present case, introducing an approximate

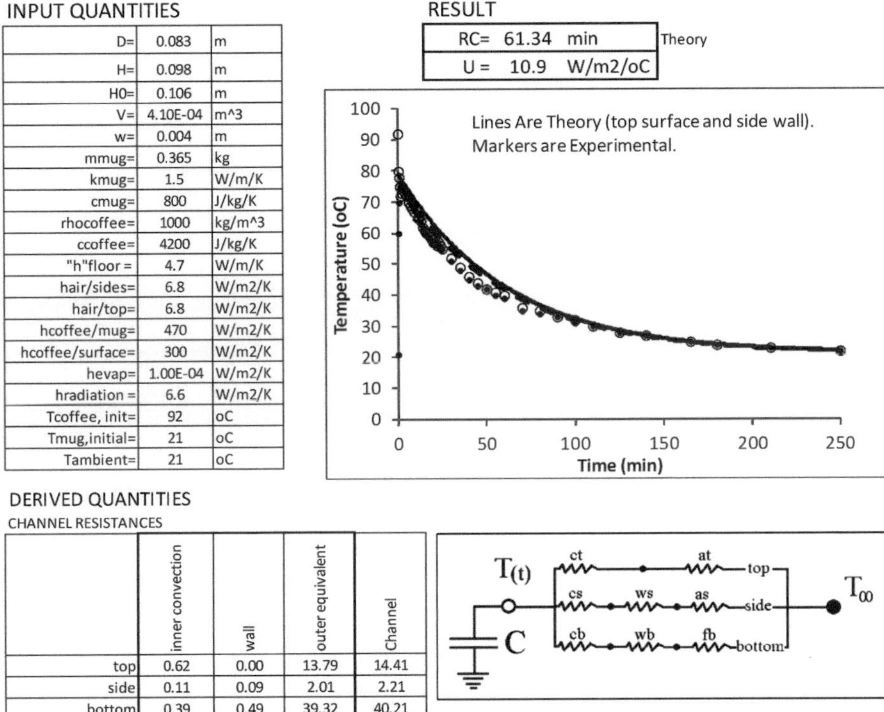

INPUT QUANTITIES

D=	0.083	m
H=	0.098	m
HO=	0.106	m
V=	4.10E-04	m^3
w=	0.004	m
mmug=	0.365	kg
kmug=	1.5	W/m/K
cmug=	800	J/kg/K
rhocoffee=	1000	kg/m^3
ccoffee=	4200	J/kg/K
"h"floor =	4.7	W/m/K
hair/sides=	6.8	W/m2/K
hair/top=	6.8	W/m2/K
hcoffee/mug=	470	W/m2/K
hcoffee/surface=	300	W/m2/K
hevap=	1.00E-04	W/m2/K
hradiation =	6.6	W/m2/K
Tcoffee, init=	92	oC
Tmug,initial=	21	oC
Tambient=	21	oC

DERIVED QUANTITIES

CHANNEL RESISTANCES

	inner convection	wall	outer equivalent	Channel
top	0.62	0.00	13.79	14.41
side	0.11	0.09	2.01	2.21
bottom	0.39	0.49	39.32	40.21

Fig. 4.8 Model 2: add radiation effects

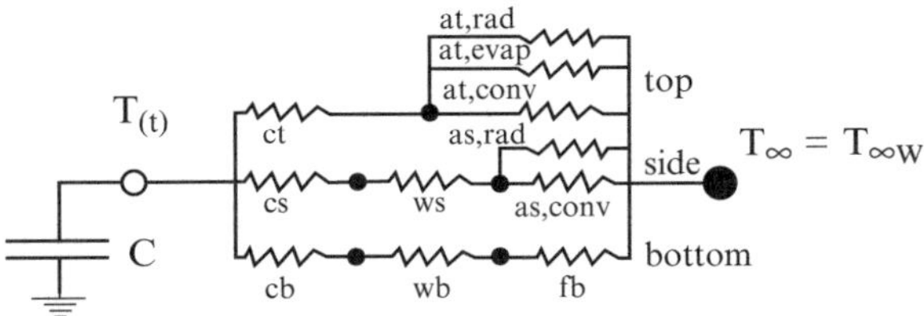

Fig. 4.9 Resistance network with evaporation included. A valuable exercise is to write in pencil the values from Fig. 4.4 directly onto this sketch, visualizing the heat transfer being represented

evaporative channel increases the overall heat transfer coefficient from 10.9 to 11.9 W/m^2 °C, a 9.1 % increase. That is considered to be significant, but not dominant. The effect of radiation is much more important, especially since radiation acts in parallel with convection on the side wall, with its much larger surface area than the free surface.

Figure 4.11 shows results (from an analysis in Chapter 13) of the effect of temperature on the three heat transfer coefficients between the coffee-free surface

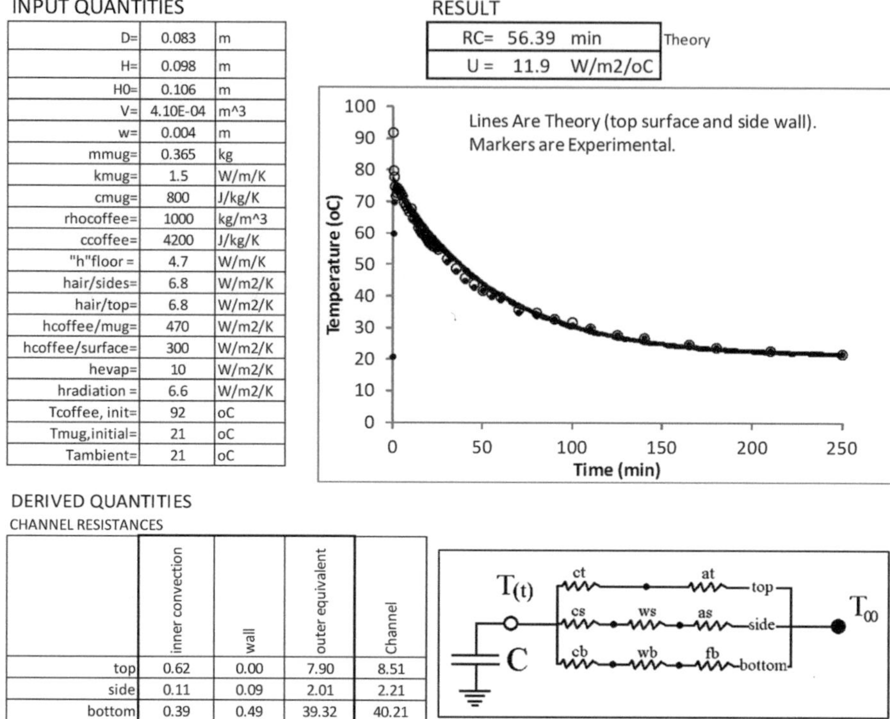

INPUT QUANTITIES

D=	0.083	m
H=	0.098	m
H0=	0.106	m
V=	4.10E-04	m^3
w=	0.004	m
mmug=	0.365	kg
kmug=	1.5	W/m/K
cmug=	800	J/kg/K
rhocoffee=	1000	kg/m^3
ccoffee=	4200	J/kg/K
"h"floor =	4.7	W/m/K
hair/sides=	6.8	W/m2/K
hair/top=	6.8	W/m2/K
hcoffee/mug=	470	W/m2/K
hcoffee/surface=	300	W/m2/K
hevap=	10	W/m2/K
hradiation =	6.6	W/m2/K
Tcoffee, init=	92	oC
Tmug,initial=	21	oC
Tambient=	21	oC

RESULT

RC=	56.39	min	Theory
U =	11.9	W/m2/oC	

DERIVED QUANTITIES

CHANNEL RESISTANCES

	inner convection	wall	outer equivalent	Channel
top	0.62	0.00	7.90	8.51
side	0.11	0.09	2.01	2.21
bottom	0.39	0.49	39.32	40.21

Fig. 4.10 Model 3: add evaporation effects

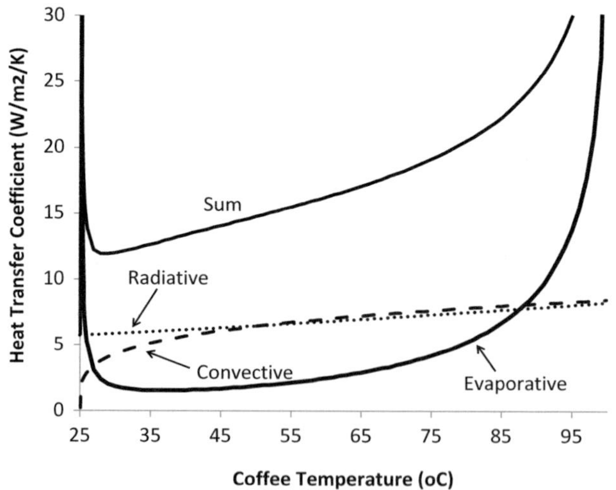

Fig. 4.11 Results of a more detailed estimate of the heat transfer coefficients on the coffee-free surface. The ambient temperature is 25 °C

and ambient. For natural convection, the convective heat transfer coefficient is a function of the temperature difference between the solid and air surface. It is actually zero when the coffee and ambient are at the same temperature, increases sharply with temperature difference, then gradually increases. The radiative coefficient is a function of both coffee and ambient temperatures, and does not show a sharp drop as the coffee temperature approaches ambient. Radiation becomes the dominant mode of heat transfer as the coffee approaches thermal equilibrium with ambient.

The evaporative term increases sharply as the coffee temperature approaches the air temperature, and is an artifact of the use of the resistance model. In fact, modeling evaporation as a dependent current source makes more sense in this case. However, for coffee temperatures above 27 °C in this case, the evaporation coefficient is generally lower than both convection and radiation until approximately 90 °C, above which the evaporation effect increases dramatically. This real effect is due to the rapid increase in vapor pressure as the boiling point is approached from below. The main point here is to argue that evaporative effects are considered secondary to radiation effects, unless the coffee is within a few degrees of its boiling point. As with radiation, for applications in which evaporative effects dominate, other resources can be consulted.

4.2   Heisler and Gurney–Lurie Charts from a 1-Node Lumped Capacity Analysis

In traditional heat transfer textbooks, exact solutions to transient problems of homogeneous objects (i.e., uniform thermal properties of density, specific heat, and thermal conductivity) that have a spatially one dimensional geometry (rectangular, cylindrical, or spherical coordinates) are presented in generalized plots of center temperature vs. time, temperature vs. position, and total heat transferred vs. time. These problems (referred to as either Heisler or Gurney–Lurie charts) are classified as "conduction with convection boundary conditions" and they are all basic heating (or cooling) events of finite-sized objects at an initial temperature suddenly placed in a fluid environment at a different temperature.

The lumped capacity model approach developed in Chaps. 1, 2, and 3 provides a good first approximation to this class of problems and is formally developed here. Furthermore, the two key dimensionless parameters, Biot number and Fourier number, naturally emerge. The Biot number gives information about the spatial distribution of temperature, that is, whether the object heats or cools uniformly (low Biot number), or whether the surface temperature approaches the environment temperature while the center lags behind (high Biot number). The Fourier number is a dimensionless time, where real time is normalized by a characteristic time it would take the interior of the object to adjust if its surface temperature were to suddenly be changed (i.e., a high Biot number event). These variables are developed naturally from a generalized lumped capacity analysis and developed here for a specific manufacturing-type application.

4.2.1 Example: Quenching a Solid Rod

In a manufacturing process, a stainless steel rod (k = 16 W/m/°C, ρ = 7,820 kg/m^3, c = 490 J/kg/°C) of length L (into the page) and diameter D at an initial temperature of $T_{initial}$ = 300 °C is placed in a quenching oil bath at T_{fluid} = 30 °C in order to harden and cool it (Fig. 4.12). Using a lumped capacity model and assuming a uniform convection coefficient on the surface of h = 300 W/m^2/°C and negligible radiation determine the time required to cool the bulk of the material to 100 °C and the surface temperature at that time for diameters of 1 mm, 1 cm, 10 cm, and 1 m.

Model Development

A general analysis will first be performed, and then applied to the specific case in question. A thermal RC network superimposed on the sketch lumps the entire heat capacity into a single node. A conductive resistance between the bulk interior of the rod and its surface acts in series with a convective resistance between the surface and the surrounding oil. Note that the capacitive-carrying node (open symbol) is not placed at the geometric center for conceptual reasons, as the numerator of the conductive resistance is an average distance between the interior and the surface. A convective resistance between the surface of the object and the surrounding fluid takes places at a well-defined surface. If radiation is important, it acts in parallel with convection, and can be combined into a single convective/radiative resistance (if the fluid and radiation wall temperatures can be considered to be equal).

For constant resistance and capacitances, the solution for the (dimensionless) temperature as a function of time is:

$$\theta \equiv \frac{T - T_{fluid}}{T_{initial} - T_{fluid}} = e^{\frac{-t}{(R_k + R_h)C}}$$

This relationship can be used for any object of any shape and construction for which estimates can be made of the total capacitance (the object need not be uniform), a conductive resistance between the interior and the surface and a convective resistance at the surface. Conceptually, the temperature represents an average temperature of the object, and in general, it is recognized that the temperature is not uniform in the object.

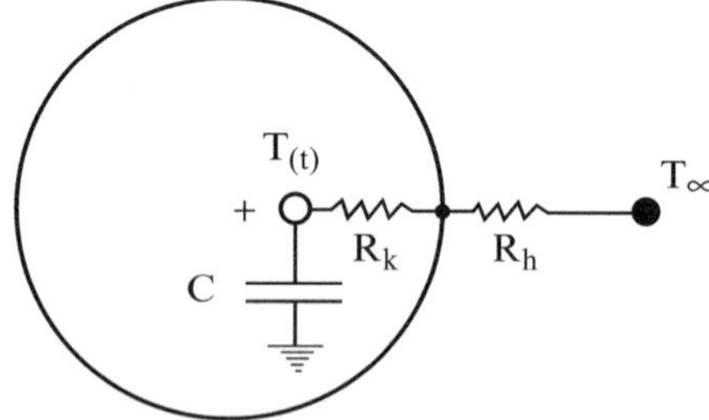

Fig. 4.12 Schematic and superimposed RC network for a solid rod undergoing a cooling or heating event

At any time, the surface temperature is related to the average temperature by a voltage divider relationship:

$$\frac{T - T_{fluid}}{R_k + R_h} = \frac{T_{surface} - T_{fluid}}{R_h}$$

Rearranging:

$$\frac{T_{surface} - T_{fluid}}{T - T_{fluid}} = \frac{R_h}{R_k + R_h} = \frac{1}{{R_k}/{R_h} + 1} = \frac{1}{Bi + 1}$$

where the Biot number is introduced as the ratio of the conductive to convective resistance. This relationship is independent of time. Rearranging for the surface temperature:

$$T_{surface} = T_{fluid} + \frac{T - T_{fluid}}{Bi + 1}$$

The convective resistance is approximated as:

$$R_h = \frac{1}{(\text{Convection Coef.})(\text{Surface Area exposed to fluid})} = \frac{1}{hA_s}$$

The conductive resistance is approximated as:

$$R_k = \frac{\text{Ave. Distance bet. Capacitive Node and Surface}}{(\text{Thermal Conductivity})(\text{Ave. Area bet. Capacitive Node and Surface})} = \frac{\overline{D}}{k\overline{A}}$$

There is more uncertainty in precisely defining the conductive resistance compared to the convective, but a reasonable first approximation of very complicated cases can usually be made.

The Biot number can therefore be expressed as:

$$Bi = \frac{R_k}{R_h} = \left(\frac{h\overline{D}}{k}\right)\left(\frac{A_s}{\overline{A}}\right)$$

Note that the Biot number is proportional to the average distance between the interior node and surface. For a given material in a given convective environment, the larger the object, the larger the Biot number.

Consider the limiting cases. For large Biot number ($Bi \gg 1$), in which case it is more difficult for heat to flow by conduction through the material than it is for heat to flow from the surface to the adjacent fluid, the surface temperature approaches the fluid temperature. For the case of small Biot number ($Bi \ll 1$), the surface temperature equals the average temperature. That is, the temperature in the object is uniform. Heat flows easily from the interior to the surface, and is "hung up" at the surface.

This analysis demonstrates that the Biot number yields insight into the spatial temperature distribution in a solid. All else being equal, small objects tend to heat or cool uniformly, while large objects tend to have larger thermal gradients within them.

For a homogeneous object, the total capacitance is $C = \rho c V$, where V is the total volume of the object and the average temperature can be expressed as:

$$\theta = \frac{T - T_{fluid}}{T_{initial} - T_{fluid}} = e^{\frac{-t}{\rho c V R_k \left(1 + {R_h}/{R_k}\right)}} = e^{-\left(\frac{Bi}{Bi+1}\right)\left(\frac{k\overline{A}t}{\rho c V \overline{D}}\right)}$$

$$\theta = e^{-\left(\frac{Bi}{Bi+1}\right)Fo}$$

where the Fourier number (Fo) has been introduced as a dimensionless time:

$$Fo = \underbrace{\frac{k\overline{A}t}{\rho c V\overline{D}}}_{} \underbrace{\frac{\alpha t}{\frac{V\overline{D}}{\overline{A}}}}_{}$$

The Fourier number is a measure of how far along a cooling or heating event an object is. For short time ($Fo \ll 1$), the dimensionless temperature is close to unity, or the average temperature is close to the initial temperature. For long time ($Fo \gg 1$), the average temperature approaches ambient.

The effect of the shape of the object is manifested in the Biot and Fourier numbers. There is a certain degree of latitude in precisely defining the average conduction distance and average conduction area, consistent with lumping the entire solid into a single node. The Heisler and Gurney–Lurie charts are developed for the three classic coordinate frames that allow for exact theoretical analysis, namely, Cartesian (wall), cylindrical and spherical. Interpretations of these cases in which geometric averages are used are summarized in Fig. 4.13. Notice that the Cartesian geometry (Fig. 4.14) consists of a wall (surface area of width × height) subjected to the same fluid environment on both sides. The centerline acts as an insulated boundary, and only half of the wall needs be considered. In the detailed solution to Workshop 4.1, several different interpretations of the conductive resistance for the cylindrical case are developed.

This dimensionless temperature (θ) vs. dimensionless time (Fourier number) is plotted in Fig. 4.15 for fixed values of the Biot number. The Heisler and Gurney–Lurie charts contain plots of the centerline temperature in this form. The spatial variation is displayed as the dimensionless surface temperature vs. the inverse Biot number (Fig. 4.16). The Heisler and Gurney–Lurie charts are of similar form, but show the temperature at various locations relative to the centerline. In this analysis, only the surface temperature relative to the average temperature is directly available. Note that these "cooling curves" become independent of Biot number for large

Geometry	Solid Wall (height H, width W) of thickness 2L subject to symmetrical cooling (or heating)	Solid rod with diameter D	Solid sphere with diameter D
Coordinate System	Cartesian	Cylindrical	Spherical
Natural dimension	L (half width)	$r = D/2$ (radius)	$r = D/2$ (radius)
Conduction distance $\overline{D}$	$L/2$	$r/2$	$r/2$
Conduction Area $\overline{A}$	$Width \times Height$	$\pi r \cdot Length$	$2\pi r^2$
Convection Area, A_s	$Width \times Height$	$2\pi r \cdot Length$	$4\pi r^2$
Volume, V	$L \times Width \times Height$	$\pi r^2 \cdot Length$	$4\pi r^3/3$
Biot #	$\dfrac{hL}{2k}$	$\dfrac{hr}{k}$	$\dfrac{hr}{k}$
Fourier #	$\dfrac{2\alpha t}{L^2}$	$\dfrac{2\alpha t}{r^2}$	$\dfrac{3\alpha t}{r^2}$

Fig. 4.13 Biot and Fourier numbers for classic geometries. Note that $\alpha = k/\rho/c$

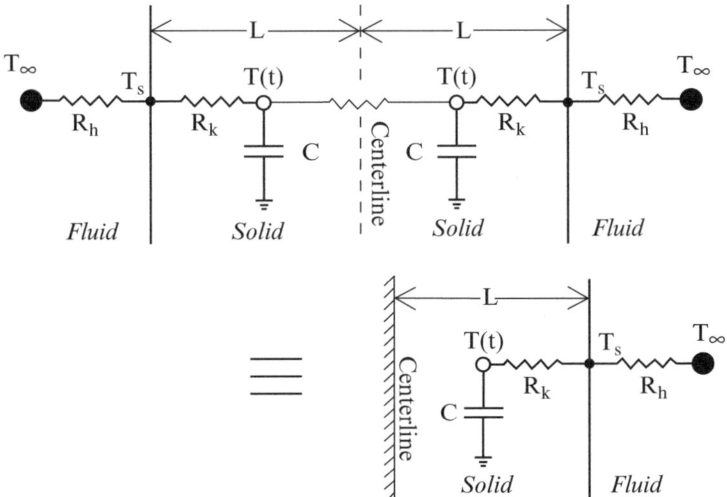

Fig. 4.14 RC network of a wall subjected to the same fluid on both sides. *Top* network shows a node on each side, and could be used as a 2-node model (next chapter) if the fluid temperatures were different on either side. For the symmetric case, the centerline acts as an insulated boundary due to symmetry and the equivalent RC network (*bottom*) is considered in this section. The *bottom* schematic applies directly to cylinders and spheres, where the centerline is a point (of zero surface area)

Biot number, in which case the surface temperature is close to the fluid temperature, and therefore the cooling rate is controlled by the rate of conduction to the interior. Low Biot number (high conductive, relative to convection) events result in uniform temperatures within the solid, and the cooling or heating is controlled by the convection.

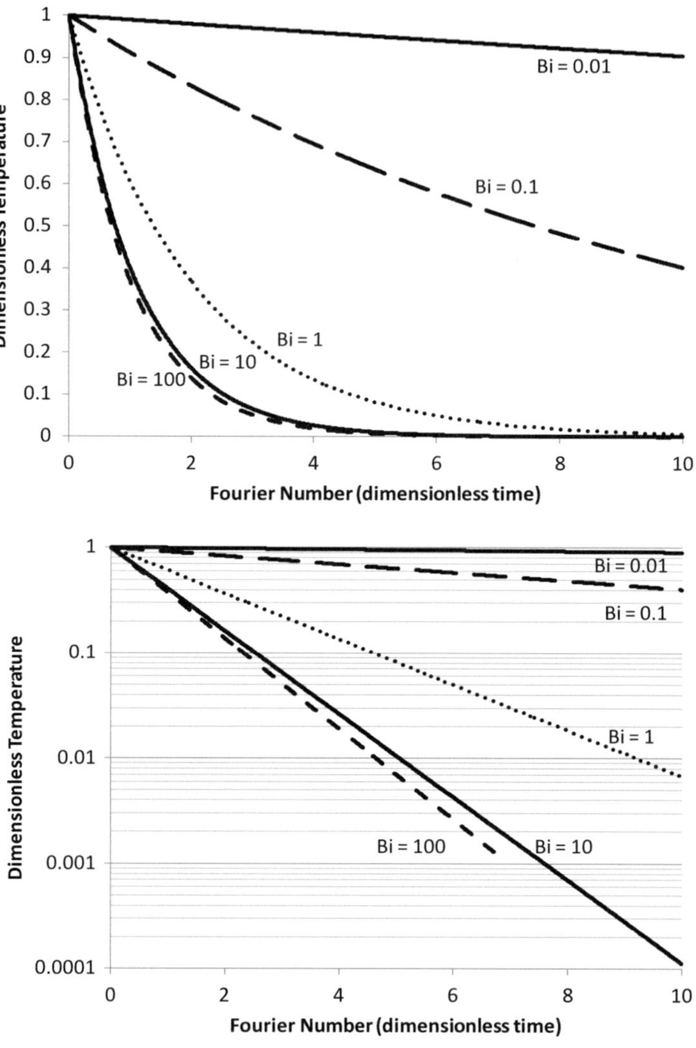

Fig. 4.15 Dimensionless temperature vs. Fourier number for various Biot numbers. Linear axes (*top*) and semi-logy axes (*bottom*)

4.2.2 Application to Quenching a Solid Rod

For a solid rod of diameter D, length L, the volume is:

$$V = \pi D^2 L \big/ 4$$

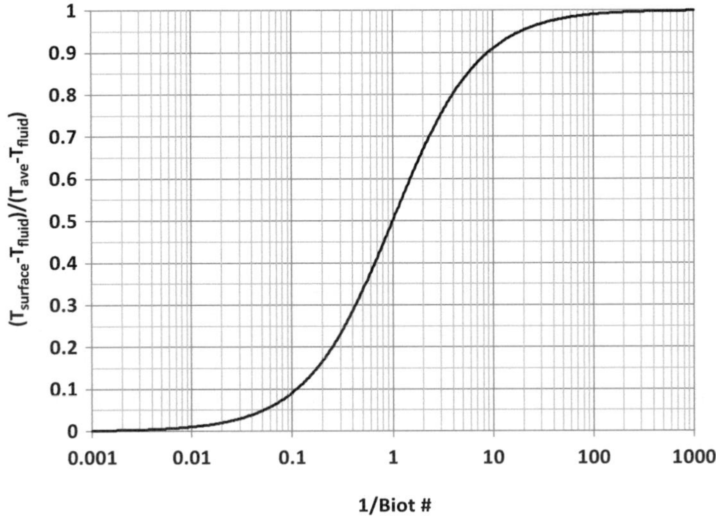

Fig. 4.16 Dimensionless surface temperature plotted against inverse Biot number

The surface area is

$$A_s = \pi D L$$

The average distance (for the conduction distance) is taken to be half the radius (quarter the diameter) and the average area is taken to be the geometric average between the surface and the centerline:

$$\overline{D} = {}^{D}\!/_{4}$$

$$\overline{A} = \frac{1}{2}(A_s + 0) = \frac{\pi D L}{2}$$

A consideration of other conduction models is developed in the detailed solution to the workshop presented at the end of this chapter.

The Biot and Fourier numbers, therefore, are:

$$Bi = \frac{R_k}{R_h} = \left(\frac{h^{D\!/_4}}{k}\right)\left(\frac{2A_s}{A_s + 0}\right)$$

$$Bi = \frac{hD}{2k}$$

$$Fo = \frac{k\overline{A}t}{\rho c V \overline{D}} = \frac{kt\pi DL/2}{\rho c \frac{\pi D^2 L}{4}\frac{D}{4}}$$

$$Fo = \frac{8\alpha t}{D^2}$$

For the cooling event defined, the dimensionless temperature is specified as:

$$\theta = \frac{T - T_{fluid}}{T_{initial} - T_{fluid}} = \frac{100 - 30}{300 - 30} = 0.259$$

Solving the cooling expression for the Fourier number:

$$Fo = -\frac{(Bi + 1)}{Bi}\ln\theta$$

And for the time:

$$t = \frac{D^2 Fo}{8\alpha} = -\left(\frac{\rho c D^2}{8k}\right)\left(\frac{2k}{hD}\right)(1 + Bi)\ln\theta$$

$$t = -\left(\frac{\rho c D}{4h}\right)\left(1 + \frac{hD}{2k}\right)\ln\theta$$

The cooling time increases linearly with diameter for low Biot numbers and increases with the diameter squared for large Biot number events.

Results for the four diameter cases are summarized in Fig. 4.17. The cooling time increases strongly with cylinder diameter. The temperature is nearly uniform for small diameters, and large internal temperature gradients are apparent for large diameters.

INPUT PARAMETERS

k	16	W/m/K
ρ	7820	kg/m^3
c	490	J/kg/K
h	300	W/m^2/K
Tinitial	300	oC
Tfluid	30	oC
Tfinal	100	oC

DERIVED PARAMETERS

α	4.18E-06	m^2/sec
θ	0.259	

RESULTS

D (m)	0.001	0.01	0.1	1
Biot #	0.009375	0.09375	0.9375	9.375
Fourier #	145	15.7	2.79	1.49
cooling time (sec)	4.35	47.1	835	44,722
Surface temperature (oC)	99.3	94.0	66.1	36.7

Fig. 4.17 Spreadsheet solution for the quenching solid cylinder problem

4.3   Thermal Boundary Layers in Thick Walls

In the 1-node fixed lumped capacity model considered in the previous section, the capacitance encompassed the entire solid object. However, at the instant the fluid temperature is changed, there is a region near the surface that responds immediately, while the interior is unaffected for a period of time. That is, there is a time delay between when the thermal signal at the surface occurs and when the interior responds. This section addresses that deficiency by introducing variable-sized nodes. The heating or cooling of a finite-sized object subjected to a sudden change at its surface can therefore be analyzed in two distinct phases: a thermally thick phase (phase I) and a lumped heating (or cooling) phase (phase II).

This phenomenon introduces the notion of a thermal boundary layer or a penetration layer. A thermal disturbance at the surface propagates, or penetrates, into the solid. The size of this boundary layer region grows as time progresses. A thermally thick wall is defined as one in which the physical thickness of the wall is larger than the penetration distance. Whether a given wall behaves as thermally thick or not depends on the time duration after a change at the surface occurs.

While this chapter is focused on thermal boundary layers growing with time in a solid, the concept of a thermal boundary layer is at the heart of modeling thermal boundary layers growing in a fluid, and gives rise to an understanding of convection coefficients. Two combination thermal properties combinations will naturally appear; thermal diffusivity and thermal inertia. The thermal diffusivity ($\alpha = k/\rho/c$) determines how thick a boundary layer is. The thermal inertia ($\Gamma = \sqrt{k\rho c}$) determines the heat transfer rate from the surface to the interior.

This section is long, and readers anxious to develop numerical methods can skip directly to Chap. 5 from here, or skim the remainder of this chapter, especially the plots.

4.3.1   Suddenly Heated Thermally Thick Wall

The first variable-sized node case is one in which the surface temperature of a thick wall is suddenly changed to a new value by some means and held there. The basic growth of a penetration layer naturally emerges. An application of this case is the so-called contact temperature, that is, the temperature that occurs at the interface of two solid objects at different temperatures when suddenly placed in contact.

4.3.1.1 Example: Cooking a Hamburger

A 1 in. (0.0254 m) thick hamburger is initially at a uniform temperature $T_{initial} = 0\,°C$ (i.e., just finished thawing). The hamburger is to be cooked by placing it on a hot skillet in a manner such that the temperature of the hamburger in contact with the skillet will be at a fixed temperature $T_{surface} = 200\,°C$. Determine the time it takes for the center of the hamburger to begin to warm (at which point the hamburger is flipped).

Model Development: 1-Node Solution—Variable Capacitance

A 1-node model where the temperature of the node is known, but its size changes can be developed to reveal the underlying relationship between space and time in this type of thermal boundary layer. A characteristic penetration depth is determined that can be useful in constructing a more detailed grid, among other things.

The RC network is shown for this 1-node model in Fig. 4.18. In this model, the size of the grid changes with time. Two nodes (surface and interior) are shown at time t and a future time $t + \Delta t$. All the capacitance of the boundary layer is lumped into the interior node. In this model, these nodal temperatures are known and constant in time, with the node at the interior of the solid representing a growing region of space as the thermal wave spreads away from the surface. The temperature of the interior node is taken to be the geometric average temperature between the surface and initial solid temperature $\left(T_{ave} = \left(T_{surface} + T_{initial}\right)/2\right)$. The distance δ is interpreted to be the distance from the surface to the region of influence at a given time, and is the depth of penetration of the thermal signal. The average difference between the surface and a location inside the region defined by node some fraction, "f" of the total size associated with the node. A value of $f = \frac{1}{2}$ is consistent with the conceptual model. However, the exact solution will be used to "calibrate" and will be shown to be somewhat less than $\frac{1}{2}$.

The capacity-carrying interior node does not represent a closed thermodynamic system. Rather, because the size increases, the boundary at the leading edge of the node moves into the wall at an instantaneous speed of $d\delta/dt$. Mass crosses this boundary at a rate $\dot{m} = \rho \frac{d\delta}{dt} A$, and carries enthalpy (internal energy plus flow work) into the node at a rate of $\dot{m}c_p\left(T_{initial} - T_{ref}\right)$, represented by a (variable) current source, with T_{ref} being a reference temperature assigned a thermal energy value of zero.

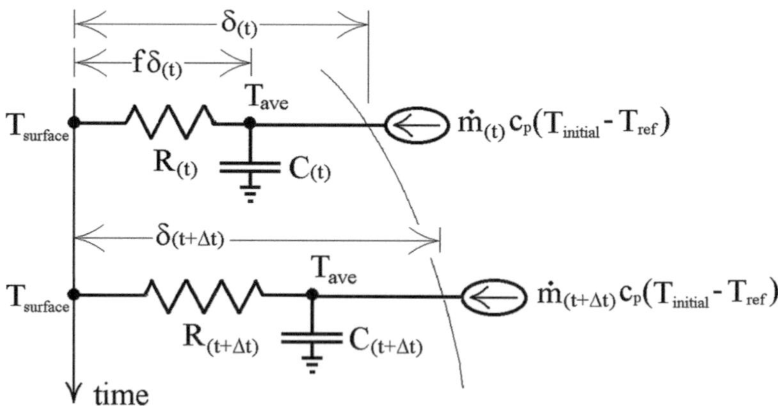

Fig. 4.18 1 Node model (with variable R and C) at time t and a short time later, $t + \Delta t$

The energy balance for the interior node is:

$$\frac{dE}{dt} = \frac{T_{surface} - T_{ave}}{R} + \dot{m}c_p\left(T_{initial} - T_{ref}\right)$$

which states that the rate of change of energy stored in the node equals the rate at which heat enters from the surface plus the rate at which energy enters at the leading edge. The energy stored in the node is changing because of its size (not its temperature, which is constant). The conductive thermal resistance also changes with time as the thermal wave propagates into the interior, spreading out. The rate of change of energy stored is:

$$\frac{dE}{dt} = \frac{d}{dt}\iiint\limits_{CV} \rho\left(\int\limits_{T_{ref}}^{T_{ave}} c_v dT\right) dV = \frac{d}{dt}\left[\rho c_v\left(T_{ave} - T_{ref}\right)A\delta\right]$$

For a solid, $c = c_v = c_p$. Assuming that the size of the node is proportional to the penetration distance (δ) and expressing the conductive resistance as the conductive distance ($f\delta$, where f is constant to be determined) divided by the thermal conductivity of the solid (k_s) and contact surface area (A):

$$\frac{d}{dt}\left[\rho c\left(T_{ave} - T_{ref}\right)A\delta\right] = \frac{T_{surface} - T_{ave}}{\frac{f\delta}{kA}} + \rho c_p\frac{d\delta}{dt}\left(T_{initial} - T_{ref}\right)A$$

Recognizing that only the depth δ changes with time, and rearranging and grouping factors (note that the reference temperature cancels):

$$\frac{d\delta}{dt} = \left(\frac{k}{\rho c}\right)\left(\frac{T_{surface} - T_{ave}}{T_{ave} - T_{initial}}\right)\frac{1}{f\delta}$$

With the average temperature is taken to be the geometric average $(T_{surface} + T_{initial})/2$, the temperature ratio $(T_{surface} - T_{ave})/(T_{ave} - T_{initial})$ equals unity and the governing differential equation becomes

$$\frac{d\delta}{dt} = \frac{\alpha}{f\delta}$$

where the combined thermal property $\alpha = k/\rho/c$ called the "thermal diffusivity" with SI units m^2/s has been introduced. The left hand side represents the rate of change of energy stored associated with the growing size of the node. The right hand side represents the heat transfer rate into the nodal volume and decreases as the penetration depth increases. As the penetration layer grows, it becomes self-insulating.

Separating variables and applying integration limits:

$$f \int_0^\delta \delta d\delta = \int_0^t \alpha dt$$

Performing the integration yields a simple (but not simplistic) expression for the thermal penetration depth:

$$\delta = \sqrt{(2/f)\alpha t}$$

This expression shows that the penetration depth increases with the square root of time. Also, the higher the thermal diffusivity of the material, the faster it grows. The penetration depth increases with the square root of the thermal conductivity and reflects competing effects. The conduction heat transfer rate is proportional to the thermal conductivity, but the thicker the boundary layer, the greater the self-insulating effect. The higher the combination ρc (which represents energy required to change the temperature per unit volume), the harder it is for the thermal disturbance to penetrate, with a square root dependency.

Expressing the penetration depth in terms of time:

$$t = \frac{f\delta^2}{2\alpha}$$

This form shows that the time it takes for a thermal "disturbance" to propagate a distance δ increases with the distance squared. Doubling the distance from the surface increases the time it would take for penetration by a factor of four. This result is due to the self-insulating property of the penetration layer.

Note that the growth of the thermal boundary layer does not depend on the magnitude of the step change in surface temperature that initiates it. The leading edge of the boundary layer can be envisioned as a signal that something has happened at the surface, and it does not matter what that something is.

The rate of heat transfer from the surface to the interior is given by:

$$q_c = \frac{T_{surface} - T_{ave}}{R} = \frac{kA}{f\delta}\left(T_{surface} - T_{ave}\right) = \frac{kA}{\sqrt{2f\alpha t}}\left(T_{surface} - T_{ave}\right)$$

Expressing the thermal diffusivity in terms of its components ($\alpha = k/\rho/c$), and recognizing that the temperature difference between the surface and the average temperature is half of the overall temperature difference between the surface and infinity ($T_{surface} - T_{ave}) = (T_{surface} - T_{initial})/2$:

$$q_c = \left[\sqrt{\frac{1}{8f}}\right]\left(\sqrt{\frac{k\rho c}{t}}\right)A\left(T_{surface} - T_{initial}\right)$$

The material property formed by the combination $\Gamma = \sqrt{k\rho c}$ is termed the "thermal inertia" or the "thermal hardness." The rate of heat transfer decreases with time (inverse square root dependency), again demonstrating the self-insulating nature. At the moment the surface temperature changes ($t = 0$), the heat transfer rate is infinite, a result of the temporal discontinuity posed in the engineering problem statement. As time passes, the wave penetrates deeper into the material, reducing the temperature gradient that drives it.

The exact solution to this problem is given by:

$$q_c = \left(\frac{1}{\sqrt{\pi}}\right)\left(\sqrt{\frac{k\rho c}{t}}\right) A\left(T_{surface} - T_{initial}\right)$$

This exact solution differs from the 1-node model only in the value of the constant. The 1-node model reproduces the functional dependency on material properties and time exactly. In this case, the 1-node solution can be "calibrated" against the exact solution by a suitable choice of the average conductive distance (which defines "f") such that:

$$\sqrt{\frac{1}{8f}} = \frac{1}{\sqrt{\pi}}$$

Squaring both sides and rearranging:

$$f = \frac{\pi}{8} = 0.39$$

This value, somewhat less than half the size of the node, reflects that a more exact solution would not predict a linear temperature distribution, but one skewed toward the surface.

Table 4.1 shows material properties for several representative materials, sorted in order of increasing thermal diffusivity. Liquid water and air are included as the only fluids. A plot of the penetration distance as a function of thermal diffusivity shown for these representative materials at various elapsed times is shown in Fig. 4.19. Each boundary layer grows with a square root dependency on time. The higher the thermal diffusivity (α), the faster it grows.

4.3.2 Application to Cooking a Hamburger

Results of this model to the cooking hamburger are shown in Figs. 4.20 and 4.21. The penetration depth is shown as a function of time until it reaches the centerline in Fig. 4.20, at which point the hamburger can be flipped. Not modeled yet is the response of the cooked portion after flipping (best done with a more detailed numerical model), during which time the temperature within the hamburger redistributes. It is apparent that in the first minute, the thermal wave penetrations

Table 4.1 Room temperature thermal properties of selected substances (sorted by thermal diffusivity)

Substance	ρ (kg/m^3)	C_p (J/kg/°C)	k (W/m/°C)	$\alpha = k/\rho c_p$ (m^2/s)	$(k\rho c_p)^{0.5}$ (J/m^2/°C/s$^{0.5}$)
Wood	420	2,720	0.11	9.63E−08	354
Meat	950	3,520	0.4	1.20E−07	1,157
H$_2$O(l)	1,000	4,200	0.57	1.36E−07	1,547
Glass	2,700	840	0.78	3.44E−07	1,330
Brick	1,600	840	0.69	5.13E−07	963
Steel	7,800	465	54	1.49E−05	13,995
Cast Iron	7,200	447	52	1.62E−05	12,937
Air	1.23	1,005	0.023	1.86E−05	5.33
Aluminum	2,707	896	204	8.41E−05	22,244
Copper	8,954	383	386	1.13E−04	36,383

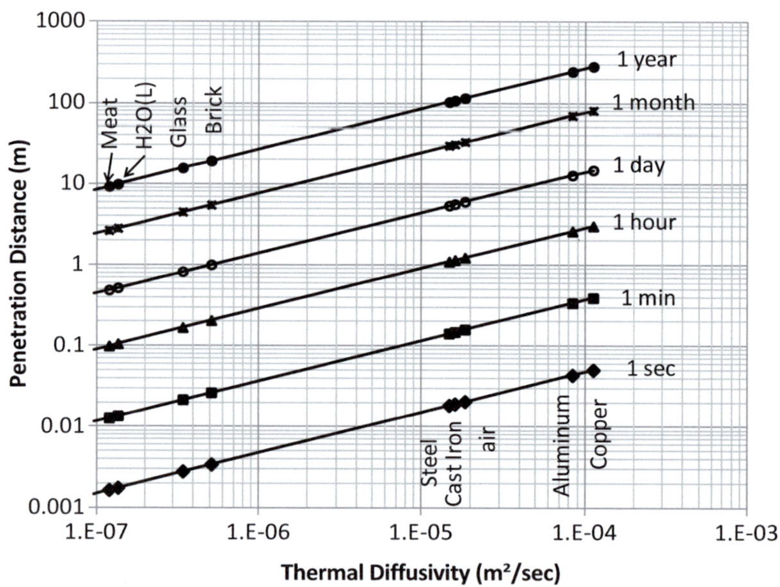

Fig. 4.19 Penetration depths as a function of thermal diffusivity for selected materials

to approximately 0.5 cm. It takes two more minutes to propagate to 1.0 cm, and almost 6 min to reach the centerline (1.27 cm).

Figure 4.21 shows temperature plotted against position at 1 min intervals. The vertical dashed line shows the hamburger at the instant it is placed on the grill. After 1 min, the penetration layer has reached 0.5 cm, and the straight line between the fixed surface temperature and the edge of the penetration layer gives a first approximation to the temperature distribution in the solid at that time. The penetration layer

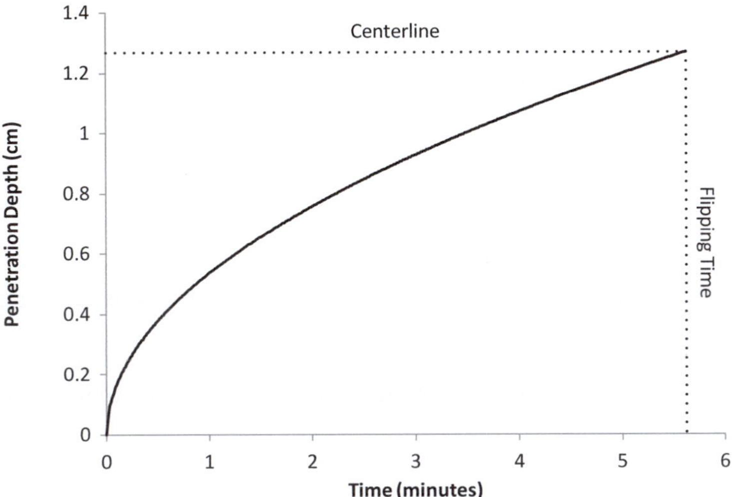

Fig. 4.20 Penetration depth vs. time for a cooking hamburger

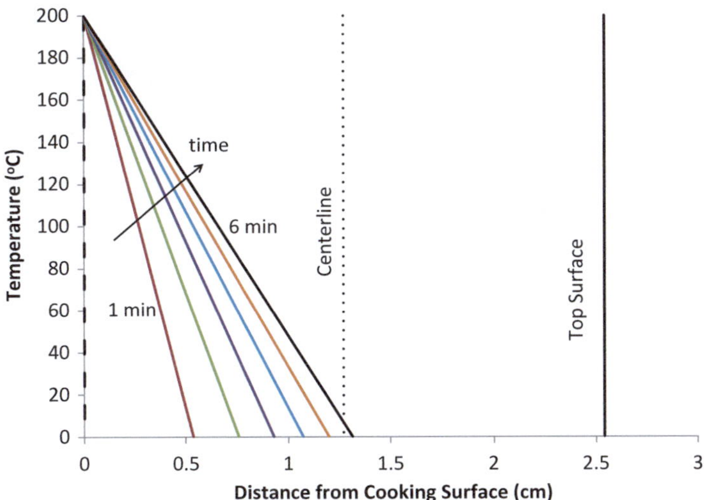

Fig. 4.21 Temperature profiles for cooking hamburgers. *Straight lines* from fixed surface temperature to the initial temperature at the penetration depth is shown at 1 min intervals

continues to grow, but the spacing between lines decreases. The temperature gradient driving the event (Fourier's law of conduction) decreases with time.

This response, of course, is consistent with experience. If a cooking hamburger is cut in half at some point to check, there will be a cooked layer near the cooking surface, and a raw hamburger above it.

4.3.2.1 Example: Contact Temperature

A block of solid material A (with density ρ_A, specific heat c_A, thermal conductivity k_A), initially at a uniform temperature T_A, is suddenly brought into good thermal contact with material B (with density ρ_B, specific heat c_B, thermal conductivity k_B), at a different temperature T_B. Determine the temperature at the point of contact, the temperature distribution in the wall and the heat transfer rate at the wall as functions of time.

Figure 4.22 shows an RC network for this problem. Both materials are assumed to experience a suddenly heated wall event, with the temperature at the surface suddenly changed to the contact temperature value, yet to be determined. No assumption has been made about whether the contact temperature changes with time, but it will be shown to be invariant.

Since the interface is a surface area, with no capacitance associated with it (and the ability to store energy), the rate at which heat enters from one side must equal the rate of heat transfer to the other:

$$q_A = q_B$$

Applying the result from the suddenly heated wall problem, choosing signs such that if A is hotter than B, heat flows from the interior of A (at T_A) to the contact point (at T_c) and from the contact point to the interior of B (at T_B) or vice versa:

$$\left[\sqrt{\frac{1}{8f}}\right]\left(\sqrt{\frac{k_A\rho_A c_A}{t}}\right)A(T_A - T_c) = \left[\sqrt{\frac{1}{8f}}\right]\left(\sqrt{\frac{k_B\rho_B c_B}{t}}\right)A(T_c - T_B)$$

Cancelling terms and rearranging:

$$\frac{T_A - T_c}{T_c - T_B} = \sqrt{\frac{k_B\rho_B c_B}{k_A\rho_A c_A}} = \frac{\Gamma_B}{\Gamma_A}$$

where the thermal inertia, $\Gamma \equiv \sqrt{k\rho c}$, naturally appears.

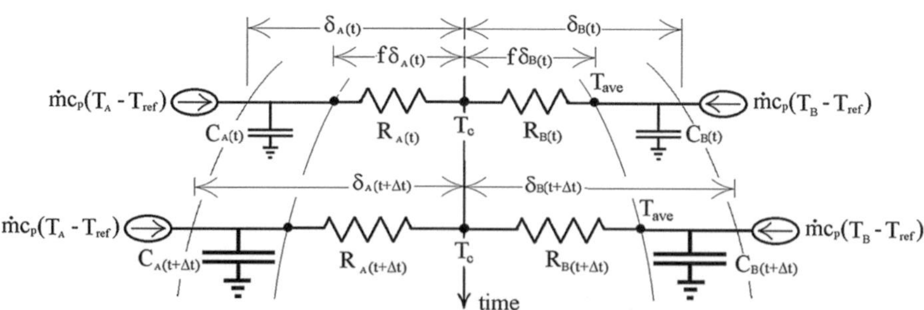

Fig. 4.22 1-Node model for two material (A and B) at different initial temperatures brought into contact, with variable R and C at two instants of time

This expression can be rearranged to solve for the contact temperature:

$$T_c = \frac{\Gamma_A T_A + \Gamma_B T_B}{\Gamma_A + \Gamma_B}$$

which shows that the contact temperature is an average temperature weighted by the thermal inertia. The higher the thermal inertia, the harder it is to change its surface temperature, from which the term "thermal inertia" is based. Notice that there is no time dependency, and therefore the contact temperature is immediately established and maintained constant as time progresses, as long as both materials exhibit thermally thick behavior. The rate of heat transfer from A to B (or vice versa) decreases with time as a thermal boundary layer grows in both materials (at different rates due to different thermal diffusivities), but the contact temperature does not change with time.

The temperature difference ratio on the left hand side indicates that the contact temperature resides between A and B. Whether it is closer to A or to B depends on the ratio of the thermal inertia values. Consider the limits. If A has a much higher inertia than B ($\Gamma_A \gg \Gamma_B$ and the temperature ratio is therefore small), then T_c is close to T_A. If A and B have equal inertias, then T_c is right in the middle.

A table of contact temperatures for a variety of material combinations is shown in Fig. 4.23. In this case, the hotter material is at 100 °C and the colder material is at 0 °C. The temperature of the contact point is listed in the corresponding box. For example, if a hot brick is placed on a cold steel plate, the contact temperature will be 6.4 °C. That temperature is much closer to the steel because the thermal inertia of steel is more than ten times that of the brick.

A good classroom demonstration is to pass around two room temperature objects of similar shape but made of materials of very different thermal inertia. The objects are known to be at the same temperature, say room temperature. Each student (at a warmer temperature than the objects) will grab the objects, one in each hand, and will think that one is hotter than the other. The sensation of hot or cold depends on the contact temperature, which will be different, even though the objects are known to be at the same temperature. Note that thermal properties of human flesh are close to that of water, so choosing a brick or other ceramic (a ceramic mug will do nicely) and a metal will give a good difference.

$(k\rho c_p)^{0.5}$	****Hot Material (at 100°C)****							
	$(k\rho c_p)^{0.5}$	5.33	354.00	963	1,330	1,547	13,995	22,244	36,383
	Cold Material at 0°C	Air	Wood	Brick	Glass	H2O(l)	Steel	Aluminum	Copper
5.33	Air	50.0	98.5	99.4	99.6	99.7	100.0	100.0	100.0
354.00	Wood	1.48	50.0	73.1	79.0	81.4	97.5	98.4	99.0
963	Brick	0.55	26.9	50.0	58.0	61.6	93.6	95.9	97.4
1,330	Glass	0.40	21.0	42.0	50.0	53.8	91.3	94.4	96.5
1,547	H2O(l)	0.34	18.6	38.4	46.2	50.0	90.0	93.5	95.9
13,995	Steel	0.04	2.5	6.44	8.68	10.0	50.0	61.4	72.2
22,244	Aluminum	0.02	1.6	4.15	5.64	6.50	38.6	50.0	62.1
36,383	Copper	0.01	1.0	2.58	3.53	4.08	27.8	37.9	50.0

Fig. 4.23 Contact temperatures for combinations of representative materials

4.3.3   Suddenly Heated Thermally Thick Cylinder

A solid cylinder of radius R_0 at an initial temperature $T_{initial}$ has its surface temperature suddenly changed to $T_{surface}$. Determine the penetration distance as a function of time.

In this case, a variable capacitance node is used (Fig. 4.18), except that the conduction area as well as the conduction distance changes with time as the thermal boundary layer grows inwardly from the surface toward the centerline. The conduction from the outer layers is focused somewhat toward a decreasing surface area at the leading edge of the penetration layer.

In this case, the conductive resistance is taken to be the exact solution for conduction between cylindrical shell with outer radius of the pipe surface (R_0) and an inner surface at a radius a fraction of the penetration depth from the surface ($R_0 - f\delta$). The volume of the nodal element is the exact value (rather than the thin-wall approximation).

The nodal balance on the variable capacity node is:

$$\frac{dE}{dt} = \frac{T_{surface} - T_{ave}}{R_k} + \dot{m}c\left(T_{initial} - T_{ref}\right)$$

$$\frac{d}{dt}\left[\rho c\left(\pi R_0^2 - \pi(R_0 - \delta)^2\right)L\left(T_{ave} - T_{ref}\right)\right]$$

$$= \frac{2\pi kL\left(T_{surface} - T_{ave}\right)}{\ln\left(\frac{R_0}{R_0 - f\delta}\right)} + \rho c 2\pi(R_0 - \delta)\frac{d\delta}{dt}\left(T_{initial} - T_{ref}\right)$$

Expanding and separating variables:

$$(R_0 - \delta)\ln\left(\frac{R_0}{R_0 - f\delta}\right)d\delta = \alpha dt$$

Dimensionless parameters will now be introduced. There is an obvious characteristic dimension for which to normalize the penetration depth, namely, the cylinder radius. Time can be normalized by a characteristic time (t*) to be determined:

$$\hat{\delta} = \frac{\delta}{R_0}$$

$$\hat{t} = \frac{t}{t^*}$$

Inserting these variables into the governing equation and rearranging:

$$(1 - \hat{\delta})\ln\left(1 - f\hat{\delta}\right)d\hat{\delta} = -\left(\frac{\alpha t^*}{R_0^2}\right)d\hat{t}$$

It is now possible to define the characteristic time for this problem by setting the coefficient on the right hand side equal to a numerical value. That is, let

$$t^* = \frac{R_0^2}{2\alpha}$$

Note that the same characteristic time as the finite cylinder is used (with the 2 in the denominator) to make the dimensionless time the same as the Fourier number derived previously for a cylinder with a fixed size node.

The governing differential equation for the penetration depth as a function of time is:

$$(1 - \hat{\delta})\ln(1 - f\hat{\delta})d\hat{\delta} = -d\hat{t}/2$$

A good exercise in integrating by parts, not shown, yields the following solution, with an initial condition that the penetration depth is zero at the initial time:

$$\frac{-(1 - f\hat{\delta})}{f^2}\left[2(1 - f)(\ln(1 - f\hat{\delta}) - 1) - (1 - f\hat{\delta})\left(\ln(1 - f\hat{\delta}) - \frac{1}{2}\right)\right] - \frac{3}{2f^2} + \frac{2}{f} = \hat{t}$$

This function is plotted in Fig. 4.24 for two values of f. The value 0.4 is based on the experience from calibrating the thick wall problem, and the value 0.5 is the geometric average value. This result indicates that the initial thick wall heating (or cooling) stage of a large Biot number (so that the surface temperature equals the fluid temperature) solid cylinder will occur until the dimensionless time

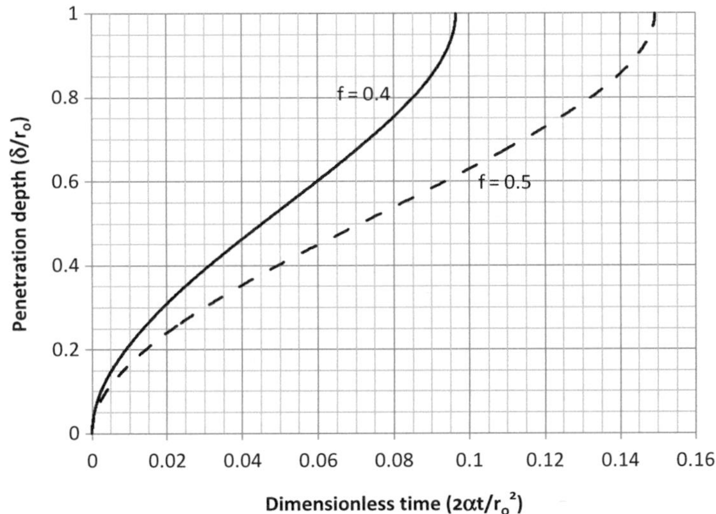

Fig. 4.24 Dimensionless penetration depth as a function of dimensionless time for the suddenly heated cylinder

(Fourier number) equals approximately 0.1. For longer times, the fixed capacity node is more appropriate as the cylinder gradually warms (or cools) as a lumped thermal mass.

4.3.4 Thermally Thick Wall Suddenly Exposed to a Hot (or Cold) Fluid

A variation of a suddenly heated wall is one in which a thick wall (with density ρ, specific heat c, thermal conductivity k) at an initial temperature $T_{initial}$ is suddenly exposed to a fluid with a constant convective/radiative coefficient h at a temperature T_{fluid}. This case is more in line with the cases covered by the Heisler and Gurnie–Lurie charts than the case where the surface temperature is changed to a known value. It is more difficult to solve because the average temperature in the penetration layer changes with time as the layer grows. Another variation (not considered here) would involve subjecting the surface to a constant heat flux, with or without a simultaneous convection to the fluid.

4.3.4.1 Example: Boiling and Baking Boneless Chicken Breasts

While chefs might cringe at the thought, boneless chicken breasts can be cooked by dropping them into boiling water. Alternatively, chicken breasts can be baked in an oven. A series of photos of boneless chicken breast (approximately 3.8 cm thick) after being boiled for different times and cut in half is shown in Fig. 4.25. The evolution of a cooked layer is revealed. Even though the shape of the chicken breast is clearly not an infinitely thick wall, for short enough times, the growth of a nearly uniform thickness of cooked layer from the surface is apparent.

Both of these cooking scenarios can be reasonably modeled as the placement of a solid object into a fluid (water for boiling, air for baking) at a different temperature. In modeling this behavior, the only difference between cooking methods is the fluid temperature (100 °C for boiling, 177 °C for baking at 350 °F) and the convection coefficient assumed (500 W/m^2/K for boiling, 15 W/m^2/K for baking). Both cases are treated as a thick wall at an initial temperature (0 °C) suddenly exposed to a fluid at different temperatures. A general treatment is developed (using dimensionless parameters) and then applied to the cooking chicken case.

A thermal resistance model for this case is shown in Fig. 4.26 at time t, and a short time later, $t + \Delta t$. In this model, heat flows from the ambient fluid to the wall surface through a convective/radiative resistance, and from the surface to the interior (to the capacity-carrying node) through a conductive resistance in series. The size of the capacity-carrying node increases, and therefore the leading edge grows by conduction into a region at the initial temperature, as before. Mass crosses this boundary, and the energy flux is modeled as a current source. The speed of the leading edge is the rate of change of the boundary layer thickness. The temperature of the capacity-carrying node is at the geometric average of the surface temperature and the initial temperature $(T_{ave} = (T_{surface} + T_{initial})/2)$. The only difference between this model and the case of a wall whose surface temperature is suddenly changed is that the surface temperature in this case is not fixed.

Fig. 4.25 Photographs
of similarly sized
(approximately 3.8 cm thick)
boneless chicken breasts cut
in half after boiling for
different times

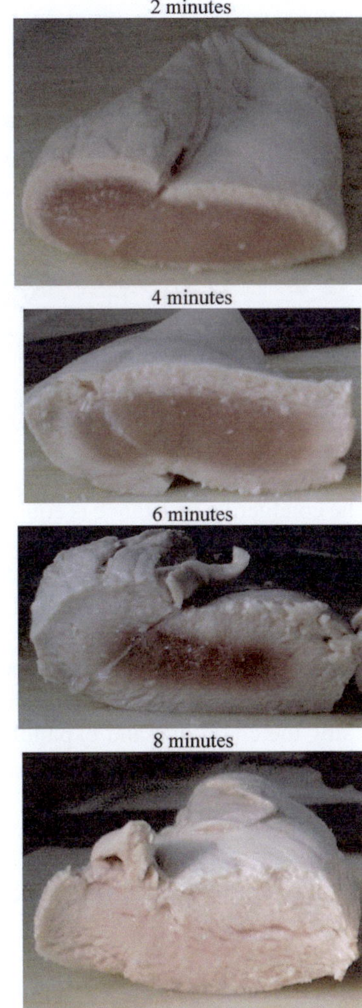

An energy balance applied to the variable capacity-carrying node is:

$$\frac{d}{dt}\left[\rho c A \delta \left(T_{ave} - T_{ref}\right)\right] = \frac{T_{surface} - T_{ave}}{R_k} + \dot{m}c\left(T_{initial} - T_{ref}\right)$$

The current source represents the enthalpy flux into the node (the flow of internal
energy plus the flow work). The mass flux is related to the rate of change of the
boundary layer thickness:

$$\dot{m} = \rho V A = \rho \frac{d\delta}{dt} A$$

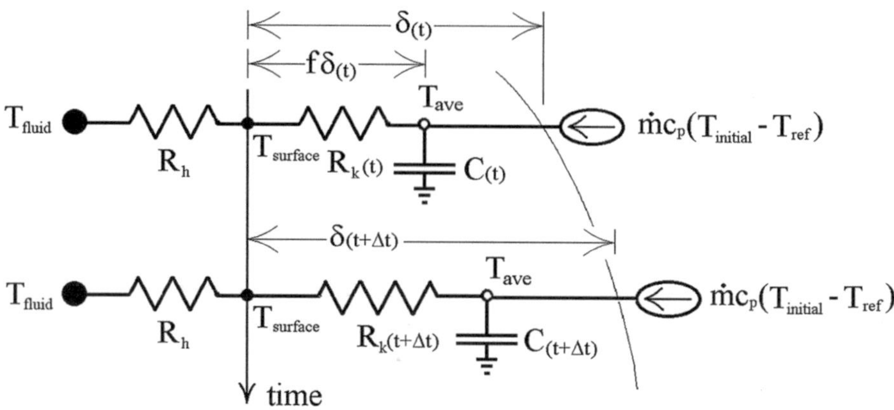

Fig. 4.26 RC network for a thermally thick wall suddenly subjected to a hot (or cold) fluid

The conductive resistance is equal to the conductive distance divided by the thermal conductivity and area. The conductive resistance is taken to be a fraction (f) of the thickness of the boundary layer. Conceptually, f is approximately ½ and could be set to that value or "calibrated" against the published exact solution for this problem.

The energy stored in the capacitive-carrying node depends on the size of the node and its average temperature, both of which change with time in this case. Using the chain rule and expanding all terms:

$$\rho c A \left[\left(T_{ave} - T_{ref} \right) \frac{d\delta}{dt} + \delta \frac{d\overline{T}}{dt} \right] = \frac{kA \left(T_{surface} - T_{ave} \right)}{f\delta} + \rho c A \frac{d\delta}{dt} \left(T_{initial} - T_{ref} \right)$$

The reference temperature term cancels on both sides. Rearranging:

$$\left(T_{ave} - T_{initial} \right) \frac{d\delta}{dt} + \delta \frac{dT_{ave}}{dt} = \frac{(k/\rho c) \left(T_{surface} - T_{ave} \right)}{f\delta}$$

The first term represents the change in energy associated with the growing size of the node, and the second term represents the change in energy due to the gradually changing surface temperature. Dividing through by the temperature change (recognizing that the average temperature is midway between the surface and the initial temperatures), and multiplying through by δ and dt:

$$\delta d\delta + \delta^2 \frac{d \left(T_{surface} - T_{initial} \right)}{\left(T_{surface} - T_{initial} \right)} = \frac{\alpha}{f} dt$$

Dimensionless variables reveal the underlying relationships between space and time in this problem. A dimensionless surface temperature is introduced that is defined to have a value of unity initially, and to decay to zero as the surface temperature approaches the fluid temperature as time goes to infinity:

$$\theta_s = \frac{T_{surface} - T_{fluid}}{T_{initial} - T_{fluid}}$$

The boundary layer thickness in normalized by a characteristic dimension (δ^*) whose value will be determined through the course of the dimensional analysis:

$$\hat{\delta} = \frac{\delta}{\delta^*}$$

Similarly, time is normalized by a characteristic time (t*) whose value will be determined:

$$\hat{t} = \frac{t}{t^*}$$

The energy balance becomes, after manipulation:

$$\hat{\delta}\,d\hat{\delta} + \hat{\delta}^2 \frac{d\theta_s}{1 - \theta_s} = \frac{\alpha t^*}{f\delta^{*2}}d\hat{t}$$

The dimensionless parameter on the right side can be set equal to unity, revealing the fundamental relationship between space and time:

$$\frac{\alpha t^*}{f\delta^{*2}} = 1$$

Finally, the parameter-free energy balance is:

$$\hat{\delta}\,d\hat{\delta} + \hat{\delta}^2 \frac{d\theta_s}{1 - \theta_s} = d\hat{t}$$

Again, both the size and the average temperature of the boundary layer are changing with time. This equation cannot be integrated, yet, as there are two dependent variables (penetration distance and surface temperature).

Figure 4.26 shows that the surface temperature represents a voltage divider between the ambient fluid and the interior:

$$\frac{T_{fluid} - T_{surface}}{R_h} = \frac{T_{surface} - T_{ave}}{R_k}$$

Rearranging:

$$\frac{R_k}{R_h} = \frac{T_{surface} - \overline{T}}{T_{fluid} - T_{surface}}$$

$$\frac{f\delta/kA}{1/hA} = \frac{T_{surface} - \left(T_{surface} + T_{initial}\right)/2}{T_{fluid} - T_{surface}}$$

Rearranging:

$$2f\frac{h\delta}{k} = \frac{T_{surface} - T_{initial}}{T_{fluid} - T_{surface}}$$

Introducing the dimensionless variables:

$$2f\frac{h\delta^*}{k}\hat{\delta} = \frac{T_{fluid} + \theta_s\left(T_{initial} - T_{fluid}\right) - T_{initial}}{T_{fluid} - \left(T_{fluid} + \theta_s\left(T_{initial} - T_{fluid}\right)\right)} = \frac{1 - \theta_s}{\theta_s}$$

The parameter $h\delta^*/k$ is a natural Biot number for this problem, and can be used to define the characteristic dimension by setting:

$$2f\frac{h\delta^*}{k} = 1$$

The characteristic penetration distance is therefore naturally defined as:

$$\delta^* = \frac{k}{2fh} \sim \frac{k}{h}$$

Returning to the characteristic time and inserting the characteristic penetration distance: $t^* = \frac{k^2}{\alpha 4fh^2} = \frac{k\rho c}{4fh^2}$

The surface boundary condition yields the following simple relationship between the penetration distance and surface temperature:

$$\hat{\delta} = \frac{1 - \theta_s}{\theta_s}$$

The rate of change is:

$$d\hat{\delta} = \frac{-d\theta_s}{\theta_s^2}$$

Eliminating the boundary layer thickness in the energy equation:

$$\frac{-(1-\theta_s)}{\theta_s^3}\,d\theta_s + \left(\frac{1-\theta_s}{\theta_s}\right)^2 \frac{d\theta_s}{1-\theta_s} = d\hat{t}$$

$$\left(\frac{1}{\theta_s} - \frac{1}{\theta_s^3}\right)d\theta_s = d\hat{t}$$

Integrating:

$$\ln\theta_s + \frac{1}{2\theta_s^2} = \hat{t} + c$$

The initial condition is that the dimensionless temperature equals unity at time $t = 0$, and therefore the constant of integration is $c = \frac{1}{2}$.

The final relationship between surface temperature and time is:

$$\ln\theta_s + \frac{1}{2}\left(\frac{1}{\theta_s^2} - 1\right) = \hat{t}$$

The surface temperature cannot be expressed analytically in terms of time, but the time is expressed in terms of the surface temperature. Therefore, a closed-form solution is obtained, and the boundary layer thickness and surface temperature are simply related. A plot of these functions is given in Fig. 4.27. The linear scale shows the penetration depth increasing with a vertical slope initially, and a gradually decaying slope. The surface temperature drops quickly at early time, and continues to drop (i.e., approach the fluid temperature) with time, but with a decreasing rate. After one time constant ($t = t^*$), the dimensionless surface temperature is 0.47. After ten time constants, it is 0.20 and after 100 time constants it is 0.07.

Temperature profiles (temperature vs. position at various times) are shown in Fig. 4.28. For each series, a straight line is drawn from the surface temperature to the penetration distance at that point. The penetration layer grows much like the case of a fixed surface temperature, but in addition, the surface temperature gradually approaches the fluid temperature. There is a rapid change initially, and the spacing between fixed temperature lines decreases as time increases. There is no steady-state solution for the truly infinitely thick case.

4.3.4.2 Application to Cooking Chicken

A spreadsheet was used to demonstrate the response of cooking chicken to the boiling and baking cooking methods. The input and derived parameter values are shown in Fig. 4.29. The material properties are the same. The convection coefficients are different, namely 500 W/m^2/K for boiling, 15 W/m^2/K for baking, taken to be representative values for those environments. Also, the fluid

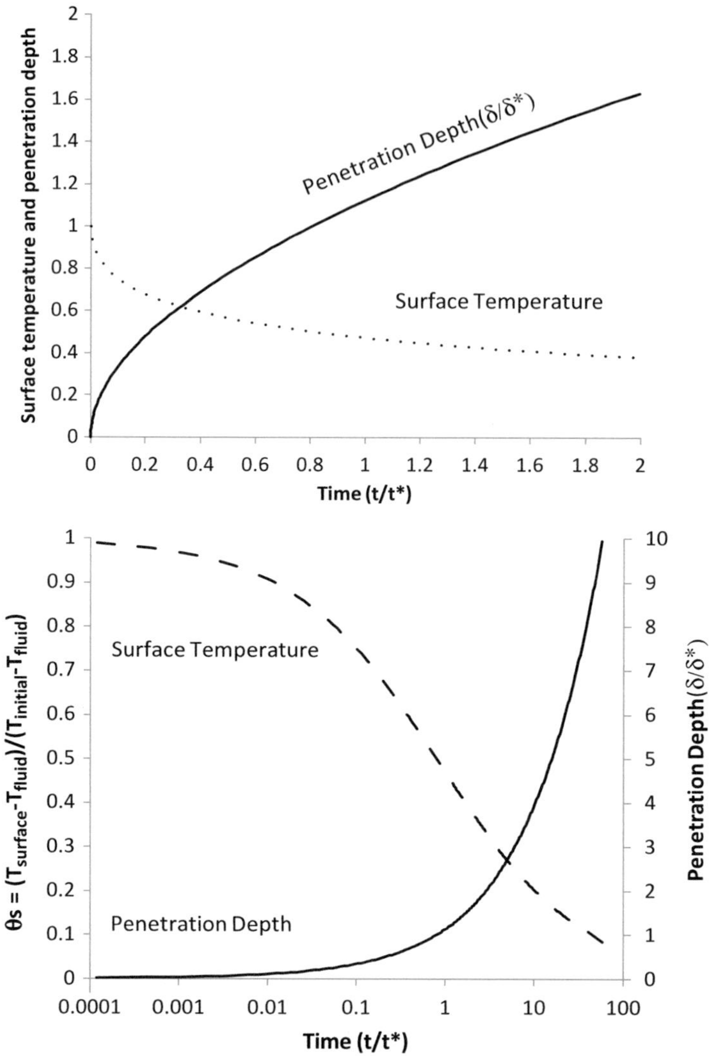

Fig. 4.27 Surface temperature and penetration distance as a function of time (all dimensionless). The *top* figure is on linear axes and the *bottom* on semi-log x axes

temperatures are different, namely 100 °C for boiling and 177 °C for baking. The dimensionless parameters for these cases are the characteristic time ($t^* = 2.7$ s for boiling and 2,972 s for baking) and characteristic penetration depth ($\delta^* = 0.8$ mm for boiling and 26.7 mm for baking). These differences are due to the convection coefficient. The higher convection coefficient for boiling means the heating of the outer layers takes place quicker, but penetrates a shorter distance.

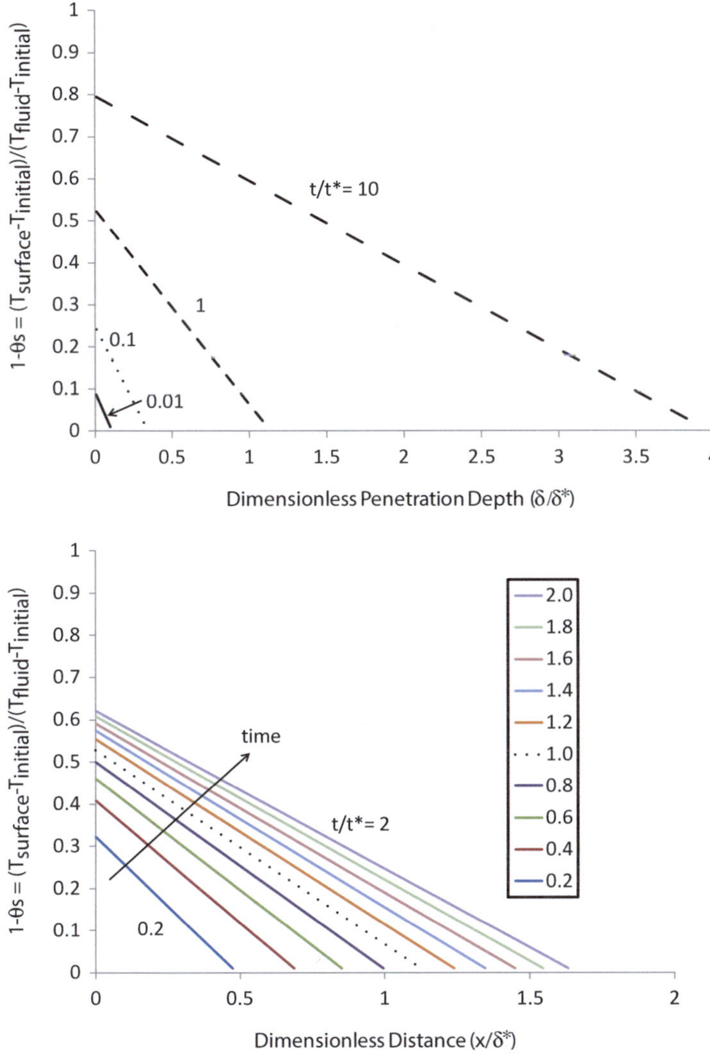

Fig. 4.28 Temperature profiles (temperature vs. distance from surface) at various times. The *top* plot shows times increasing by an order of magnitude. The *bottom* plot shows lines with equal increments (0.2) up to two time constants. The *dotted line* is for t = t*. The surface temperature is presented as 1-θs to correspond to a dimensional case in which the fluid temperature is higher than the initial temperature. For each curve, a *straight line* from the surface to the leading edge is shown. More detailed models in future chapters will capture the curvature

The surface temperatures of the two cooking methods are shown in Fig. 4.30 for 20 min of cooking. The safe cooking temperature (165 °F = 74.2 °C) is indicated as a dotted line. Boiling causes the surface temperature to rise above the safe cooking temperature in a very short time, and then gradually approach (the fluid temperature). On the other hand, with baking, the surface temperature increases slowly and

INPUT PARAMS

ksolid	0.4	W/m/K
rhosolid	950	kg/m^3
csolid	3520	J/kg/K
kfluid	0.023	W/m/K
h	500	W/m2/K
Tinfinity	0	oC
Tfluid	100	oC

DERIVED PARAMS

t*	2.68	sec
δ*	0.00080	m

INPUT PARAMS

ksolid	0.4	W/m/K
rhosolid	950	kg/m^3
csolid	3520	J/kg/K
kfluid	0.023	W/m/K
h	15	W/m2/K
Tinfinity	0	oC
Tfluid	177	oC

DERIVED PARAMS

t*	2972	sec
δ*	0.0267	m

Fig. 4.29 Parameter values for boiling (*left*) and baking (*right*)

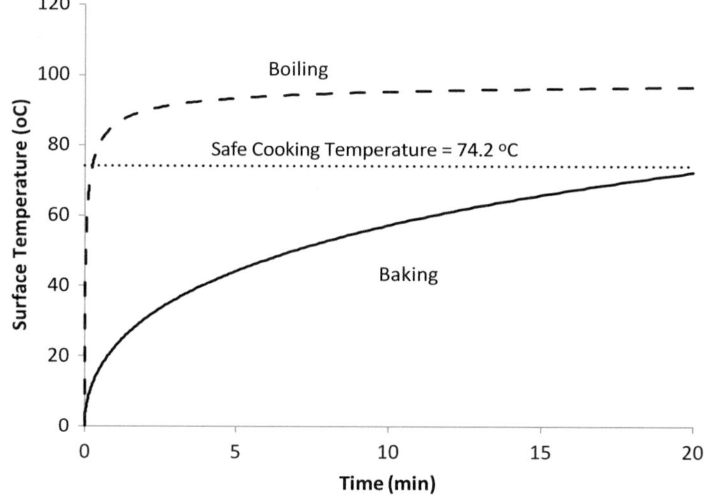

Fig. 4.30 Surface temperatures for boiling and baking chicken

after 20 min is near the safe cooking temperature, but still far from the fluid temperature (177 °C). That is, the surface layers are overcooked during boiling, while the interior is still raw (making the chef cringe).

The corresponding depth of penetration of the surface layer is shown in Fig. 4.31. The penetration depth grows somewhat faster in the boiling case, but not dramatically different, since it is the conduction within the solid that is the primary factor in the propagation of the thermal wave forward. Higher convection driving the surface does have a noticeable influence on penetration depth, however.

The temperature distribution within the solid is shown in Figs. 4.32 (for boiling) and 4.33 (for baking) for the early times. Figures 4.34 and 4.35 show the

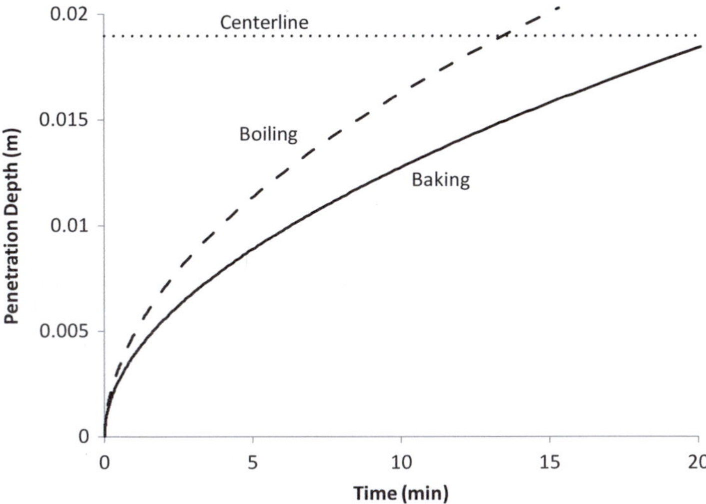

Fig. 4.31 Penetration depths for boiling and baking chicken. The centerline of a 3.8 cm thick sample is shown

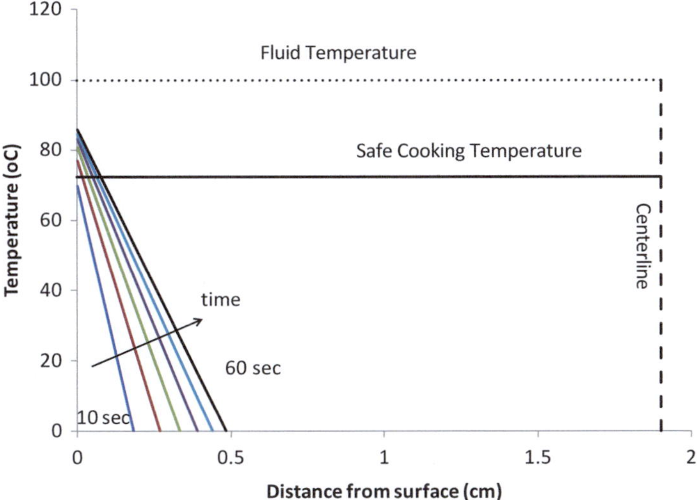

Fig. 4.32 Temperature distributions for boiling for the first minute of cooking (lines at 10 s intervals)

temperature distribution for later times. For boiling, the surface temperature exceeds the safe cooking temperature within the first 10 s of cooking, but does not penetrate very far. For baking, the temperature is more uniform, but cooking still proceeds by heating an outer layer inward.

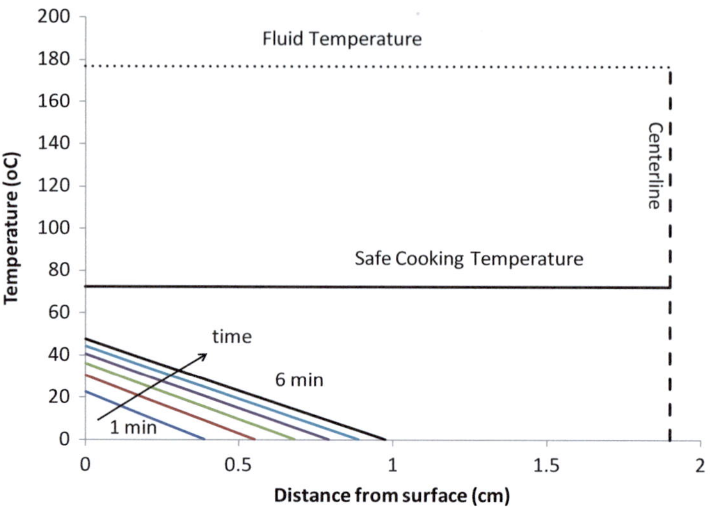

Fig. 4.33 Temperature distributions for baking for the first 6 min (lines at 1 min intervals)

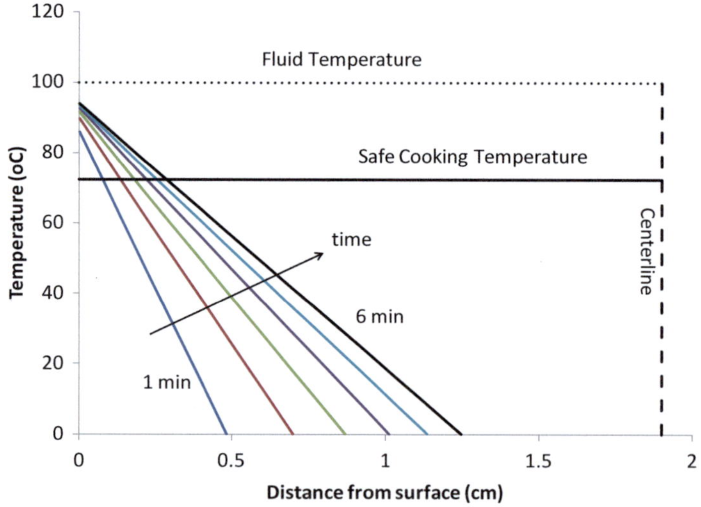

Fig. 4.34 Temperature distributions for boiling for 6 min (lines at 1 min intervals)

These results are all consistent with experience, but hopefully now with a deeper fundamental grasp. The slower something is cooked, which is controlled by the convection environment, the more uniform the cooking will be, as conduction to the interior limits how quickly it cooks in the center.

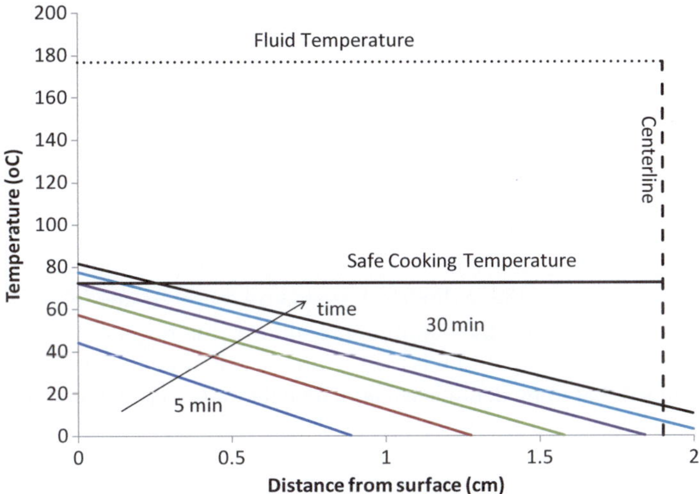

Fig. 4.35 Temperature distributions for baking for 30 min (lines at 5 min intervals)

The focus of this analysis is on the early cooking times, when the cooking object exhibits thermally thick behavior. When the penetration depth reaches the centerline, the object is no longer thermally thick, and the fixed size 1-node model (or more detailed numerical analysis) is more appropriate.

4.3.4.3 Heisler and Gurnie–Lurie Charts Revisited

Close inspection of the published Heisler or Gurnie–Lurie charts at low Fourier number (dimensionless time) shows an offset of approximately 0.1. That is, the centerline temperature remains at the initial temperature for some time before an exponential decay occurs. This behavior reflects the transition from an initial thick wall phase, where the penetration layer grows with time, to a lumped heating (or cooling) stage. That transition occurs when the penetration distance reaches the centerline of the object.

The difference between the Heisler charts (dimensionless temperature vs. time) and the closed formula predicted by the fixed size 1-node model (Fig. 4.15) is that the former is a plot of the centerline temperature, while the latter is a plot of the average temperature. The Heisler charts, therefore, include a separate plot of the average temperature (a spatial integral) in the form of a dimensionless total heat transferred. That plot is not necessary with the 1-node model which predicts the time dependence of the average temperature. Furthermore, the average temperature of a finite-sized object during the thick wall phase can be calculated from the 1-node model, even though all the action is in the growing penetration layer. Therefore, there is really no need to attempt to adjust the 1-node model to account for the initial phase.

In short, for applications of heating (or cooling) objects subjected to sudden changes in their environments, the 1-node model can be used to quickly and in closed-form estimate the response. If it is established that the scenario falls into a thick wall category, that model can be used. The 1-node model is not restricted to any geometry, or to homogeneous objects. Informed estimates of conductive and convective/radiative resistances are needed. The chapters on multidimensional steady-state conduction will provide additional tools for that. For cases where the change in the environment is not sudden, or there are additional changes, numerical methods that address time dependency (i.e., Euler method) can be used.

Two examples that extend the method to situations not covered in the Heisler or Gurnie–Lurie charts are considered in detail next. In one, an internal heat generation term is added to the cylindrical geometry case to model the effect of microwaving. In the other, the cooling mug of coffee is revisited, with a focus on the initial moments after pouring hot coffee into a room temperature mug.

If it is judged that more detail is needed, then the next step is to begin the process of breaking physical space into more than 1-node. The next chapters develop that methodology.

4.4 Cooling Mug of Coffee Revisited: Mug Heating Phases

An application that ties together several concepts from this chapter and shows the versatility of the 1-node models is to analyze the response of a room temperature mug immediately after hot coffee is poured into it. There is an initial heating phase during which time the mug acts as a thermally thick wall, followed by a phase where the mug is simultaneously heated by the coffee and cooled by the air until it reaches a quasi-steady state temperature.

Imagine it takes no time at all to pour hot coffee into a room temperature mug. Imagine also that the coffee behaves like a solid, and does not move after it is poured. The interface between the mug and coffee will immediately attain an intermediate contact temperature, and a thermal wave will propagate into both the coffee and the mug (but at different rates, due to different thermal properties). The outer temperature of the mug remains at its initial temperature during this phase, and the variable-sized nodal analysis applies. Once the penetration depth reaches the outer surface of the mug, the fixed size nodal analysis applies. This behavior is developed in this section.

In this analysis, the average temperature of the coffee is assumed to remain fixed at the boiling point of water (100 °C). This assumption is equivalent to assuming that the coffee has an infinite capacitance, so the heat transferred to the mug does not change the coffee temperature. In the next chapter, this assumption will be relaxed, and the coffee will cool simultaneously with the mug being heated. That problem, however, cannot be solved in closed form, and a numerical simulation is developed for that problem.

4.4.1   Phase 1: Thermally Thick Mug

During the initial phase of heating, the thin-wall approximation (mug thickness much less than the inner diameter) is invoked, so that the mug behaves like a thick wall of constant-area. A good exercise would be to apply the cylindrical model instead. Also, an alternative model to treating the coffee as a solid during this phase would be to assume a representative value of convection (say, 400 $W/m^2/K$).

Table 4.2 lists thermal properties assumed for a typical ceramic and for coffee at nominal room temperature values (the thermal properties are assumed to not change significantly with temperature).

The contact temperature, that is, the temperature that is achieved at the instant of the event is therefore:

$$T_c = \frac{\Gamma_A T_A + \Gamma_B T_B}{\Gamma_A + \Gamma_B} = \frac{(1,587)(100) + (1,166)(25)}{(1,587) + (1,166)}$$

$$T_c = 68.2\,°C$$

Notice that this temperature is different than the 87.7 °C that was calculated for the starting temperature of the main cooling event. The latter thermodynamic calculation depends on the finite amount of coffee and mug and is a fundamentally different phenomenon than the contact behavior which is independent of material size.

Immediately after the coffee is poured, the contact temperature of 68.2 °C is established and a thermal boundary layer grows in both the ceramic and the coffee. The underlying model is valid as long as the material is thermally thick. That criterion is met as long as the thickness of the growing boundary layer is less than the physical thickness of the material (w):

$$\delta < w$$

Inserting the suddenly heated wall result for the boundary layer thickness as a function of time:

$$\delta = \sqrt{(16/\pi)\alpha t} < w$$

Table 4.2 Property values for coffee/mug problem at nominal room temperature

Property	Symbol	Unit	Coffee (A)	Ceramic (B)
Thermal conductivity	k	W/m/K	0.6	1.0
Density	ρ	kg/m^3	1,000	1,700
Specific heat	c	J/kg/K	4,200	800
Thermal diffusivity	$\alpha \equiv k/\rho/c$	m^2/s	$1.43(10)^{-7}$	$7.35(10)^{-7}$
Thermal inertia	$\Gamma = \sqrt{k\rho c}$	$J/s^{0.5}/m^2/K$	1,587	1,166

Solving for the time restriction:

$$t < \frac{\pi}{16} \frac{w^2}{\alpha}$$

For the coffee mug:

$$t < \left(\frac{\pi}{16}\right) \frac{(0.00398\,\text{m})^2}{7.35(10)^{-7}\,\text{m}^2/\text{s}}$$

$$t < 4.2\,s$$

That seems like a reasonable value. It can be estimated empirically by quickly pouring hot coffee into a ceramic mug, holding the mug and counting how long it takes until it starts to feel warm.

Figure 4.36 shows the results for this initial phase of the coffee cooling process. Temperature is plotted against position at 1 s increments of time for the first 5 s. The details of construction of this plot are discussed in Appendix 2. During this time, the mug behaves as if it were infinitely thick. The temperature at the inner wall is brought instantly to the contact temperature (68.2 °C) and it stays there. Rapid growth of the two penetration layers occurs in the first second, due to the extremely high temperature gradient, with a thermal boundary layer growing approximately 2 mm in the mug, and about 1 mm in the coffee. The difference between coffee and

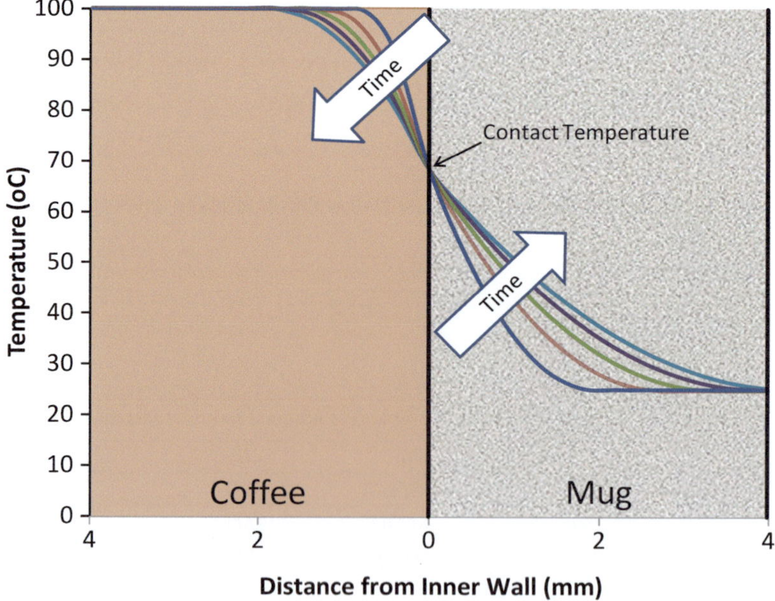

Fig. 4.36 Temperature profiles for the first 5 s of the coffee cooling example. Temperature is plotted against position at 1 s increments. Each curve is a parabolic fit (explained in Appendix 2)

mug penetration depths is a direct result of the thermal diffusivity being approximately four times as great in the mug as in the coffee. The leading edge of the boundary layer advances with time, but the incremental change (i.e., space between curves) decreases with time (because the temperature gradient at the wall that drives the heat decreases).

The coffee in this model is treated as a hypothetical solid. Of course, the coffee flows, introducing another mechanism for heat transport. There are actually two distinct mechanisms for liquid movement. The coffee was just poured, so it is "churning." Even if that effect settles (but in 10 s, that's not likely to be complete), there is a natural convective mechanism. The coffee adjacent to the mug will be colder, and hence slightly more dense, than the coffee in the middle and will tend to fall along the wall, to be replaced from the top. A mild circulation pattern will be induced that profoundly changes the thermal behavior. Convection coefficients allow engineering models to incorporate these effects. If incorporated, these effects would increase the effectiveness of heat transfer on the coffee side, and result in a somewhat higher contact temperature. Also, the contact temperature would drift.

Note that during this initial mug heating phase, the outer wall of the mug remains at ambient temperature, which means that heat is not being lost by the coffee/mug combination. This initial phase involves redistribution within the original system boundary.

4.4.2 Phase II: Mug as Lumped Capacity Node

The thermally thick variable capacitance 1-node model that produced these results breaks down as soon as the leading edge reaches the outer wall of the mug. There is a shift to a new phase, and a new model, occurs. A conventional fixed size model is shown in Fig. 4.37. Treating the mug itself as a 1-node model in contact with coffee at a fixed temperature on one side and air on the other, the mug will continue to warm to an intermediate temperature between the air and the coffee. This process establishes a quasi-steady balance between the mug and coffee, and they subsequently cool together in a third phase, namely, the main cooling event analyzed in Chap. 3.

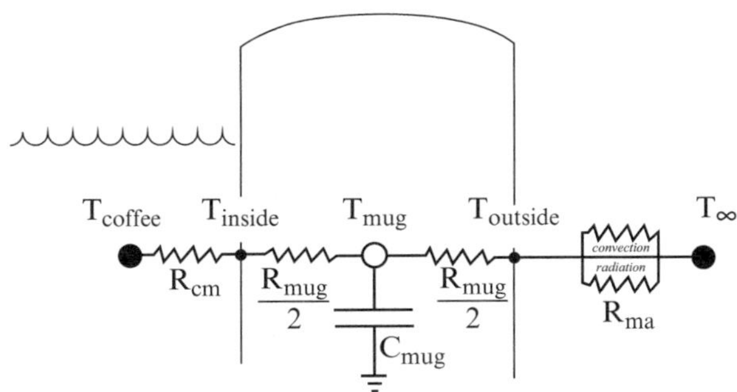

Fig. 4.37 RC network for phase II mug heating

The nodal energy balance applied to the mug node is:

$$C_m \frac{dT_m}{dt} = \frac{T_{coffee} - T_{mug}}{R_{cm} + R_m/2} + \frac{T_{air} - T_{mug}}{R_m/2 + R_{ma}}$$

The solution to this problem, with the coffee (and air) temperature fixed is:

$$\frac{T_{mug} - T_{ss}}{T_{mug,initial} - T_{ss}} = e^{\frac{-t}{RC}}$$

where the steady-state mug temperature is given by:

$$T_{ss} = \frac{\frac{T_{coffee}}{(R_{cm}+R_m/2)} + \frac{T_{air}}{(R_{ma}+R_m/2)}}{\frac{1}{(R_{cm}+R_m/2)} + \frac{1}{(R_{ma}+R_m/2)}}$$

And the time constant is:

$$RC = \frac{C_{mug}}{\left(\frac{1}{R_{cm}+R_m/2}\right) + \left(\frac{1}{R_m/2+R_{wa}}\right)}$$

Input and derived parameter values for the phase II mug heating are listed in Table 4.3.

The time constant of 23.2 s is long compared to the 4.2 s it took for the initial thermally thick boundary layer phase, and short compared to the subsequent cooling of the coffee/mug assembly. Experimentally, the mug temperature peaked at about a minute, giving good order of magnitude agreement.

The model predicts the temperature of the mug at its centerline (T). The temperatures at the inner and outer walls (T_i and T_o) of the mug are related by a voltage divider-type calculation. The inner wall temperature is obtained by equating the heat rate from the coffee to the center with the heat rate from the coffee to the inner wall:

$$q_{coffee-->mugcenter} = q_{coffee-->innerwall}$$

$$\frac{T_{coffee} - T}{R_{cm} + R_m/2} = \frac{T_{coffee} - T_i}{R_{cm}}$$

Solving for the inner wall temperature:

$$T_i = T_{coffee} - \left(\frac{R_{cm}}{R_{cm} + R_m/2}\right)(T_{coffee} - T)$$

A similar analysis between the mug center and ambient air yields an expression for the outer wall temperature:

Table 4.3 Parameter values for phase II mug heating

Input parameters			
Phase I time	4.22	s	
Tmug, initial	46.6	°C	Average mug temperature (end of phase I)
Tair	25	°C	Ambient temperature
Tcoffee	100	°C	Fixed coffee temperature
w	0.003975	m	Mug wall thickness
D	0.08255	m	Mug inner diameter
H	0.098	m	Mug inner height
hma	13.9	W/m^2/K	Outer convective/radiative coefficient
hcm	406	W/m^2/K	Inner convective coefficient
kmug	1.0		Thermal conductivity of mug
rhomug	1,700	kg/m^3	Density of mug
cmug	800	J/kg/K	Specific heat of mug
Derived parameters			
Cmug	144.0	J/°C/W	Capacitance of mug wall
Rcm	0.097	°C/W	Inner convective resistance
Rm	0.149	°C/W	Mug wall conductive resistance
Rma	2.58	°C/W	Outer convective/radiative resistance
R	0.161	°C/W	Equivalent resistance
RC	23.2	s	Time constant
Tss	95.5	°C	Steady-state mug temperature

$$T_o = T_\infty + \left(\frac{R_{m\infty}}{R_{m\infty} + R_m/2} \right)(T - T_\infty)$$

Figure 4.38 shows temperature profiles during the second phase of heating of the mug at 10 s increments. The dashed line shows the temperature profile at equilibrium. The coffee temperature is fixed in this model at 100 °C. Within the mug wall, straight lines are drawn from the inner surface to the center and from the center to the outer wall. In moving from phase I to II model, there is a discontinuous jump on the outer wall surface from ambient (25 °C) to the value given by this second phase of mug heating (38.7 °C). There is a similar discontinuous jump from the contact temperature used in the infinite mug phase (68.2 °C) to the value of this model (69.8 °C), but the change there is (coincidentally) less obvious. The thicknesses of the fluid thermal boundary layers shown are estimates from a future chapter and are meant to be illustrative. This model indicates that a concave up shape persists for the temperature profile in the mug wall as the wall heats, and eventually attains a linear temperature distribution in the mug at equilibrium. The model is focused on the heating of the mug in the initial stages. The gradual decrease in coffee temperature with time follows with the next phase of the cooling, which is taken up further in the next chapter.

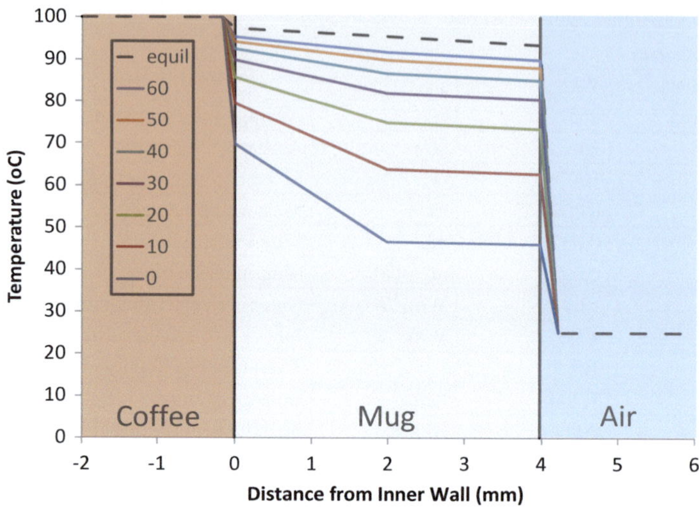

Fig. 4.38 Phase II heating of mug. Temperatures are plotted against distance from the inner wall of the mug at 10 s increments of time (from the start of phase II at 4.2 s). The dashed line shows the temperature profile at quasi-equilibrium (steady-state, with the coffee and air temperatures fixed). The thermal boundary layer thicknesses in the fluids (coffee and air) are estimates based on models of future chapters

Appendix 1. Detailed Calculation of Individual Coffee Mug Thermal Resistances

In this appendix, the formulas used to calculate the individual resistance values are detailed. Figures 4.1 and 4.9 are repeated here for reference. Input parameter values were listed in Figs. 4.4 and 4.10. The rationale for choosing values is described, much of the detail of which is developed in later chapters.

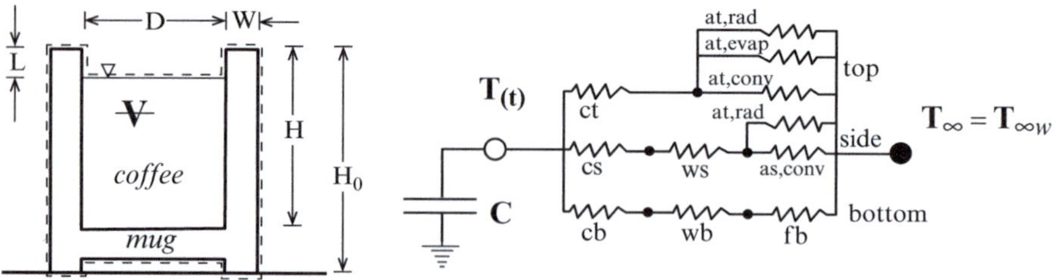

Top Channel Resistors

Free Surface Convective Resistance, Liquid Side: R_{ct}

For this resistance, the coffee/air interface at the free surface is treated like a convective resistance. On the liquid side, the resistance is:

$$R_{ct} = \frac{1}{h_{coffee/surface}A_{surface}} = \frac{1}{h_{coffee/surface}\left(\pi D^2/4\right)}$$

The value chosen for convection coefficient (300 W/m^2/K) is representative of a natural convection for liquids.

Free Surface Convective Resistance, Air Side: R_{at}

There are three resistances in series: convective, radiative, and evaporative. Each of them is modeled as:

$$R_{at} = \frac{1}{h_{effective}A_{surface}} = \frac{1}{h_{effective}\left(\pi D^2/4\right)}$$

The values chosen (6.8, 6.6, and 10 W/m^2/K for convective, radiative, and evaporative) were chosen based on detailed Nusselt number and radiation coefficient analyses (methods developed in later chapters) for a coffee temperature of 70 °C.

Side Channel Resistors

Inner Side Wall Convective Resistance: R_{cs}

The heat transfer between the coffee and the side wall is by the convective mode, and radiation is not important for two reasons: the inner surface does NOT fall into the model for a resistance model because the inner wall "sees" itself. Also, because the liquid convection coefficient is over an order of magnitude larger than typical air values, convection dominates radiation. The thermal resistance is therefore:

$$R_{cs} = \frac{1}{h_{coffee/side}A_{InnerSurface}} = \frac{1}{h_{coffee/side}\left(\pi DL\right)}$$

The value for $h_{coffee/side}$ (470 W/m^2/K) was obtained from a Nusselt number calculation for a coffee temperature of 70 °C.

Side Wall Conductive Resistance: R_{ws}

This resistance represents a conductive resistance across the side wall. Following the general outline, two surfaces are identified, and average distance between the surfaces and average areas calculated. For the side wall, the surface 1 is defined as the inner submerged surface. Its area is:

$$A_1 = \pi D(H - L) = 0.0198 \ \text{m}^2$$

Surface 2 is defined as the total surface area of the outer wall exposed to air, including the top rim and the inner surface area above the coffee surface. This 2D conduction problem is discussed in some detail in a future chapter. The area is:

$$A_2 = \pi(D + 2w)H_0 + \pi DL + \frac{\pi}{4}\left((D + 2w)^2 - D^2\right) = 0.0372\,\text{m}^2$$

The outer area is 88 % greater than the inner one. The geometric average area is:

$$A_{12} = (A_1 + A_2)/2 = 0.0285\,\text{m}^2$$

The average distance between surfaces 1 and 2 (Δx_{12}) can be taken to be simply the wall thickness, w, although that underestimates the true distance. The thermal resistance is:

$$R_{ws} = \frac{\Delta x_{12}}{k_{12}A_{12}} = \frac{w}{k_{12}(A_1 + A_2)/2}$$

The thermal conductivity k_{12} is something you "look up" for the material in question, unless there is more specific information available for a material in question. In this case, an Internet search (January 30, 2011) of "thermal conductivity of ceramic mug" yielded:

http://www.askmehelpdesk.com/physics/why-ceramic-mug-cools-liquid-faster-98327.html

$$k_{12} = 15\,\text{W}/\text{m}/\text{K}$$

http://www.madsci.org/posts/archives/2004-12/1102383405.Eg.r.html

$$k_{12} = 0.5 - 20\,\text{W}/\text{m}/\text{K}$$

A good website for material property values is "Engineering Toolbox" http://www.engineeringtoolbox.com/.

Searching in material properties tabs for thermal conductivity... http://www.engineeringtoolbox.com/thermal-conductivity-d_429.html.

The thermal conductivity for porcelain, and the value used here, is published as:

$$k_{12} = 1.5\,\text{W}/\text{m}/\text{K}$$

Such a wide range of uncertainty in a material property is not uncommon. Engineering judgment often comes into play.

Outer Side Wall Convective Resistance: R_{as}

The exposed outer surface of the side wall, including the inner area above the coffee level and the top of the rim (A_2 from the conduction analysis), exchanges heat by convection in parallel with radiation with ambient air. Technically, the inner portion "sees" the coffee and itself in addition to ambient, but this is a small part of the total exchange so this effect is not considered. The resistances are:

$$R_{as,conv} = \frac{1}{h_{air/sides}A_2} \quad \text{and} \quad R_{as,rad} = \frac{1}{h_{radiation}A_2}$$

The values for $h_{air/sides}$ (6.8) and $h_{radiation}$ (6.6) are obtained from a Nusselt and Kirchhoff analysis for a coffee temperature of 70 °C.

Bottom Channel Resistors

Inner Bottom Wall Convective Resistance: R_{cb}

The only difference between this resistance and the inner side wall is the exposed area:

$$R_{cm} = \frac{1}{h_{cm}(\pi D^2)/4}$$

where h_{cm} is taken as the same natural convection value as between the coffee and the sides.

Bottom Wall Conductive Resistance: R_{fb}

For the conductive resistance through the mug bottom, consider the shape to be a truncated cone, with surface 1 as the inside surface of the mug, and surface 2 to be the surface area in contact with the floor. The area is taken to be the geometric average:

$$A_{12} = \frac{1}{2}\left[\frac{\pi D^2}{4} + \frac{\pi(D+2w)^2}{4}\right]$$

The thermal resistance is therefore:

$$R_{wb} = \frac{\Delta x_{12}}{k_{12}A_{12}} = \frac{w}{k_{12}A_{12}}$$

Conductive Resistance Across the Floor: R_{fb}

Heat transfer between the floor of the mug and its surroundings is difficult to precisely define without more information as to the nature of the table on which the mug rests. There is an air gap trapped beneath the bottom rim. If (a big "if") natural convection effects are sufficiently suppressed so that the air is stationary, then a conductive resistance between the mug bottom and the table surface can be estimated. A radiative channel between flat plates exists between the floor and the bottom.

Then there is a conductive series resistance from the table surface toward the interior of the table. To estimate this resistance, assume the table is large and take a long range view so that the heat transfer emanates outward spherically. The inner surface (surface 1) is taken to be a hemisphere with a radius such that the surface area of the hemisphere equals the surface area of the floor underneath the mug. That is:

$$2\pi r_1^2 = \pi(D + 2w)^2/4$$

or

$$r_1 = (D + 2w)/\sqrt{8}$$

The rate of heat transfer across a hemispherical shell of inner radius r_1 and outer radius r_2 (from 1 to 2) is

$$q_{12} = -k2\pi r^2 \frac{dT}{dr}$$

Separating variables and integrating:

$$\int_{T_1}^{T_2} dT = \frac{-q_{12}}{2\pi k} \int_{r_1}^{r_2} \frac{dr}{r^2}$$

$$T_2 - T_1 = \frac{-q_{12}}{2\pi k} \left[\frac{r^{-1}}{-1}\right]_{r_1}^{r_2} = \frac{q_{12}}{2\pi k}\left(\frac{1}{r_2} - \frac{1}{r_1}\right)$$

$$q_{12} = \frac{(T_1 - T_2)}{\frac{1}{2\pi k}\left(\frac{1}{r_1} - \frac{1}{r_2}\right)}$$

In this case, surface 1 represents the floor surface (subscript "b" for bottom), which is modeled as a hemisphere of radius r_1 (that gives the same area as that of the floor under the mug), and surface 2 represents infinity (so $1/r_2 \rightarrow 0$ and $T_2 = T_\infty$ is ambient). The heat transfer rate, defining the conductive resistance, is:

$$q_{1\infty} = \frac{T_1 - T_\infty}{R_{1\infty}} = 2\pi k r_1(T_1 - T_\infty)$$

$$R_{b\infty} = \frac{1}{2\pi k r_1} = \frac{\sqrt{2}}{\pi k(D + 2w)}$$

The following table details results for the floor resistance estimate. This analysis could use a couple of schematics, but the text in the opening paragraph of this section contains a word picture. In the end, the result suggests that the total resistance through the bottom of the mug (40.2 °C/W) is much higher than between the free surface and ambient (8.51 °C/W), which is less than that between the side wall and ambient (2.21 °C/W). That means the bottom surface behaves as a nearly insulated surface.

Detail of floor resistance			
D	0.083	m	Inner diameter of mug
hrim	0.004	m	Height of air gap
wrim	0.001	m	Width of rim in contact with floor
kai r	0.023	W/m/K	Thermal conductivity of air
kfloor	0.17	W/m/K	Thermal conductivity of table (Oak value)
kmug	1.5	W/m/K	Thermal conductivity of mug
Aair gap	0.00541	m^2	Area of air gap
Arim	0.000264	m^2	Area of rim contact
emiss	0.95		Emissivity of both surfaces
hrad	6.8		Radiation coefficient (between 70 and 25 °C plates)
Rbc,air	32.1	°C/W	Across air gap
Rbc,gap radiation	27.2	°C/W	Across air gap
Rrim	10.1		Through rim
Rbinf	31.2	°C/W	Infinite floor
Rfloor	37.1	°C/W	Effective floor contact resistance
UAfloor	0.027	W/°C	Effective floor contact conductance
Ufloor	4.7	$W/m^2/K$	Effective floor contact coefficient

Appendix 2. Parabolic Fits for Plotting Temperature Profiles

This appendix demonstrates a common technique for approximate methods for boundary layer analysis. If three conditions are specified at two locations (the wall and the free stream), then the shape of a function can be fit to a parabola, as detailed here. The three conditions here are that the temperature at the wall is fixed at the contact temperature, and at the edge of a boundary layer, the temperature is fixed at the initial value, and the temperature gradient is set to zero.

For this case, the heat transfer rate between the two solids will not be constrained, and therefore the heat transfer rate between them will not match. A fourth condition would be required to match the heat transfer rate between the two solids, and the temperature could be fit to a cubic. A good exercise (with a fair degree of additional algebra) would be to develop that cubic fit.

For the purposes of demonstrating the temperature profiles (snapshots of temperature plotted against distance from the wall surface at fixed times), a simple parabolic fit that fixes the temperature as the wall surface, the temperature at the edge of the boundary layer and forces the slope to be zero at the edge of the boundary layer. That is, with "x" the distance from the wall, an assumed parabola is:

$$T_{(x)} = a + bx + cx^2$$

where the coefficients (a, b, c) are determined from constraints imposed on the fit. At the wall surface (x = 0), the temperature is fixed at T_c:

$$T_{(0)} = T_c = a + b(0) + c(0)^2$$

or simply, $a = T_c$.

At the edge of the boundary layer, $x = \delta$, the temperature is set to ambient:

$$T_{(\delta)} = T_\infty = T_c + b\delta + c\delta^2$$

where the result for coefficient "a" has been inserted. Also, the slope at the edge of the boundary layer is set to zero:

$$\left.\frac{dT}{dx}\right|_{x=\delta} = 0 = b + 2c\delta$$

Multiplying this result by d and subtracting from the previous:

$$T_{(\delta)} = T_\infty = T_c - c\delta^2$$

$$c = \frac{T_c - T_\infty}{\delta^2}$$

Inserting this into the third constraint:

$$b = -2c\delta = \frac{-2(T_c - T_\infty)}{\delta}$$

A curve fit that yields an approximate shape for the temperature profile therefore is:

$$T_{(x)} = T_c - 2(T_c - T_\infty)\frac{x}{\delta} + (T_c - T_\infty)\left(\frac{x}{\delta}\right)^2$$

A somewhat more compact and dimensionless form of this expression is:

$$\frac{T_{(x)} - T_\infty}{T_c - T_\infty} = \left(\frac{x}{\delta}\right)\left(\frac{x}{\delta} - 2\right)$$

This formula does not give the exact shape of the temperature profile, but an approximate fit. The time dependency of the profile is contained in the boundary layer thickness, δ, which grows with time.

The inner wall of the mug is at the contact temperature, and the temperature falls within the thermal boundary layer, eventually reaching ambient. There is no single mug temperature. However, an average mug temperature can be obtained by integration (whose value would yield the correct total energy stored, assuming constant specific heat):

$$T_{ave}w = \int_{x=0}^{w} T\,dx$$

where the coordinate x is measured from the inner wall of the mug outward. Within the boundary layer ($x < \delta_m$), the temperature profile was fit to a parabola to generate the plot of Fig. 4.36, and outside the boundary layer ($x > \delta_m$), the temperature is ambient. Therefore, the integration can be broken into two pieces:

$$T_{ave}w = \int_{x=0}^{\delta_m} T_{(x)}\,dx + \int_{\delta_m}^{w} T_\infty\,dx$$

The parabolic fit for temperature is: $T_{(x)} = T_c - 2(T_c - T_\infty)\frac{x}{\delta_m} + (T_c - T_\infty)\left(\frac{x}{\delta_m}\right)^2$

Performing the integrations:

$$T_{ave}w = \left[T_c - 2(T_c - T_\infty)\frac{x^2}{2\delta_m} + (T_c - T_\infty)\frac{x^3}{3\delta_m^2} \right]\Bigg|_0^{\delta_m} + T_\infty[x]\big|_{\delta_m}^{w}$$

Doing the algebra: $T_{ave} = T_\infty + \frac{\delta_m}{3w}(T_c - T_\infty)$

Inserting the expression for the boundary layer thickness as a function of time:

$$T_{ave} = T_\infty + \frac{4\sqrt{\alpha_m t}}{3w\sqrt{\pi}}(T_c - T_\infty)$$

Similarly, the average temperature of the coffee can be calculated during this time:

$$T_{ave,coffee}\pi R^2 = \int_{r=0}^{R} T 2\pi r\,dr$$

where in this case, the cylindrical shape is considered, so the integration is conducted over a cylindrical shell from the centerline of the coffee ($r = 0$) to the inner wall of the mug ($r = R$). The coordinates r and x are related simply by $r = R + x$. Breaking the integral into two parts, inside and outside the thermal boundary layer:

$$T_{ave,coffee}\pi R^2 = \int_{r=0}^{R-\delta_c} T_{coffee,init} 2\pi r\,dr + \int_{R-\delta_c}^{R} T 2\pi r\,dr$$

Since the radius (R = 32 mm) is much larger than the maximum thermal boundary layer thickness ($d_{c,max}$ = 2.7 mm), the average coffee temperature changes very little during the initial heating, while the mug has increased from ambient (25 °C) to 39.4 °C.

Workshop 4.1. Cooking 1-Node Hot Dogs

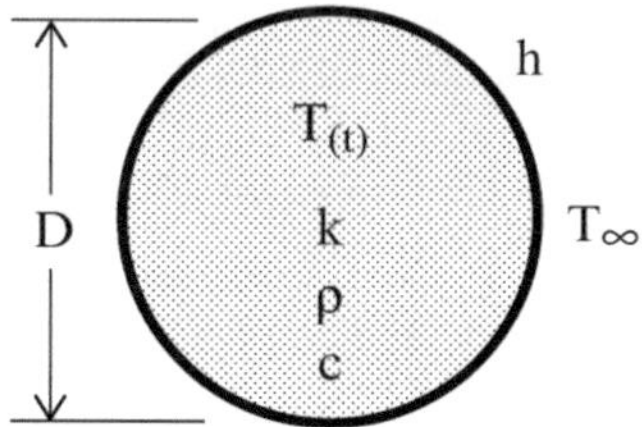

A hot dog of diameter D (=11/16″), length L is to be cooked by several methods (boiling baking, barbequing, microwaving). The hot dog is initially at 5 °C and is placed into a cooking environment with a fluid temperature T_∞. It is considered to be cooked when the average temperature exceeds 70 °C. Estimate the cooking times and the surface and center temperatures at that time for each method. Develop a 1-node extended model, where the total thermal capacitance of the hot dog is lumped into a single node and a thermal channel with a conductive resistance from the interior to the surface is in series with a convective resistance between the outer surface and ambient. Use input parameters from the input parameter tables shown.

INPUT PARAMS

D=	0.0175	m, diameter
L=	0.12	m, length
k=	0.6	W/m/K, thermal conductivity
rho=	1000	kg/m^3, density
c=	4200	J/kg/K, specific heat
Tinitial =	5	oC, initial temperature
Tfinal=	70	oC, final cooked temperature

Cook Method	Tinf (oC)	h (W/m2/K)	Qdot/V (W/m^3)
boiling	100	800	0
baking	177	15	0
bbq	500	25	0
microwaving	25	15	1.00E+07

For the convection coefficient, the value for boiling water (800 W/m²/K) is on the high end of natural convection in liquid water. There is expected to be significant fluid motion induced by the boiling process. For baking (at 350 °F = 177 °C) and microwaving, a value of 15 W/m^2/K is typical of natural convection in air, in parallel with a radiative coefficient. The value for barbequing (25 W/m^2/K) is taken to be somewhat higher due to the strong updraft caused by the fire.

For the barbequing and microwaving cases, some of these input parameters were chosen after the fact, to give cooling times consistent with experience. For barbequing, the fluid temperature is taken to be 500 °C. In practice, raising the distance between the coals and the food lowers the fluid temperature in contact with the food. There is an important radiation effect from the coals that is not considered here as it does not fit into either radiation model. For the microwaving case, the value of the rate of heat generation is given in Watts absorbed per unit volume, and the value $(1(10)^7 \, \text{W/m}^3)$ is chosen to be one which gives a cooking time consistent with empirical experience. That's cheating, a little... Or, rather, it is using experimental results and a theoretical model to evaluate an approximation for the effective rate of absorption of microwave energy, for which no attempt to model from first principles is made here.

Note: A detailed model development follows. Conduct your own analysis before referring to it.

Model Development
A general 1-node model with a heat source (for the microwave case) and a thermal channel between the capacity-carrying node and ambient with a conductive and convective/radiative channel in series is shown in Fig. 4.39.

The energy balance applied to this general RC network is:

$$C\frac{dT}{dt} = \dot{Q} + \frac{T_\infty - T}{R_k + R_h}$$

The solution (with constant parameter values) has been given in Chap. 1:

$$\frac{T - T_{SS}}{T_0 - T_{SS}} = e^{\frac{-t}{C(R_k + R_h)}}$$

where $T_{ss} = T_\infty + \dot{Q}(R_k + R_h)$ is the steady-state temperature and T_0 is the initial temperature. Note that the microwave case is the only one for which there is internal heat generation, and there is a difference between the steady-state and ambient fluid temperatures.

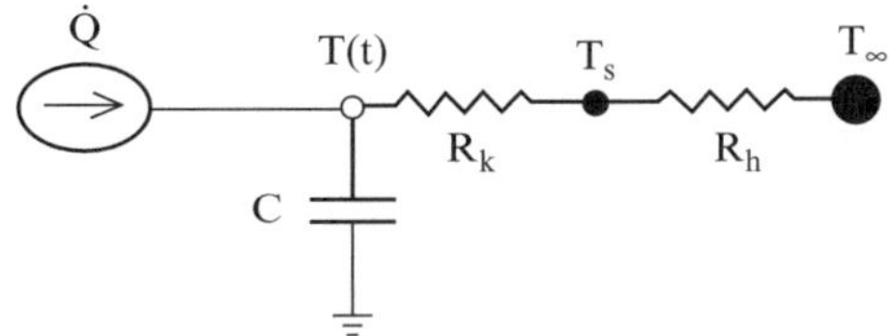

Fig. 4.39 General 1-node model applicable to hot dog model

The surface temperature is directly related to the average hot dog temperature through the voltage divider channel between the capacity-carrying node and the outer hot dog surface:

$$\frac{T - T_\infty}{R_k + R_h} = \frac{T_s - T_\infty}{R_h} = \frac{T - T_s}{R_k}$$

$$T_s = T_\infty + \frac{R_h(T - T_\infty)}{R_k + R_h} = T - \frac{R_k(T - T_\infty)}{R_k + R_h}$$

While not essential, it is useful to define Biot and Fourier numbers:

$$Bi = R_k/R_h \ \ \text{Gives SPATIAL INFO} \quad Fo = t/CR_k \ \ \text{Gives TEMPORAL INFO}$$

$$\frac{T - T_{SS}}{T_0 - T_{SS}} = e^{\overline{\frac{-t}{CR_k\left(1 + {R_h}/{R_k}\right)}}} = e^{-\left(\frac{Bi}{1+Bi}\right)Fo}$$

$$T_s = T_\infty + \frac{(T - T_\infty)}{Bi + 1} = T - \frac{Bi(T - T_\infty)}{Bi + 1}$$

Consider the Biot number limits:

For $Bi \gg 1$, there is good convection and Ts $\to$ T$_\infty$, that is, the hot dog cooks on the edges first, and the center lags behind.

For $Bi \ll 1$, there is good conduction, and Ts $\to$ T$_{ave}$, that is, the hot dog cooks with a uniform temperature.

Evaluating the capacitance is unambiguous:

$$C = \rho c L \pi D^2 / 4$$

Similarly, evaluating the convective resistance is unambiguous because convection occurs at a single, specific surface:

$$R_h = 1/(h\pi DL)$$

The convection coefficient is considered to be an average value. For the barbequing case, it is known from experience that the hot dog cooks more on the bottom side. That effect is probably due to radiation heat coming from the hot coals, an effect that could be accounted for by a parallel radiation channel acting on the bottom surface (but from a different temperature than the hot gases flowing past it, requiring a new RC network be drawn). But think of the hot dog as being turned frequently during the heating, and the higher convection coefficient value takes the nonuniform heating effect into account.

Evaluation of the conductive resistance requires some more thought because conduction occurs between two surfaces that have different surface area. As suggested by the sketch in Fig. 4.40, the cylindrical hot dog is modeled as an equivalent wall, with one "surface" (of zero surface area) being the centerline, and

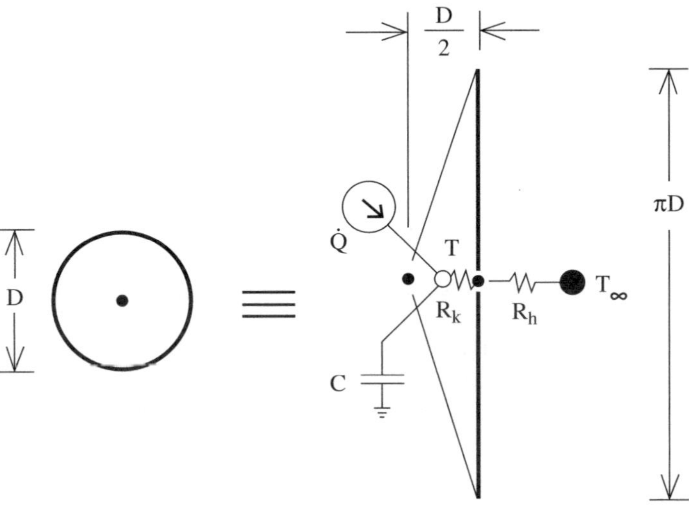

Fig. 4.40 Conceptual model for estimating hot dog conductive resistance

Table 4.4 Summary of conduction models for 1-node hot dog problem

Model	Node placement	R_k	Biot number	Fourier number
1	r = 0	$\dfrac{D/2)}{k(0+\pi DL)/2} = \dfrac{1}{\pi kL}$	$\dfrac{hD}{k}$	$\dfrac{4t}{D^2}\left(\dfrac{k}{\rho c}\right)$
2	r = D/4 (midway between surface and center)	$\dfrac{D/2-D/4}{\pi kL(D/2+D/4)/2} = \dfrac{1}{1.5\pi kL}$	$\dfrac{hD}{k}\left(\dfrac{1}{1.5}\right)$	$\dfrac{4t}{D^2}\left(\dfrac{k}{\rho c}\right)\left(\dfrac{1}{1.5}\right)$
3	r = D/3 (centroid of wedge)	$\dfrac{D/2-D/3}{k(\pi LD/3+\pi LD/2)/2} = \dfrac{1}{2.5\pi kL}$	$\dfrac{hD}{k}\left(\dfrac{1}{2.5}\right)$	$\dfrac{4t}{D^2}\left(\dfrac{k}{\rho c}\right)\left(\dfrac{1}{2.5}\right)$
4	r = D/3 (centroid of wedge, cylindrical shell)	$\dfrac{\ln\left(\frac{D/2}{D/3}\right)}{2\pi kL} = \dfrac{\ln(1.5)}{2\pi kL} = \dfrac{1}{4.93\pi kL}$	$\dfrac{hD}{k}\left(\dfrac{1}{4.93}\right)$	$\dfrac{4t}{D^2}\left(\dfrac{k}{\rho c}\right)\left(\dfrac{1}{4.93}\right)$

the other surface being the outer surface of the hot dog, with surface area πDL (with L the length of the hot dog), a distance $D/2$ away. The capacity-carrying node is placed at some radius between the centerline and the surface.

The general form of a conductive resistance is:

$$R_k = \frac{1}{k}\int_{x_1}^{x_2} \frac{dx}{A}$$

which has the general form of an average distance between surfaces (Δx) divided by the thermal conductivity, and an average area. A few models (summarized in Table 4.4) are considered before obtaining results for the base case hot dog.

MODEL 1: The most straightforward conduction model (for conductive resistance between the hot dog surface and the capacity-carrying node) is to "place" the capacity-carrying node at the centerline of the hot dog (at $r = 0$), and use the basic estimate for R_k as the distance between surfaces dividing by the thermal conductivity times average surface area.

MODEL 2: The centerline in a cylindrical geometry acts like an insulated boundary because the surface area approaches zero with the radius. Conceptually, heat conducted from the surface into the interior does not require it to conduct all the way to the centerline. Therefore, placement of the capacity-carrying node at the centerline will overestimate the conductive resistance. This model places the capacity-carrying node at the midpoint between the surface and the centerline (at $r = D/4$). The conductive resistance is based on the distance between the surface and node, and on the average areas. When executed, the functional dependencies are the same as Model 1, but the Biot number is reduced by a factor of 2/3.
MODEL 3: This model places the capacity-carrying node at the centroid of a wedge of the hot dog (at $r = D/3$), rather than at the midpoint between the centerline and center. The centroid is more representative of the solid than the geometric center. The Biot number is 2/5 of the value of Model 1.

MODEL 4: The model places the capacity-carrying node at the centroid of a wedge, but uses the exact formula for conduction between two surfaces at different radial positions, that is, across a cylindrical shell. The Biot number is reduced by a factor of 4.93 compared to Model 1.

Model 4 is considered to be conceptually the most accurate for this case. However, the important take-home message is that all models give the same functional dependency. Moving forward, methods for breaking the hot dog into multiple nodes will be developed, and the differences between models for conduction will be less important.

MODEL 4 results:

Spreadsheet results using conduction Model 4 are summarized in Fig. 4.41 and plots of the average hot dog temperature and the surface temperature are shown in Figs. 4.42, 4.43, 4.44, and 4.45. The conductive resistance and capacitance are independent of the method of cooking and are calculated in the top section (derived parameters) of a worksheet. This block refers to the input parameter block of the problem statement. The convective resistance depends on the method of cooking. Cooking in boiling water results in a large (but not effectively infinite) Biot number (4.73) event, which means conduction through the hot dog is slow compared to the rate at which heat is transferred by convection from the water to the surface. There is a large spatial gradient, with the surface temperature much closer to the fluid temperature than the core temperature. The other three cases are low Biot number cases (but not effectively zero), which means that conduction is effective relative to convection and the hot dog cooks nearly uniformly. Comparing the boiling and baking cases, the cooking time is faster in boiling water, despite the fact that the

DERIVED PARAMS

rnode=	0.00583	m
Rk=	0.896	oC/W
C=	121.2	J/oC

Cook Method	Rh	Bi = Rk/Rh	RC (sec)	Teq
boiling	0.189	4.730	131.6	100
baking	10.105	0.089	1333.7	177
bbq	6.063	0.148	843.7	500
microwaving	10.105	0.089	1333.7	3200

RESULTS

Cook Method	Cook Time (min)	Tsurface (oC)
boiling	2.53	94.76
baking	10.55	78.72
bbq	1.98	125.38
microwaving	0.46	66.33

Fig. 4.41 Results of 1-node hot dog workshop for conduction Model 4

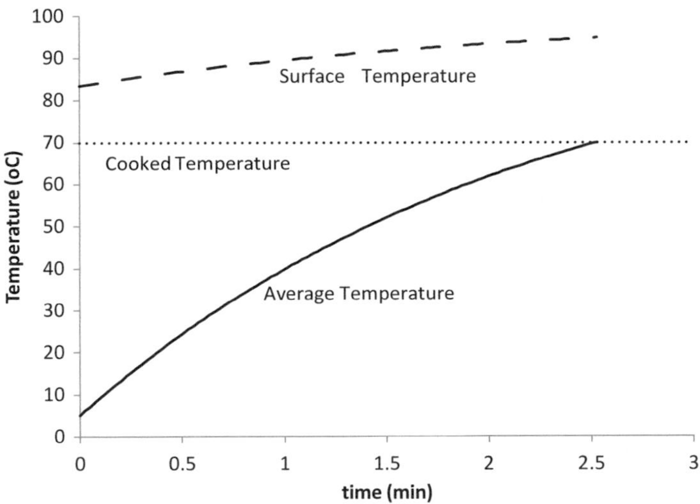

Fig. 4.42 Boiling: Biot number $= 4.73$, $T_{inf} = 100\ °C$

fluid temperature (100 °C) is lower than that in an oven (177 °C). But at the time the hot dog is cooked, the surface temperature in water is 95 °C (25 °C above the hot dog average, and only 5 °C below the fluid temperature) while the surface temperature is 79 °C for baking (9 °C above the hot dog, and almost 100 °C below the fluid temperature).

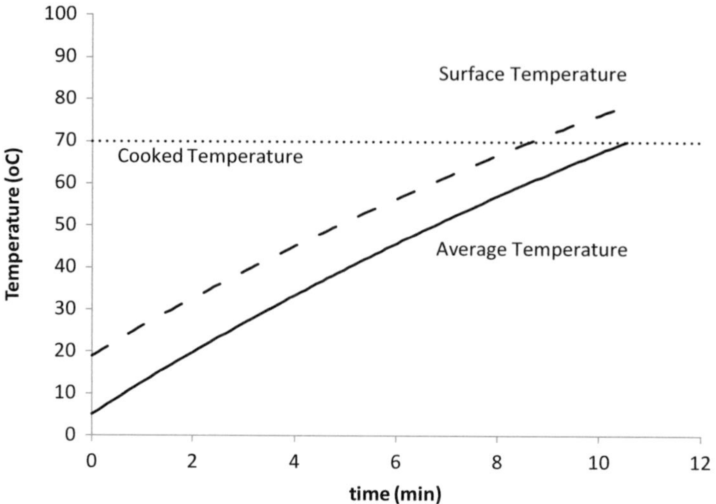

Fig. 4.43 Baking: Biot number $= 0.089$, $T_{inf} = 177\ °C$

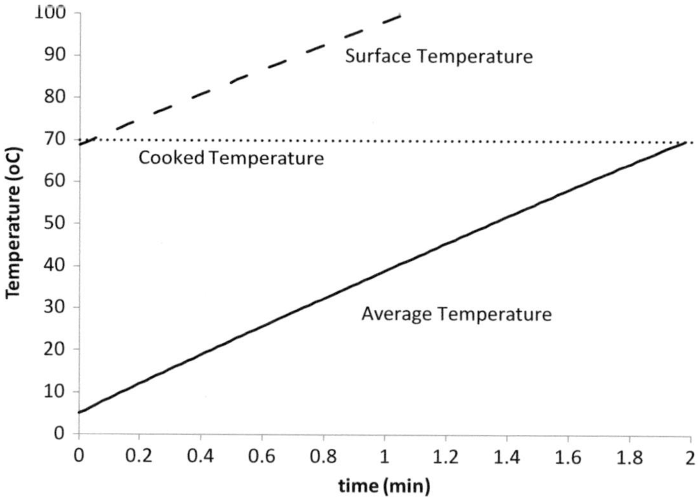

Fig. 4.44 Barbeque, Biot number $= 0.148$, $T_{inf} = 500\ °C$

For barbequing, the cooking time is shorter than boiling, because the fluid temperature (500 °C) is so high. Meanwhile, the surface temperature is 125 °C. In this case, despite the low Biot number, the hot dog appears to heat nonuniformly, but that is merely because the fluid temperature is so high. The surface temperature is closer to the average hot dog temperature than it is to the fluid temperature.

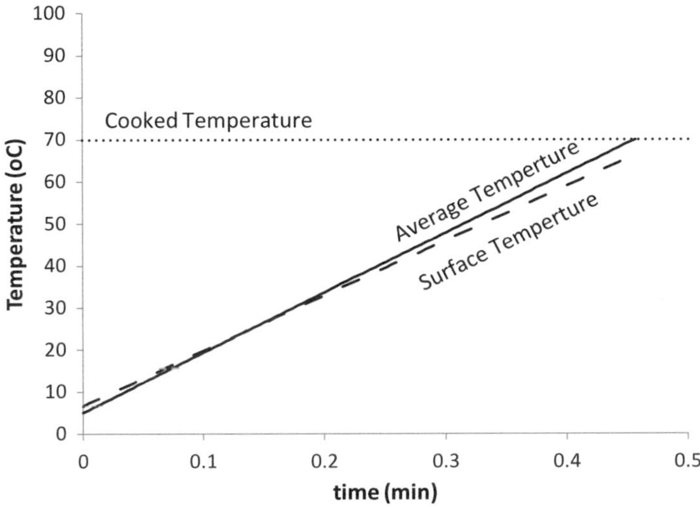

Fig. 4.45 Microwave, Biot number $= 0.089$, $T_{inf} = 25\ °C$

The fluid temperature chosen for this case is not well characterized. Nevertheless, it is well known from experience that it is a function of the vertical distance between the source of the fire and the hot dog. It burns on the outside if placed too close.

Finally, microwaving yields the shortest cooking times (but to me, a hot dog in a microwave is too rubbery…). The equilibrium (steady-state) temperature of 3,200 °C represents the temperature the hot dog would achieve in theory if the microwave were left on indefinitely. Of course, something would happen to the hot dog to change the model long before it reached that point.

The surface temperature at the microwave cooking time is lower than surface temperature. That is, heat flows outward from the core of the hot dog to the surface. Heat is absorbed uniformly throughout the interior, yet the temperature is not uniform. In order for heat to flow from the interior to the surface, and ultimately to the ambient, there must be a temperature gradient driving it. From Fig. 4.45, at the start of the cooking event, the surface temperature is higher than average. That is because the hot dog is taken from the refrigerator, and it is being warmed by the air temperature at the same time it is being heated from within. Once the hot dog average temperature exceeds the air temperature, the core is hotter than the surface.

Workshop 4.2. Beverage Chilling Methods

Beverages are commonly purchased at room temperature, but are consumed cold. This workshop develops a means to predict the time it takes to chill a room temperature beverage to a desirable drinking temperature. A standard beverage chilling time is defined here as the time it takes to chill a beverage from 25 to 5 °C using a specific cooling method. The Cooper Cooler™ is a consumer appliance developed at the Cooper Union that is designed to do so rapidly. The user prepares an ice/water slurry in a reservoir and the Cooper Cooler sprays a cold recycled jet of water on a horizontal beverage while is it rotated along its natural axis. For this workshop, five different chilling methods are considered: three conventional modes and two Cooper Cooler modes.

1. Refrigerator: The beverage is placed in a standard household refrigerator.
2. Freezer: The beverage is placed in a freezer.
3. Ice/water bath: The beverage is placed in a cooler filled with ice, and enough water to completely immerse the beverage.
4. Cooper Cooler™ Spray-only Mode: A jet of cold water is sprayed onto a stationary, horizontal beverage.
5. Cooper Cooler™ Spray and Spin Mode: A jet of cold water is sprayed onto a rotating horizontal container.

Model Development

All five chilling methods are modeled with the same RC network which the reader can draw based on the following description. The capacitance of the beverage and container are lumped into a single node, T_{bev}. For heat to be transferred from the beverage to the coolant (at $T_{coolant}$), it must flow from the beverage to the inner wall of the container (a convective resistance), across the container wall (conductive), and from the outer wall to the coolant (convective/radiative). In other words, there is a single thermal channel from the beverage to the coolant with three resistances in series. The differences in chilling times are due to different values of the various capacitances (for different beverage types), thermal resistances, and coolant temperature.

The governing nodal equation is:

$$C_{bev} \frac{dT}{dt} = \frac{T_{coolant} - T_{bev}}{R_{eq}}$$

where R_{eq} is the equivalent thermal resistance (sum of three series resistances). The solution to this now familiar governing first-order equation is:

$$\frac{T_{bev} - T_{coolant}}{T_{initial} - T_{coolant}} = e^{\frac{-t}{RC_{bev}}}$$

This solution can be used, once estimates of R and C are made, to prepare plots of beverage temperature vs. time. To determine the standard chilling time (t_{chill}), the natural logarithm of both sides is taken and the beverage temperature set to the desired final temperature:

$$t_{chill} = -RC_{bev} \ln\left(\frac{5 - T_{coolant}}{25 - T_{coolant}}\right)$$

A spreadsheet template for a specific beverage type is shown in Fig. 4.46. Blocks for input parameters, derived parameters, and results are shown in which suggested values

BEVERAGE TYPE

12 oz. Glass Bottle

INPUT PARAMETERS

Vbev	**0.355**	liters	beverage volume
Douter	**0.07**	m	outer diameter
w	**0.003**	m	wall thickness
kwall	**0.81**	W/m/K	wall thermal conductivity
rhowall	**2800**	kg/m^3	wall density
cpwall	**800**	J/kg/K	wall specific heat
cbev	**4200**	J/kg/K	beverage specific heat
rhobev	**1000**	kg/m^3	beverage density
Tinitial	**25**	oC	initial beverage temperature
Tfinal	**5**	oC	final beverage temperature

DERIVED PARAMETERS

L		m	equivalent cylinder length
Cbev		J/oC	Beverage Capacitance
Ccontainer		J/oC	Container Capacitance

RESULTS

COOLING METHOD	Tcoolant (oC)	hinner (W/m2/K)	houter (W/m2/K)	Cooling Area (m^2)	Rinner (oC/W)	Rwall (oC/W)	Router (oC/W)	RC (min)	Cooling Time (min)
Refrigerator	4	300	15						
Freezer	-10	300	15						
Ice/Water Bath	0	300	300						
Cooper Cooler Spray-Only	1	300	2500						
Cooper Cooler Spray and Spin	2	1100	2500						

Fig. 4.46 Spreadsheet templates for beverage chilling workshop

for the case of a 12 oz. bottled beverage are listed. Input values considered to be common for different beverage types are listed in the results block. Blank boxes are left for the formulas needed to implement the method. The beverage is considered to have a clearly identified outer diameter and treated as an equivalent cylinder to obtain the length. A thin-wall approximation is suggested in which the inner and outer areas are treated as approximately equal when thermal resistances are calculated.

The coolant temperatures are taken to be those typical of a refrigerator, freezer, and the melting point of water for the three conventional cooling methods. For the Cooper Cooling methods, a coolant temperature of 1 °C is suggested for spray only, and 2 °C for spray and spin. In practice, the coolant is recycled from an ice/water bath, and heat added to the coolant from the beverage warms it somewhat, even as the ice in the bath cools it. The modeling of this process involves more than one node and is discussed later. The coolant temperature is warmer for the spray and spin because of the higher heat transfer rates from the beverage.

The value for inner heat transfer coefficient is taken to be a typical value for the natural convection of fluid with nominal water properties (300 $W/m^2/K$) for all stationary beverages. When the beverage is rotated, there is mixing inside which produces a sort of wind chill on the inside and the value used (1,500 $W/m^2/K$) is representative of a mild forced convection with water. The outer convection/radiative coefficient is taken to be representative of natural convection in air, with a parallel radiation channel, for the refrigerator and freezer. For the ice/water bath, the outer coefficient is representative of natural convection in water. A value representative of a forced convection with water (stronger than for the inside) is listed for the two cases, where a jet of cold water is sprayed.

For the three conventional cooling methods (refrigerator, freezer, ice/water bath), the coolant is considered to be in contact with the entire surface area of the beverage (sides, top, and bottom) whereas for the two Cooper Cooling cases, only the side surface is considered to be in good thermal contact with the coolant.

It is suggested that the case of a 12 oz. canned beverage be compared with the 12 oz. bottled beverage, in which case the only difference is the thermal resistance of the container wall. Also, the cases of a 0.5, 1, and 2 L plastic bottles be considered, where both the thermal capacitance and surface area vary. It is also suggested that plots of temperature vs. time be prepared with all five cooling methods on the same set of axes for a given beverage. Include the inner and outer wall temperatures. A final suggestion is to conduct some experiments and compare results. Notice that in an experiment, the natural dependent variable to set is the chilling time, and this analysis can help inform an intelligent choice for that time for a single experiment. Too short a time, and the temperature difference between initial and final temperature will be small and experimental uncertainty will be high. Too long a time, and the beverage will be too close to the final equilibrium to obtain a reasonable dimensionless temperature.

Few-Node Transient Models 5

Try this experiment. Fill a ceramic mug with boiling water and immediately grab its side. Close your eyes and concentrate on what it feels like. You've just experienced the mug heating phase, the topic of this chapter.

In the cooling mug of coffee problem, the mug is initially at ambient temperature when hot coffee is poured into it. Figure 5.1 shows the measured outside wall and coffee-free surface temperatures (as symbols), and the prediction of the 1-node lumped capacity model for the coffee from Chap. 4 (solid line), the top surface (dotted line), and the outer side wall (dashed line). The 1-node extended model is capable of distinguishing the temperature of the coffee (think average bulk coffee temperature) from the various surface locations (two of which are measured). However, it misses the distinction between the mug and the coffee during the first few minutes. It predicts that the outside wall of the mug is hot at the instant the coffee is poured. However, it takes some time for that to happen.

In the 1-node model (coffee and mug as a single node), the capacitance of the coffee and mug are lumped together, and their temperatures must rise or fall together. In essence, the initial thermal mixing of coffee and mug is assumed to take place instantaneously, and the initial temperature for the ensuing cooling process is obtained from the energy balance detailed in Chap. 2 (in which only the capacitances of the mug and coffee come into play). The initial period where the mug rises as the coffee falls cannot be observed.

In Chap. 4, a variable-sized 1-node model was developed that yielded the basic response of the mug to being suddenly exposed to a hot fluid. However, the bulk temperature of the fluid was fixed. The initial cooling of the coffee as the mug is being heated is not captured in that model.

In order to observe the separate behavior of the coffee and the mug, the capacitance of the coffee and the mug must be treated separately, creating a 2-node model. The goal of this chapter is to develop a general scheme for doing so, and apply it to the mug of coffee problem.

© Springer International Publishing Switzerland 2015   163
G. Sidebotham, *Heat Transfer Modeling: An Inductive Approach*,
DOI 10.1007/978-3-319-14514-3_5

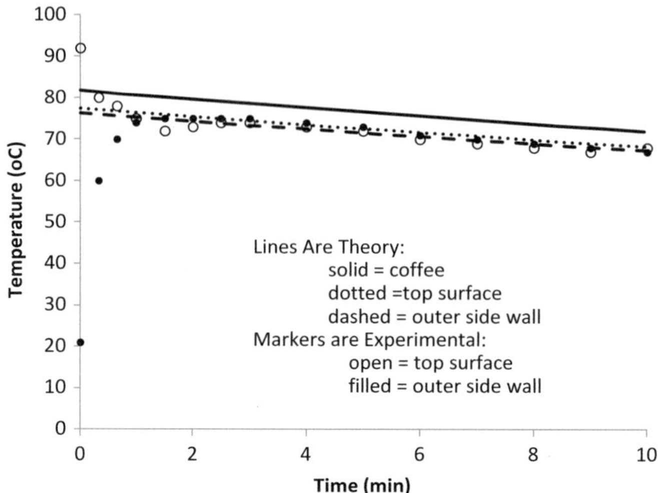

Fig. 5.1 1-Node lumped capacity model predictions during mug heating phase

The term "Few-Node Model" is loosely defined as one in which space is discretized into clearly identifiable regions that can be given clear names (as opposed to numbers). In the next chapter, called "multi-node," space will be discretized into finer elements in a way that there is no clear sense of how many nodes there should be beforehand. Once a spreadsheet is set up to solve for a particular number of nodes, analyzing the same problem but for a different number of nodes is not possible without setting up a new sheet. For solution approaches that set the number of nodes as an input parameter, MATLAB or other programming code can be written that allows the user to simply change this input parameter, and run the code right away.

This chapter closes with a 3-node analysis of a cooling mug of coffee into which Joulies™ are placed (http://www.joulies.com), which involves simulating a solid absorbing heat from the coffee with an inner core material that changes phase. Coffee Joulies placed into a cup of hot coffee dramatically reduces the time it takes the coffee to cool to a suitable drinking temperature range, then releases stored latent energy, holding the coffee in a suitable range for a while. They represent an engineering solution to the problem of *"why do I have to wait to drink my coffee?"*

5.1   A Generic 2-Node RC Network

Figure 5.2 shows an RC network with two nodes that have capacitance associated with them (T_1 and T_2). Initially, both nodes are at ambient (T_∞, held by a battery above ground). The switch (s) is momentarily closed, long enough to charge capacitor C_1 to a starting point (T_0), while T_2 is initially undisturbed, and then released. Subsequently, heat flows directly to ambient through $R_{1\infty}$ and

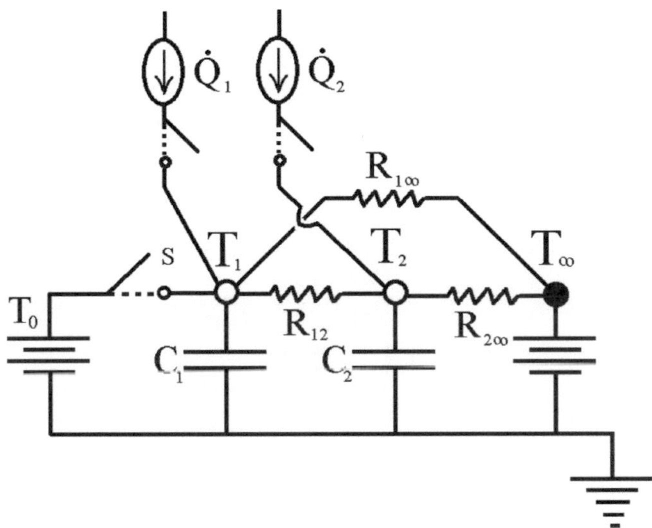

Fig. 5.2 RC network for a 2-node system

simultaneously to the second node (T_2), whose capacitor begins to charge. As soon as T_2 rises above ambient (if $T_0 > T_\infty$), heat flows from T_2 to ambient. If T_0 is less than T_∞, then everything flows in the opposite direction. Both nodes can also receive heat from separate heat sources.

This model represents a more complicated model of the cooling mug of coffee. The difference between this model and the previous lumped model is that instead of lumping the capacitance of the mug and coffee together, each component is given its own capacitance. The direct channel from T_1 to T_∞ (through $R_{1\infty}$) represents the heat transfer from the coffee to ambient through the free surface. The channel between T_1 and ambient through T_2 represents heat transfer from the coffee to the mug through the sides and bottom.

The three thermal resistances are shown as equivalent resistances of a combination of series and parallel channels. Through the use of voltage divider type calculations, intermediate temperatures can be related to the nodal temperatures as desired. For example, T_2 represents the average temperature of the mug, while the inner wall temperature is between the coffee and mug temperatures, and the outer wall is between the mug and ambient. As with the 1-node lumped model approach, more spatial detail can be gleaned by appropriate selection of thermal resistances.

In this model, the radiation and convection ambient temperatures are considered equal. It is relatively easy to modify the RC network to account for the effect of different fluid and radiation wall temperatures if needed, or to introduce a correction to either radiation or convection as detailed previously.

5.2   Nodal Equations

To solve this system, a nodal balance for both capacitor-carrying nodes is required. Figure 5.3 isolates node 1. The energy balance for this node, being careful about signs, is:

$$\begin{pmatrix} \text{Rate of Change} \\ \text{of Energy Stored} \\ \text{in Node 1} \end{pmatrix} = \begin{pmatrix} \text{Heat Flow into} \\ \text{Node 1 from} \\ \text{Heat Source 1} \end{pmatrix} + \begin{pmatrix} \text{Heat Flow into} \\ \text{Node 1} \\ \text{from Node 2} \end{pmatrix}$$

$$+ \begin{pmatrix} \text{Heat Flow into} \\ \text{Node 1} \\ \text{from Ambient} \end{pmatrix}$$

$$C_1 \frac{dT_1}{dt} = \dot{Q}_1 + \frac{T_2 - T_1}{R_{12}} + \frac{T_\infty - T_1}{R_{1\infty}} \tag{5.1}$$

This differential equation cannot be integrated directly because it depends on T_2, which is an unknown variable, and therefore is shown with an open symbol.

Figure 5.4 isolates T_2. The energy balance for this node is:

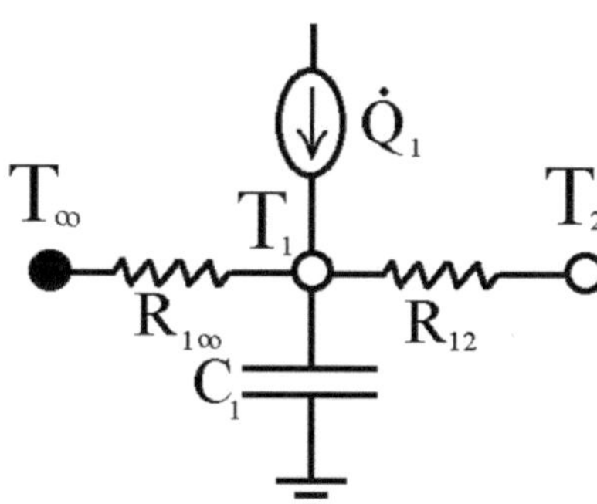

Fig. 5.3 Isolation of node 1

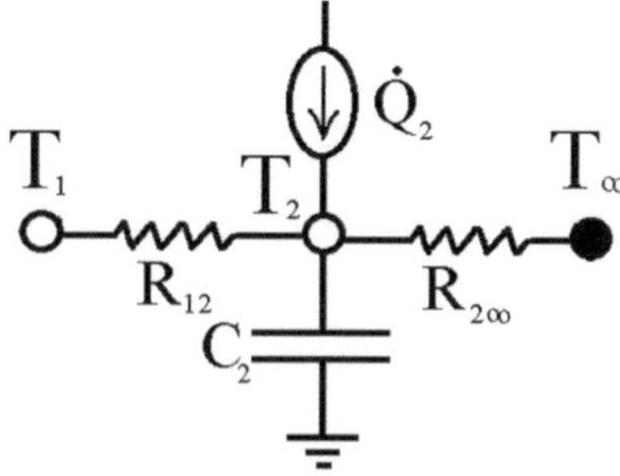

Fig. 5.4 Isolation of node 2

$$\left(\begin{array}{c}\text{Rate of Change}\\\text{of Energy Stored}\\\text{in Node 2}\end{array}\right) = \left(\begin{array}{c}\text{Heat Flow into}\\\text{Node 2 from}\\\text{Heat Source 2}\end{array}\right) + \left(\begin{array}{c}\text{Heat Flow into}\\\text{Node 2}\\\text{from Node 1}\end{array}\right)$$

$$+ \left(\begin{array}{c}\text{Heat Flow into}\\\text{Node 2}\\\text{from Ambient}\end{array}\right)$$

$$C_2 \frac{dT_2}{dt} = \dot{Q}_2 + \frac{T_1 - T_2}{R_{12}} + \frac{T_\infty - T_2}{R_{2\infty}} \tag{5.2}$$

Equations (5.1) and (5.2) represent a coupled system of first-order differential equations. If all resistance and capacitance values are constants, the system is linear, and there are analytical techniques for solving them (LaPlace transforms, for example). However, the objective of this text is to build toward more complex models for which analytic solutions do not exist. Numerical methods, on the other hand, can treat these more general complex cases, and are developed immediately. Two-node problems using the explicit method can generally be solved equally well with spreadsheets or writing code in MATLAB or other programming language.

5.2.1   Explicit Numerical Formulation

For the explicit method, a forward finite difference approximation is used to approximate the derivatives with respect to time:

$$\frac{dT_1}{dt} = \frac{T_1^{(p+1)} - T_1^{(p)}}{\Delta t^{(p)}}$$

$$\frac{dT_2}{dt} = \frac{T_2^{(p+1)} - T_2^{(p)}}{\Delta t^{(p)}}$$

Inserting these approximations into the two nodal equations and rearranging them:

$$T_1^{(p+1)} = T_1^{(p)} + \frac{\Delta t^{(p)}}{C_1^{(p)}} \left[\dot{Q}_1^{(p)} + \frac{T_2^{(p)} - T_1^{(p)}}{R_{12}^{(p)}} + \frac{T_\infty^{(p)} - T_1^{(p)}}{R_{1\infty}^{(p)}} \right] \tag{5.3}$$

$$T_2^{(p+1)} = T_2^{(p)} + \frac{\Delta t^{(p)}}{C_2^{(p)}} \left[\dot{Q}_2^{(p)} + \frac{T_1^{(p)} - T_2^{(p)}}{R_{12}^{(p)}} + \frac{T_\infty^{(p)} - T_2^{(p)}}{R_{2\infty}^{(p)}} \right] \tag{5.4}$$

On the right hand side, all variables and parameters are evaluated at the present time. A superscript "p" has been applied to everything (time step, R and C values) indicating that all these may change as time marches forward.

There are two underlying time constants for this system, one for each node, that represent a first-order response of each node to an equilibrium value if the neighboring nodes were fixed (detailed in Appendix 1). For example, in Fig. 5.3, if T_2 is held fixed, T_1 will approach a value that is intermediate to T_2 and T_∞, with an equivalent resistance resembling a parallel path. That is, a time constant for node 1 is:

$$\tau_1^{(p)} = (RC)_{1,equiv}^{(p)} = \frac{C_1^{(p)}}{\dfrac{1}{R_{12}^{(p)}} + \dfrac{1}{R_{1\infty}^{(p)}}}$$

For node 2:

$$\tau_2^{(p)} = (RC)_{2,equiv}^{(p)} = \frac{C_2^{(p)}}{\dfrac{1}{R_{12}^{(p)}} + \dfrac{1}{R_{2\infty}^{(p)}}}$$

Note that the heat sources are not associated with the inherent response time. A Second Law limit for the system is defined by limiting the time step taken at each step to the smaller of the two time constants. If a larger step size is used, the temperature will overshoot this "meta-equilibrium" value. As was seen in Chap. 1, there is also a numerical stability limit for the step size, which, if exceeded, would cause the system temperature to grow without bounds.

A common misuse of this algorithm is to use a different step size for each node, an egregious conceptual error. The same step size must be used for both nodes when marching in time. On the other hand, it is possible for either the capacitance or especially one or more of the thermal resistors to be temperature and/or time dependent (hence the superscripts "p"), and the response time of one or both of the nodes will change with time. A variable step-size algorithm can be used in those cases.

It may sound confusing, but "the maximum allowable step size for the method is the minimum time constant." That is:

$$\Delta t_{max}^{(p)} = \min\left(\tau_1^{(p)}, \tau_2^{(p)}\right)$$

In setting up a computer simulation, it is often useful to set a parameter (called "f" for example), that is, set to be a fraction of the maximum allowable step size, and therefore setting the time step equal to a fraction of the maximum allowable step size:

$$\Delta t^{(p)} = f \cdot \Delta t_{max}^{(p)} = f \cdot \min\left(\tau_1^{(p)}, \tau_2^{(p)}\right)$$

This approach places a constraint on time (the independent variable). In some applications, it is advantageous to invoke a variable time step algorithm by placing a constraint on one or both of the temperatures (the dependent variables), and calculate the required time. One approach is to fix the maximum change in temperature. Another is to fix the fractional change in temperature.

5.3   Coffee Mug: Explicit Numerical Solution with Constant Coefficients

To apply this general model to the specific case of a cooling mug of coffee, the three thermal resistances and the two capacitances must be evaluated. Figure 5.5 shows a more detailed RC network. The two temperatures (coffee and mug, shown as open circles) represent the main capacity-carrying nodes. However, using voltage divider formulas, the temperatures of the free surface, inside wall of the mug and outside wall of the mug can be determined (shown as small filled circles). The free surface and outside wall surface are the most closely related conceptually to the measured temperatures to which the model predictions will be compared. Ambient temperature is a fixed temperature (large filled circle) that could easily be made to vary with time in a prescribed way.

Convection and radiation act in parallel at the outer surface of the side wall. For the free surface, an additional parallel channel is shown, namely, an evaporative channel. As discussed in Chap. 3, evaporation effects are probably dominant when the coffee is near the boiling point, but negligible thereafter. A single constant heat transfer coefficient is chosen to include all these effects. The side and bottom channels from Chap. 3 are treated as a single channel from the coffee to the outside surface of the mug. There, a side channel and a parallel bottom channel connect to ambient. More capacity-carrying nodes would be required to further distinguish the bottom of the mug from the side.

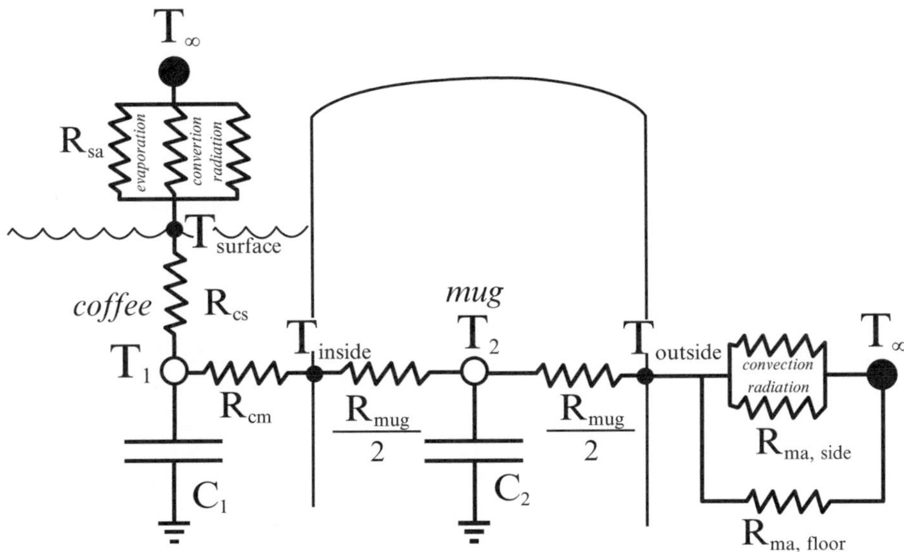

Fig. 5.5 RC network of mug of coffee detailing thermal channels

Detailed derivations of all these individual thermal resistances are included in Appendix 2. There are many subtle points and different ways to construct an RC network such as this one, and there could be debate as to how to define specific formulas. It is suggested that time be taken to define all these individual resistances as a group in-class and/or homework exercise before consulting the appendix. The strategy taken here is, given the constraint of only two capacity-carrying nodes, to model the problem in some detail, and then compare the results with the experimental data. As will be seen, there is good general agreement, but some discrepancy. The temptation to adjust input parameter values to obtain better agreement is resisted, as the modeling methods developed are intended to be applied to situations where experimental data does not exist.

5.3.1 Spreadsheet Program Description

The explicit method was implemented in an Excel workbook. The construction of this worksheet is described in some detail, as a case study for a general methodology for using spreadsheets for this type of problem.

A worksheet named "params" contains a block of input parameter values (which are input as numbers, not formulas). Figure 5.6 shows this block of cells. The values listed are those that apply to the experiment detailed in Chap. 2. The user can change any of these parameter values to simulate other scenarios. In this implementation, all heat transfer coefficients (convective, radiative, and evaporative) are treated as constants, and are listed as input quantities. For the parallel channels at the side wall and free surface, a single heat transfer

D=	0.08255	m	Inner diameter of mug
H=	0.098	m	Inner height of mug
H0=	0.106	m	Outer height of mug
V=	4.10E-04	m^3	Coffee Volume
w=	0.003975	m	Mug thickness
mmug=	0.335	kg	Mug mass
kmug=	1	W/m/K	Mug thermal conductivity
rhomug=	1700	kg/m^3	Mug density
cmug=	800	J/kg/K	Mug specific heat
rhocoffee=	1000	kg/m^3	Coffee density
ccoffee=	4200	J/kg/K	Coffee specific heat
hair/sides=	13.2	W/m2/K	hrad+ hconv outer heat transfer coefficient
hair/top=	23.2	W/m2/K	hrad + hconv + hevap heat transfer coefficient for free surface
hfloor=	4.7	W/m2/K	effective floor coefficient
hcoffee/mug=	470	W/m2/K	convection coefficient between coffee and mug
hcoffee/surface=	470	W/m2/K	convection coefficient between coffee and free surface
Tcoffee, init=	92	oC	Temperature of coffee when poured
Tambient=	21	oC	Ambient temperature
Tmug,initial=	21	oC	Mug initial temperature

Fig. 5.6 Input parameter values for 2-node mug of coffee problem

coefficient that represents the sum of the individual coefficients is used. The values chosen are based on estimates evaluated at a coffee temperature of 70 °C from analyses from future chapters.

In constructing a spreadsheet based on a problem statement, there are generally fixed input parameters, and then many other parameter values (quantities that do not change with time) that are obtained from formulas based on them. It is a good practice to keep the fixed values separate from the derived values, especially for conducting parametric studies (i.e., changing a fixed input to observe its effect on the system response). It is easy to tell the difference. Any spreadsheet cell that contains a number is an input parameter. Any cell that contains a formula is a derived parameter.

In the same worksheet, quantities derived from the input quantities that do not change with time are shown in Fig. 5.7 in three different blocks. There are dimensions, capacitances, and surface areas that will be used as inputs to the individual thermal resistances, and the individual resistances are combined into series and parallel channels to form the three main equivalent resistances in the 2-node simulation. Detailed explanations of all these calculations are given in Appendix 2.

The conductive resistance across the mug wall is taken to be between the surface area of contact between the coffee and the mug and the surface area of the mug exposed to ambient (including the bottom, through a different mechanism). As was

DERIVED QUANTITIES

H - L =	0.0766	m	Coffee depth
Cmug=	268	J/oC	Capacitance of mug
Ccoffee=	1722	J/oC	Capacitance of coffee
Afreesurface =	0.0054	m^2	coffee free surface area
Ainside=	0.0252	m^2	contact area between coffee and mug
Aoutside=	0.0432	m^2	contact area between mug and environment
Aave=	0.0342	m^2	averaged mug area

INDIVIDUAL RESISTANCES

Rcs=	0.3975	oC/W	convection between coffee and free surface
Rsinf=	8.0536	oC/W	conv/rad/evap between environment and free surface
Rcm=	0.0844	oC/W	convection between coffee and mug
Rm=	0.1162	oC/W	conduction across mug
Rma,side=	2.0605	oC/W	conv/rad between environment and mug
Rma,floor=	33.0762	oC/W	conduction between mug bottom and floor

MODEL RESISTANCES

R1inf=	8.4511	oC/W	between coffee node and environment
R12=	0.1425	oC/W	between coffee and mug nodes
R2inf=	1.9978	oC/W	between mug node and environment

Fig. 5.7 Derived parameter values for 2-node mug of coffee problem

done with the 1-node model, the heat transfer from the inside surface area of the mug above the free surface to ambient is included in the side channel (not the top channel). The conductive resistance across the mug wall could be further improved by considering a parallel channel through the rim (a future chapter on heat transfer fins). Either approach could be used in either the 1-node or 2-node model and highlights some of the art of heat transfer modeling.

An important feature of the detailed RC network is how the total conductive resistance across the mug wall is split into two, with the mug capacitance placed in the middle. This placement allows for spatial detail in the temperature field. It also adds numerical stability, although that may not be clear at this point.

Much can be inferred from a consideration of the relative magnitudes of the three model resistances. Heat flows easily between the coffee and mug due to the low value of R_{12} compared to the other two resistors. Heat leaves the side wall more easily than it does through the top, despite the higher net coefficient (reflecting the inclusion of evaporation), a trend attributed to the larger surface area of the side wall.

The basic structure of the 2-node simulation is shown in the worksheet "2-nodeMug" (Fig. 5.8). The top block includes the main inputs to the general 2-node simulation. These cells refer to the appropriate cells in the "params" worksheet, except for the cell that contains the user-defined parameter "f", which has the value 0.2 chosen for this implementation, a choice based on experience from Chap. 1 with the general consideration of the Euler method for a 1-node problem. This numerical parameter is used to determine the (constant) step size used in the simulation. An input value of 1.0 would represent the maximum Second Law step size of the system.

The time constants for nodes 1 and 2 are calculated in a separate block below the inputs. It is apparent that the characteristic response time of the coffee in this case (241 s) is about eight times as large as that of the mug (35.6 s). An "if-then-else" statement is used to calculate the step size (Dt = 7.12 s) used in the simulation, which is the numerical parameter "f" multiplied by the smaller of the two time constants.

The simulation is implemented in the next block. The first row contains the initial time (0), and initial coffee and mug temperatures (92 and 21). The next row contains the recursion formulas for time (add the fixed time step to the previous time and convert from seconds to minutes), coffee temperature and mug temperature (T_1 and T_2, from Eqs. (5.3) and (5.4)), which implement the explicit method. Once written, this second row is copied down as many additional rows are needed to simulate the desired total elapsed time. For this case, 2,020 rows are needed to simulate 4 h of elapsed time. If a smaller step size were chosen (by choosing f = 0.1 instead of 0.2, for instance), that number of rows would simulate less time (2 h instead of 4).

Another worksheet was written to extract additional information used to analyze the response of the system. The structure of this worksheet ("2NodeMugDetailed") is shown in Fig. 5.9. The top block (not shown) is the same as the top of Fig. 5.8, but several additional quantities are calculated from the two main capacitive-bearing nodes (coffee and mug) in the simulation. A row of spatial locations (x) is added. For the two fluids (coffee and air), a thin boundary layer is added for the purposes of demonstrating the relationship between surface temperature and fluid temperature

Fig. 5.8 Structure of worksheet that implements the fixed step-size explicit numerical method (Euler method)

INPUTS

T01=	92	oC
T02=	21	oC
Tinf=	21	oC
R1inf=	8.451099	oC/W
R12	0.142467	oC/W
R2inf=	1.997784	oC/W
C1	1722	J/oC
C2	268	J/oC
f=	0.2	fraction of maximum step size

TIME CONSTANTS

(RC)1=	241.2608	sec
(RC)2=	35.63957	sec
Dt=	7.127915	sec

time(min)	T1(oC)	T2(oC)
0	92	21
0.118799	89.90234	34.25477
0.237597	88.25178	44.46698
0.356396	86.94669	52.32861
0.475194	85.90857	58.37427
0.593993	85.07678	63.017
0.712791	84.40445	66.57589
0.83159	83.8554	69.2975
0.950389	83.40164	71.37228
1.069187	83.02156	72.94739

x(cm)=	4.0775	4.1275	4.32625	4.525	4.575		Heat Transfer Rates (Watts)				
time(min)	T1(oC)	Tinner(oC)	T2(oC)	Touter(oC)	Tinf (oC)=	Tsurface (oC)	Coffee to Mug	Coffee to Ambient	Mug to Ambient	dT1/dt	dT2/dt
0.000	92	49.95431	21	21	21	88.66018	498.3615	8.401274	0	-0.29429	1.859558
0.119	89.90234	56.94825	34.25477	33.89128	21	86.6612	390.6001	8.153063	6.634735	-0.23156	1.432707
0.238	88.25178	62.32274	44.46698	43.82345	21	85.08827	307.3332	7.957755	11.7465	-0.1831	1.102935
0.356	86.94669	66.44611	52.32861	51.46949	21	83.84457	242.9904	7.803327	15.68168	-0.14564	0.848167
0.475	85.90857	69.60296	58.37427	57.34935	21	82.85529	193.2681	7.680489	18.70786	-0.11669	0.651344
0.594	85.07678	72.01314	63.017	61.86476	21	82.06262	154.8415	7.582065	21.0318	-0.09432	0.49929
0.713	84.40445	73.84651	66.57589	65.32606	21	81.42193	125.1418	7.50251	22.81322	-0.07703	0.381823
0.832	83.8554	75.23431	69.2975	67.97303	21	80.8987	102.1845	7.437541	24.17553	-0.06366	0.291078
0.950	83.40164	76.27794	71.37228	69.99091	21	80.46628	84.4362	7.383849	25.21407	-0.05332	0.220978
1.069	83.02156	77.05571	72.94739	71.52283	21	80.10408	70.7124	7.338875	26.0025	-0.04533	0.166828

Fig. 5.9 Structure of worksheet in which additional detail is extracted (i.e., intermediate temperatures and heat transfer rates)

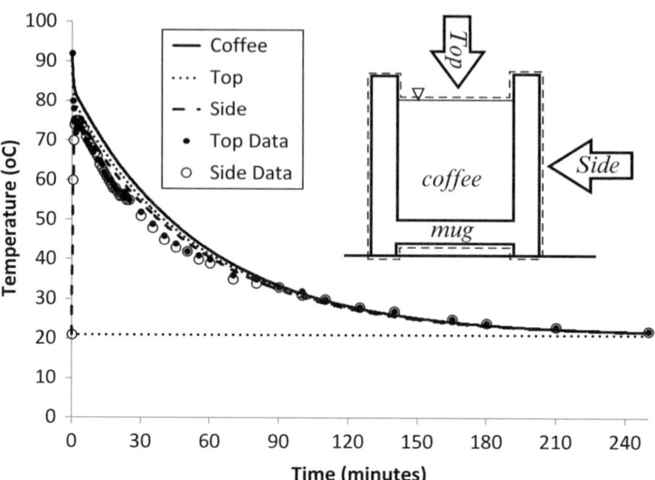

Fig. 5.10 2-Node model prediction for entire cooling event on linear axes

across a convective thermal boundary layer. Columns for the inner and outer mug surface temperatures are added and the voltage divider formulas implemented. A column for the free surface temperature is added. Finally, the individual heat transfer rates between nodes are calculated (individual terms from governing equations).

5.3.2   Model Prediction Results

Figures 5.10 (linear axes) and 5.11 (semi-logx axes) show the temperature vs. time predictions from the 2-node model (shown as lines) compared to the experimental data (shown as markers) for the entire cooling event using the parameter values from the previous section. Figure 5.12 focuses on the mug heating phase (early time) on linear axes. The line for coffee (solid line) shows node 1 at a slightly higher temperature than the surface and outer wall. The dotted line which corresponds with the filled circles shows the free surface temperature. The dashed line which corresponds with the open circles shows the outer mug wall temperature. The mug temperature itself (node 2) is not shown in these plots, but only the two non-capacity-carrying nodes that correspond to the measured quantities.

The linear axes (Fig. 5.10) show the entire event, and the initial mug heating period is not clearly visible. The coffee, top (free) surface of the coffee, and mug side wall all fall together. The model prediction lags the experimental for the first hour or so, but then the final approach to equilibrium gives good agreement between theory and experiment. However, this model assumes constant coefficients, and the effect of a variable coefficient cannot be seen. Recall that the constant values chosen were based on a temperature of 70 °C, and more

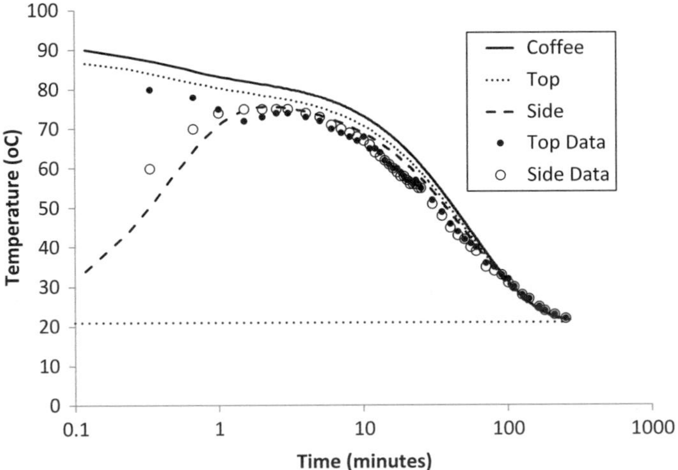

Fig. 5.11 2-Node model prediction for entire cooling event on semi-logx axes

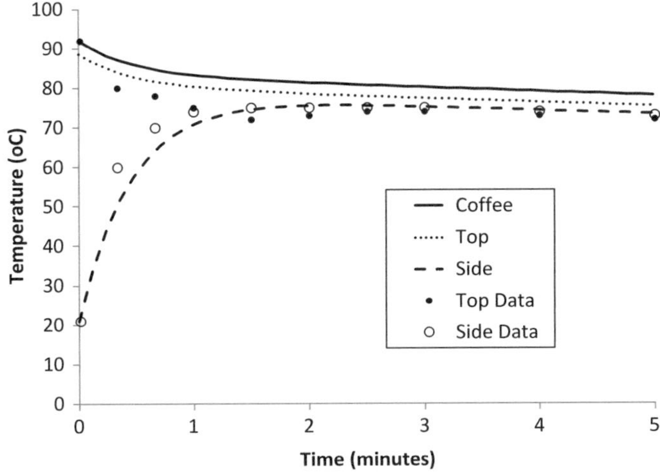

Fig. 5.12 2-Node prediction for the cooling mug of coffee for the first 5 min

sophisticated models would show the resistance values increasing with time (as the temperature falls).

The semi-logx axes (Fig. 5.11) show the entire event, but the behavior during the initial mug heating phase is apparent in this view as well. The focus on the first 5 min of the event on linear axes (Fig. 5.12) shows the undistorted shape of the temperature response, with the same trends observed on the semi-logx scale. In the first minute, the model predicts a slower rise of the outer side wall than measured, and then excellent agreement between 1 and 10 min. The predicted top (free)

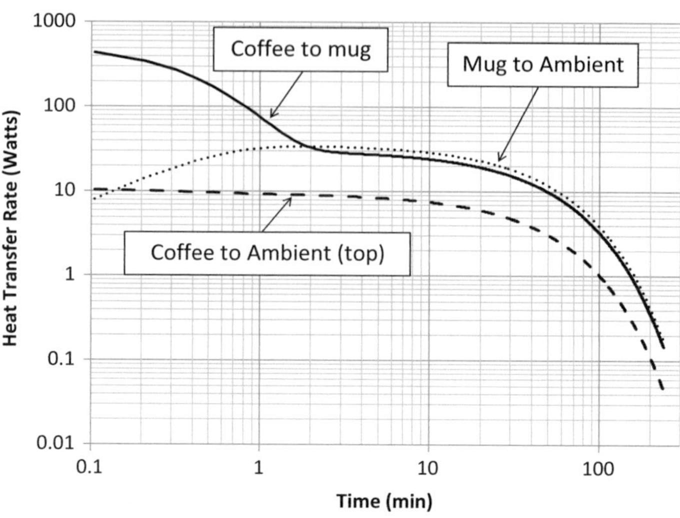

Fig. 5.13 Heat transfer rates between nodes for the entire event (log–log axes)

surface of the coffee is noticeably higher than the experimental value throughout. Use of a higher value for thermal resistance between the coffee and the free surface (R_{cs}) would give better agreement (but not the "dip" in the data, which may or may not be a repeatable trend, and only a single experiment was conducted). Inclusion of this resistance represents a deviation from the classic modes of heat transfer because the free surface represents a boundary between two immiscible fluids, not between a fluid and a solid. Nevertheless, application of Newton's Law of Cooling with a convection coefficient as if the fluid/fluid boundary behaved as a fluid/solid boundary (on either side) is a plausible modeling tool to reflect a mechanism for heat transfer from a bulk fluid to an interface.

The heat transfer rates between nodes are shown as a function of time in Fig. 5.13. During the mug heating phase (the first minute), the rate of heat transfer from the coffee to the mug is very high. The heat transfer rate from the mug to ambient rises as the outside wall temperature rises during the first minute or two. After that, the rate of heat transfer from the mug to ambient is slightly higher than from the coffee to the mug (the difference is the change in energy stored in the mug wall). The rate of heat loss through the top surface is always small, and this effect can be attributed to the smaller surface area, despite the higher effective heat transfer coefficient value input there. The heat transfer rate between the side and bottom channel are not distinguished, but the thermal resistance of the bottom channel is much larger than that of the side wall (due to smaller area and lower effective heat transfer coefficient).

A main point of presenting this heat transfer plot is to demonstrate how much more information can be gleaned from a theoretical analysis of a problem (if one chooses to look for it) compared to experiment. The experiment has the advantage of being a measurement (keeping experimental errors always in mind), but

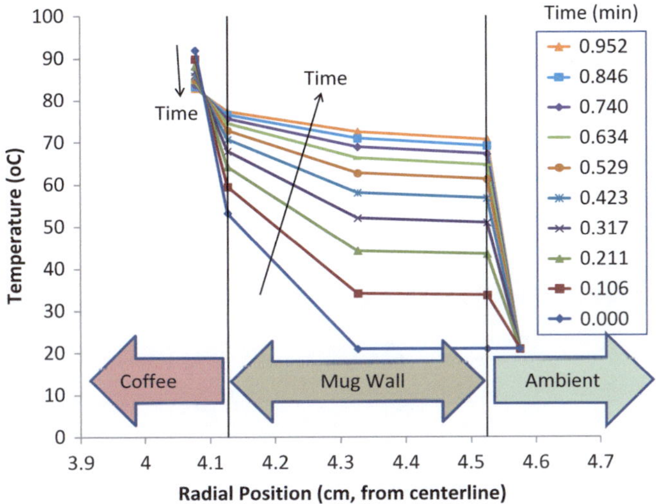

Fig. 5.14 Temperature as a function of side wall position for the first minute of the simulation (mug heating phase)

generally limited information. Theory and experiment make a great team. For example, an IR detector was used that can only yield surface temperatures, and the model shows the side and surface temperatures with excellent agreement. This experiment does not show that the bulk of the coffee is actually slightly higher the temperature at the free surface and at the mug surface. A thermocouple array might show that detail (a great project for a course on experimentation). The heat transfer rates cannot be measured directly at all.

The spatial temperature variation in the side wall as time elapses is demonstrated in Figs. 5.14 (for the mug heating phase) and 5.15 (for the coffee/mug cooling phase), which show temperature as a function of radial position at fixed times. The fluid temperatures (coffee and ambient air) are shown with somewhat arbitrary 0.05 cm thermal boundary layer. Straight lines are drawn between points with markers that indicate specific locations from the model. Arrows are used to indicate the trends with time.

The first plot (Fig. 5.14) shows these temperatures for the first ten discrete times of the simulation which focuses on the first minute of the event. At the initial time, where the coffee and mug temperatures are set to their initial values (92 °C and 21 °C, respectively), the model predicts (through the voltage divider calculation) that the initial temperature on the inner wall surface does not equal the initial mug temperature, but is between the mug and the coffee. Its value (49.95 from Fig. 5.9) is roughly halfway between the mug and coffee temperatures because the convective resistance between the coffee and mug (0.0844 °C/W) is coincidentally of the same order as that of the conductive resistance across the mug wall (0.116/ 2 = 0.0553 °C/W). After one step in time, the mug temperature rises to about 35 °C and the outer surface wall temperature rises with it. The outer surface

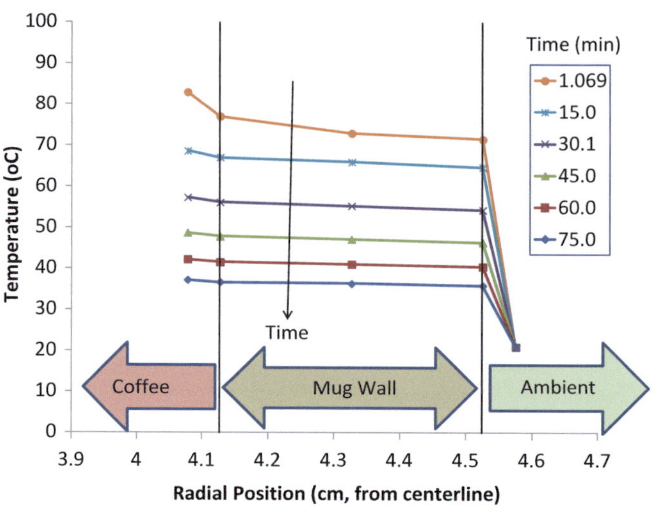

Fig. 5.15 Temperature as a function of side wall position at 15 min time intervals (after initial mug heating)

temperature is between the mug and ambient, but is much closer to the mug because the convective resistance to ambient (2.06 °C/W) is much larger than the conductive resistance across the mug (0.0553 °C/W), a Biot number concept. During this time, the inner wall temperature rises about 5 °C as the coffee temperature drops about 1 °C. Each step in time reveals a similar qualitative change, but with reduced magnitude temperature jumps.

Experience from the variable-sized 1-node model from Chap. 4 would suggest that it takes approximately 4 s for the initial thermal wave to traverse the mug wall, and that time is short compared to the step size used in this simulation. Therefore, the use of a fixed size mug node is justified.

The concavity in the temperature profile across the mug is an indication of the degree to which it is important to assign separate capacitance to the mug (as opposed to lumping it together with the coffee). If heat transfer rate from the inner wall to the mug exceeds that from the mug to the outer wall, the temperature profile will be concave up. The difference is the rate of change of the energy stored in the mug. After 1 min, the concavity essentially vanishes (the temperature is approximately linear across the mug wall), meaning that the NET heat transfer rate to the mug is negligible. They are not exactly equal, as shown in Fig. 5.13. A high heat transfer rate enters from the inside and exits the outside at almost exactly the same rate. This behavior is an example of "quasi-steady state" behavior, and why it is justified to lump the capacitance of the mug and coffee together in order to simulate the response AFTER the mug attains this quasi-equilibrium. The heat transfer rate at any time (after 1 min) is the same as it would be if the slowly changing coffee temperature were maintained by some means at a constant value. Recalling Fig. 5.13, these trends are also made apparent by considering the heat transfer rates between the coffee and mug, and between the mug and ambient.

Once this quasi-steady state equilibrium for the mug is established, the temperature remains approximately linear across the mug as the coffee and mug cool (Fig. 5.15). During this time, the coffee and mug respond together, and this 2-node model provides justification for lumping their capacitance together during the cooling phase. However, the 1-node lumped model is not capable of revealing the key physics during the mug heating phase. In this particular case, the coffee/mug cooling phase dominates the entire event, and therefore a 1-node model does a good job of simulating the main event, while missing detail at the beginning.

A really good problem statement would be to conduct a similar experiment and analysis with a Styrofoam cup. The main difference would be a lower mug capacitance and a higher conductive resistance across the mug wall. The mug outer side wall would be much lower than the coffee-free surface in this case. Keeping the same geometry, the mug mass (from a measurement), specific heat and the mug thermal conductivity (from a literature search) could be changed to their appropriate values. The time constant for the mug will be much smaller than the ceramic mug, and this problem is becoming numerically "stiff" where the time step restriction on a fast responding node requires many more time steps to simulate the overall response than would be required for the slower responding node.

5.3.3   Numerical Accuracy and Stability Issues

An important distinction to be made in evaluating heat transfer models is between the physics of the model and the mathematics of its simulation. The discussion of the previous section was entirely about the physics of the model, and it was assumed (without formal statement) that the plots presented were from an accurate implementation of the model. An inherent challenge and potential danger of numerical analysis is generating output that "looks right" but may in fact be unacceptably and unnecessarily inaccurate.

Figure 5.16 shows the effect of step size used in the simulation defined as a fraction (f) of the maximum value allowed by the Second Law of Thermodynamics. Three plots show the predicted coffee and mug temperatures for different values of "f". The bottom right compares the mug temperature directly for these cases, plus an additional two values. Symbols with straight lines are used to indicate the discretization of time. A value $f = 1$ (bottom left) shows the mug jumping up and overshooting the coffee temperature, which at face value, appears to violate the Second Law. However, the value $f = 1$ represents a step size in which the mug will come to thermal equilibrium with its neighbors (coffee and ambient) based on their present values, not their future values. Values of $f = 0.1$ and 0.5 reveal similar qualitative predictions, with the smaller value yielding smoother behavior, at the expense of more calculations. However, when

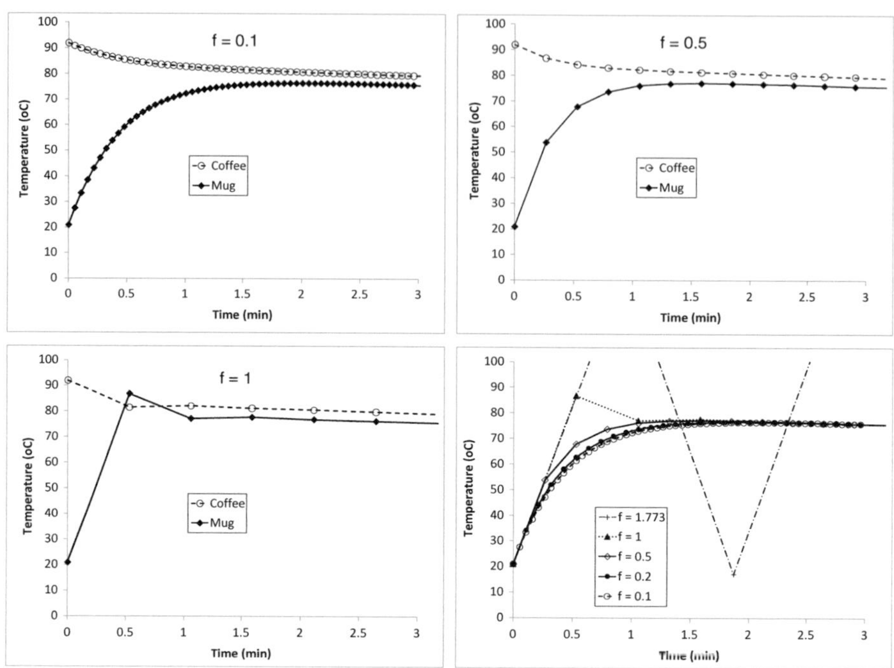

Fig. 5.16 Effect of step size (f = fraction of maximum Second Law step size) during the mug heating phase. The *bottom right* graph compares the mug temperature for 5 different values of the step size (as a fraction, f, of the maximum allowable)

compared directly (bottom right), the $f = 0.5$ case shows a noticeable difference from $f = 0.1$. An intermediate value ($f = 0.2$) is included which shows good agreement with $f = 0.1$.

This analysis shows a somewhat informal or brute force test for "grid insensitivity." To accurately simulate a given mathematical model numerically, a typical Question and Answer series might be

Q1. "What step size should be used?"
A1. "As small as it needs to be to accurately solve the mathematical problem?"
Q2. "How do you know what's small enough?"
A2. "When the prediction of two values of step size agree acceptably, the bigger one is small enough"

In this case, a value $f = 0.2$ yields a good compromise between computational expense and numerical accuracy. More formal test methods can be developed, especially as the models become spatially more complex.

Notice for this system that even the case $f = 1$ yields excellent agreement with the smaller step-size cases at the end of the mug heating phase. So that choice is acceptable from the point of view of the simulation of the entire cooling event.

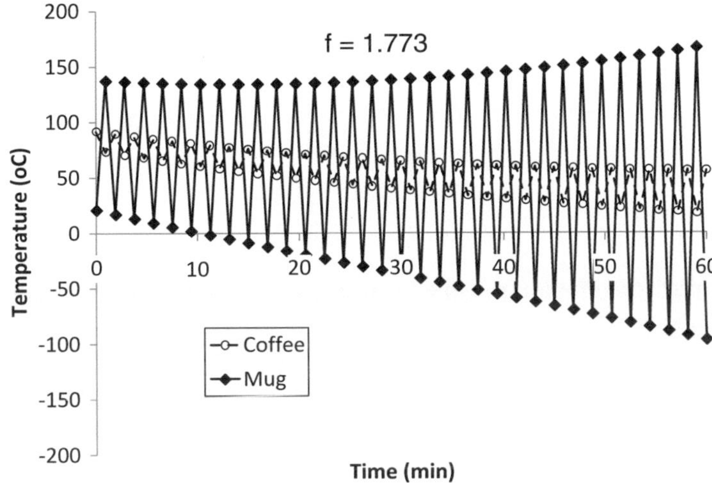

Fig. 5.17 Coffee and mug temperatures for the step size just above a critical value for numerical stability

It just misses the detail at the beginning. It can be argued that the reason for that agreement is because it merely finds the right quasi-equilibrium in a few steps. For a problem for which that quasi-equilibrium is not a good model, all bets are off.

Numerical stability limits are demonstrated in Fig. 5.17, which shows the mug and coffee temperatures for an hour with a value $f = 1.773$. This value (also one of the cases in the bottom right plot of Fig. 5.16) was chosen by trial and error, and is considered near, but above the stability limit of this explicit formulation. The mug temperature would eventually grow without bounds (numerical instability) if allowed to continue.

5.3.4   Implicit Numerical Formulation

There may come a time when the numerical stability limit of an explicit formulation becomes so restrictive as to make it untenable. For example, in a 2-node system, if the response time constant of one node is 1,000 times smaller than that of the other, a minimum of 1,000 time steps would need to be taken just to simulate an amount of time corresponding to one time constant of the slower responding node. An implicit method represents another numerical tool that overcomes this problem. It is introduced in this section, but not implemented, as the explicit method is considered adequate for this particular job. This method is not readily solved using a spreadsheet.

As was demonstrated in Chap. 1, a backward difference for the time derivative brings a quickly responding node to equilibrium if an excessive step size is used. In this case, the two backward difference approximations (for nodes 1 and 2, coffee and mug) are:

$$\frac{dT_1}{dt} = \frac{T_1^{(p)} - T_1^{(p-1)}}{\Delta t^{(p)}}$$

$$\frac{dT_2}{dt} = \frac{T_2^{(p)} - T_2^{(p-1)}}{\Delta t^{(p)}}$$

Updating present to future ($p \rightarrow p+1$), inserting these approximations into the two nodal equations yields:

$$C_1^{(p+1)} \left(\frac{T_1^{(p+1)} - T_1^{(p)}}{\Delta t^{(p+1)}} \right) = \dot{Q}_1^{(p+1)} + \frac{T_2^{(p+1)} - T_1^{(p+!)}}{R_{12}^{(p+1)}} + \frac{T_\infty^{(p+1)} - T_1^{(p+1)}}{R_{1\infty}^{(p+1)}}$$

$$C_2^{(p+1)} \left(\frac{T_2^{(p+1)} - T_2^{(p)}}{\Delta t^{(p+1)}} \right) = \dot{Q}_2^{(p+1)} + \frac{T_1^{(p+1)} - T_2^{(p+!)}}{R_{12}^{(p+1)}} + \frac{T_\infty^{(p+1)} - T_2^{(p+1)}}{R_{2\infty}^{(p+1)}}$$

All temperature variables and parameters with a superscript "p" are evaluated at the present time. If the initial condition is known, these are considered to be known quantities. A superscript "p + 1" is a future value, and is unknown. If the numerical time step (Δt), R and C parameters are functions are variable, their (unknown) future value is required. The two nodal equations constitute a system of two algebraic equations for the future values of T_1 and T_2.

If the parameters are constants, it is a linear system, and can be solved using linear algebra. The two governing equations can be written in matrix form for the two future temperatures, and then solved using Cramer's rule or Gaussian elimination, or easily inverted in a MATLAB program.

If the parameters are variables, the system is nonlinear, and an iterative numerical technique is required (as opposed to a "stepping" technique as with time). The Jacobi Method (Successive Substitution) and closely related Gauss–Seidel method are powerful methods, and involve rearranging the governing equations by "solving" for each nodal future value in terms of its neighbors, making an initial guess, and iterating until either convergence or failure of the method. If the method fails, then Newton's method is a good technique to try next. The recursion equations in the form for Gauss–Seidel are:

$$T_1^{(p+1)} = \frac{\dot{Q}_1^{(p+1)} + \dfrac{T_\infty^{(p+1)}}{R_{1\infty}^{(p+1)}} + \dfrac{T_2^{(p+1)}}{R_{12}^{(p+1)}} + \dfrac{C_1^{(p+1)} T_1^{(p)}}{\Delta t^{(p+1)}}}{\dfrac{1}{R_{1\infty}^{(p+1)}} + \dfrac{1}{R_{12}^{(p+1)}} + \dfrac{C_1^{(p+1)}}{\Delta t^{(p+1)}}}$$

$$T_2^{(p+1)} = \frac{\dot{Q}_2^{(p+1)} + \dfrac{T_\infty^{(p+1)}}{R_{2\infty}^{(p+1)}} + \dfrac{T_1^{(p+1)}}{R_{12}^{(p+1)}} + \dfrac{C_2^{(p+1)} T_2^{(p)}}{\Delta t^{(p+1)}}}{\dfrac{1}{R_{2\infty}^{(p+1)}} + \dfrac{1}{R_{12}^{(p+1)}} + \dfrac{C_2^{(p+1)}}{\Delta t^{(p+1)}}}$$

Application of the implicit method, then, involves solving a system of algebraic equations for each step in time taken. It is that extra work, requiring a nested loop (which cannot be done in a spreadsheet without using macros) that makes it more difficult to implement than the explicit method. It can be advantageous (or necessary) to use an implicit method if the value of the two time constants differ by orders of magnitude, in which case the problem is said to be "stiff." To test for stiffness, the ratio of the two time constants can be evaluated:

$$\frac{\tau_1}{\tau_2} = \left(\frac{C_1}{C_2}\right) \left[\frac{1 + \dfrac{R_{12}}{R_{2\infty}}}{1 + \dfrac{R_{12}}{R_{1\infty}}}\right]$$

If this ratio is less than, say 0.001, then the response time of node 1 is much shorter than that of 2. In this case, the response of the system is controlled by the slower moving 2, and node 1 will be in a quasi-equilibrium between 2 and the environment. On the other hand, if the ratio is greater than, say 1,000, then node 2 quickly responds, and the system is controlled by node 1.

Alternatively, if a 2-node problem is shown to be stiff, the original model can be simplified by appropriately lumping the two capacitances together, and thereby reducing the problem to a 1-node problem. Detail is compromised, and if that detail would help make engineering decisions, then lumping would not be a good simplification of the model. The coffee/mug problem is a case is point. Lumping the coffee and mug prevents investigation of the mug heating phase, but does not significantly affect the overall cooling behavior.

5.4 Coffee Joulies™: A 3-Node Model

Coffee Joulies represent an engineering solution whose objective is to shorten the time before pouring hot coffee and being able to drink it, and lengthen the time coffee stays in an acceptable drinking range. Joulies fit comfortably in your hand, and can be placed into a hot mug of coffee, or placed in a mug, and then coffee

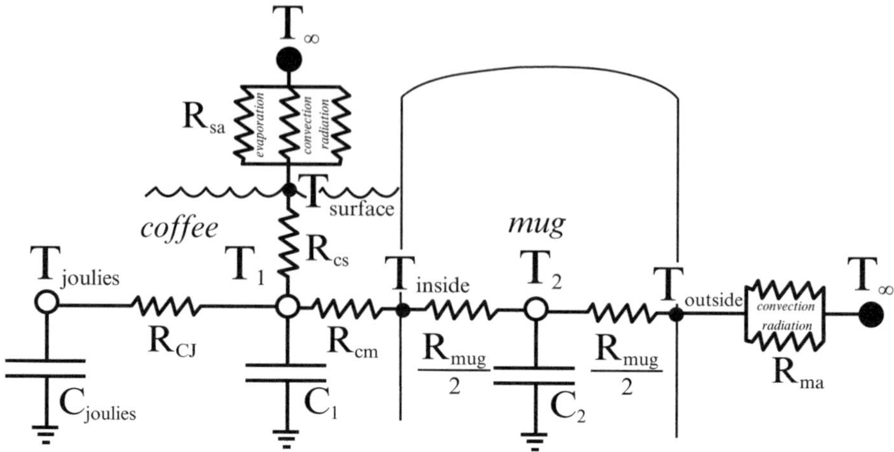

Fig. 5.18 3-Node model with Coffee Joulies in a single-stage phase of heating or cooling

poured over them. They are constructed with a shell of stainless steel, and filled with a "phase change material" that is called "wax" in this text. The requirement of the wax is that its melting point temperature be within the suitable temperature range. The Joulies store energy in the form of "latent heat."

To model the cooling of coffee with Joulies, a 3-node model consisting of coffee, mug, and Joulies captures the key effects (Fig. 5.18). The coffee and mug are the same as the 2-node model, except that the coffee has an additional thermal partner, namely, the Joulies. The Joulies are treated as a lumped element. During single phase stages (i.e., completely solid or liquid), they are modeled with a capacitance, and a thermal resistance (R_{CJ}) that consists of three resistances in series: conduction through the wax, conduction across the shell, and convection between the Joulie and the coffee.

When undergoing phase change (Fig. 5.19), the temperature remains fixed at the melting point, but the fraction of wax that is liquid and the fraction that is solid changes. There is a "quality" (x) of the solid/liquid mixture that changes. The quality, defined as the mass fraction of liquid, is analogous to the quality used in liquid/vapor phase change in traditional thermodynamic analysis. The quality is only defined during phase change, when the Joulie contains a mixture of saturated solid (further heating would cause melting) and saturated liquid (cooling would cause freezing) in equilibrium. If the Joulie is below the melting point, the solid is not saturated and the quality is not applicable (x = NA). Similarly, for temperatures above the melting point the liquid is not saturated (and x = NA).

The Coffee Joulie node is represented in Fig. 5.19 as a "rechargeable battery" given the symbol of a battery with an arrow through it. The node is filled, indicating that the temperature of the node is fixed at the melting point of the wax, but the arrow indicates that the degree of battery "charge" varies. When the wax is

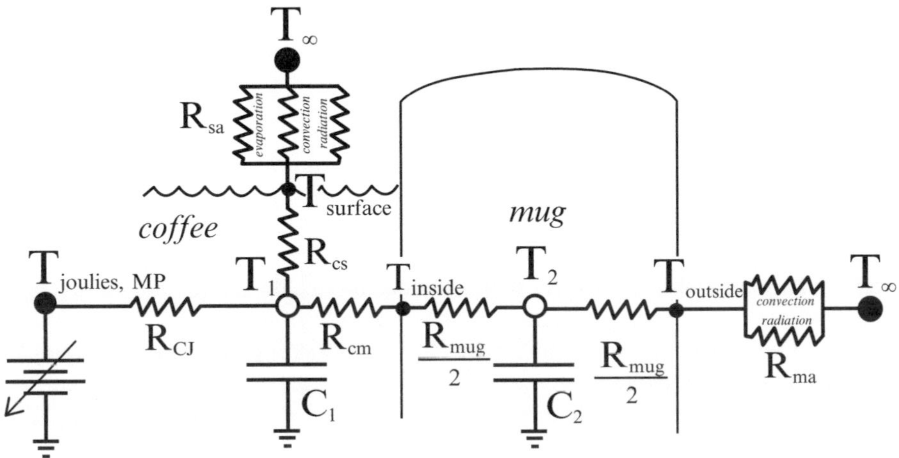

Fig. 5.19 3-Node model with Coffee Joulies undergoing phase change (melting or freezing)

completely solid, the battery is discharged, and when liquid it is fully charged. The quality "x" monitors the degree of charge.

When placed in sufficiently hot coffee, the Joulies will go through five distinct stages.

1. Heating stage: The Joulies will be heated (by the coffee) to the melting point of the wax. Meanwhile the coffee is also heating the mug wall and is being cooled itself. If the Joulies reach the melting point with the coffee still above it, the wax will become a saturated solid and stage 2 will begin.
2. Melting stage: The Joulies will continue to absorb heat from the coffee but remain at the melting point temperature of the wax. The quality (mass fraction of vapor) will increase until all the solid is melted, and the wax is a saturated liquid. If melting is complete, the coffee temperature remains above the melting point, and stage 3 will begin.
3. Superheating stage: The temperature of the Joulies rises above the melting point, and will tend to approach the coffee temperature, then fall back with it as the coffee cools. When the Joulies cool back to the melting point, the liquid will be saturated and stage 4 will begin.
4. Freezing stage: The coffee temperature drops below the Joulies and the wax begins to freeze. Heat is now transferred from the Joulies to the coffee, helping delay the coffee from cooling below a suitable drinking temperature. Once the wax has completely solidified, the final stage begins.
5. Cooling stage: The coffee, mug, and Joulies cool together, eventually attaining thermal equilibrium with ambient. Unless the coffee has been drunk by now...

5.4.1   Incorporating a Node with Phase Change

The general nodal equation for the Coffee Joulie(s) is:

$$\frac{dE_{Joulie}}{dt} = \frac{T_{coffee} - T_{Joulie}}{R_{CJ}}$$

Table 5.1 summarizes the model for the energy stored in the Joulie and its rate of change. The Joulie is considered to be constructed of a central spherical core of wax, and a spherical shell of steel. In this lumped model, the temperature of the wax and shell are not necessarily equal, but they rise and fall together.

If the Joulie is below the melting point (T_{MP}), the wax is completely solid, and its internal energy, relative to a reference temperature (T_{ref}, below the melting temperature) has been handled previously. The term "quality" has no relevance (x = NA). The total capacitance (sum of wax plus shell) is $C_{solid} = \left((mc)_{shell} + (mc)_{wax,s} \right)$. There is a similar expression for when the wax is completely liquid, the third row in Table 5.1, except the liquid wax could have a different specific heat than the solid. In the energy expression for single phases, only the temperature changes.

If the temperature equals the melting temperature and it is completely solid, the wax in the Joulie has properties of a saturated solid, and the quality has a value of 0. The internal energy (relative to the reference temperature) is $u_s = C_{solid}(T_{MP} - T_{ref})$. If the wax is at the melting point and is completely liquid, the wax has properties of a saturated liquid, and the quality has a value of 1.0. If the internal energy is between a saturated solid and liquid, the quality has a value between 0 and 1, and the internal energy of the wax can be expressed as $E_{wax} = m_{wax}[(1-x)u_s + xu_L] = m_{wax}[u_s + xu_{LS}]$ where the heat of fusion is $u_{LS} = u_L - u_S$.

In the phase changing region, the only variable is the quality, so the rate of change of energy depends on the heat of fusion and the rate of change of quality.

Table 5.1 Joulie energy, relative to a reference temperature below the melting point, stored in a Joulie and its rate of change for different temperature ranges

Temperature range	$E_{Joulie} = E_{shell} + E_{wax}$	$\dfrac{dE_{Joulie}}{dt}$
$T_{Joulie} < T_{MP}$	$\left((mc)_{shell} + (mc)_{wax,s} \right)\left(T_{Joulie} - T_{ref}\right) = C_{solid}\left(T_{Joulie} - T_{ref}\right)$	$C_{solid}\dfrac{dT_{Joulie}}{dt}$
$T_{Joulie} = T_{MP}$	$(mc)_{shell}\left(T_{MP} - T_{ref}\right) + m_{wax}[u_s + xu_{LS}]$	$m_{wax}u_{LS}\dfrac{dx}{dt}$
$T_{Joulie} > T_{MP}$	$\left[(mc)_{shell} + (mc)_{wax,l} \right]\left(T_{Joulie} - T_{MP}\right) + m_{wax}u_L$	$C_{liquid}\dfrac{dT_{Joulie}}{dt}$

5.4.2   Nodal Equations

The nodal equations (in the form for the Euler method) are therefore:

$$T_{mug}^{(p+1)} = T_{mug}^{(p)} + \frac{\Delta t^{(p)}}{C_{mug}} \left(\frac{T_\infty - T_{mug}^{(p)}}{R_{M\infty}} + \frac{T_{coffee}^{(p)} - T_{mug}^{(p)}}{R_{CM}} \right)$$

$$T_{coffee}^{(p+1)} = T_{coffee}^{(p)} + \frac{\Delta t^{(p)}}{C_{coffee}} \left(\frac{T_\infty - T_{coffee}^{(p)}}{R_{C\infty}} + \frac{T_{mug}^{(p)} - T_{coffee}^{(p)}}{R_{CM}} + \frac{T_{Joulie}^{(p)} - T_{coffee}^{(p)}}{R_{CJ}} \right)$$

The nodal equation for the Joulies depends on whether it is in the single or mixed phase. When in the single phase (solid or liquid):

$$T_{Joulie}^{(p+1)} = T_{Joulie}^{(p)} + \frac{\Delta t^{(p)}}{C_{Joulie}} \left(\frac{T_{coffee}^{(p)} - T_{Joulie}^{(p)}}{R_{CJ}} \right)$$

When in the mixed phase, the temperature of the Joulies is fixed at the melting point, and an Euler formula for the quality is:

$$x^{(p+1)} = x^{(p)} + \frac{\Delta t^{(p)}}{m_{wax} u_{LS}} \left(\frac{T_{coffee}^{(p)} - T_{MP}}{R_{CJ}} \right)$$

As always with the Euler method, there is a maximum time step for numerical stability that must be carefully monitored.

A variable time step algorithm is used during the melting and freezing stages, as indicated by a superscript (p) for the time step. Rather than input a time step to march, a step in quality (Δx) is made. Rearranging the governing equation for quality gives the required time step (which must be less than the maximum time step for the coffee and mug).

$$\Delta t^{(p)} = \left| \frac{\Delta x m_{wax} u_{LS} R_{CJ}}{T_{coffee}^{(p)} - T_{MP}} \right|$$

Absolute values are used to prevent negative time steps.

All the resistance values are equivalent resistances. A detail for the resistor between the Joulies and the coffee (R_{CJ}) is shown in Fig. 5.20, which consists of three thermal resistances in series. For the purposes of modeling, the Joulies are treated as spherical, with a diameter that reproduces the total volume of a manufactured Joulie, and an average density equal to the total mass divided by the total density (detailed in Appendix 3). The actual shape is not spherical so the surface

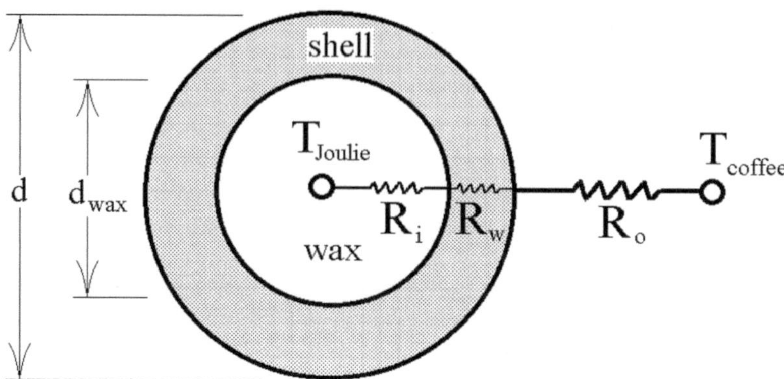

Fig. 5.20 Detail of R_{CJ}, the thermal resistance between the Joulies and coffee

to volume ratio of a manufactured Joulie is different than the one used in this simulation. The total mass (mass of the wax plus the mass of the shell) and volume are easily measured, and the equivalent shell thickness can be estimated given assumed density values. The mass-averaged specific heat is similarly estimated.

The outer resistance is modeled as a convective resistance:

$$R_o = \frac{1}{h_o\left(\pi d^2\right)}$$

An outer convection coefficient typical of a natural convection in water is assumed, acting on the outer surface area $(4\pi r^2 = \pi d^2)$ of the Joulie.

The resistance across the shell wall is a conductive resistance. Using a thin-wall approximation:

$$R_w = \frac{(d - d_{wax})/2}{k_{shell}\left(\pi d^2\right)}$$

The thermal conductivity of the shell is required.

The inner conductive resistance (between the inner shell surface and the wax) is intriguing and a good heat transfer modeling case study. A first approach is to assume a pure conduction during the Joulie heating stage, in which case a first approximation would be:

$$R_w = \frac{d/4}{k_{wax}\left(\pi d^2_{wax}/2\right)} \quad \text{(a first approximation, not used in simulation)}$$

Here, the numerator is the average distance between the center and the inner shell surface, and the denominator is taken to be the average area of those surfaces.

When evaluated (Appendix 3), the value of this resistance is much larger than the two others in this series channel. Therefore, in practice, the shell temperature will change first, and the surface of the wax will follow, then quickly enter into a stage where the wax melts. This behavior requires the shell and wax to be treated as separate nodes. However, that approach will not be pursued here (a good workshop assignment!) in the interest of gaining a reasonable prediction of the system response with Joulies. The modeling compromise is to treat the inner thermal resistance as a convective resistance:

$$R_i = \frac{1}{h_i\left(\pi d_{wax}^2\right)}$$

An inner convection coefficient typical of natural convection with a liquid is input. Actually, during phase change, the effective heat transfer coefficient would be much larger than that, as convection coefficients during phase change tend to be larger than single phase convection due to latent heat effects. During those periods, the total resistance would be controlled by the sum of the shell and outer resistances.

5.4.3   Spreadsheet Implementation

The worksheet used to calculate various parameter values that are inputs to the simulation is shown in Fig. 5.21. Many of these parameters were determined from a separate spreadsheet detailed in Appendix 3 based on basic measurements (volume and mass) and assumptions about a single Joulie. A total of 3 Joulies are added to the same mug. The mass of the wax (40.6 g for 3 Joulies) is approximately half of the total Joulie mass (86 g), the remainder being the shell. The 3 Joulies displace 62 mL of volume. Note that the capacitance of the Joulies during the heating phase ($C_{solid} = 123$ J/K) is less than half that of the mug ($C_{mug} = 268$ J/K), so it is expected that the main effect of the Joulies on coffee temperature will be due to phase change (latent heat), not the sensible energy effect that adding more thermal mass would have.

Five different "simulation" spreadsheets are written, one each for each stage of the event (Joulie heating, melting, superheating, freezing, cooling), and these sheets are linked so that the time and temperatures at the end of one stage are referenced as the initial condition of the subsequent stage. The structures of two of these sheets are shown in Figs. 5.22 and 5.23 for the heating and melting stages, respectively. Each sheet marches forward a total of 100 steps (arbitrarily chosen, and not tested to yield a time step smaller than that required of the other two nodes, mug and coffee).

During heating (Fig. 5.22), the quality is set to a value "NA," and the Joulie temperature varies. The numerical simulation marches in time, with a fraction of the maximum allowable step size (f = 0.0218..., small enough for excellent Euler accuracy) determined using "goal seek" to the value that results in the temperature

INPUT PARAMETERS

n	3		Number of Joulies
Djoulie	0.034	m	Diameter of Joulie
rhojoulie	1360	kg/m3	density of Joulie
csolid	1461	J/kg/K	specific heat of solid Joulie
cliquid	1461	J/kg/K	specific heat of liquid Joulie
rhocoffee	1000	kg/m3	coffee density
ccoffee	4200	J/kg/K	coffee specific heat
uLS	210000	J/kg	heat of fusion of Joulie
Tmp	60	C	melting point of Joulie
Tinf	25	C	ambient
Vtotal	4.10E-04	m^3	total volume (coffee plus Joulies)

DERIVED

Ajoulie	0.0109	m^2	total surface area of Joulies
Vjoulie	6.17E-05	m^3	total volume of Joulies
mjoulie	0.0840	kg	total joulie mass
mwax	0.0406	kg	mass of wax
Esat,Solid	4294	J	Joulie energy at saturated solid
Esat,Liquid	21926	J	Joulie energy at saturated liquid
Coffee Volume	3.48E-04	m^3	coffee volume

RC VALUES

Rcj	0.86	oC/W	bet. Coffee and Joulie
Rcm	0.15		bet Coffee and mug
Rminf	1.74		bet. Mug and ambient
Rcinf	6.63		bet. Coffee and ambient
Cmug	268	J/K	mug capacitance
Ccoffee	1463	J/K	coffee capacitance
Cjoulie,solid	123	J/K	Joulie solid capacitance
Cjoulie,liquid	123	J/K	Joulie liquid capacitance

Fig. 5.21 Worksheet of input parameters for the 3-node simulation

at the last step equal to the melting point. The maximum step size (inherent time constant) is controlled by the mug (dtmax,mug $= 37.2$ s) which is smaller than the time constant for the Joulies (dtmax,Joulies $= 105.2$ s). Even though the capacitance of the Joulies is half that of the mug, the thermal resistance is almost six times as large between the Joulies and coffee ($R_{CJ} = 0.86$ °C/W) as it is between the coffee and the mug ($R_{CM} = 0.15$ °C/W). This effect is one of contact area.

During melting, the Joulie temperature is set to the melting temperature and the quality varies with time. The time step (dt) varies according to the governing equation for quality and is confirmed to be lower than the inherent time constant of the mug at all times.

This workbook is not completely automatic. Any time a parameter value is changed, it is necessary to adjust the time step (through "f") at each stage to transition from stage to stage.

JOULIE HEATING PHASE — VALID SIMULATION

dtmax,Joulie	105.1924716	sec
dtmax,Coffee	184.0242627	sec
dtmax,mug	37.18518815	sec
f=	0.021782786	fraction of max step size
dt	0.80999699	sec

time (min)	Tcoffee	Tmug	Tjoulie	x	dTc/dt	dTmug/dt	dTJ/dt	dx/dt	Ejoulie
0	100	25	25	NA	-0.40755	1.855774	0.712979	NA	0
0.01349995	99.66988171	26.50317	25.57751	NA	-0.39849	1.807182	0.704351	NA	70.84444
0.0269999	99.34710938	27.96698	26.14803	NA	-0.38964	1.75983	0.695859	NA	140.8315
0.040499849	99.03150161	29.39244	26.71168	NA	-0.38101	1.713687	0.6875	NA	209.9748
0.053999799	98.72288158	30.78052	27.26855	NA	-0.3726	1.668721	0.679272	NA	278.2876
0.067499749	98.42107692	32.13218	27.81876	NA	-0.36439	1.624904	0.671173	NA	345.7828
0.080999699	98.12591956	33.44835	28.36241	NA	-0.35639	1.582206	0.663199	NA	412.4733
0.094499649	97.83724568	34.72993	28.89959	NA	-0.34858	1.540598	0.655348	NA	478.3713
0.107999599	97.55489558	35.97781	29.43042	NA	-0.34097	1.500053	0.647617	NA	543.4893
0.121499548	97.27871353	37.19285	29.95499	NA	-0.33354	1.460544	0.640005	NA	607.8392
0.134999498	97.00854776	38.37589	30.47339	NA	-0.32629	1.422045	0.632509	NA	671.4327
0.148499448	96.74425028	39.52774	30.98572	NA	-0.31923	1.384529	0.625126	NA	734.2813
0.161999398	96.48567683	40.6492	31.49207	NA	-0.31233	1.347972	0.617854	NA	796.3963
0.175499348	96.23268675	41.74105	31.99253	NA	-0.30561	1.312349	0.610692	NA	857.7888

Fig. 5.22 Spreadsheet simulation during the initial Joulie heating phase. Similar sheets are used for Joulie superheated and cooling phases. A fixed time step based on a fraction of the maximum allowable step size is used. "Goal seek" is used to calculate the correct time step so that the Joulie temperature equals the melting point at the end of the stage (100 steps chosen). An "if-then-else" statement is placed in the first row that checks, and returns "VALID SIMULATION" if the last row for Joulie temperature matches the melting point. If not, the formula returns "SIMULATION NOT VALID"

JOULIE MELTING PHASE — VALID SIMULATION

dtmax,Joulie	NA	sec
dtmax,Coffee	184.0243	sec
dtmax,mug	37.18519	sec
dx	0.01	step in x

time(min)	Tcoffee	Tmug	Tjoulie	x	dTc/dt	dTmug/dt	dTJ/dt	dx/dt	Ejoulie	dt
1.349994983	85.79192	77.1434	60	0	-0.06605	0.101952	NA	0.003528	4293.523	2.834817
1.397241934	85.6047	77.43241	60	0.01	-0.06372	0.089547	NA	0.003502	6393.523	2.855546
1.444834368	85.42275	77.68812	60	0.02	-0.06157	0.078168	NA	0.003477	8493.523	2.875983
1.492767416	85.24568	77.91293	60	0.03	-0.05959	0.067741	NA	0.003453	10593.52	2.896155
1.541036665	85.0731	78.10912	60	0.04	-0.05776	0.058195	NA	0.003429	12693.52	2.916089
1.589638147	84.90467	78.27882	60	0.05	-0.05608	0.049464	NA	0.003406	14793.52	2.935811
1.638568332	84.74004	78.42404	60	0.06	-0.05452	0.041485	NA	0.003384	16893.52	2.955347
1.687824115	84.5789	78.54664	60	0.07	-0.05309	0.034201	NA	0.003362	18993.52	2.974722
1.737402809	84.42097	78.64838	60	0.08	-0.05177	0.027557	NA	0.00334	21093.52	2.99396
1.787302135	84.26596	78.73088	60	0.09	-0.05056	0.021503	NA	0.003319	23193.52	3.013084
1.837520202	84.11364	78.79567	60	0.1	-0.04943	0.015991	NA	0.003298	25293.52	3.032118
1.888055505	83.96374	78.84416	60	0.11	-0.0484	0.010978	NA	0.003278	27393.52	3.051084
1.9389069	83.81607	78.87765	60	0.12	-0.04745	0.006424	NA	0.003257	29493.52	3.070002
1.990073602	83.67041	78.89737	60	0.13	-0.04656	0.002289	NA	0.003237	31593.52	3.088894
2.041555164	83.52658	78.90444	60	0.14	-0.04575	-0.00146	NA	0.003218	33693.52	3.107778
2.093351468	83.3844	78.89991	60	0.15	-0.045	-0.00486	NA	0.003198	35793.52	3.126674

Fig. 5.23 Spreadsheet simulation during melting phase of Joulie. A similar sheet is used for the Joulie freezing stage. The numerical step is taken with a fixed change in quality, "x", and the required time step "dt" is calculated from the governing equation for the quality. The initial conditions (time, temperatures) refer to the last row from the heating phase

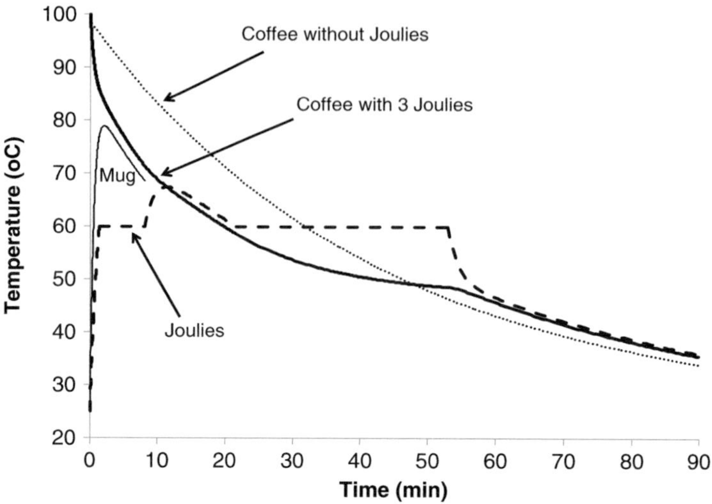

Fig. 5.24 Temperatures as a function of time for the case of a mug of coffee with 3 Joulies. A 2-node simulation (with additional coffee replacing the volume occupied by the Joulies) was conducted as a control with the same conditions

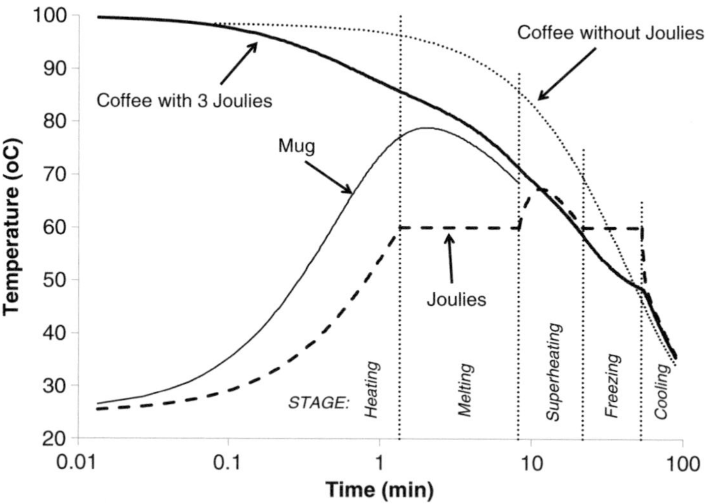

Fig. 5.25 The same data as Fig. 5.24 plotted on semi-logx axes. The boundaries for the stages undergone by the Joulies are shown as vertical dotted lines

Plots of temperature vs. time are shown in Figs. 5.24 (linear axes) and 5.25 (semi-logx axes). The 2-node coffee/mug model is shown as the control case. The mug curve is shown only during the Joulie heating and melting stages to reduce clutter in the plots. The mug follows the coffee (a few degrees lower) after that.

During the Joulie heating phase, the mug and Joulie temperatures rise quickly, the mug rising faster owing to its smaller time constant. The Joulies then reach the melting point and stay there until melting is complete. Thereafter, in the superheating phase, the Joulie temperature increases (as a liquid) toward the coffee, and then decay as the coffee continues to cool. The peak in the Joulie temperature coincides with the point where the coffee temperature equals the Joulie temperature, as there is no heat transfer during this time. After the peak, the coffee is colder than the Joulie, and heat flows from the Joulie to the coffee, returning energy stored to the coffee. Once the Joulie cools to its melting temperature, the liquid freezes to a solid until phase change is complete. The Joulie cools with the coffee after that.

Compared to the control curve (coffee without Joulies), the coffee cools quickly with Joulies, reaching a drinking temperature in a noticeably shorter time. The heat is retained for a slightly longer time, yielding a much longer time window.

For me, it's being able to drink hot coffee much sooner that sold me on the product. As I said *"I like Joulies so much, I wrote a book about it!"*

Appendix 1. Nodal Response Time

Figure 5.3, reproduced here, shows a node with a capacitance, two neighbors and heat generation. The inherent response time of this node is established by determining its response assuming its neighbors are fixed, not variable. In this case, T_1 will approach an equilibrium value that will depend on its two neighbors and the heat generation.

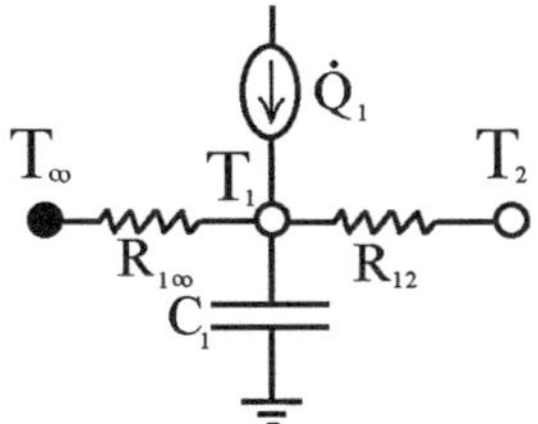

Repeating the governing nodal equation:

$$C_1 \frac{dT_1}{dt} = \dot{Q}_1 + \frac{T_2 - T_1}{R_{12}} + \frac{T_\infty - T_1}{R_{1\infty}}$$

A dimensionless temperature (ψ_1) that starts at a value of unity (1) and approaches a value of zero at equilibrium, and a dimensionless time defining a response time for node 1 ($\tau_1 = RC_{1equiv}$, to be determined) are defined:

$$\psi_1 = \frac{T_1 - T_{1eq}}{T_0 - T_{1eq}}$$

$$\hat{t} = \frac{t}{\tau_1} = \frac{t}{RC_{1equiv}}$$

Inserting into the nodal equation:

$$\frac{C_1}{\tau_1} \frac{d\left[T_{1eq} + \psi_1\left(T_0 - T_{1eq}\right)\right]}{d\hat{t}} = \dot{Q}_1 + \frac{T_{2fixed} - \left[T_{1eq} + \psi_1\left(T_0 - T_{1eq}\right)\right]}{R_{12}}$$

$$+ \frac{T_\infty - \left[T_{1eq} + \psi_2\left(T_0 - T_{1eq}\right)\right]}{R_{1\infty}}$$

Note that T_2 is given a "fixed" superscript. In the system, T_2 is a moving target, so this calculation is a hypothetical "what if." With T_2 and therefore T_{1eq} fixed, the first term under the time derivative is zero. Rearranging:

$$\frac{C_1\left(T_0 - T_{1eq}\right)}{\tau_1} \frac{d\psi_1}{d\hat{t}} = \left[\dot{Q}_1 + \frac{T_{2fixed} - T_{1eq}}{R_{12}} + \frac{T_\infty - T_{1eq}}{R_{1\infty}}\right]$$

$$- \left(T_0 - T_{1eq}\right)\psi_1 \left(\frac{1}{R_{12}} + \frac{1}{R_{1\infty}}\right)$$

The term in brackets [ ] is identically zero because the equilibrium value for T_1 is defined by setting the time derivative of the nodal equation to zero:

$$0 = \dot{Q}_1 + \frac{T_{2fixed} - T_{1eq}}{R_{12}} + \frac{T_\infty - T_{1eq}}{R_{1\infty}}$$

Then the factor $(T_0 - T_{1eq})$ cancels and the dimensionless equation is:

$$\frac{d\psi_1}{d\hat{t}} = -\frac{\tau_1}{C_1} \left(\frac{1}{R_{12}} + \frac{1}{R_{1\infty}}\right)\psi_1$$

The time constant can now be defined as:

$$\tau_1 = \frac{C_1}{\left(\frac{1}{R_{12}} + \frac{1}{R_{1\infty}}\right)} = R_{1equiv}C_1$$

and the governing equation becomes the first-order model:

$$\frac{d\psi_1}{dt} = -\psi_1$$

with solution:

$$\psi_1 = e^{-\hat{t}}$$

It is reemphasized that this temperature response is a hypothetical one, with T_2 fixed. The aim of this derivation is to determine the inherent response of a given node to its immediate environment. In a real simulation, T_1 would not approach this equilibrium value because T_2 is changing with time, unless the system itself has reached equilibrium.

Appendix 2. Detailed Thermal Resistance Formulas for Coffee Mug Problem

Figures 5.6 and 5.7 (repeated side by side) shows the input parameters (left) and derived quantities (right) from the "params" worksheet from "Ch5_2Node_Transient.xls." The formulas used in the tables on the right side (that reference the cells in the input quantities block) are detailed in this appendix.

INPUT QUANTITIES

D=	0.08255	m	Inner diameter of mug
H=	0.098	m	Inner height of mug
H0=	0.106	m	Outer height of mug
V=	4.10E-04	m^3	Coffee Volume
w=	0.003975	m	Mug thickness
mmug=	0.335	kg	Mug mass
kmug=	1	W/m/K	Mug thermal conductivity
rhomug=	1700	kg/m^3	Mug density
cmug=	800	J/kg/K	Mug specific heat
rhocoffee=	1000	kg/m^3	Coffee density
ccoffee=	4200	J/kg/K	Coffee specific heat
hair/sides=	13.2	W/m2/K	hrad+ hconv outer heat transfer coefficient
hair/top=	23.2	W/m2/K	hrad + hconv + hevap heat transfer coefficient for free surface
hfloor=	4.7	W/m2/K	effective floor coefficient
hcoffee/mug=	470	W/m2/K	convection coefficient between coffee and mug
hcoffee/surface=	470	W/m2/K	convection coefficient between coffee and free surface
Tcoffee, init=	92	oC	Temperature of coffee when poured
Tambient=	21	oC	Ambient temperature
Tmug,initial=	21	oC	Mug initial temperature

DERIVED QUANTITIES

H - L =	0.0766	m	Coffee depth
Cmug=	268	J/oC	Capacitance of mug
Ccoffee=	1722	J/oC	Capacitance of coffee
Afreesurface =	0.0054	m^2	coffee free surface area
Ainside=	0.0252	m^2	contact area between coffee and mug
Aoutside=	0.0432	m^2	contact area between mug and environment
Aave=	0.0342	m^2	averaged mug area

INDIVIDUAL RESISTANCES

Rcs=	0.3975	oC/W	convection between coffee and free surface
Rsinf=	8.0536	oC/W	conv/rad/evap between environment and free surface
Rcm=	0.0844	oC/W	convection between coffee and mug
Rm=	0.1162	oC/W	conduction across mug
Rma,side=	2.0605	oC/W	conv/rad between environment and mug
Rma,floor=	33.0762	oC/W	conduction between mug bottom and floor

MODEL RESISTANCES

R1inf=	8.4511	oC/W	between coffee node and environment
R12=	0.1425	oC/W	between coffee and mug nodes
R2inf=	1.9978	oC/W	between mug node and environment

The basic nomenclature of the geometry is repeated here, introduced previously in Chap. 2.

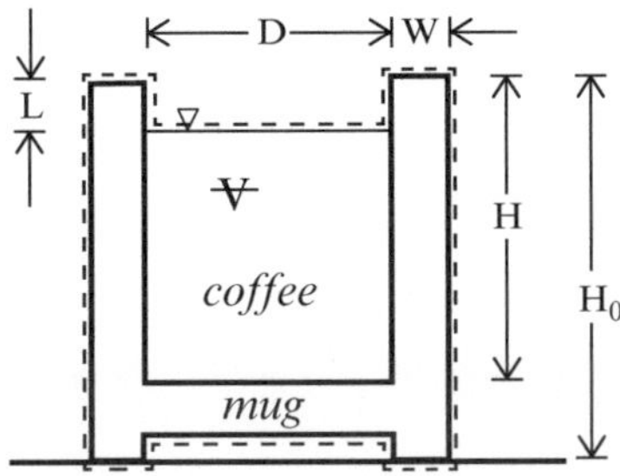

The coffee depth is:

$$H - L = \frac{\forall}{\pi D^2 / 4}$$

The mug and coffee capacitances are:

$$C_{mug} = m_{mug} c_{mug} \quad \text{and} \quad C_{coffee} = \rho_{coffee} c_{coffee} \, \forall$$

The area of the coffee-free surface is:

$$A_{free\ surface} = \pi D^2 / 4$$

The area of contact between the coffee and the mug is the sum of the bottom and sides:

$$A_{inside} = \pi D^2 / 4 + \pi D (H - L)$$

The total outside surface area of the mug (exposed to ambient) includes the outside, bottom (neglecting the bottom ridge area), the top rim surface, and the inside mug surface above the coffee level:

$$A_{outside} = Outside + Bottom + Rim + Inside$$

$$A_{outside} = \pi(D + 2w)H_o + \pi(D + 2w)^2 / 4 + \pi\left[(D + 2w)^2 - D^2\right] + \pi D L$$

The average mug surface area (used in the mug-conductive resistance) is the geometric average of the inside and outside areas:

$$A_{average} = (A_{inside} + A_{outside})/2$$

Individual Resistances

The RC network is repeated here for reference.

The convective resistance between the coffee and the mug is:

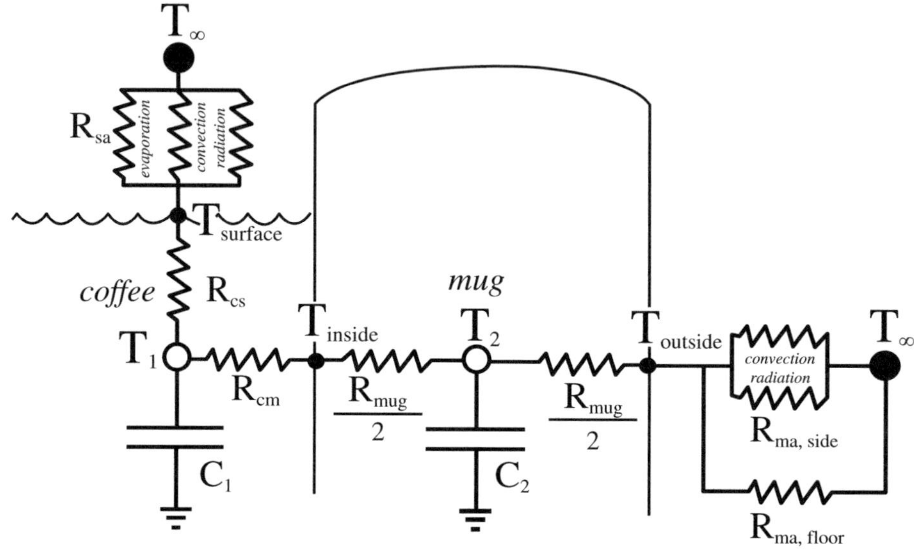

$$R_{cs} = \frac{1}{h_{coffee/surface}A_{top}}$$

The value chosen for the equivalent heat transfer coefficient ($h_{coffee/surface} = 470\ \mathrm{W/m^2/K}$) is representative of a natural convection between a liquid with water properties and a solid surface.

The equivalent convective/radiative/evaporative resistance between the coffee surface and ambient is:

$$R_{sa} = \frac{1}{h_{air/top}A_{top}}$$

The value chosen for the equivalent heat transfer coefficient is built on values of 6.8, 6.6, and 10 $\mathrm{W/m^2/K}$ for convection, radiation, and evaporation, respectively.

The convective resistance between the coffee and the mug inner surface is:

$$R_{cm} = \frac{1}{h_{cm}A_{inside}}$$

The value chosen for the convection coefficient (470 $\mathrm{W/m^2/K}$, with radiation considered negligible in comparison) is representative of natural convection between a liquid with water properties and a vertical surface.

The conductive resistance across the mug wall is:

$$R_m = \frac{w}{k_{mug}A_{average}}$$

This resistance is split into two equal parts between the capacity-carrying mug node and the mug surfaces. This model is considered reasonable for nearly full mugs. A detailed analysis of nearly empty mugs is deferred to a later chapter.

Heat transfer from the outer surface of the mug to ambient is split into two parallel channels: side and bottom. For the side, a convective/radiative equivalent resistance for those portions exposed to air is:

$$R_{ma,sides} = \frac{1}{h_{air/sides}\left(A_{outside} - \pi(D + 2w)^2/4\right)}$$

The area is the outside area subtracted by the surface area of the bottom. The equivalent heat transfer coefficient is the sum of the convective and radiative portions (8.8 and 8.6 $W/m^2/K$, respectively). For the heat transfer through the bottom:

$$R_{ma,floor} = \frac{1}{h_{floor}\pi(D + 2w)^2/4}$$

The effective heat transfer coefficient ($h_{floor} = 4.7$ $W/m^2/K$) is detailed in Chap. 4, and reflects heat that flows from the bottom of the mug to the floor it is placed on, and then conducted outward into the floor.

Appendix 3. Joulie Property Estimation

Resistance and capacitance value estimation requires assumptions about the construction of the Joulie, detailed in this appendix, and calculated in a separate worksheet (Fig. 5.26). The mass and volume of Joulies are easily measured (34.0 g and 20.6 mL). Thermal properties are obtained online (accessed 1 February 2014), with convection coefficients inside and outside entered as representative values.

The Joulies are modeled as equivalent spheres. That is, the equivalent outer diameter is calculated from the measured volume as:

$$d = \left(\frac{6V}{\pi}\right)^{1/3}$$

MEASURED VALUES

mass	0.028	kg	Single Joulie mass
Vtotal	2.06E-05	m^3	Single Joulie volume

THERMAL PROPERTIES (STEEL SHELL, PARAFFIN WAX CORE)

kshell	20	W/m/K	http://www.engineeringtoolbox.com/thermal-conductivity-metals-d_858.html
rhoshell	8000	kg/m^3	http://www.engineeringtoolbox.com/metal-alloys-densities-d_50.html
cpshell	490	J/kg/K	http://www.engineeringtoolbox.com/specific-heat-metals-d_152.html
kwax	0.25	W/m/K	http://www.engineeringtoolbox.com/thermal-conductivity-d_429.html
rhowax	721	kg/m^3	http://www.engineeringtoolbox.com/density-materials-d_1652.html
cpwax	2500	J/kg/K	http://en.wikipedia.org/wiki/Paraffin_wax
hCJ	250	W/m2/K	convection bet. Coffee and Joulie
hwax	200	W/m2/K	convection bet. Wax and shell

DERIVED PROPERTY VALUES

d	0.034	m	equivalent diameter of single Joulie
dwax	0.0330	m	diameter of wax core
Atotal	3.63E-03	m^2	surface area
Awax	3.42E-03	m2	surface area of wax
Vwax	1.88E-05	m^3	volume of wax core
tshell	5.13E-04	m	shell thickness
mwax	0.0135	kg	mass of wax
mshell	0.0145	kg	mass of shell
rhoave	1361	kg/m^3	average density
cpave	1462	J/kg	average heat capacity

THERMAL RESISTANCES OF SINGLE JOULIE

Rwax	19.31	oC/W	conductive resistance across wax
Rwax	1.46	oC/W	Convective inner wax layer
Rshell	0.01	oC/W	conductive resistance across shell
Rconv	1.10	oC/W	convective resistance
Rcj	2.57	oC/W	equivalent resistance between single Joulie and coffee

Fig. 5.26 Input spreadsheet focused on property values for a single Joulie used as inputs to the simulation input sheet

The diameter of the wax (and thereby the wall thickness) is determined by summing the total mass from wax core plus shell:

$$m_{Total} = \rho_{shell} V_{shell} + \rho_{wax} V_{wax} = \rho_{shell}(V_{Total} - V_{wax}) + \rho_{wax} V_{wax}$$
$$= \rho_{shell} V_{Total} - V_{wax}(\rho_{shell} - \rho_{wax})$$

Solving for the wax volume:

$$V_{wax} = \frac{\rho_{shell} V_{total} - m_{total}}{\rho_{shell} - \rho_{wax}}$$

$$d_{wax} = \left(\frac{6V_{wax}}{\pi}\right)^{1/3}$$

The total specific heat (during single phase stages) is:

$$C_{Joulie} = (mc)_{Total} = \rho_{shell} V_{shell} C_{shell} + \rho_{wax} V_{wax} C_{wax}$$

Yielding effective specific heat:

$$c_{ave} = \frac{\rho_{shell} V_{shell} C_{shell} + \rho_{wax} V_{wax} C_{wax}}{\rho_{shell} V_{shell} + \rho_{wax} V_{wax}}$$

For the resistances, there are two separate entries for R_{wax}, representing the thermal resistance between the wax and the inner shell wall. The first is a pure conduction value and the second is assuming a convective mode. The conduction value is much larger than the convective one and also large compared to the outer convection plus conduction across the shell wall (which is negligible). That means that the Biot number of the wax is large, and during heating, the center of the wax will respond much more slowly than its outer layers. Use of the convection resistance is considered a better way to estimate the rate at which heat is absorbed by the Joulie. A more detailed spatial model would be the way to address this issue (and a great assignment!).

These calculations are for a single Joulie. Note that when multiple Joulies are used, the total capacitance is proportional to the number of Joulies, while the thermal resistance is inversely proportional. Adding more Joulies increases the total surface area, reducing the resistance of heat flow. That is:

$$C_{3\ Joulies} = 3C_{single\ Joulie} \quad \text{while} \quad R_{3\ Joulies} = \frac{1}{3}R_{single\ Joulie}$$

I'd bet that lots of people would not think that through and assume $R_{3\ Joulies} = 3R_{1\ Joulie}$, and in so doing confuse a parallel heat transfer concept with a series effect.

Workshop 5.1. A Batch Mode Solar Water Heater

An alternative to burning fossil fuels to produce hot water for domestic needs is to heat water using solar energy. It is a common experience that placing a garden hose in direct sunlight can produce hot water. It is also a common experience that sunlight can be focused using a magnifying glass. Alternatively, a reflecting mirror can be used to focus sunlight. This workshop applies these principles to a specific arrangement in which a "solar pipe" is filled with initially cold water and placed in focused sunlight. The water is heated (without a flow of water) until the water reaches a target temperature, at which point a pump is activated (say by a solar photovoltaic cell) and the hot water is pumped out of the pipe into a separate tank, and the pipe is filled with a fresh batch of cold water, completing the batch cycle. The time it takes to complete a cycle depends on the angle and intensity of sunlight which varies throughout the course of a day.

An alternative mode, called continuous mode, is to use the same arrangement but with a continuous flow of water through the solar pipe. That problem will be studied much later, after the effects of advective flows are developed.

Problem statement: In a solar thermal application (Fig. 5.27), black polypropylene tubing ($k_p = 0.15$ W/m/°C, $\rho_p = 1{,}400$ kg/m^3, $c_p = 1{,}600$ J/kg/°C) of length $L = 3$ m, outer diameter (D_0) of 0.040 m, and wall thickness $t = 0.0028$ m is subjected to focused sunlight distributed evenly on the outer surface (concentrated with a parabolic mirror). The intensity of sunlight that enters the top plane (the solar insolation) has a maximum value of $I_{0,max} = 800$ W/m^2 (on a clear day at noon) and a solar collection width to rod diameter ratio (n) to be determined as a design parameter. An optical efficiency $\eta_{opt} = 0.9$ (defined as the ratio of the solar radiation that strikes the pipe to the collected radiation) accounts for collected solar energy not absorbed by the pipe. In the end view, the angle of the sun remains perpendicular to the collecting plane through proper orientation of the mirror (and changes gradually through the course of a year). From the side view, the angle between the incoming solar rays and the perpendicular to the pipe varies through the course of a day, from $\theta° = 90°$ at sunrise to $0°$ at noon to $90°$ at sunset. The absorption coefficient of incident radiation is $\alpha = 0.85$ (85 % of the radiation that strikes the pipe is absorbed, 15 % is reflected). At the start of a cycle, the tube is filled with liquid water ($\rho_w = 1{,}000$ kg/m^3, $c_w = 4{,}200$ J/kg/K) at a temperature $T_0 = 15$ °C. The tube is exposed to ambient at $T_\infty = 25$ °C. The objective is to raise the temperature of the water to a target of $T_{final} = 50$ °C before pumping it into a separate collection tank and starting a new cycle.

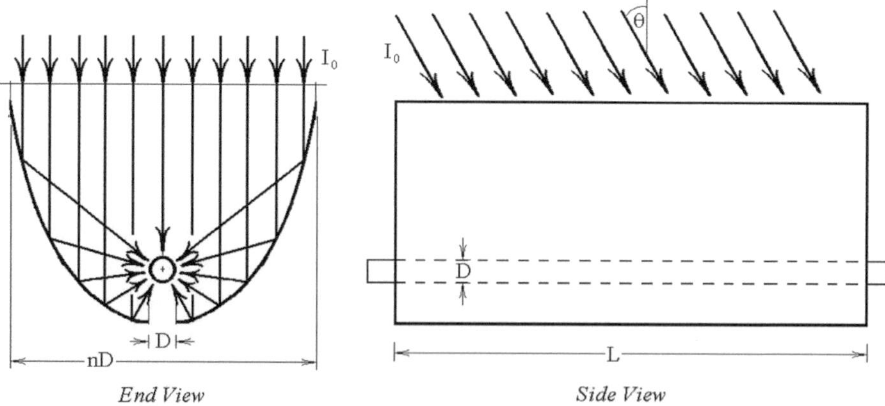

Fig. 5.27 Schematic of a solar pipe placed at the focal point of a parabolic mirror

Write an Excel spreadsheet or program code that implements an explicit numerical formulation of a 2-node model (tube and water). Assume the inner surface is exposed to the water with a constant convection coefficient of $h_i = 300$ W/m^2/°C, and the outer surface is exposed to ambient with a constant convection coefficient of $h_o = 15$ W/m^2/°C. Use the program to determine the size of the solar collection area (nDL) that would cause the outer surface of the pipe to reach its melting point ($T_{melt} = 160$ °C) at the same time the water reaches its target of 50 °C when the solar flux equals its maximum value at noon. Then determine the time it takes to heat a fresh batch of water to its target as a function of solar angle (which can be related to time of day) for fixed values of solar flux. Report a minimum intensity required to heat the water to the target temperature, the maximum intensity before the outer surface of the pipe reaches its melting point ($T_{melt} = 160$ °C), and the size of the solar collection area (nD), assuming the maximum intensity of direct sunlight is 800 W/m^2 (a typical maximum value on a clear day at noon). Prepare a table of relevant quantities and plots of

- required heating time as a function of intensity (above the minimum required)
- all nodal temperatures (including non-capacity-carrying nodes) vs. time with maximum intensity
- temperature vs. radial position at fixed times with maximum intensity

Note: A specific 2-node model is developed on the next page. It is advised that you develop your own model first before consulting that model. Then decide which model to implement.

Model Development

The 2-node thermal RC network shown in Fig. 5.28 treats node 1 as the water and node 2 as the pipe. The pipe node (2) is placed in the middle of the pipe wall. The solar absorption is modeled as a heat source absorbed on the surface (not directly on the capacity-carrying node). The surface node is in thermal contact with ambient (through a parallel convective/radiative channel with an assumed constant effective heat transfer coefficient) and the pipe center (through a conductive channel). The pipe is in thermal contact with the water (through a series conductive/convective resistance).

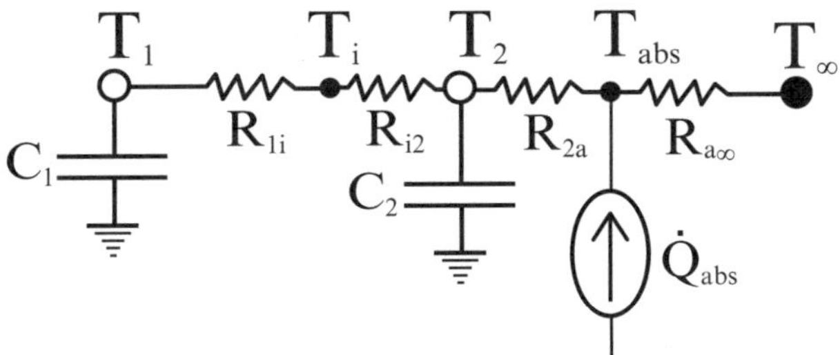

Fig. 5.28 2-Node RC network for a solar pipe in batch mode

The nodal equations are (water $= 1$, pipe $= 2$, pipe outer surface $=$ abs):

$$C_1 \frac{dT_1}{dt} = \frac{T_2 - T_1}{R_{1i} + R_{i2}}$$

$$C_2 \frac{dT_2}{dt} = \frac{T_1 - T_2}{R_{1i} + R_{i2}} + \frac{T_{abs} - T_2}{R_{2a}}$$

The solar flux is absorbed at the outer surface, not at the center, and a nodal balance on the surface node is:

$$0 = \dot{Q}_{abs} + \frac{T_2 - T_{abs}}{R_{2a}} + \frac{T_\infty - T_{abs}}{R_{a\infty}}$$

Expressing the time derivatives with a forward difference approximation and rearranging for the recursion relations, the future values (superscript "p + 1", after a time step Δt) are explicitly expressed in terms of present values (superscript "p"):

$$T_1^{(p+1)} = T_1^{(p)} + \frac{\Delta t}{C_1} \left(\frac{T_2^{(p)} - T_1^{(p)}}{R_{1i} + R_{i2}} \right)$$

$$T_2^{(p+1)} = T_2^{(p)} + \frac{\Delta t}{C_2} \left[\left(\frac{T_1^{(p)} - T_2^{(p)}}{R_{1i} + R_{i2}} \right) + \left(\frac{T_{abs}^{(p)} - T_2^{(p)}}{R_{2a}} \right) \right]$$

After rearrangement, the temperature at the absorbing surface is expressed as:

$$T_{abs}^{(p)} = \left(\dot{Q}_{abs} + \frac{T_2^{(p)}}{R_{2a}} + \frac{T_\infty}{R_{a\infty}} \right) \bigg/ \left(\frac{1}{R_{2a}} + \frac{1}{R_{a\infty}} \right)$$

The time constants for the two nodes are given by:

$$\tau_1 = (R_{1i} + R_{i2})C_1$$

and

$$\tau_2 = \frac{C_2}{\dfrac{1}{R_{1i} + R_{i2}} + \dfrac{1}{R_{2a} + R_{a\infty}}}$$

The time step can be chosen to be a fraction of the smallest of these.

The rate of heat absorption is defined by:

$$\dot{Q}_{abs} = \begin{pmatrix} \text{fraction of} \\ \text{collected radiation} \\ \text{that strikes pipe} \end{pmatrix} \begin{pmatrix} \text{fraction of} \\ \text{radiation that strikes} \\ \text{pipe that is absorbed} \end{pmatrix} \begin{pmatrix} \text{Rate of Solar} \\ \text{Radiation} \\ \text{Collected} \end{pmatrix}$$

$$\dot{Q}_{abs} = (\eta_{opt})(\alpha)(nDLI_0 \cos \theta)$$

Workshop 5.2. Cooper Cooling with a Finite Reservoir

In Workshop 4.2, various methods of chilling a beverage were considered, the fastest of which involved the process of Cooper Cooling™ in which a horizontal beverage is rotated along its axis while being sprayed with a cold jet of water. In that problem statement, the temperature of the water jet was a fixed input value, a result that would occur if the source of the water was a large (infinite) reservoir of ice and water, and/or the water heated by the beverage was not returned and recycled. However, the first marketed consumer product that incorporated the Cooper Cooling process has a finite reservoir. Jet temperature data and a schematic of the process are shown in Fig. 5.29, taken from Jaime Cachiero's Master's Thesis. The cold water jet is drawn from an ice/water bath and is pumpcd onto the beverage. Heat transferred from the beverage is added to the jet before the jet is returned to the ice water bath. Furthermore, a large portion of the electrical input to the pump is added to the recirculating water. In the reservoir, the water is in contact with ice, which removes heat from the water jet. Since the beverage temperature changes with time, the rate at which heat is added to the jet changes. In practice, during the first four or so beverages, the jet temperature fluctuates around approximately 2 °C (the input value used for Workshop 4.2). The goal of this workshop is to develop a 3-node model (beverage, reservoir, ice) that predicts this behavior and can be used to explore other scenarios.

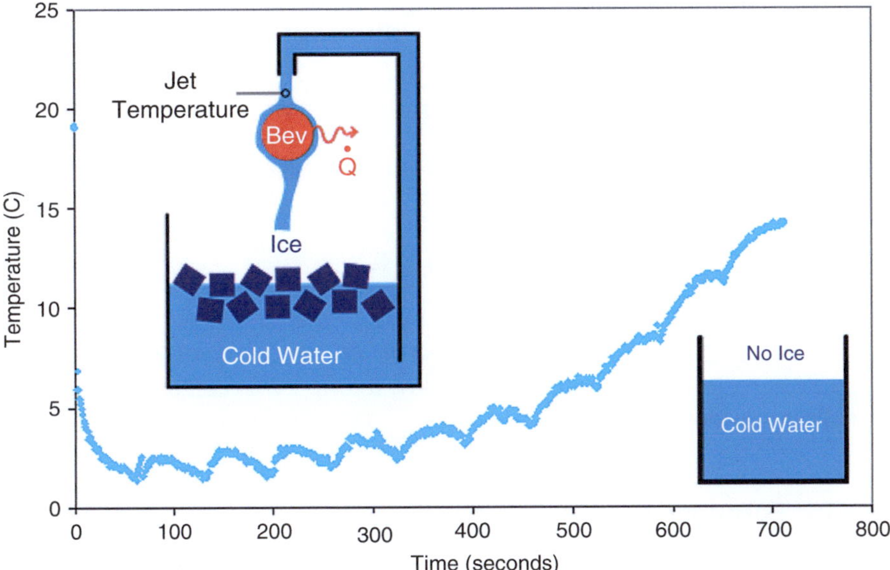

Fig. 5.29 Schematic of Cooper Cooling with a Finite Reservoir. The data shows the measured temperature of the water jet before it is used to chill a horizontal rotating beverage. A warm beverage is added every 60 s (copyright Wei Dai, reproduced with permission)

Note: A specific 3-node model is developed on the next page. It is advised that you develop your own model first before consulting that model. Then decide which model to implement.

Model Development

It is good practice to draw a thermal resistance network for this problem in which capacitance is assigned to the beverage and reservoir, and a rechargeable battery at the melting point of the reservoir (T_{mp}) is used to represent the ice. The mass of the ice, but not its temperature, changes as it melts. Notice that if a reservoir fluid other than water is used (like salt water), the melting point can differ from that of water. Also, the capacitance of the reservoir changes with time in a predictable way as ice melts. The composition of the reservoir might change if a reservoir other than water were used.

The governing equation for the beverage is the same as that in Workshop 4.2, namely:

$$C_{bev} \frac{dT_{bev}}{dt} = \frac{T_{res} - T_{bev}}{R_{bev}}$$

The difference in this model is that the reservoir temperature is not a fixed input, and this differential equation cannot be integrated directly.

The liquid water in the system is considered to be at an average temperature (T_{res}) that exchanges heat with the beverage (at T_{bev}), the ice (at T_{mp}, the melting point of the reservoir), and ambient (at T_∞) and picks up electrical energy that is converted to thermal energy in the pump (VI_{pump}):

$$C_{res} \frac{dT_{res}}{dt} = \frac{T_{bev} - T_{res}}{R_{bev}} + \frac{T_{mp} - T_{rex}}{R_{ice}} + \frac{T_\infty - T_{res}}{R_\infty} + VI_{pump}$$

The energy stored in the ice is in the form of latent (phase) energy, and an energy balance on the ice is given by:

$$L_f \frac{dM_{ice}}{dt} = \frac{T_{res} - T_{mp}}{R_{ice}}$$

L_f is the heat of fusion of ice (330 kJ/kg) and the thermal resistance between the ice and the reservoir (R_{ice}) is one of external flow forced convection on a surface area that changes with time.

Explicit Formulation

The nodal equations written in the form of an explicit formulation are with superscripts used to indicate all values that change with time:

$$T_{bev}^{(p+1)} = T_{bev}^{(p)} + \frac{\Delta t^{(p)}}{C_{bev}}\left(\frac{T_{res}^{(p)} - T_{bev}^{(p)}}{R_{bev}}\right)$$

$$T_{res}^{(p+1)} = T_{res}^{(p)} + \frac{\Delta t^{(p)}}{C_{res}^{(p)}}\left(\frac{T_{bev}^{(p)} - T_{res}^{(p)}}{R_{bev}} + \frac{T_{mp} - T_{res}^{(p)}}{R_{ice}^{(p)}} + \frac{T_\infty - T_{res}^{(p)}}{R_\infty} + VI_{pump}\right)$$

$$M_{ice}^{(p+1)} = M_{ice}^{(p)} + \frac{\Delta t^{(p)}}{L_v}\left(\frac{T_{res}^{(p)} - T_{mp}}{R_{ice}^{(p)}}\right)$$

The reservoir capacitance equals the specific heat of the reservoir times the mass of the reservoir, which equals the initial mass of the reservoir ($M_{res,0}$) plus the mass of ice that has melted (the initial mass of ice less the remaining ice, $M_{ice,0} - M_{ice}$):

$$C_{res}^{(p)} = c_{res}\left(M_{res,0} + M_{ice,0} - M_{ice}^{(p)}\right)$$

The thermal resistance that models the melting of the ice requires an assumption about the shape of the ice, and how that changes. For example, if ice cubes (actual cubes of dimension "w") are used, and it is assumed that they remain as cubes (even though in practice the edges become rounded), then the surface area can be related to the mass. That is, if a total of n_{cubes} are added initially of dimension w_0, the initial mass of ice is given by:

$$M_{ice,0} = n_{cubes}w_0^3$$

As the ice melts, the number of cubes does not change, but their size does, so that the remaining mass of ice can be related to the ice cube size by:

$$M_{ice} = n_{cubes}w^3$$

The thermal resistance between the ice and the reservoir is given by:

$$R_{ice}^{(p)} = \frac{1}{h_{ice}A_{ice}^{(p)}} = \frac{1}{h_{ice}n_{cubes}6(w^{(p)})^2}$$

Notice that the thermal resistance increases as the ice cubes become smaller. A value of 1,500 W/m^2/K would be reasonable for the convection coefficient in this case.

The values for the beverage capacitance and thermal resistance can be taken from Workshop 4.2, and simulations with different beverages can be conducted. The electrical input to the pump can be varied, but $VI_{pump} = 20$ W is a good starting place.

There are a variety of scenarios that can be used to run a simulation. For example, a prechilling period can be conducted in which case a number of ice cubes are placed into an input mass of reservoir at an initial temperature, without actively chilling a beverage, and letting the initial ice/water reservoir come to equilibrium before adding beverages. A decision about how long to chill each individual beverage can be made and that simply involves resetting the temperature of the beverage. Chilling consecutive beverages can be simulated by either fixing the chilling time or allowing the system to operate until the beverage reaches a target temperature. In the former case, the final temperature of each beverage will be different. In the latter case, the total time required to chill each beverage will be different. Eventually, the reservoir will not be able to chill a beverage to that target.

Additional Workshop Ideas

Workshop 5.2: Solve the 2-node mug of coffee problem using a variable step-size algorithm that fixes the maximum change in temperature (absolute value) of either the mug or coffee.

Workshop 5.3: Solve the 2-node mug of coffee problem using a variable step-size algorithm that fixes the maximum fractional change in temperature of either the mug or coffee.

Workshop 5.4: Solve the 2-node mug of coffee problem using the backward implicit method with a fixed time step.

Workshop 5.5: Solve the 2-node mug of coffee problem using a Styrofoam cup.

Workshop 5.6: Solve the 2-node mug of coffee problem using the explicit numerical scheme with variable thermal resistances on the side channel. Assume an emissivity of 0.9 for the radiation coefficient, and use the following formulas (derived in a future chapter) for the convection coefficients. For convection between the outer surface of the mug and ambient:

$$h_{air} = \left(1.5646 - 0.0015 T_{film,air}\right) \frac{\left|\left(T_{mug} - T_{ambient}\right)\right|^{0.25}}{H_0^{0.25}}$$

where $T_{film,air}$ is the average of the mug and ambient. For convection between the inner surface of the mug and the coffee:

$$h_{coffee} = \left(32.38 T_{film,coffee}^{0.441}\right) \frac{\left|\left(T_{coffee} - T_{mug}\right)\right|^{0.25}}{H_0^{0.25}}$$

where $T_{film,coffee}$ is the average of the coffee and mug. Compare this model with the model any of the constant convection coefficient models (and the data).

Workshop 5.7: A hybrid implicit method

For a stiff problem ($\tau_2 \ll \tau_1$), a legitimate approach is to use a forward difference approximation for the node with the larger response time (i.e., coffee) and a backward difference for the node with the shorter response time (mug). The future value of T_1 is obtained (from before) as:

$$T_1^{(p+1)} = T_1^{(p)} + \frac{\Delta t^{(p)}}{C_1^{(p)}} \left[\dot{Q}_1^{(p)} + \frac{T_2^{(p)} - T_1^{(p)}}{R_{12}^{(p)}} + \frac{T_\infty^{(p)} - T_1^{(p)}}{R_{1\infty}^{(p)}} \right]$$

The backward difference equation for T_2 is rearranged as:

$$T_2^{(p+1)} \left(\frac{C_2^{(p+1)}}{\Delta t^{(p+1)}} + \frac{1}{R_{12}^{(p+1)}} + \frac{1}{R_{2\infty}^{(p+1)}} \right) = \dot{Q}_2^{(p+1)} + \frac{C_2^{(p+1)} T_2^{(p)}}{\Delta t^{(p+1)}} + \frac{T_1^{(p+1)}}{R_{12}^{(p+1)}} + \frac{T_\infty^{(p+1)}}{R_{2\infty}^{(p+1)}}$$

Since T_1 changes slowly compared with T_2, then for one time step, $T_1^{(p+1)} \approx T_1^{(p)}$. Furthermore, when the coefficients vary with time, they will do so slowly, and the R and C values can be evaluated at the present, not the future time. Therefore, a legitimate explicit prediction of the future temperature of T_2 that is inherently stable numerically (let $\Delta t \to \infty$) is:

$$T_2^{(p+1)} = \frac{\dot{Q}_2^{(p)} + \dfrac{C_2^{(p)} T_2^{(p)}}{\Delta t^{(p)}} + \dfrac{T_1^{(p)}}{R_{12}^{(p)}} + \dfrac{T_\infty^{(p)}}{R_{2\infty}^{(p)}}}{\dfrac{C_2^{(p)}}{\Delta t^{(p)}} + \dfrac{1}{R_{12}^{(p)}} + \dfrac{1}{R_{2\infty}^{(p)}}}$$

For this scheme, the time step can be chosen to be a value greater than the Second Law restriction on a full explicit scheme.

Workshop 5.8: 2-node Electric Water Heater:

An electrically resisting element made of nichrome (thermal conductivity $k_{element} = 11.0$ W/m/K, density $r_{element} = 8{,}400$ kg/m^3, specific heat $c_{element} = 450$ J/kg/K) of diameter $d = 0.015$ m with a total length to be determined is designed to be immersed in 1.0 L of liquid water in a cylindrical container of inner diameter $D = 0.15$ m with a thin wall (negligible thermal capacity and conductive resistance). The water is initially at 15 °C and electrical power at the rate of $Q = 1{,}000$ W is delivered to the heating element.

Conduct a 2-node analysis (element and water). Assume that the heat transfer coefficient between the element and the water, and between the water and the container is $h_{element,water} = h_{water,container} = 400$ W/m^2/K, the convective/radiative coefficient between the outer wall and ambient is $h_{container,air} = 15$ W/m^2/K.

Determine:

1. The length of the heating element needed so that the water temperature is 90 °C at the point the water starts to boil (i.e., reaches 100 °C) in the thermal boundary layer between the element and the water and the corresponding time required to boil. Note: if heated beyond this point, most of the heat will go toward boiling the water in that boundary layer, and not to raising the temperature of the bulk of the liquid. That is to say, it is difficult to obtain water hotter than 90 °C for this heating element at this heating rate.
2. The time to start to boil and the final water temperature obtained for this length heating element, but with a heating power of 500 W.
3. The minimum power that would yield 90 °C water and the corresponding time and the temperature of the element at that time.

Multi-Node Transients

6

This chapter builds on the few-node numerical approach, detailing the side wall of the mug of coffee problem, which is where the majority of heat is lost (60 % of the surface area exposed to ambient). Results are generated only for the mug heating phase, as computationally, simulation of the entire cooling event would require several large spreadsheets. Conceptually, multi-node problems involve a simple extension of the principles developed for the 2-node mug of coffee model. For example, adding Coffee Joulies™ added a third node. This type of problem is generally solvable with a spreadsheet approach for an explicit formulation (but not for an implicit one).

However, a subtle distinction is made between a few-node and a multi-node problem. In programming a multi-node simulation, some or all of the recursion formulas can be written in terms of an input numerical parameter, say "n". For example, the mug side wall will be split into "n" nodes. The governing equations will be derived in terms of "n". However, for spreadsheet solutions, as detailed in this chapter, a commitment to a specific number must be made at the outset (set up the number of columns). Technically, then, this is a "few-node" approach, but a spreadsheet implementation is developed here to demonstrate the details of the method.

Numerical modeling in heat transfer involves discretization of both space and time. In the 1-node model, space was discretized by lumping all the mass of the coffee/mug system into a single node. If the thermal resistances are taken as constants, the governing ODE is easy to integrate and it is not necessary to discretize time. For variable resistances (natural convection and radiation boundaries, developed in future chapters), time can be discretized by turning the time derivative into an approximate algebraic expression using finite difference approximations, and the transient problem can be simulated by appropriately stepping forward in time. In the 2-node model, space was discretized into two lumped pieces; coffee and mug. Time was discretized using a finite difference approximation. This chapter introduces a more general approach by breaking space into many discrete nodes.

© Springer International Publishing Switzerland 2015

G. Sidebotham, *Heat Transfer Modeling: An Inductive Approach*,

DOI 10.1007/978-3-319-14514-3_6

The model developed is spatially one-dimensional and is focused primarily on the heat transfer process across the mug side wall. Edge effects near the rim and base are not modeled. The heat loss through the free surface and bottom are modeled as they were for the extended lumped model. An extension to 2D and 3D is relatively straightforward, conceptually. In this model, the convection and radiation coefficients depend on temperature, and therefore vary with time. The convection formulas are developed in a future chapter, but those results are used here, much the same way that constant values of h have been used throughout. The radiation coefficients have been developed already.

Results for the coffee/mug problem are presented. The model implementation was done with a rather large Excel file, written so that all parameters could be changed to simulate other problems that fit the same basic model of a cylindrical container holding a liquid at a temperature different from a constant ambient temperature. There are good modeling lessons at play. This more detailed model requires more considerable effort, both in setup and in computation, than the 2-node model, yet the key results are virtually identical. That is not to say more detailed models are never needed. It is just demonstrated here that they are overkill for the cooling mug of coffee example.

Problem Statement

The first part of the problem statement, the description of the mug of coffee and its dimensions, is exactly the same as the original one. The objective stated is different:

A volume $V = 0.41$ L of coffee at 1 atm, 92 °C is quickly poured into a room temperature (23 °C) ceramic mug with mass of $m_{mug} = 0.335$ kg, inner diameter $D = 0.083$ m, inner height $H = 0.098$ m, and nominal wall thickness $w = 0.0040$ m. The mug of coffee is placed on a table in a still room. Determine the temperature distribution in the mug side wall as a function of time.

6.1   Multi-Node Model Development

Figure 6.1 shows an RC network for the mug of coffee. The coffee itself is modeled with a single node, with three parallel channels for heat flow: top (free) surface, bottom, and side wall. The side wall has been broken up into a total of "n" nodes, where n will be set to 8 for the case study of the mug of coffee implemented in an Excel file. Each node represents a cylindrical shell and has a capacitance associated with it. The sum of all "n" capacitances is the total capacitance of the side wall of the mug. Heat transfer out of the top rim and bottom of the side wall is neglected. Thermal resistances between each node represent a conductive thermal channel between them. The coffee (at temperature $T(t)$) has a capacitance C that consists of the capacitance of the coffee plus the capacitance of the bottom wall lumped into it. This portion of the mug capacitance is easy to account for by lumping it with the coffee, rather than including it with the mug capacitance as was done in the 2-node mug of coffee example.

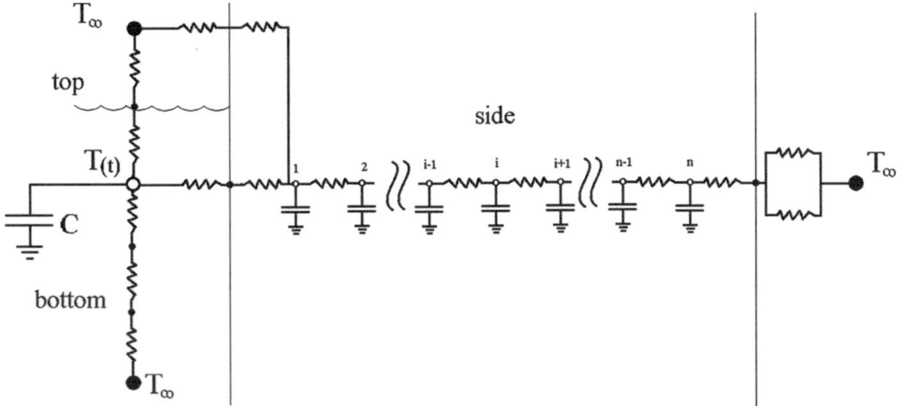

Fig. 6.1 RC network for the mug side wall discretized into n-nodes

The two end nodes (1 and n) are not placed on the mug surface, but rather are placed inside the volume of the section of the mug it represents. Therefore, the inner and outer surfaces of the mug are shown as filled circles. These are surfaces, not volumes, and therefore have no capacitance associated with them. Node 1 has two parallel channels: one between the node and the coffee and the other between the node and the air inside the rim. Each of these channels consists of a convective and a conductive resistance in series. This detail is a result of a "define volumes, then place nodes" or finite volume approach, as opposed to a "place nodes, then define volume" approach typical of a finite difference method. It gives additional information (really the accuracy of $n+2$ nodes) plus added numerical stability for the explicit method to be used. Notice that the distance between the inner surface and node 1 is approximately half the distance between node 1 and 2, so these resistance values differ by a factor of about 2. It is not exactly 2 due to the radial geometry. The last node (n) has a conductive resistance to the outer surface in series with a convective/radiative parallel channel.

6.2   Defining the Grid (i.e., Discretization of Space)

There are different ways that the nodal volumes can be defined. There is not necessarily a "right" way to do it, although the method should be self-consistent. The approach taken here is to define all the nodes as cylindrical shells of equal height (from the floor to the rim, H_0) and equal capacitance. As a result, the spacing between nodes decreases from the inside to the outside due to the radial geometry. An alternative approach would be to space the nodes equally, and then their volumes would increase with radial position. In this control volume approach to the spatial discretization, the locations of the element boundaries are first defined by the chosen volumes, and then each node is placed in the center of the volumes

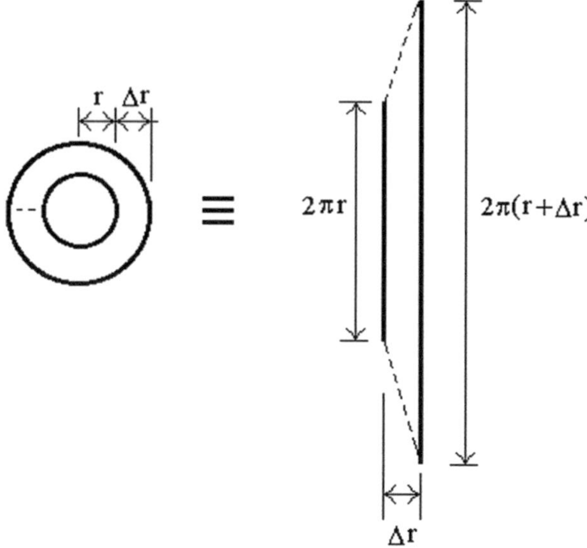

Fig. 6.2 Top view of a cylindrical shell and its equivalent as a rectangular wall of variable width

inside. Conceptually, the node could be placed at the centroid of the nodal element (for non-square elements).

Figure 6.2 shows a top view of a cylindrical shell imagined to be cut and opened up to as a wall of uniform thickness (Δr) and a width that varies from $2\pi r$ to $2\pi(r + \Delta r)$. The height (into the page) is also uniform (H_0).

Figure 6.3 defines the geometry for node 1. The distances to the radial position of the boundary of the element are shown above (r_{1-} and r_{1+}), and the distance to the node is shown below (r_1). The distance to the inner boundary is simply the radial distance from the centerline to the inner surface of the mug:

$$r_{1-} = \frac{D}{2}$$

In this particular approach, the volumes of each element are constrained to be equal to each other, namely the total volume of the mug side wall divided by the number of nodes (n):

$$V_{node} = \frac{\pi\left[(D + 2w)^2 - D^2\right]H_0}{4n}$$

The height of the node is taken to be the measured outer height of the mug (H_0). Conceptually, that would be the inner height plus the mug thickness if the mug were uniformly thick. However, the actual mug under investigation has a bottom rim, and the exterior height is therefore a little larger. The distance to the outer boundary of node 1 is also the inner boundary of node 2 and is determined by defining its volume:

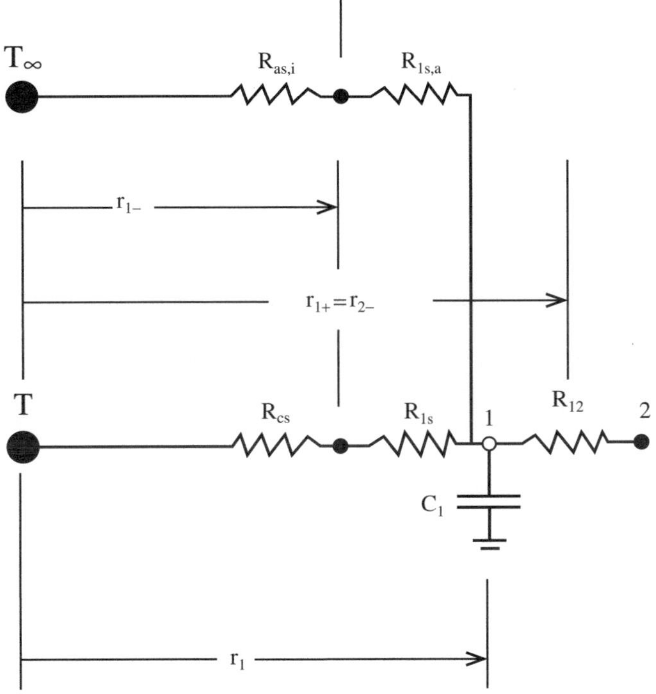

Fig. 6.3 Geometry of node 1, the inner surface node

$$V_{node} = \pi\left[r_{2-}^2 - r_{1-}^2\right]H_o$$

Solving for the radial position of the boundary between node 1 and 2:

$$r_{2-} = \sqrt{r_{1-}^2 + \frac{V_{node}}{\pi H_o}}$$

The node itself is placed at the centroid of the volume, defined by:

$$r_1 A = \int_{r_{1-}}^{r_{2-}} r\,dA = \int_{r_{1-}}^{r_{2-}} r(2\pi r\,dr) = 2\pi\frac{r^3}{3}\bigg|_{r1-}^{2-} = \frac{2\pi}{3}\left(r_{2-}^3 - r_{1-}^3\right)$$

$$r_1 = \frac{2\left(r_{2-}^3 - r_{1-}^3\right)}{3\left(r_{2-}^2 - r_{1-}^2\right)}$$

Alternatively, simply placing the node halfway between would actually be just as good. . .

Consider next node 2. Its volume is the same as node 1 (V_{node}) and is defined by:

$$V_{node} = \pi\left[r_{3-}^2 - r_{2-}^2\right]H_o$$

Inserting the result for r_{2-} from node 1 analysis:

$$r_{3-}^2 = \frac{V_{node}}{\pi H_o} + r_{2-}^2 = \frac{V_{node}}{\pi H_o} + \frac{V_{node}}{\pi H_o} + r_{1-}^2$$

$$r_{2+} = r_{3-} = \sqrt{\frac{2V_{node}}{\pi H_o} + r_{1-}^2}$$

The location of node 2 (the centroid) is:

$$r_2 = \frac{2\left(r_{3-}^3 - r_{2-}^3\right)}{3\left(r_{3-}^2 - r_{2-}^2\right)}$$

The results for node 3 are:

$$r_{3+} = r_{4-} = \sqrt{\frac{3V_{node}}{\pi H_o} + r_{1-}^2}$$

and

$$r_3 = \frac{2\left(r_{4-}^3 - r_{3-}^3\right)}{3\left(r_{4-}^2 - r_{3-}^2\right)}$$

For the general interior node "i":

$$r_{i+} = r_{(i+1)-} = \sqrt{\frac{iV_{node}}{\pi H_o} + r_{1-}^2}$$

The location of the centroid is:

$$r_i = \frac{2\left(r_{(i+1)-}^3 - r_{i-}^3\right)}{3\left(r_{(i+1)-}^2 - r_{i-}^2\right)}$$

For the last node, n, the distance to the outer boundary is the outer radius of the mug:

$$r_{n+} = \frac{D}{2} + w$$

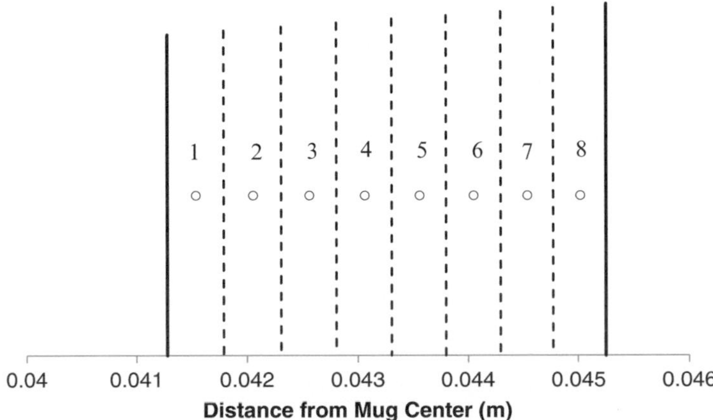

Fig. 6.4 Locations of nodes and boundaries for an 8-node model applied to mug of coffee problem

The distance to the centroid is:

$$r_n = \frac{2\left((D/2 + w)^3 - r_{n-}^3\right)}{3\left((D/2 + w)^2 - r_{n-}^2\right)}$$

Figure 6.4 shows the precise radial locations of all boundaries of each nodal element (to scale, horizontally) as dotted lines, except for the inner and outer surfaces (shown as solid lines) for the 8-node case applied to the mug of coffee problem. The length of each vertical line is proportional to the surface area of the boundary. The area increases with radial position because of the increasing circumference. The height of each "wall" is the same, however. To the naked eye, the nodes may appear equally spaced, but they are in fact closer together as the distance from the mug center increases. In this case, the mug wall thickness (0.004 m) is small compared to the inner radius (0.041 m), making a thin-wall approximation (not taken here) reasonable. The surface area of inner boundary (line length) increases slightly with radial distance.

6.3 Nodal Equations for Explicit Method (Euler)

In general, the transient thermal energy balance for a fixed volume element "i" in thermal contact with neighbors "j" is:

$$C_i \frac{dT_i}{dt} = \dot{Q}_i + \sum_j \frac{T_j - T_i}{R_{ij}}$$

$\dot{Q}_i$ represents heat generation, which could come from an internal chemical reaction (conversion of chemical to thermal energy), from the absorption of radiation, from I^2R heating (conversion of electrical to thermal energy), from microwaving, etc. In the problem at hand, there are no heat sources. R_{ij} represents the thermal resistance between node "i" and its neighbor in question "j". The detailed engineering work is in estimating the resistance values (which may be variable).

Discretizing time using a forward finite difference approximation:

$$C_i^{(p)} \frac{T_i^{(p+1)} - T_i^{(p)}}{\Delta t^{(p)}} = \dot{Q}_i^{(p)} + \sum_j \frac{T_j^{(p)} - T_i^{(p)}}{R_{ij}^{(p)}}$$

Rearranging for the future temperature of node "i":

$$T_i^{(p+1)} = T_i^{(p)} + \frac{\Delta t^{(p)}}{C_i^{(p)}} \left(\dot{Q}_i^{(p)} + \sum_j \frac{T_j^{(p)} - T_i^{(p)}}{R_{ij}^{(p)}} \right)$$

For a system with n total capacity-carrying temperature nodes (as opposed to nodes fixed by a battery, or a surface node), there are n equations. Starting with an initial temperature, a numerical step size is chosen, and all future temperatures are calculated.

Recall from Chap. 1 that the explicit formulation places a strict limit on the numerical step size chosen. Each node has a different time constant associated with it which represents how quickly it responds to its immediate neighbors. From Chap. 1, the time constant for a node toward an equilibrium value with its neighbors at a fixed temperature is given by:

$$\tau_i^{(p)} = \frac{C_i^{(p)}}{\sum_j \dfrac{1}{R_{ij}^{(p)}}}$$

The maximum step size for the system that can be used in an explicit formulation is the smallest of these:

$$\Delta t_{\max}^{(p)} = \min \tau_i^{(p)}$$

If a step size larger than this limit is used, the node with the minimum time constant will overshoot its equilibrium value.

The nodal equations for all nine variable temperature nodes (coffee plus node 1, six interior nodes, and node 8) are:

For the coffee:

$$T_{coffee}^{(p+1)} = T_{coffee}^{(p)} + \frac{\Delta t^{(p)}}{C_{coffee}^{(p)}} \left(\frac{T_1^{(p)} - T_{coffee}^{(p)}}{R_{cs} + R_{s1}} + \frac{T_\infty^{(p)} - T_{coffee}^{(p)}}{R_{top}} + \frac{T_\infty^{(p)} - T_{coffee}^{(p)}}{R_{bottom}} \right)$$

For node 1:

$$T_1^{(p+1)} = T_1^{(p)} + \frac{\Delta t^{(p)}}{C_1^{(p)}} \left(\frac{T_{coffee}^{(p)} - T_1^{(p)}}{R_{cs} + R_{s1}} + \frac{T_2^{(p)} - T_1^{(p)}}{R_{12}} + \frac{T_\infty - T_1^{(p)}}{R_{as,i} + R_{s1,a}} \right)$$

For interior nodes i (nodes 2–7):

$$T_i^{(p+1)} = T_i^{(p)} + \frac{\Delta t^{(p)}}{C_i^{(p)}} \left(\frac{T_{i-1}^{(p)} - T_i^{(p)}}{R_{i,i-1}} + \frac{T_{i+1}^{(p)} - T_i^{(p)}}{R_{i,i+1}} \right)$$

For node 8:

$$T_8^{(p+1)} = T_8^{(p)} + \frac{\Delta t^{(p)}}{C_8^{(p)}} \left(\frac{T_7^{(p)} - T_8^{(p)}}{R_{78}} + \frac{T_\infty - T_8^{(p)}}{R_{8s} + R_{s\infty}} \right)$$

The resistance $R_{s\infty}$ is the equivalent resistance of the parallel convective/radiative channel between the outer surface and ambient. The two surface temperatures (surface nodes 0 and 9, which have no capacitance) are determined from a voltage divider analysis. For the inner surface:

$$q_{coffee,1} = \frac{T_{coffee} - T_1}{R_{cs} + R_{s1}} = \frac{T_{coffee} - T_0}{R_{cs}}$$

Solving for the surface temperature:

$$T_0 = T_{coffee} - \frac{R_{cs}}{R_{cs} + R_{s1}} \left(T_{coffee} - T_1 \right)$$

For the outer surface:

$$q_{8\infty} = \frac{T_8 - T_\infty}{R_{8s} + R_{s\infty}} = \frac{T_9 - T_\infty}{R_{s\infty}}$$

$$T_9 = T_\infty + \frac{R_{s\infty}}{R_{9s} + R_{s\infty}} \left(T_8 - T_\infty \right)$$

The final step before implementing the method in a computer simulation is to define the capacitances and thermal resistances in terms of system parameters, detailed in the following paragraphs. Results for the coffee/mug problem are listed in Table 6.1, and the various entries are discussed as they arise.

Table 6.1 Nodal geometry, capacitance, resistance, and step-size restriction at the start of the simulation

Quantity	Symbol	unit	Node										
			coffee	"o"	1	2	3	4	5	6	7	8	"9"
inner boundary radius	r-	m	NA	NA	0.0413	0.0418	0.0423	0.0428	0.0433	0.0438	0.0443	0.0448	NA
node radius	r	m	NA	0.0413	0.0415	0.0420	0.0426	0.0431	0.0436	0.0440	0.0445	0.0450	0.0453
Inner Boundary Area	A-	m^2	NA	NA	0.0199	0.0278	0.0282	0.0285	0.0288	0.0292	0.0295	0.0298	0.0301
Capacitance	C	J/°C	1757	0	29.2	29.2	29.2	29.2	29.2	29.2	29.2	29.2	0
Resistance (to the right)	R+	°C/W	VARIES	0.0131	0.0185	0.0180	0.0176	0.0172	0.0168	0.0165	0.0161	0.0079	VARIES
Maximum step size	dtmax	sec	22.871	NA	0.413	0.266	0.260	0.254	0.248	0.243	0.238	0.469	NA

Varies with time. Initial value is shown, which yields a minimum step size

The shaded cells indicate that these step sizes vary with time and must be confirmed to not control the system at a later time

6.3.1   Capacitances

The capacitance of each node is the same in this particular implementation:

$$C_{node} = \rho_{mug} c_{mug} V_{node}$$

Each node has a capacitance of 29.2 J/°C, and the total capacitance of the mug side wall is $8(29.2) = 234$ J/°C, which is 87.3 % of the total capacitance of the mug (the remainder at the bottom which is lumped with the coffee).

The capacitance for the coffee is that calculated for the 1-node cases, plus the capacitance of the bottom of the mug:

$$C_{coffee} = \rho_{coffee} c_{coffee} V_{coffee} + \left\lfloor m_{mug} c_{mug} V_{mug} - n * C_{node} \right\rfloor$$

The capacitance of the coffee, 1,757 J/°C, is (not surprisingly) much larger than the capacitance of each node in the mug wall. Inclusion of the bottom of the mug has a small (2.9 %) effect on the coffee capacitance for this case study and could be neglected. However, it is easy to incorporate it, and it might be more important for other applications, so it is included.

6.3.2   Resistances

All the resistance values are shown in Table 6.1 (except for the resistance between node 1 and the air inside the rim of the mug, which is treated separately). They are derived one at a time, from left to right. Starting with the convective thermal resistance between the coffee and the inner wall of the mug:

$$R_{cs} = \frac{1}{h_{cs} A_{r1-}} = \frac{1}{h_{cs} \pi D (H - L)}$$

In this expression, H is the depth of the coffee on the inside and L is the vertical distance of the air column on the inside of the mug, making $H - L$ the vertical dimension of the contact area between the coffee and the inner surface of the mug.

The convection coefficient, h_{cs}, results from a natural convection process between the coffee and the inner wall of the mug. From the Nusselt number correlation of a future chapter:

$$h_{cs} = \left(32.38\,T_{film}^{0.441}\right)\frac{\left|\left(T_{coffee} - T_1\right)\right|^{0.25}}{(H - L)^{0.25}}$$

The film temperature is the average of T_{coffee} and T_1.

The thermal resistance between the inner surface (node "0") and the first elemental node (node 1) is a conductive resistance and is given by:

$$R_{s1} = \frac{r_1 - r_{1-}}{k_{mug}\left(\frac{A_{r1-}+A_{r1}}{2}\right)} = \frac{r_1 - r_{1-}}{\pi k_{mug}(r_{1-} + r_1)(H - L)}$$

The numerator is the spacing between nodes 0 and 1, and the area is the geometric average between the two. This thermal resistance is in series with the convective resistance, and it is conceptually half that of a thermal resistance between two interior nodes (from Table 6.1, $R_{01} = 0.0094\ °C/W$, while $R_{12} = 0.0185\ °C/W$).

For the resistance between node 1 and ambient through the inner surface of the mug rim:

$$R_{as,i} + R_{s1,a} = \frac{1}{h_{eff}\pi DL} + \frac{r_1 - r_{1-}}{\pi k_{mug}(r_{1-} + r_1)L}$$

In this expression, h_{eff} represents the sum of the convective and radiative coefficients between the surface and the ambient. The same value as calculated for the outer surface will be used. There are admittedly several inconsistencies in how this thermal channel is treated. The philosophy is that the exposed surface area in question is a significant (but not dominant) channel for heat loss, and it is accounted for in a reasonable approximation by the approach taken. Modeling is an art.

Continuing left to right, the conductive resistance between nodes 1 and 2 is:

$$R_{12} = \frac{r_2 - r_1}{k_{mug}2\pi r_{2-}H_0}$$

Generally, between nodes "i" and "i + 1":

$$R_{i,i+1} = \frac{r_{i+1} - r_i}{k_{mug}2\pi r_{(i+1)-}H_0}$$

It would be just as easy to use the exact conductive formula for a cylindrical geometry, but the more general approach is taken here. From Table 6.1, the conductive resistance between adjacent interior nodes decreases with radial position. There are two reasons for this trend. First, the nodes are closer together, easing conduction. Second, the surface area across which heat is conducting is larger, again easing conduction.

The conductive thermal resistance between the last element node ("n" in general, "8" in this implementation) and the surface node ("s" in general, "9" for this implementation) is:

$$R_{s,n} = R_{9,8} = \frac{r_9 - r_8}{k_{mug}\left(\frac{A_{r8+}+A_{r8}}{2}\right)} = \frac{r_9 - r_8}{\pi k_{mug}(r_9 + r_8)H_0}$$

Finally, the convective resistance between the outer wall surface ("s" = "9") and the ambient air is:

$$R_{s,\infty} = \frac{1}{h_{s\infty}A_{r9}} = \frac{1}{h_{s\infty}\pi(D + 2w)H_0}$$

The 1-node models taught that the overall system response is very sensitive to the choice of convection and radiation coefficients between the outer wall and ambient. This sensitivity is due to the large surface area (a parallel resistance concept), and because this combined resistance is the largest resistance in the side channel (a series resistance concept). Therefore, extra attention is warranted in modeling these parameters. In this numerical simulation, it is relatively easy to account for heat transfer coefficients that vary with temperature (and therefore time) by calculating them at each time step. Conceptually, the temperature used should be for the slightly different node 9, not 8, but that would require an iterative loop, making a spreadsheet simulation impossible.

For parallel convective and radiative resistances acting on a surface of area A, with coefficients h_{conv} and h_{rad}, the equivalent resistance is:

$$\frac{1}{R} = \frac{1}{R_{conv}} + \frac{1}{R_{rad}} = h_{conv}A + h_{rad}A = h_{eff}A$$

That is, the convective and radiative coefficients add together to form an effective heat transfer coefficient. If the ambient air immediately adjacent to the mug and the walls of the room are at significantly different temperatures, then the model would have to be adjusted by adding a separate channel for radiation and convection.

The convective coefficient in this case is modeled as a natural convection of air for a vertical cylinder of height H_0, with fluid properties evaluated at the film temperature. The simplified formula for h_{as} from the Nusselt number correlation (with temperature in °C and height in m) is:

$$h_{air} = \left(1.5646 - 0.0015T_{film}\right)\frac{|(T_8 - T_\infty)|^{0.25}}{H_o^{0.25}}$$

The leading term represents the combined fluid properties and their relatively mild temperature dependency.

The radiative coefficient, with an emissivity ε as derived in Chap. 3, is:

$$h_{rad} = \varepsilon\sigma\left[(T_8 + 273)^2 + (T_\infty + 273)^2\right]\left[(T_8 + 273) + (T_\infty + 273)\right]$$

where the temperatures are in Celsius. Again, technically speaking, the surface temperature (T_9) should be used. However, that would require an iterative solution for each time step, and that would make a spreadsheet implementation problematic.

6.3.3   Step-Size Restriction

In an explicit formulation, each capacitive node has a step-size restriction equal to its characteristic response time, $R_{eq}C$, where R_{eq} is the equivalent parallel resistance of all its neighbors. The step size used to march forward in time cannot exceed the smallest of these. That is, the maximum step size that can be used in an explicit formulation is the minimum time constant of any individual node.

For nodes that have variable resistances, the maximum step size will also vary with time. In Table 6.1, the shaded cells contain maximum step sizes that vary throughout the simulation. The values shown are minimum values, based on early time, when the convection and radiation coefficients are highest (giving lowest resistances, and shorter response times). It is apparent that these variable values are not going to control the step size for this application. It is conceivable that they could, however, for other input parameter values, and could be checked at each step. It is relatively easy to implement a variable step-size algorithm, in which the step size chosen at a given step is a user input-defined fraction of the maximum step size within the system of equations.

In this case, node 7 clearly limits the step size used in the system formulation to a value of 0.238 s. Contrast that restriction with the 2-node numerical simulation, in which case the mug as a lumped node yielded a step-size restriction of 31.8 s.

A comment on the computational requirements is in order. In order to simulate a fixed total time ($t_{simulation}$), the computations per simulation can be expressed verbally as:

$$\frac{Computations}{Simulation} = \left(\frac{Computations}{Step}\right) * \left(\frac{Steps}{time}\right) * \left(\frac{time}{Simulation}\right)$$

The computations per step is a fixed number of times the total number of nodes. That is:

$$\left(\frac{Computations}{Step}\right) \sim n$$

The steps per simulation is the inverse of the step size:

$$\left(\frac{Steps}{Simulation}\right) = \left(\frac{Step}{time}\right) * \left(\frac{time}{Simulation}\right) = \left(\frac{1}{\Delta t}\right) * t_{simulation}$$

The step size $= R_{eff,i} \, C_i$

$C \sim 1/n$ (the total capacitance divided by number of elements)

$R \sim 1/n$ (the distance between nodes is the total distance divided by number of elements)

Therefore:

$\Delta t \sim 1/n^2$

$$\frac{Computations}{Simulation} \sim (n) * \left(n^2\right) * \left(t_{simulation}\right)$$

That is, if the number of nodes is doubled, the computational requirement increases by a factor of 8.

6.4   Spreadsheet Implementation

This model was implemented into a spreadsheet simulation (Table 6.2) using a 9-node model (eight nodes in the mug, plus the coffee) and stepping in time with the maximum step size of 0.238 s. Initial conditions and fixed parameters are shaded. The maximum step size (dtMax) refers to the appropriate cell in Table 6.1, and the numerical input parameter "f", set to 1, is the fraction of the maximum step size used to define dt, the step size taken. It will be demonstrated why a smaller step size than that is not required for numerical accuracy for this case. In this model, three heat transfer coefficients depend on surface temperature (h_{coffee}, $h_{conv, \, air}$, and $h_{rad, \, air}$) and therefore must be calculated at each time step. They are placed to the right, and then the appropriate thermal resistance values are calculated. In the second row, the Euler formulas are entered (for coffee and nodes 1–8), and then all equations are copied down for as many rows as is desired.

Temperature/time plots are presented followed by a series of plots of temperature vs. radial position at fixed times. The heat transfer coefficients are plotted to

Table 6.2 Structure of spreadsheet simulation showing the first 3 s

	dtMax=	0.2375
	f =	1
	dt (sec) =	0.2375

time(s)	time(min)	coffee	"0"	1	2	3	4	5	6	7	8	"9"	hcoffee	Rcs	hconv	hrad	heff	Rinf	Tsurface
0.00	0.0000	92.00	36.29	21.00	21.00	21.00	21.00	21.00	21.00	21.00	21.00	21.00	1058	0.048	0.00	5.19	5.19	6.40	87.05
0.24	0.0040	91.84	43.72	30.53	21.00	21.00	21.00	21.00	21.00	21.00	21.00	21.00	1057	0.048	0.00	5.19	5.19	6.40	86.90
0.48	0.0079	91.70	46.81	34.55	25.20	21.00	21.00	21.00	21.00	21.00	21.00	21.00	1053	0.048	0.00	5.19	5.19	6.40	86.77
0.71	0.0119	91.57	49.50	38.07	27.42	22.89	21.00	21.00	21.00	21.00	21.00	21.00	1048	0.048	0.00	5.19	5.19	6.40	86.65
0.95	0.0158	91.45	51.35	40.51	30.07	24.06	21.88	21.00	21.00	21.00	21.00	21.00	1043	0.048	0.00	5.19	5.19	6.40	86.54
1.19	0.0198	91.34	52.99	42.67	31.96	25.76	22.47	21.41	21.00	21.00	21.00	21.00	1038	0.048	0.00	5.19	5.19	6.40	86.43
1.43	0.0238	91.23	54.29	44.38	33.88	27.04	23.49	21.71	21.20	21.00	21.00	21.00	1034	0.049	0.00	5.19	5.19	6.40	86.33
1.66	0.0277	91.12	55.45	45.93	35.42	28.49	24.29	22.31	21.35	21.10	21.00	21.00	1029	0.049	0.00	5.19	5.19	6.40	86.23
1.90	0.0317	91.02	56.43	47.23	36.92	29.67	25.29	22.78	21.69	21.17	21.05	21.05	1025	0.049	1.30	5.19	6.48	5.12	86.14
2.14	0.0356	90.93	57.32	48.42	38.19	30.92	26.13	23.44	21.96	21.37	21.11	21.11	1021	0.049	1.59	5.19	6.77	4.90	86.05
2.38	0.0396	90.83	58.09	49.45	39.41	31.99	27.07	24.00	22.38	21.53	21.24	21.24	1017	0.049	1.92	5.19	7.11	4.67	85.96
2.61	0.0435	90.74	58.80	50.41	40.48	33.06	27.89	24.67	22.74	21.80	21.39	21.39	1014	0.050	2.16	5.20	7.36	4.51	85.88
2.85	0.0475	90.65	59.44	51.26	41.51	34.02	28.76	25.26	23.21	22.06	21.60	21.60	1010	0.050	2.41	5.20	7.61	4.36	85.79
3.09	0.0515	90.57	60.03	52.05	42.42	34.97	29.54	25.92	23.63	22.40	21.83	21.83	1007	0.050	2.61	5.21	7.82	4.24	85.71

The column group to the right (hconv, hrad, heff, Rinf, Tsurface) is labelled **AIR SIDE**.

The shaded cells contain initial conductions and formulas are embedded in all other cells observe both their values as well as their change over time. Finally, a consideration of the numerical issues is presented, demonstrating both spatial and temporal grid effects.

The workbook written is rather large (15 MB) and it only simulates 36 min of the entire event. The number of time steps required to simulate 36 min is:

$$Number\ of\ steps = \frac{Time\ elapsed}{Step\ size} = \frac{36\,min}{0.238\,s} * \left|\frac{60\,s}{min}\right| = 9,075$$

63,025 steps would be required to simulate the entire 250 event, resulting in a spreadsheet of over 100 MB. In short, this simulation taxes the limit of spreadsheet applications, and a shift to a computer program may be warranted.

The input and derived parameters are listed in Fig. 6.5.

6.5   Temperature/Time Plots

Figure 6.6 shows the predicted temperature vs. time for locations that correspond to the measurements at the mug-free surface and mug side wall during the first 10 min. The predicted rise of the outer side wall temperature and fall of the free surface both lag from the measured values, somewhat, but in general, the agreement is considered to be quite good. The 2-node model (which used fixed resistance values) gave similar results. However, the 9-node model gives more detail, as shown in the following paragraphs.

A plot of all nine nodal temperatures, plus the inner and outer surface (non-capacitive nodes) vs. time is presented in Fig. 6.7, without the experimental data. The closer each node is to the inner surface, the earlier its temperature rises. The same plots are shown in Fig. 6.8 on semi-logx axes. This plot more clearly shows how it takes time before the first "signal" that the inner wall has been exposed to hot liquid reaches a given location. That is, a thermal "wave" propagates outward from the inner wall. The initial conditions cannot be shown on the latter plot (time equals zero cannot be plotted on a logarithmic scale).

6.6   Radial Temperature Profiles

A series of similar plots that show the temperature vs. radial position at fixed times is presented in Figs. 6.9, 6.10, 6.11, 6.12. On the coffee and air side, hypothetical thermal boundary layers are shown using several columns and an exponential fit for illustrative purposes.

The first plot shows the first 3 s, during which time the outer wall never experiences a change. During this period of time, the wall of the mug behaves as if it were infinitely thick. In traditional heat transfer texts, analytic solutions of heat transfer into an infinitely thick wall are presented. Those solutions are for flat walls with constant convection coefficients, whereas this one is for a radial geometry with

INPUT QUANTITIES

D=	0.08255	m	Inner diameter of mug
H=	0.098	m	Inner depth of mug
HO=	0.106	m	outer height of mug
V=	4.10E-04	m^3	coffee volume
w=	0.003975	m	mug thickness
mmug=	0.335	kg	mass of mug
kmug=	1	W/m/K	thermal conductivity of mug
cmug=	800	J/kg/K	specific heat of mug
rhocoffee=	1000	kg/m^3	density of coffee
ccoffee=	4200	J/kg/K	specific heat of coffee
hair/top=	30	W/m2/K	effective heat transfer coefficient between surface and air
hcoffee/surface=	400	W/m2/K	effective heat transfer coefficient between coffee and top surface
hfloor=	7.91	W/m^2/K	effective heat transfer coefficient between coffee and floor ambient
Tcoffee, init=	92	oC	initial coffee temperature
Tmug,initial=	21	oC	initial mug temperature
Tambient=	21	oC	ambient temperature
hair,outside	15	W/m2/K	representative value (not used in simulation)
hcoffee/mug=	400	W/m2/K	representative value (not used in simulation)

DERIVED QUANTITIES

H-L=	0.076605	m	depth of coffee
Vmug=	0.000131	m^3	volume of mug
rhomug=	2548.312	kg/m^3	density of mug
Cmug=	268	J/oC	total capacity of mug
Ccoffee=	1.72E+03	J/oC	capacity of coffee
Atop,bottom	0.005352	m^2	surface area of top and bottom

RESISTANCES

	inner convection	wall	outer convection	Channel
top	0.47	0.00	6.23	6.70
side	variable	NA	variable	variable
bottom	0.47	0.74	23.62	24.83

Fig. 6.5 Input and derived fixed parameter values

variable convection coefficients (which would be impossible to solve analytically). This simulation shows the basic response, and gives a sense that it is in fact possible to have an infinitely thick wall response, provided the elapsed time is sufficiently short to justify it. So there is a relationship between space and time in transient conduction problems. Short time is like long distance.

The first 30 s sees the temperature of the outer wall increasing with time, and retaining a general concave up shape. This problem would be extremely hard, if not impossible, to solve analytically. The first 3 min shows the mug wall reaching a maximum, with a subsequent gradual falling of the coffee and mug wall together

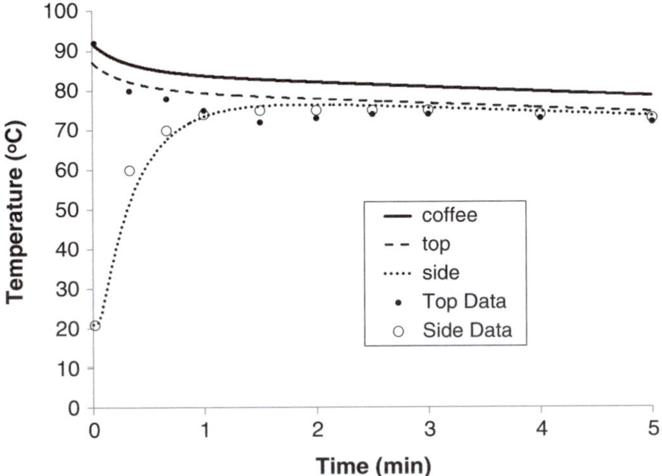

Fig. 6.6 9-Node model predictions during mug heating phase

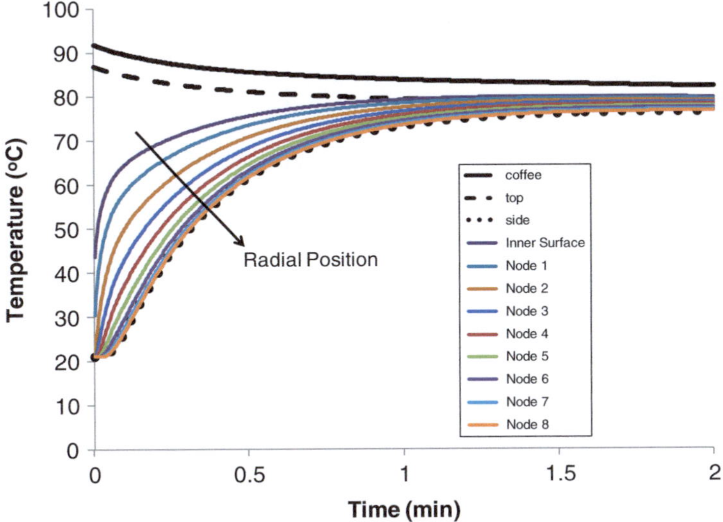

Fig. 6.7 Nodal temperatures vs. time for the first 2 min

during the first 30 min. This behavior maintains itself until the entire event is over, when the mug reaches equilibrium. During this period, there a very mild temperature drop from the coffee to the inner surface, another mild drop across the mug wall, and then a large drop from the outer surface to ambient. This behavior is due to the relative resistance values of the side channel.

What this behavior suggests, rather strongly, is that using eight nodes in the coffee mug is total overkill to do this particular job. The main objective here is to

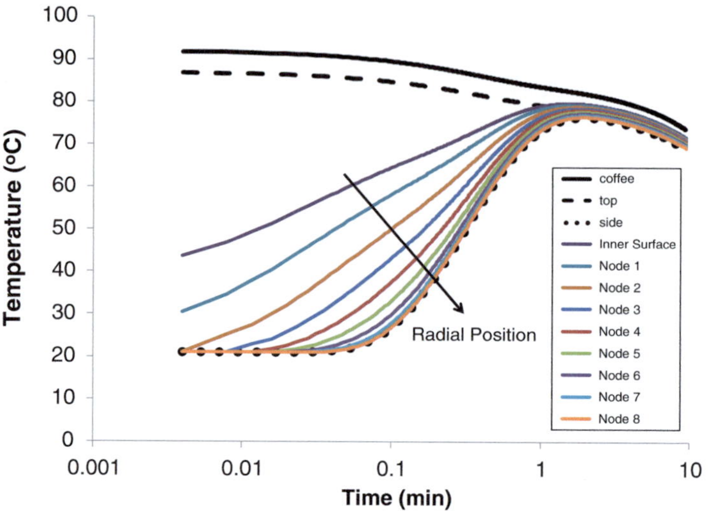

Fig. 6.8 Nodal temperatures vs. time for the first 10 min on semi-logx axes

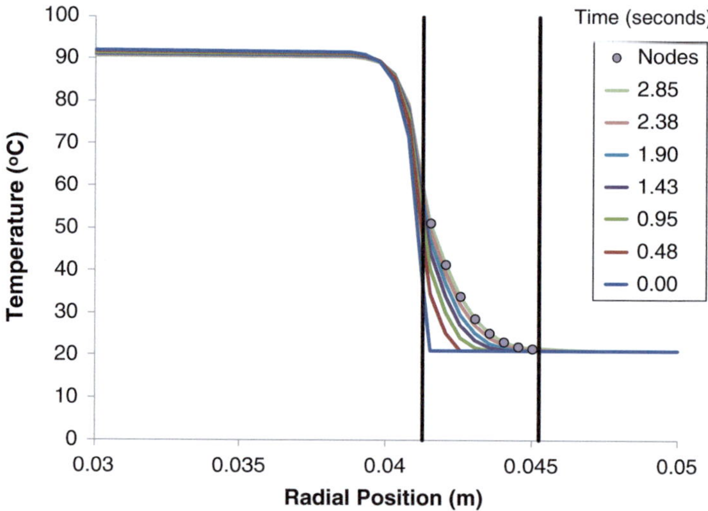

Fig. 6.9 Temperature vs. radial position for the first 3 s. *Symbols* shown for the latest curve to indicate nodal positions

teach the method, but good engineering judgment would suggest that the 2-node model (with the finite volume approach to yield surface behavior) is adequate for this application. And the extended lumped model is sufficient for the entire event (but not for the early mug heating phase).

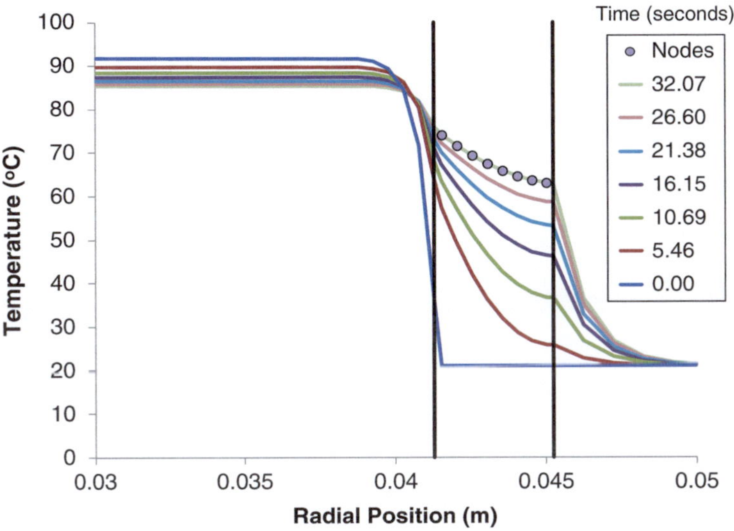

Fig. 6.10 Temperature vs. radial position for the first 30 s. *Symbols* shown for the latest curve to indicate nodal positions

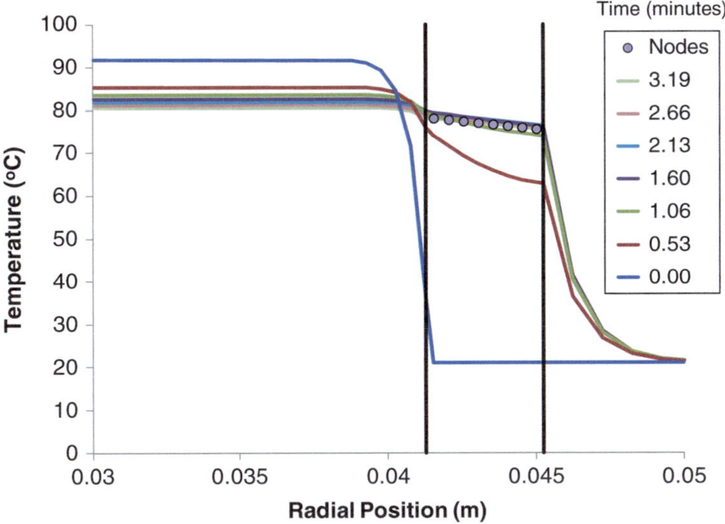

Fig. 6.11 Temperature vs. radial position for the first 3 min. *Symbols* shown for the latest curve to indicate nodal positions

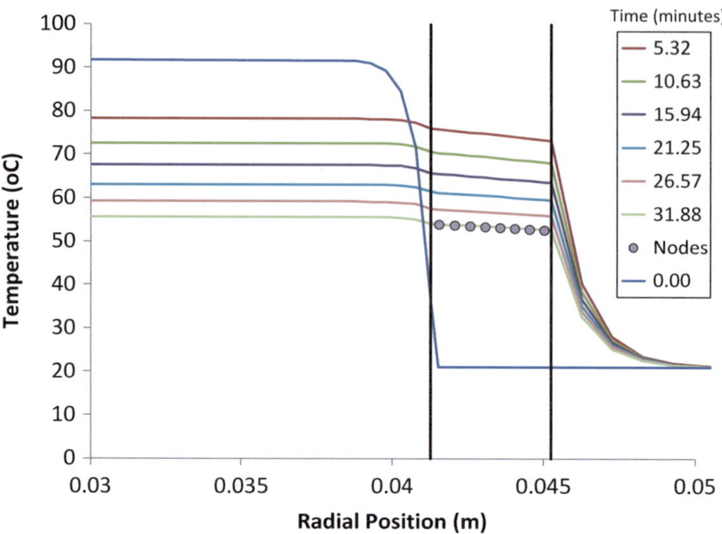

Fig. 6.12 Temperature vs. radial position for the first 30 min. *Symbols* shown for the latest curve to indicate nodal positions

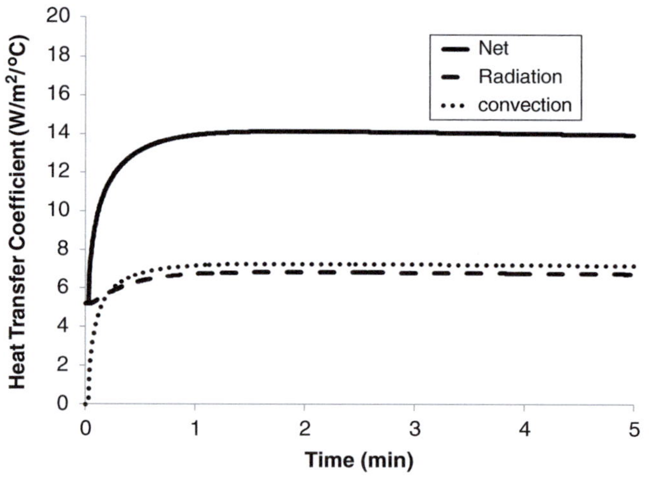

Fig. 6.13 Heat transfer coefficents on the outer surface of the mug vs. time for the first 5 min. If simulated for the entire cooling event, the convection coefficient will go to zero, while the radiation coefficient will remain finite

6.7 Heat Transfer Coefficients

The heat transfer coefficients are shown as functions of time for the first 3 min on the air side in Fig. 6.13 and on the coffee side in Fig. 6.14. At time zero, when the outer wall is at ambient, the natural convection coefficient is zero, according to

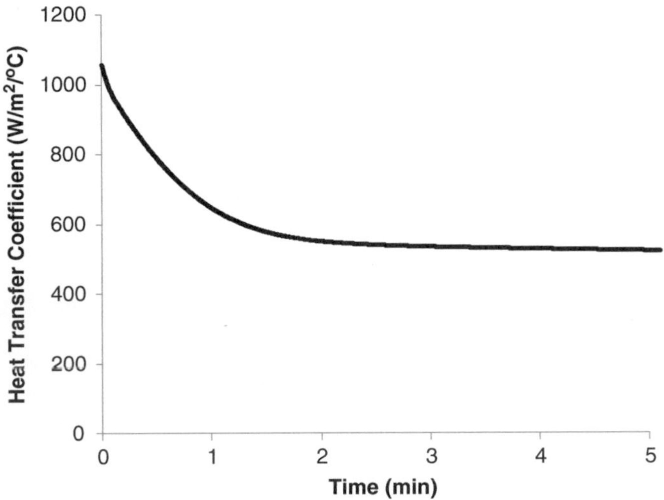

Fig. 6.14 Heat transfer coefficients on the inner surface of the mug vs. time for the first 5 min

the model. However, the radiation coefficient is not zero initially, and would control the overall heat transfer rate during periods of time where the solid surface temperatures are nearly equal to the adjacent fluid. This behavior emphasizes the importance of including radiation when natural convection predominates. As the outside surface wall temperature rises in the first 30 s, the convection coefficient rises and becomes almost equal to the radiative. The effective heat transfer coefficient is the sum of the two and equals approximately 14 $W/m^2/°C$.

In contrast, the convection coefficient on the water side starts high because the mug and coffee are at very different temperatures early. The inner mug wall surface temperature rises rapidly initially, and therefore the temperature difference driving the natural convection inside decreases. The convection coefficient inside therefore starts at a value just above 1,000 $W/m^2/°C$ and falls rapidly to a value of approximately 500 $W/m^2/°C$.

As time proceeds, the two natural convection coefficient values gradually fall as both the mug and coffee approach ambient. The model should really build in a lower cap to the convection coefficients.

6.8 Numerical Stability and Accuracy

Figures 6.15 and 6.16 show the same plots as Figs. 6.8 and 6.9, respectively, but with the step size chosen to be only 10 % larger than the minimum step size (0.262 s as opposed to 0.238 s). It is apparent that an instability arises, and it starts at node 7, and emanates from there. This instability reveals the biggest problem with an explicit formulation, especially as the spatial grid is made finer.

On the other hand, a comparison of the model prediction for node 7 is made when the step size is reduced by an order of magnitude in Fig. 6.17. There are

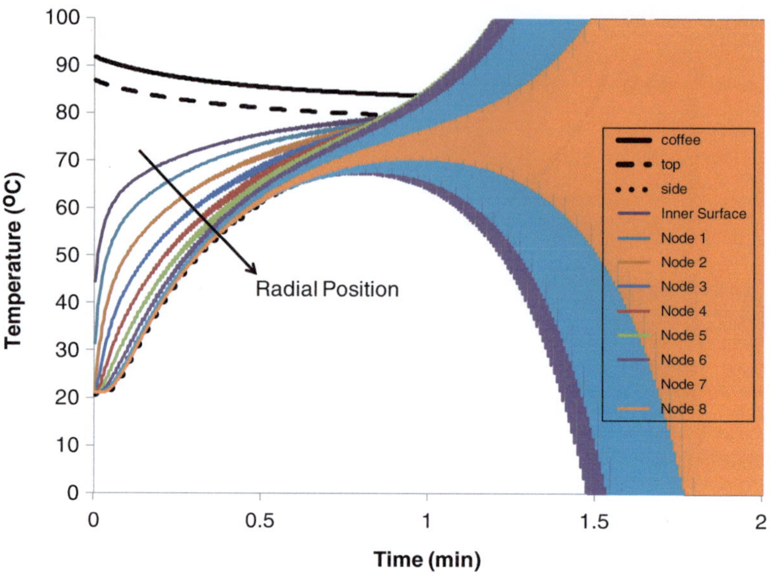

Fig. 6.15 Simulation with a step size of 1.1 times the maximum allowable step

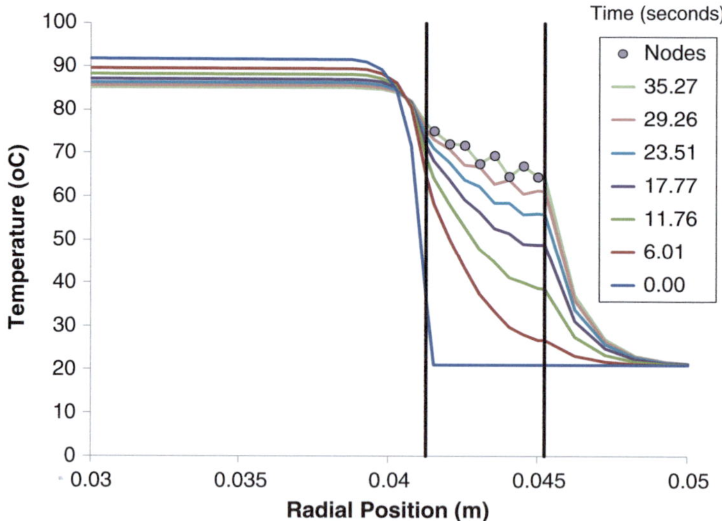

Fig. 6.16 Radial profile for first 30 s with a step size 1.1 times the minimum

actually two indistinguishable curves plotted, one when $f = 1$ and the other for when $f = 0.1$, a reduction in step size of an order of magnitude. This plot demonstrates temporal grid insensitivity exists when the maximum allowable step size is used in this case.

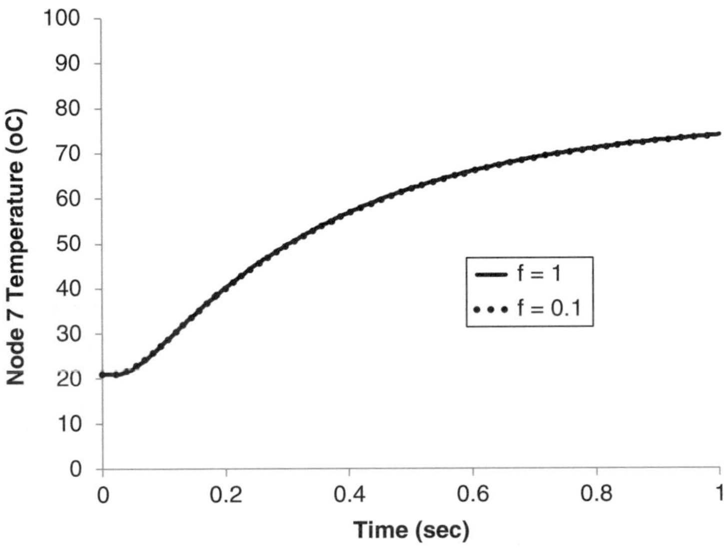

Fig. 6.17 Effect of step size on node 7. The two curves are indistinguishable

6.8.1 Implicit Method

The general governing equation (assuming constant capacitances, but variable thermal resistances) for node "i" in thermal contact with nodes "j" written as a backward-difference approximation, and upgrading (p) to (p + 1) is:

$$C_i \frac{T_i^{(p+1)} - T_i^{(p)}}{\Delta t^{(p)}} = \dot{Q}_i^{(p+1)} + \sum_j \frac{T_j^{(p+1)} - T_i^{(p+1)}}{R_{ij}^{(p+1)}}$$

Isolating all terms for node "i" evaluated in the future (p + 1):

$$T_i^{(p+1)} = \frac{\dot{Q}_i^{(p+1)} + \sum_j \dfrac{T_j^{(p+1)}}{R_{ij}^{(p+1)}} + \dfrac{C_i T_i^{(p)}}{\Delta t^{(p)}}}{\dfrac{C_i}{\Delta t} + \sum_j \dfrac{1}{R_{ij}^{(p+1)}}}$$

This expression can be used as a recursion relation for an iterative procedure, called, in general the Method of Successive Substitution. The method is developed more formally in later chapters, and outlined here.

An initial "guess" for all the unknown future temperatures (and the associated thermal resistances) is made, which for transient problems is logically the present value, and then an iterative loop that calculates an "improved guess" for all the temperatures is made, tests for convergence, and then either executes another iteration or exits the iteration loop and saves the temperatures. Testing for convergence requires the change in all temperatures between subsequent iterations be calculated, and setting the largest change to be less than a defined criterion.

This procedure requires a nested loop that cannot be executed in a spreadsheet (without using macros). The outer loop steps forward in time, the inner loop is an iterative loop needed to solve for the temperatures one step into the future.

Workshop 6.1. Transient Heat Transfer Fins

A "heat sink" (Fig. 6.18) is mounted onto a 6 cm × 6 cm (w × w) electronic component that generates 250 W (q) when operating. The heat sink has a 0.6 cm thick (b) base plate that fits exactly onto the component. Sixteen (M) rectangular "fins" of width 6 cm (w), thickness 0.10 cm (t), and length 4 cm (L) are mounted onto the base plate with an air gap of 0.275 cm (g) between each fin. The fins are cooled by a fan (not shown) that provides sufficient ventilation to maintain the air temperature in the gaps at ambient temperature ($T_\infty = 25$ °C). Compare the transient performance of heat sinks of copper, aluminum, and steel, with thermal properties listed in the following table:

Material	k (W/m/K)	ρ (kg/m^3)	c (J/kg/K)
Copper	400	8,900	390
Aluminum	200	2,700	910
Steel	20	7,900	490

Write a spreadsheet or program code to simulate the thermal response when the component is powered on, allowed to reach an effective steady-state, and then turned off and allowed to cool until it is effectively back to ambient. Treat the base plate as a uniform temperature lumped node (neglect its conductive resistance) that receives the generated heat and transfers it to the fins, which are assumed to behave identically. Analyze a single fin by breaking it into "n" equal sized nodes and

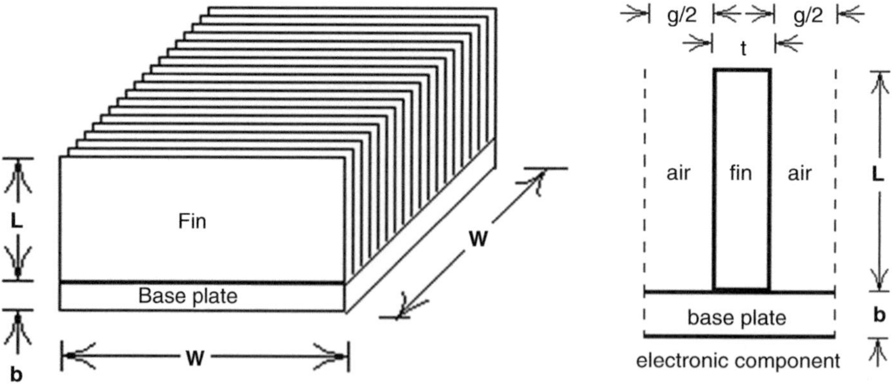

Fig. 6.18 Schematic of heat sink (*left*) and an end view of an isolated fin (*right*)

performing a transient analysis using the explicit method. Assume the fin (and base plate) is exposed to ambient with a fixed heat transfer coefficient h. Radiation is neglected in this case because the fins "see" each other.

The main objectives of this exercise are to successfully implement an "n-node" transient analysis, and to gain an understanding of the transient and steady-state response of heat transfer fins. At a minimum, prepare and discuss well-formatted plots of base plate temperature vs. time (with all materials on the same plot), steady-state temperature vs. position (distance from the base plate, all three materials). Also, consider temperature vs. position at fixed times for aluminum. Conduct a study of the effect of h (range from 10 to 100 W/m^2/K to simulate natural convection to a "strong crosswind" using aluminum.

Note: A specific 5-node model is developed on the next pages. It is advised that you develop your own model first before consulting that model. Then decide which model to implement.

Model Development

A 5-node RC network (base plate segment plus n = 4 equal sized nodes of the fin) is shown in Fig. 6.19 for a single fin. The term "longitudinal" will be used to designate a distance along the fin length (L or Δx), and "lateral" will refer to across the fin (t).

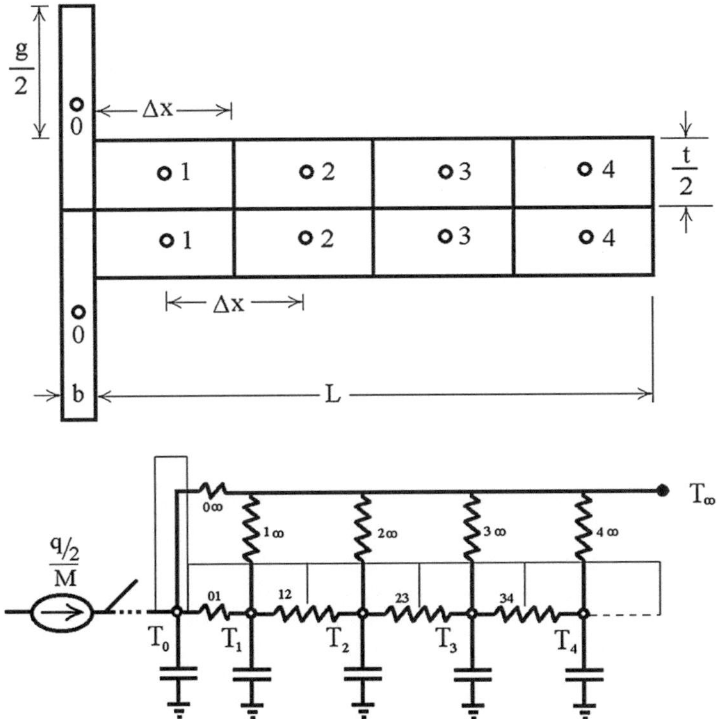

Fig. 6.19 Schematics for 5-node transient fin problem. The *top* figure defines the nodal volumes used. The *bottom* shows the RC network (the *light lines* define the nodal volumes)

The fin is broken into "n" (4 in this schematic) equally sized longitudinal elements, with a recognized symmetry so that only half of the fin need be considered. The base plate node is just that section associated with the fin being analyzed. Capacitances are assigned to each node and thermal channels between them are shown in the RC network. The symmetry has been used on the fin, accounting for the "q/2" being delivered to each fin, with M being the total number of fins and q the total heat dissipated.

The equations will be developed with a general form with the fin being broken into "n" equally sized nodes. Therefore, the length of each node, which also equals the distance between them is given by $\Delta x = L/n$. The distance between the base plate (node 0) and node 1 of the fin is $\Delta x/2$.

The capacitances are all equal to

$$C_i = \rho c w \left(\frac{t}{2}\right) \Delta x = \rho c w \left(\frac{t}{2}\right) \left(\frac{L}{n}\right)$$

The resistance between the base plate directly to the air is by:

$$R_{0\infty} = \frac{1}{hwg/2}$$

The conductive resistance between the base plate and node 1 is:

$$R_{0\infty} = \frac{\Delta x/2}{kwt/2}$$

Notice that the distance from the base plate to node 1 is half of element length. The conductive resistances between the other nodes are:

$$R_{i,i+1} = \frac{\Delta x}{kwt/2}$$

The resistances between each node and ambient is given by a series conduction and convection resistance. For interior node "i":

$$R_{i,i+1} = \frac{1}{h\Delta x} + \frac{t/4}{k\Delta x} = \frac{1}{h\Delta x}\left(1 + \frac{ht/4}{k}\right) = \frac{1}{h\Delta x}(1 + Bi)$$

where a Biot number has been introduced (the ratio of lateral conductive to convective resistance). The end node ("n") has a different surface area, and its resistance consists of two parallel channels (lateral and longitudinal) each of which has a conductive and convective resistance in series:

$$\frac{1}{R_{n\infty}} = \frac{1}{\frac{1}{h\Delta x} + \frac{t/4}{k\Delta x}} + \frac{1}{\frac{1}{ht/2} + \frac{\Delta x/2}{kt/2}}$$

This resistance could be expressed in terms of the Bi defined above and the ratio of fin thickness to element length. It is common in fin analysis to neglect the heat going out the end in comparison to that going out the sides, but is included for completeness, and practice in defining resistance values.

Using the explicit numerical method, the nodal equations are:

Base Plate:

$$T_0^{(p+1)} = T_0^{(p)} + \frac{\Delta t}{C_0} \left[\frac{q}{2M} + \frac{T_1^{(p)} - T_0^{(p)}}{R_{01}} + \frac{T_\infty - T_0^{(p)}}{R_{0\infty}} \right]$$

Node $i = 1$ to $n-1$:

$$T_i^{(p+1)} = T_i^{(p)} + \frac{\Delta t}{C_i} \left[\frac{T_{i-1}^{(p)} - T_i^{(p)}}{R_{i,i-1}} + \frac{T_{i+1}^{(p)} - T_i^{(p)}}{R_{i,i+1}} + \frac{T_\infty - T_i^{(p)}}{R_{i\infty}(1 + Bi_i)} \right]$$

Node n:

$$T_n^{(p+1)} = T_n^{(p)} + \frac{\Delta t}{C_n} \left[\frac{T_{n-1}^{(p)} - T_n^{(p)}}{R_{n-1,n}} + \frac{T_\infty - T_n^{(p)}}{R_{n\infty}(1 + Bi_n)} \right]$$

In these expressions, the Biot numbers represent the ratio of the conductive to convective resistance in the direction perpendicular to the direction of the primary heat flow and can be shown to be negligible for these cases. The time constants are:

$$\tau_0 = C_0 \left/ \left(\frac{1}{R_{01}} + \frac{1}{R_{0\infty}} \right) \right.$$

$$\tau_i = C_i \left/ \left(\frac{1}{R_{i,i-1}} + \frac{1}{R_{i,i+1}} + \frac{1}{R_{i\infty}(1 + Bi_i)} \right) \right.$$

$$\tau_n = C_n \left/ \left(\frac{1}{R_{n,n-1}} + \frac{1}{R_{n\infty}(1 + Bi_n)} \right) \right.$$

The maximum step size for numerical stability is the smallest time constant. Choose the step size to be a fraction (f) of this limit.

Part III

Steady-State Conduction

Heat Transfer Fins (and Handles)

7

In this chapter, a lumped capacity approach to steady-state conduction problems is developed, using the important practical application of heat transfer fins (and handles). The approach is applied to the coffee/mug problem, this time with half a mug, and the heat transfer into the rim and out to ambient is detailed.

7.1 Motivation for Fins

Extended surfaces are widely used as either heat transfer fins (devices designed to increase the overall rate of heat transfer between a solid and a fluid) or handles (something you can grab without burning your hand). For fin applications, since the rate of heat transfer by convection (and possibly radiation in parallel) is proportional to the exposed surface area, the basic idea is to increase that exposed surface area, as suggested in Fig. 7.1. However, because the temperature in the fin decreases with the distance from the wall (x), there is a limit to how far the fin can extend and still be effective. A "perfect" fin is one in which the temperature of the fin is the same everywhere as its base. In a real fin, the average temperature of the fin exposed to the fluid will be less than that, and therefore the fin transfers heat at a lower rate. If the fin is long enough, the temperature at the end of the fin approaches the ambient temperature. That's a handle you can grab.

An exact analysis of the steady-state 1D heat transfer associated with an "extended surface" is included in virtually all heat transfer textbooks for the case where the cross-sectional area is constant with distance from the wall (i.e., the area of the fin base (t*d) is projected outward from the base). The result for the more general transient case (the steady-state being a limiting case of the transient solution) is summarized in Appendix 1. This analysis also includes the inherent 2D lateral heat transfer that is required for heat to flow from the interior of the fin toward the surface. Appendix 2 develops the governing differential equation that could be the starting point for a finite difference numerical method for the more

© Springer International Publishing Switzerland 2015
G. Sidebotham, *Heat Transfer Modeling: An Inductive Approach,*
DOI 10.1007/978-3-319-14514-3_7

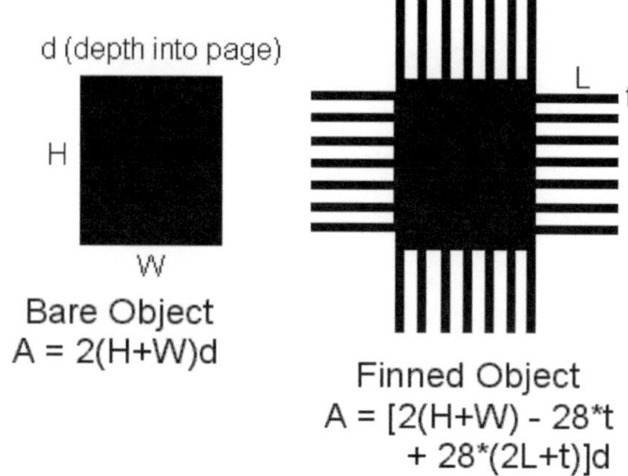

Fig. 7.1 Motivation for heat transfer fins. The bare object is a rectangular rod of length d (*left*). Rectangular heat transfer fins (length L, thickness t, and depth d) are added to increase the exposed surface area

general case with a non-constant-area fin. This method is not formally developed here, however.

The approach taken in the main body of the chapter is to develop a finite element approach. First, the fin is considered as a single lumped element, a 1-node model. This approach yields a simple closed-form solution that is surprisingly versatile and captures all the key functional dependencies of the exact solution. Next, a multi-element approach ("few node") is developed considering the fin broken into three equal elements. This method is easily extended to two or three dimensions. Finally, a detailed methodology of the 1-node model is applied to the case of the rim of a mug of coffee.

Engineering Problem Statement

An aluminum ($k = 200$ W/m/K) rectangular fin with a thickness t ($=0.001$ m), length L ($=0.1$ m), and width w ($=0.25$ cm) is attached to a wall (Fig. 7.2). The base of the fin is maintained at a temperature ($T_o = 100\ °C$). The fin is exposed to a fluid at T_∞ ($=0\ °C$) with an effective convection/radiation coefficient h ($=50$ W/m^2/K). Determine

- The rate of heat transfer from the base wall, q_{ss}.
- The fin efficiency, η, defined as the rate of heat transfer normalized by that of a perfect fin (defined shortly).
- The fin effectiveness, ε, defined as the rate of heat transfer divided by the rate of heat transfer that would occur from the base if there were no fin attached.

By specifying k and especially h and implying that the fluid temperature is not affected by the fin (outside a thermal boundary layer), the accuracy of this model to a "real life" fin will always be suspect. The main goal here is to compare the accuracy of the numerical methods on a problem statement that is well posed so a comparison with an exact solution is possible. The main advantage of the numerical method will be in addressing problems that do NOT have an exact solution.

Fig. 7.2 Schematic of a constant-area fin attached to a wall. A rectangular fin is shown, but the shape of the cross-section can be anything, as long as it does not vary with distance from the wall (if it is to be a constant-area fin)

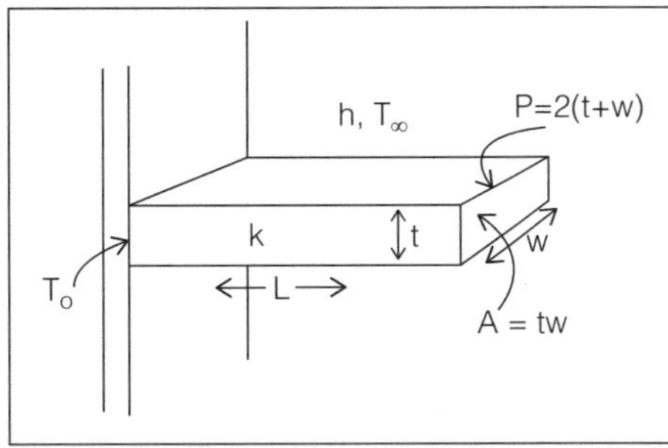

7.1.1   Perfect Fin

The heat transfer rate from a perfect fin, in which the temperature of the fin equals the base temperature everywhere, is governed by Newton's Law of Cooling:

$$q_{perfect} = h(PL + A)(T_0 - T_\infty)$$

P is the perimeter of the fin, PL is the area of the sides, and A is the area of the exposed end (and is the base area projected a distance L from the wall). Note that a constant-area fin can be any shape (area and perimeter) and is not restricted to a rectangular geometry.

7.1.2   Corrected Length

In a fin, heat flows out the sides and out the end. In most cases, the heat out the sides is dominant and inclusion of the end complicates the mathematics. Therefore, it is common to define a corrected length in which the heat out the end is accounted for in the term PL, as suggested in Fig. 7.3. The added length adds the same area as the end. That is, $P\Delta L = A$, and the corrected length is $L_c = L + \Delta L = L + A/P$. Therefore, the heat transfer rate from a perfect fin with a corrected length is $q_{perfect} = h(PL_c)(T_0 - T_\infty)$.

7.2   1-Node Fin Analysis

Two 1-node models will be developed; one for fins, and one for handles. The concepts of "thermally short" for fins and "thermally long" for handles will naturally emerge.

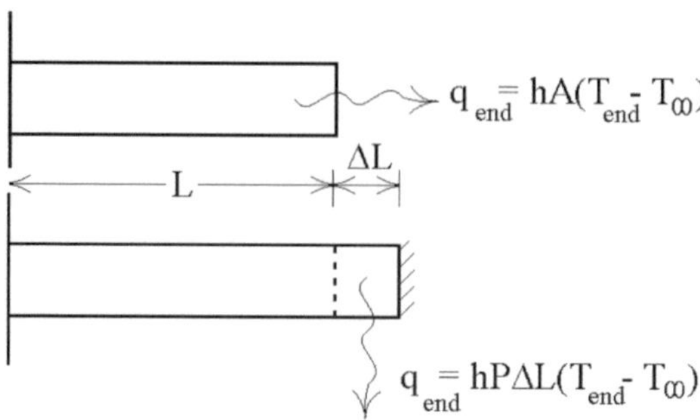

Fig. 7.3 Schematic of corrected length. The *top* fin has heat transfer out the end. The *bottom* fin has an added length (ΔL) that adds surface area to the sides and has an insulated (adiabatic) end. It's not that insulation is placed on the end; it is that heat out the end of the longer fin is neglected

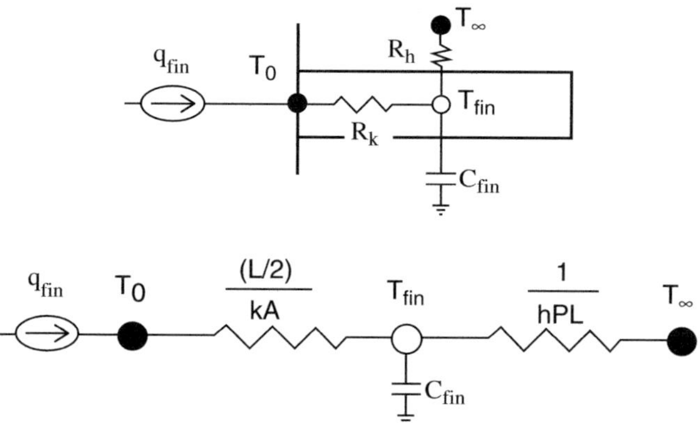

Fig. 7.4 Schematic and thermal resistance network of a thermally short fin represented by a single node. This model is called the "thermally short" fin model, for reasons which will become apparent. The *top* network is expressed in general terms of conductive and convective resistances. The *bottom* network is defined for a constant-area fin

7.2.1 Thermally Short Fin Model

The resistance network of Fig. 7.4 considers the fin as a single element, represented by a node. This simple network captures the fundamental heat transfer process associated with fins. Heat (shown as a current source) must flow from the base (at a temperature T_0) into the fin by conduction and from the fin to the fluid by convection (and radiation in parallel). Alternatively, a battery could be used to maintain a fixed T_0, as is conventionally presented. However, since the heat transfer rate into the fin is the key objective of the analysis, a current source that supplies that heat is more appropriate. Also, fins are often called "heat sinks" and a current source fits that language better than a battery.

The average temperature of the fin is represented by T_{fin}. In this model, the distance used in the conductive resistance is the half length (the distance between the two nodes), while the surface area of the convection is based on the full length (times the perimeter P). To be complete, the corrected length could be used in the convective resistance (but conceptually, not in the conduction), but is excluded here to keep the final result cleaner. A thermal capacitance is shown, but at steady-state, the net heat flow rate into the capacitor is zero. A good review exercise is to analyze a transient fin using a 1-node lumped transient. For a fin undergoing a transient heating or cooling (an important practical case when there is a cycling on and off of some device, for example), a closed-form transient solution using the previous 1-node methods works fine. The objective here is to introduce steady-state analysis techniques, however.

At steady-state, the rate of heat transfer is determined in one step, namely:

$$q_{ss} = \frac{T_0 - T_\infty}{R_k + R_h}$$

If a given application is posed with a specified base temperature (the conventional approach), then this form can be used to determine the heat transfer rate. If a given problem is posed with a specified heat transfer rate (for example, to "dissipate" heat from a computer chip), then this equation can be rearranged to solve for the required operating temperature of the base:

$$T_0 = T_\infty + (R_k + R_h)q_{ss}$$

A perfect fin will have no conductive resistance, and therefore the fin efficiency can be expressed as:

$$\eta = \frac{q_{ss}}{q_{perfect}} = \frac{\frac{T_0 - T_\infty}{R_k + R_h}}{\frac{T_0 - T_\infty}{R_h}} = \frac{R_h}{R_k + R_h} = \frac{1}{R_k/R_h + 1}$$

The ratio of the conductive and convective resistances is shown to be the key to the thermal parameter of a fin. The average fin temperature can be expressed using a voltage divider analysis:

$$\frac{T_{Fin} - T_\infty}{T_0 - T_\infty} = \frac{R_h}{R_h + R_k} = \eta$$

For fins with high efficiency, the fin temperature is close to the wall temperature. For fins with low efficiency, the fin temperature is close to ambient.

The thermal resistances can be estimated for a wide variety of complicated shapes, generally by placing the node at the centroid of the fin. For the case of the constant-area fin defined by a base area A, and base perimeter P:

$$q_{ss} = \frac{T_0 - T_\infty}{\frac{L}{2kA} + \frac{1}{hPL}} = \frac{hPL(T_0 - T_\infty)}{\frac{hPL^2}{2kA} + 1} = \frac{hPL(T_0 - T_\infty)}{\left(\frac{L}{L^*}\right)^2 + 1} = \frac{hPL(T_0 - T_\infty)}{\hat{L}^2 + 1}$$

where a thermal length L* has been defined as:

$$L^* = \sqrt{\frac{2kA}{hP}}$$

and a dimensionless fin length as the ratio of the actual fin length to the thermal length:

$$\hat{L} = \frac{L}{L^*}$$

The fin temperature is expressed as:

$$T_{fin} = T_0 - \left(\frac{\hat{L}^2}{\hat{L}^2 + 1}\right)(T_0 - T_\infty)$$

Notice the limits:

$$\text{For} \quad \hat{L} \ll 1, \quad T_{fin} = T_0$$

that is, if the actual length of the fin (L) is much less than the thermal length of the fin (L*), then the fin temperature equals that of the base. Such a fin can be called a "thermally short" fin, and will behave well as a fin, but lousy as a handle (you'll burn your hand if you grab the end). Conversely:

$$\text{For} \quad \hat{L} \gg 1, \quad T_{fin} = T_\infty$$

That is, if the actual length is much longer than the thermal length, the fin will be at the ambient fluid temperature. Such a "thermally long" fin will be a lousy fin, but an excellent handle. An alternative 1-node model will be developed for thermally long fins shortly.

$$\text{For} \ \hat{L} = 1, \quad T_{fin} = (T_0 + T_\infty)/2$$

That is, the fin temperature is the average of the base and ambient temperature. In a revised model tailored to thermally long fins, this result is used to determine the distance over which the temperature actually changes.

The fin efficiency is:

$$\eta = \frac{q_{ss}}{q_{perfect}} = \frac{hPL(T_0 - T_\infty)/(\hat{L}^2 + 1)}{hPL(T_0 - T_\infty)} = \frac{1}{\hat{L}^2 + 1}$$

The fin efficiency equals 1 for thermally short fins and decreases monotonically with fin length.

The fin effectiveness is:

$$\varepsilon = \frac{q_{ss}}{q_{No\ Fin}} = \frac{hPL(T_0 - T_\infty)/(\hat{L}^2 + 1)}{hA(T_0 - T_\infty)} = \frac{PL/A}{\hat{L}^2 + 1}$$

For thermally short fins, the fin effectiveness increases with length. For thermally long fins, the fin effectiveness decreases with length. Conceptually, once the fin is thermally long, added extra length does not change the heat transfer rate, and therefore this model does not capture that effect. The alternative long fin model, to which attention now turns, will address that deficiency.

7.2.2   Thermally Long Fin Model for Handles

The fin model just developed can be applied to a fin of any length. However, if the fin is long compared to its thermal length ($L > L^*$), then the area near the end is not thermally involved, and defining the convective resistance based on the total surface area is problematic, conceptually. An alternative thermal resistance model, shown in Fig. 7.5, recognizes that there is a region adjacent to the base whose average temperature is known to be halfway between the base and ambient temperature. The size of that region is not known a priori, however, and is to be determined.

In this model, it is recognized that the actual temperature at the end of the fin is closer to the fluid temperature than the base temperature. Therefore the fin is broken into two segments. The segment adjacent to the wall (on the left) is fixed at a temperature midway between the wall and fluid temperatures. The boundary between this segment and the rest of the fin is considered to be adiabatic. That is, heat flows into the fin from the base, and out to the fluid through the sides of the first segment, but not further out along the fin. The distance from the wall to the first node is left unknown, and expressed as a fraction, f, of the

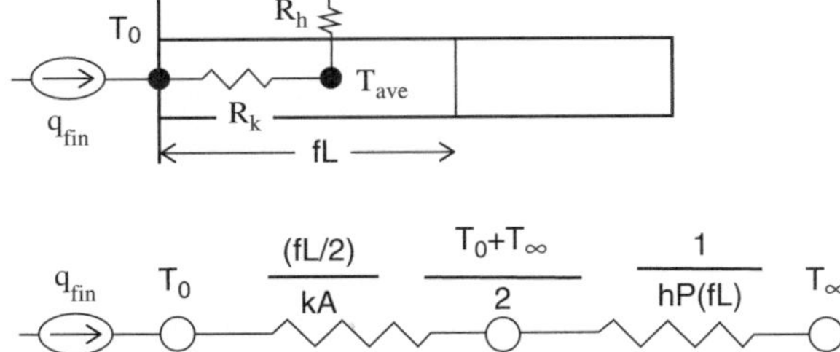

Fig. 7.5 Schematic and thermal resistance network for a thermally long fin at steady-state. The factor "f" is a fraction of the total length, whose value is solved for

total length. The temperature of the fin will be midway between the wall and the fluid when the convective and conduction resistances are equal. That is:

$$\frac{fL}{2kA} = \frac{1}{hPfL}$$

Solving for "f":

$$f = \frac{1}{L}\sqrt{\frac{2kA}{hP}} = \frac{L^*}{L} = \frac{1}{\hat{L}}$$

The heat transfer rate, recognizing the two resistances are equal, is:

$$q_{ss} = \frac{T_0 - T_\infty}{\dfrac{2}{hPfL}} = \frac{T_0 - T_\infty}{\dfrac{2\sqrt{hP}}{hP\sqrt{2kA}}} = \frac{\sqrt{hPkA}}{\sqrt{2}}(T_0 - T_\infty)$$

The exact solution for a long fin has exactly the same functional dependency, but without the factor $2^{1/2}$ in the denominator. It would be relatively easy to "adjust" the 1-node model to make the agreement exact.

The 1-node model captures the right physics, and the characteristic length L* is the same as that in the exact solution. Furthermore, if the basic logic is understood, it is easy to solve for fins with non-constant-area or other complicated geometry. Notice that the heat transfer rate is independent of the actual fin length. Adding length to a thermally long fin will merely add material that is already close to ambient. That's why a thermally long fin is a good handle. The fin efficiency is:

$$\eta = \frac{\sqrt{hPkA/2}(T_0 - T_\infty)}{hPL(T_0 - T_\infty)} = \frac{1}{2L}\sqrt{\frac{2kA}{hP}} = \frac{1}{2\hat{L}}$$

Both thermally short and thermally long models yield a fin efficiency of 0.5 when the dimensionless fin length equals unity, so the two models smoothly transition with length.

The exact solution (Appendix 1) for a constant-area fin, with a corrected length, is:

$$\eta_{Exact} = \frac{\tanh\left(\hat{L}\sqrt{2}\right)}{\hat{L}\sqrt{2}} = \frac{\tanh(mL)}{mL}$$

where the factor "m" found in traditional heat transfer literature has been introduced and is the inverse of L* (with a factor of $2^{1/2}$).

The efficiency for both the long and short fin model results are plotted versus dimensionless length and compared to the exact solution in Fig. 7.6. The functional dependency (shape of the curve) is basically right, although the predicted heat transfer rates are somewhat less than the exact result (lower by $2^{1/2}$ to be precise).

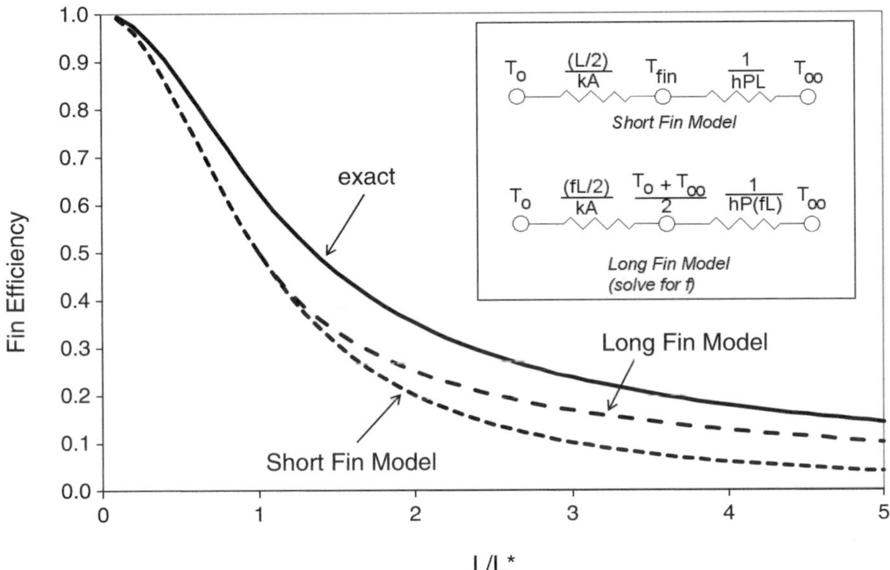

Fig. 7.6 Fin efficiency as a function of dimensionless fin length for the two 1-node models compared to the exact solution

Of most importance, the emergence and meaning of "m" is apparent in the simple 1-node model. A sound understanding of this parameter is all that is really needed for fin design. It is tempting to use this feedback to alter the model to give better quantitative agreement. However, the definition of R_k and R_h used here seems the most conceptually correct.

Figure 7.7 shows a spreadsheet programmed to solve for the 1-node rectangular fin. The numerical results are those of the problem statement given at the start of this section. An "if-then-else" statement is used for the heat transfer rate and fin efficiency to select the appropriate model (thermally short or thermally long). The user can change any of the seven input parameters, and an approximate solution for any rectangular fin follows.

7.2.3   Thick Fins

The basic flow path of heat in a fin is by conduction from the base of the fin to the interior of the fin, and lateral convection to the environment. However, a thick fin is defined as one in which the lateral conduction from the interior to the convective surface is important. In this case, a series resistance can be added to the convective fin resistance. That is, for a rectangular fin:

Fig. 7.7 Spreadsheet for 1-node model

INPUT PARAMETERS

t	0.001	m, fin thickness
w	0.25	m, fin width
L	0.1	m, fin length
k	200	W/m/K, fin thermal conductivity
h	50	W/m2/K, convection coef.
T0	100	oC, wall temperature
Tinf	0	oC, ambient temperature

DERIVED PARAMETERS

P	0.502	m, perimeter
A	0.0003	m2, base and fin area
Lc	0.1005	m, corrected length
L*	0.0631	m, thermal length
Rk	1	K/W, 1-node conductive resistance
Rh	0.3964	K/W
Lhat	1.5843	dimensionless fin length

RESULTS

q1node	79.215	Watts
ε	63.372	fin effectiveness
η,1node	0.3156	fin efficiency, 1node model
η,exact	0.4363	fin efficiency, exact model

$$R_h = \frac{1}{2}\left(\frac{(t/4)}{kwL} + \frac{1}{hwL}\right) = \frac{1}{2Lw}\left(\frac{ht/4}{k} + 1\right) = \frac{1}{2Lw}(B+1)$$

A lateral Biot number ($B = ht/4/k$) has been defined and shown to give a correction to the convective resistance. A similar procedure can be conducted for alternative shapes. Note that the lateral conduction distance of t/4 has been used. Conceptually, the centerline of the fin is an insulated surface, by symmetry, and the node is placed at the centroid of one half of the fin. The factor ½ reflects that there are two surfaces.

7.2.4 Orientation of Fins

When a fan or some other means of forcing a fluid flow across a fin, the orientation of the fin is not generally important. However, when there is no imposed flow (natural convection), the performance of a fin can be greatly affected by its orientation. For example, for a heat sink with many rectangular fins attached to a base plate, there is only one orientation that will work well without a forced convection, namely, the one with the base plate vertical and the fins oriented vertically, as shown in Fig. 7.8. For all the other orientations, the flow will not effectively be generated for the interior fins (Fig. 7.9).

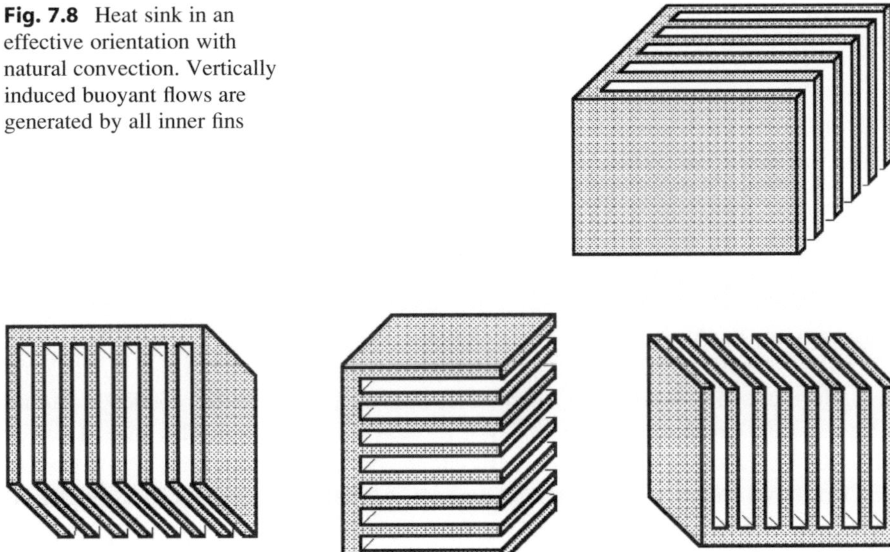

Fig. 7.8 Heat sink in an effective orientation with natural convection. Vertically induced buoyant flows are generated by all inner fins

Fig. 7.9 Ineffective orientations of heat sinks with natural convection. Vertically induced flows are blocked for inner fins

7.3   Half a Mug of Coffee

Here's another coffee story twist. My mom not only likes her coffee as hot as you can get it, she usually says "just half a cup," or "just an inch." This section considers the effect of coffee volume on the cooling time with a 1-node lumped transient analysis, but with a fin analysis applied to that portion of the mug wall that is above the coffee level.

The spreadsheet used for the 1-node lumped capacity model from Chap. 4 gives a first approximation by simply changing the input coffee volume. However, the model for the side channel was developed with the tacit understanding that the surface area exposed on the inside of the mug was small compared to the total exposed surface area. Therefore, a more detailed resistance model is developed here for the 1-node mug of coffee in which the exposed rim is treated like a heat transfer fin.

The main point of this chapter has already been made. Therefore, the next section, and the detailed appendices, provides detailed numerical and analytical applications, but those can be skipped if that level of detail is not of interest. The final results are worth reviewing, however, if all the details are not of interest.

Fig. 7.10 Schematic of a
partially filled mug of coffee

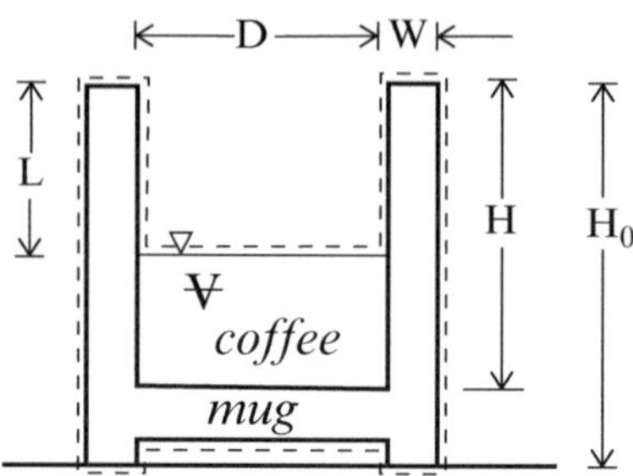

Engineering Problem Statement

Coffee (Fig. 7.10) at 1 atm, 95 °C is quickly poured into a room temperature (23 °C)
ceramic mug with mass of $m_{mug} = 0.335$ kg, inner diameter D = 0.083 m, inner
height H = 0.098 m, and nominal wall thickness w = 0.0040 m. The mug of coffee
is placed on a table in a still room. Estimate the time it takes for the water to cool to
70 °C (above which it is too hot) and to 50 °C (defined as "tepid"). Consider all
ranges of coffee amounts indicated by the distance between the top of the mug and
the coffee surface (L).

7.3.1   Pure Lumped Capacity Result

The computer programs written to solve the lumped capacity model can handle this
scenario by merely changing a single input parameter, the volume of coffee. The
thermal resistance between the system boundary (the mug outer surface plus the
free surface of the coffee) and ambient air ($R_{s\infty}$) is defined, using a heat transfer
coefficient (h) applied to the entire surface, as:

$$R = \frac{1}{hA_{surface}}$$

The capacitance is the sum of the coffee and mug. Table 7.1 compares the full, half,
and quarter mug cases for the pure lumped model, in which an effective heat
transfer coefficient $h_{as} = 15$ W/m^2/K is applied to the entire exposed surface area.
The exposed area increases (and thermal resistance decreases) as the volume of
coffee is reduced from a full cup to a half to a quarter due to the inner surface of the
rim. The capacitance (mug plus coffee) decreases. Therefore, the time constant is
reduced due to both a smaller thermal mass and an enhanced surface heat transfer.
The initial temperature (after the coffee and mug come to thermal equilibrium)
decreases as well due to a higher percentage of capacitance being attributed to the
initially cold mug. In fact, for a quarter mug, the coffee temperature after pouring is
below 70 °C as soon as the mug and coffee thermally equilibrate.

The pure lumped model assumes that the entire surface of the mug is at the bulk
temperature of the coffee. Is that a good model? For the full cup, it seems a
reasonable first approximation, since the external resistance was shown to dominate

Table 7.1 Effect of coffee volume on cooling time: pure lumped capacity analysis with $h = 15$ W/m^2/K applied to entire exposed surface

Quantity	Symbol	Unit	Full mug	Half a mug	Quarter mug
Coffee volume	V	L	0.500	0.250	0.125
Total exposed surface area	$A_{surface}$	cm^2	452.2	573.3	633.9
Resistance	R	°C/W	1.47	1.16	1.05
Capacitance	C	J/°C	2,392	1,342	817
Time constant	$\tau = RC$	min	58.6	25.9	14.3
Initial temperature	T_0	°C	83.3	76.6	66.6
Time to cool to 70 °C	$t_{too\ hot}$	min	14.6	3.4	NA
Time to cool from 70 °C (or initial temperature) to 50 °C	$t_{opportunity}$	min	32.5	14.4	6.85

the side channel. Furthermore, probing a hot mug of coffee with your fingertips confirms it to be uniformly hot. This situation would not exist for a Styrofoam mug, in which case the conductive resistance across the wall could not be considered a secondary effect. But for half a mug, the very top rim feels a little cooler than the side. For a quarter cup, the top of the rim feels like it might not even be warm. So something very different is happening in the mug above the level of the coffee compared to what is below it. Heat flows directly across the mug wall where it is exposed to coffee. Above that, heat must be flowing upward into the portion of the rim exposed to air, and then transferring heat to the air on both sides. The rim is acting like a fin.

For the purposes of the following series of analyses, the mug is divided into two parts. The "rim" of the mug is defined to be that portion of the mug above the level of coffee. The "mug" is what is left below. The heat transfer physics that takes place in the rim portion is very different than the conduction across the mug portion. Heat is transferred to the air by the rim, but in order to do so, heat must first conduct into the rim, parallel to the wall of the rim. It is intuitive that there is a vertical temperature gradient in the rim.

7.3.2   Modified Lumped Capacity Model

The lumped capacity model developed in Chap. 4 accounts for the rim effect in an approximate way. In that model, the surface area in R_{as} (between the mug outer surface and air) includes the inner surface exposed to air. Also, the conductive resistance between the inner mug wall and outer wall (a portion of which includes the inner wall above the level of the coffee) uses the geometric average of these two areas. Conceptually, this model is reasonable when the rim effects are small, that is, when the mug is nearly full.

Figure 7.11 shows a modified 1-node RC network designed for partially full mugs. The top and bottom channels for heat flow are the same as before. The side channel,

Fig. 7.11 Extended 1-node thermal resistance network of a half a mug of coffee

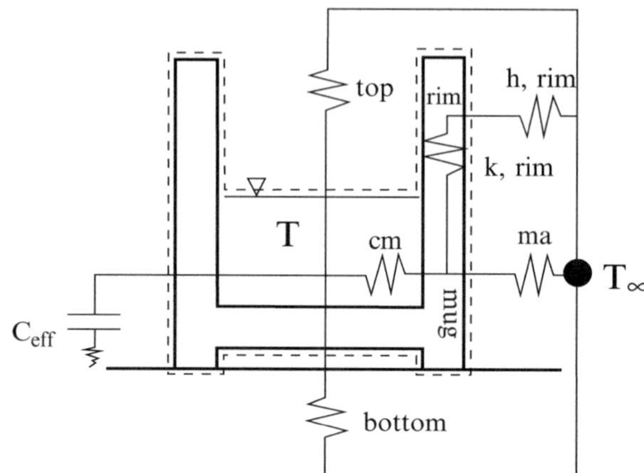

however, branches into two parallel channels. Heat enters the mug wall along the submerged surface and is split into a channel directly across the mug wall and a channel that flows vertically through the mug itself into the rim, and from the rim to ambient. The rim, while not necessarily designed this way, acts like a heat transfer fin.

The effective capacitance of the system is modified to account for the fact that the rim is not fully involved. The total capacitance in the model is the sum of the coffee capacitance, the mug capacitance below the level of the coffee, and a fraction, f, of the capacitance of the rim: $C_{eff} = \left(C_{coffee}\right) + \left(C_{mug} - C_{rim}\right) + \left(fC_{rim}\right)$

The value of "f" is determined from a subsequent analysis of the rim branch, where two different models are developed, depending on the thermal length of the rim.

The equivalent resistance of the entire system is given by a parallel network with three branches:

$$R = \frac{1}{\frac{1}{R_{top}} + \frac{1}{R_{side}} + \frac{1}{R_{bottom}}}$$

The equivalent resistance of the side channel is given by:

$$R_{side} = R_{cm} + \frac{1}{\frac{1}{R_{ma}} + \frac{1}{R_{k,rkm}+R_{h,rim}}}$$

The resistance between the coffee and the non-rim portion of the mug consists of a convective resistance between the coffee and the mug inner wall in series with a conductive resistance between the mug inner surface and center of the wall of the mug:

$$R_{cm} = \frac{1}{h_{coffee}\pi D(H_0 - L)} + \frac{(w/2)}{k_{mug}\pi(D + w/2)(H_0 - L)}$$

In this expression, the height of the area is taken to be the outside mug height (H_0) minus the rim height (L), which seems "wrong" but is used to account for edge effects on the floor. It is instructive to express this resistance and those to follow in terms of the convective resistance and a factor that has the ratio of conductive to convective resistance, a form of Biot number:

$$R_{cm} = \frac{1}{h_{coffee}\pi D(H_0 - L)}\left(1 + \frac{h_{coffee}(w/2)D}{k_{mug}(D + w/2)}\right)$$

Defining a Biot number to account for the inner portion, $B_{coffee} = h_{coffee}(w/2)/k_{mug}$:

$$R_{cm} = \frac{1}{h_{coffee}\pi D(H_0 - L)}\left(1 + \frac{B_{coffee}}{(1 + w/2D)}\right)$$

The geometric factor in the denominator of the Biot number arises from the different area associated with the convective resistance and the conductive resistance. Defining a modified Biot number (B') that incorporates the geometric factor:

$$B'_{coffee} = \frac{B_{coffee}}{(1 + w/2D)}$$

The resistance between the coffee and the mug is thus:

$$R_{cm} = \frac{1 + B'_{coffee}}{h_{coffee}\pi D(H_0 - L)}$$

Similarly, the resistance between the non-rim portion of the mug and ambient consists of a conductive resistance between the mug wall center and the outer surface in series with a convective/radiative resistance between the outer surface and ambient:

$$R_{ma} = \frac{(w/2)}{k_{mug}\pi(D + 3w/2)(H_0 - L)} + \frac{1}{h_{ma}\pi(D + 2w)(H_0 - L)}$$

In terms of the Biot number for the outer portion, $B_{air} = h_{air}(w/2)/k_{mug}$:

$$R_{ma} = \frac{1}{h_{air}\pi(D + 2w)(H_0 - L)}\left(1 + \frac{B_{air}(1 + 2w/D)}{(1 + {}^{3}w/_{2D})}\right)$$

Introducing a modified Biot number:

$$B'_{air} = \frac{B_{air}(1 + 2w/D)}{(1 + 3w/2D)}$$

The resistance between the mug and ambient air is:

$$R_{ma} = \frac{1 + B'_{air}}{h_{air}\pi(D + 2w)(H_0 - L)}$$

7.3.3   Rim Details

Since the mug and rim capacitances have been lumped into the coffee, the initial mug heating process cannot be observed in this model. In effect, the mug and the rim are both considered to be in a quasi-steady state equilibrium. They are changing with time, but not independently from the coffee temperature. The emphasis in this model is on estimating the effective thermal resistances associated with the rim, which involves the following complex web of individual conductive and convective resistances in series and parallel channels.

7.3.3.1 Rim Convective Resistance, $R_{h,rim}$

Heat leaves the rim by convection in parallel with thermal radiation all along its surface (inner, top edge, and outer) as shown in Fig. 7.12. In series with that, lateral (horizontal) conduction can be accounted for by introducing resistance from the centerline of the mug wall to the inner or outer surface. The rim branch convective resistance is related to system parameters as:

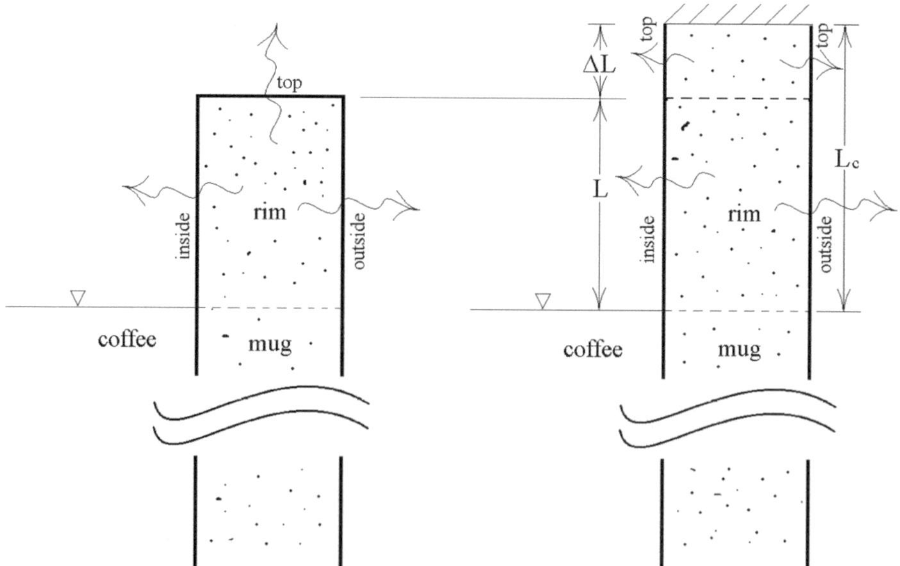

Fig. 7.12 Schematic showing the use of a corrected length to account for heat transfer from the top surface of the rim by attributing it to the sides

$$R_{h,rim} = \cfrac{1}{\cfrac{1}{R_{InnerSurface}} + \cfrac{1}{R_{OuterSurface}} + \cfrac{1}{R_{TopSurface}}}$$

There are several subtle differences between the inner and outer surfaces:

- The convective coefficient can be different due to the different type of natural convection taking place.
- The air temperature inside the mug is warm (and humid), and not at ambient temperature as implied by the resistance network, in which all the rim surface effects are combined into a single resistance.
- The radiative coefficient can be different since the inner wall "sees" a hot coffee surface looking down, the mug inner wall looking across and ambient temperature looking up.

In this model, all of these effects could be accounted for, however, by suitable choice of the inner convective/radiative coefficient (and/or a modified resistance network).

7.3.3.2 Corrected Length

Most of the action is anticipated to take place at the inner and outer wall surfaces. Rather than merely neglecting heat leaving the top, it can be accounted for by introducing a corrected length to the rim, sketched in Fig. 7.12. The top surface of the mug is imagined to be cut down its center and bent upward on both sides. The new top "surface" is insulated (no heat transfer leaves it), indicated with hatch marks. The corrected length, L_c, to be used for estimating convective heat loss from the rim, is merely the sum of the actual length, L, and the added length, ΔL: $L_c = L + \Delta L$. For this transient analysis, the added length is of a massless material in order to preserve the correct thermal capacitance.

The added length, ΔL, adds precisely the area of the top surface to the inner and outer wall:

$$A_{Top} = \left(\begin{array}{c} \text{Area Added} \\ \text{to Inner Wall} \end{array} \right) + \left(\begin{array}{c} \text{Area Added} \\ \text{to Outer Wall} \end{array} \right)$$

Expressed in terms of the geometry of the mug:

$$\pi \left[(D + 2w)^2 - D^2 \right] = \Delta L \pi D + \Delta L \pi (D + 2w)$$

Solving for the added length:

$$\Delta L = \frac{\left[(D + 2w)^2 - D^2\right]}{2(D + w)}$$

For thin walls $(2w \ll D)$, the analysis yields a simpler result from which the controlling physics is clearer:

$$A_{Top} = \begin{pmatrix} \text{Area Added} \\ \text{to Inner Wall} \end{pmatrix} + \begin{pmatrix} \text{Area Added} \\ \text{to Outer Wall} \end{pmatrix}$$

Expressed in terms of system geometry:

$$\pi D w = (\pi D \Delta L) + (\pi D \Delta L)$$

Solving for ΔL:

$$\Delta L_{ThinWall} = \frac{w}{2}$$

It is much easier to see the controlling physics in this thin wall limit than for the thick wall.

The thin-wall approximation and equal convection coefficients (h_{ma}) on the inside and outside surfaces will be used going forward. Lateral conduction effects could be included, giving rise to a form of Biot number as a correction factor to the convective resistance term.

The convective resistance for a thin-walled rim is given by:

$$R_{h,rim} = \frac{1}{h_{ma}A_{RimSurface}} = \frac{1}{h_{ma}2\pi D L_c}$$

The "2" arises in the denominator because there are two surfaces, inner and outer.

7.3.3.3 Rim Conductive Resistance

The rim branch conductive resistance is related to system parameters as:

$$R_{k,rim} = \frac{(\text{Average Conduction Distance})}{(\text{Conductivity})(\text{Conduction Area})} = \frac{(L/2)}{k_{mug}\frac{\pi}{4}\left[(D + 2w)^2 - D^2\right]}$$

The numerator is the distance to the centroid of the area $(L/2)$, which represents the average distance heat must conduct from the base of the rim. Again, the corrected length should not be used here. The denominator is the thermal conductivity times the area of the base of the rim.

For thin walls, this expression simplifies to:

$$R_{k,rim,THINWALL} = \frac{(\text{Average Conduction Distance})}{(\text{Conductivity})(\text{Conduction Area})} = \frac{(L/2)}{k_{mug}\pi Dw}$$

The determination of whether a given rim has thermally short or long behavior can be made by comparing the magnitude of the conductive thermal resistances between the base of the rim and the bulk of the rim ($R_{k,rim}$) with the convective/radiative resistance between the surface of the rim and the air ($R_{h,rim}$). For thermally short fins, the conductive resistance is less than the convective, and the average temperature of the rim is closer to the mug than to ambient. The thermal length is defined by equating the convective and conductive resistances:

$$\frac{(L^*/2)}{k_{mug}\pi Dw} = \frac{1}{h_{ma}2\pi DL^*}$$

Solving for the thermal length:

$$L^* = \sqrt{\frac{k_{mug}w}{h_{ma}}}$$

For the test case:

$$L^* = 0.017 \text{ m} = 1.7 \text{ cm}$$

If the actual length of the rim is shorter than the thermal length, then the rim is thermally short, otherwise, it is thermally long. The appropriate thermal resistances to use are summarized in the following table:

Rim resistance	Thermally short rim: L < L*	Thermally long rim: L > L*
$R_{k,rim}$	$\frac{(L/2)}{k_{mug}\pi Dw}$	$\frac{1}{2\pi D\sqrt{k_{mug}wh_{ma}}}$
$R_{h,rim}$	$\frac{1}{h_{ma}2\pi DL_c}$	$\frac{1}{2\pi D\sqrt{k_{mug}wh_{ma}}}$

7.3.3.4 Effective Rim Capacitance

Since not all of the capacitance is involved, the total lumped capacitance can be reduced appropriately to reflect that the rim temperature is between the mug temperature and ambient. For the capacitance for thermally short cases, the fraction, "f" of the capacitive action of the rim is defined (by instinct) to be the temperature ratio:

$$f_{short} = \frac{T_{rim} - T_\infty}{T_{mug} - T_\infty} = \frac{1}{\left(\frac{R_{k,rim}}{R_{h,rim}}\right) + 1}$$

For thermally long fins, the capacitance fraction "f" is defined by the ratio of half the thermal length divided by the actual length:

$$f_{long} = \frac{L^*}{2L}$$

In summary, the method outlined above for the "1-node Fin" is a versatile technique that can be used to approximate heat transfer rates and temperatures from any so-called extended surface. The mathematics is algebra based, and often yields simple closed-form relationships between geometric properties and thermal properties (k and h). The tricky part is in identifying resistances, and relating them to system parameters. This general method can be useful as a design tool to help assess performance of engineered heat transfer devices.

7.3.4　Results of Extended 1-Node Model

The coffee/mug problem is revisited as a 1-node transient solution but with the rim treated as a fin, using the resistance network of Fig. 7.11. In coding this problem, the thin-wall approximations are used, as is justified in this case ($hw/k \sim 0.05 \ll 1$). The thin-wall approximations yield simple closed-form expressions that more clearly highlight the underlying physics in the model. A main spreadsheet was used to perform the calculations, and a separate spreadsheet was used to produce results for the full range of coffee volume. Table 7.2 lists the input parameters used. The coffee volume was changed to generate a table of results. Table 7.3 lists all derived quantities, including individual resistances that make up the three thermal channels (not all of which are shown explicitly in Fig. 7.11), the three channel equivalent resistances, and the final results for effective capacitance, equivalent resistance, and the time constant. Note that there is an "if-then-else" statement for the fin resistance that uses either the short fin or long fin model as needed. The long fin model is used for this set of input parameter values because the actual rim length (L = 6.06 cm) is longer than the thermal length of the rim ($L^* = 1.7$ cm)

The behavior of the rim as a fin is demonstrated in Fig. 7.13 which shows the thermal resistance of the fin (the sum of the conductive and convective resistances in series) and the fin efficiency as a function of coffee volume. When the mug is nearly full, there is a high thermal resistance due to the low exposed surface area (the top edge is always present, even when filled to capacity). As the coffee volume is reduced, the thermal resistance drops rapidly as more surface area is exposed, and the conduction resistance into the rim is lower than the convective resistance (the short fin model). Below approximately 0.45 L volume, the rim resistance remains constant because the conductive resistance exceeds the convective, and the long fin model applies. The fin efficiency drops continuously as coffee volume is reduced. The efficiency is used to calculate the effective capacitance of the rim, considered next.

Figure 7.14 shows the thermal capacitance of the coffee, the mug, and the calculated initial temperature of the coffee/mug combination after the mug heating phase (during which time "thermal mixing" of the mug and coffee occur). The coffee capacitance is proportional to coffee volume, while the effective mug capacitance depends on the portion of the rim that is involved as a fin. This effective

Table 7.2 Input parameters used for partially filled mug 1-node analysis

Input quantities			
D	0.08255	m	Mug inner diameter
H	0.098	m	Mug inner height
H0	0.106	m	Mug outer height
V	2.00E−04	m^3	Coffee volume
w	0.003975	m	Mug thickness
mmug	0.335	kg	Mug mass
kmug	1	W/m/K	Mug thermal conductivity
rhomug	1,700	kg/m^3	Mug density
cmug	800	J/kg/K	Coffee density
rhocoffee	1,000	kg/m^3	Mug-specific heat
ccoffee	4,200	J/kg/K	Coffee-specific heat
hair/outside	13.9	$W/m^2/K$	Outside convection coefficient
hair/top	20	$W/m^2/K$	Convection coefficient between free surface and ambient
hcoffee/mug	400	$W/m^2/K$	Convection coefficient between coffee and mug
hcoffee/surface	400	$W/m^2/K$	Convection coefficient between coffee and free surface
hfloor	10	$W/m^2/K$	Effective coefficient of table
Tcoffee, init	95	°C	Initial coffee temperature
Tmug,initial	23	°C	Initial mug temperature
Tambient	23	°C	Ambient temperature

mug capacity is simply the total capacitance of the rim multiplied by the fin efficiency, plus the capacitance of the mug below the coffee level.

The initial coffee/mug temperature after the mug heating phase remains high as the coffee volume is decreased and decreases markedly below approximately 0.2 L as the effective capacitance of the mug becomes comparable to that of the coffee.

The thermal resistances of the three main channels are shown in Fig. 7.15. The top and bottom resistances are independent of the coffee volume because the top and bottom surface areas do not change. On the other hand, the side channel equivalent resistance decreases sharply with coffee volume, the primary cause of which is the increase in surface area across the sides. The parallel channel of heat into the rim reduces the equivalent side resistance somewhat, but since the rim resistance is always greater than approximately 16 °C/W (Fig. 7.13), the rim heat transfer does not have a dominant effect. Overall, the net equivalent resistance has a mild dependency on coffee volume.

Finally, the response time of the partially filled mug with the rim treated as a heat transfer fin, the time needed for the coffee to cool to drinkable temperature (70 °C) and the time to cool to a tepid temperature (50 °C) are shown in Fig. 7.16.

Table 7.3 Derived quantities and final results for partially filled mug 1-node model

Derived quantities			
H-L	0.0374	m	Depth of coffee
L	0.0606	m	Rim length
Crim	89.1	J/°C	Total capacitance of rim
Crim,effective	6.4		Effective capacitance of rim
Cmug	178.9	J/°C	Capacitance of mug (below rim)
Ccoffee	840	J/°C	Capacitance of coffee
Atop,bottom	0.0054	m^2	Top and bottom surface area
Aoutside	0.0106	m^2	Outside surface area below coffee level
Amiddleside	0.0102	m^2	Average surface area below coffee level
Ainside	0.0097	m^2	Inside surface area below coffee level
Tinitial	82.0	°C	Initial temperature (after mug heating)
rim conduction area	0.0011	m^2	Base area of rim (as fin)
rim surface area	0.0340	m^2	Convection area of rim (as fin)
L*	0.0169	m	Fin thermal length
fin efficiency	0.0722		Fin efficiency
Individual resistances			
Rtop,innerconv	0.47	°C/W	Between coffee and free surface
Rtop,outerconv	9.34	°C/W	Between free surface and ambient
Rbottom,innerconv	0.47	°C/W	Between coffee and bottom surface
Rbottom, wall	0.74	°C/W	Across bottom wall
Rbottom,floor	18.68	°C/W	Between bottom and table
Rcoffee, innersurface	0.26	°C/W	Between coffee and inner side surface
Rsidewall	0.39	°C/W	Across side wall (below coffee level)
Rside,outerconv	6.77	°C/W	Between ambient and outer side wall (below coffee)
Rcm	0.45	°C/W	Coffee to mid plane of side wall
Rma	6.97	°C/W	Mid plane of side wall to ambient
Rk,rim SHORT	28.06	°C/W	Conduction into rim
Rh,rim SHORT	2.11	°C/W	Convection from rim to ambient
Rh,rim = Rk,rim_LONG	8.20	°C/W	Long model resistances
Rfin	16.40	°C/W	Long fin model resistance
Channel resistances			
Rtop	9.81	°C/W	Equivalent top channel resistance
Rbottom	19.89	°C/W	Equivalent bottom channel resistance
Rside	5.34	°C/W	Equivalent side channel resistance
Results			
Ceff	1,025.3	J/°C	Effective total capacitance
Requiv	2.95	°C/W	Effective resistance
RC	3,021.5	s	Time constant
RC	50.4	min	Time constant

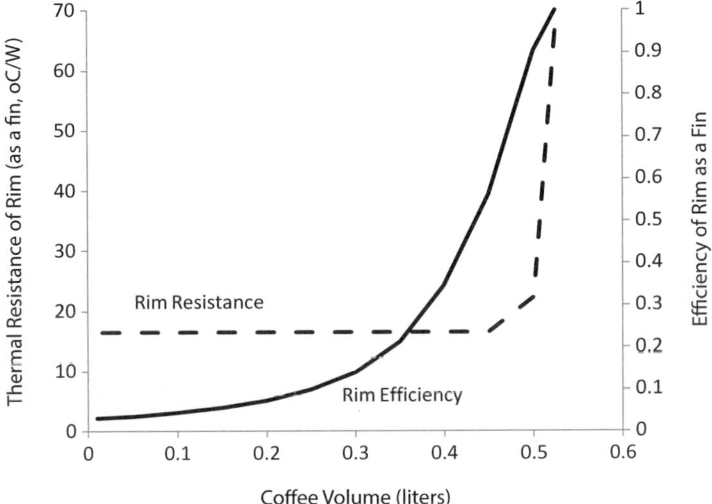

Fig. 7.13 Thermal resistance of the rim (acting as a fin) and the efficiency of the rim as a function of coffee volume

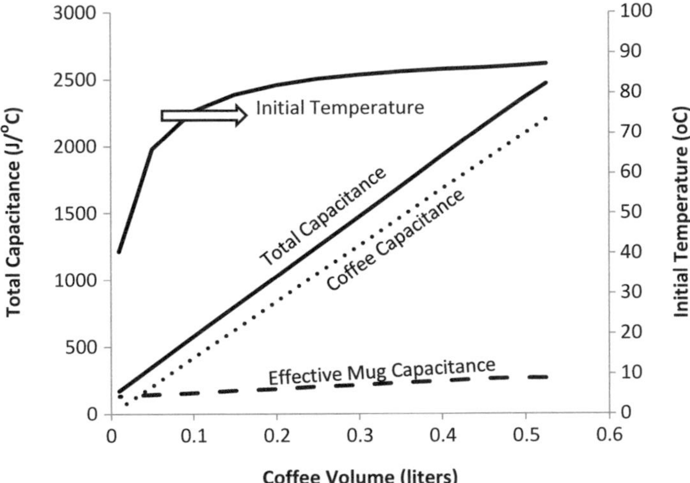

Fig. 7.14 Thermal capacitance of coffee and mug and initial temperature (after mug heating phase) as a function of coffee volume. 0.525 L constitutes a mug filled to capacity

The response and therefore cooling times increase with coffee volume. The increased capacitance has a stronger effect than the decrease in thermal resistance, since the capacitance increases with the volume, while the thermal resistance decreases with the surface area.

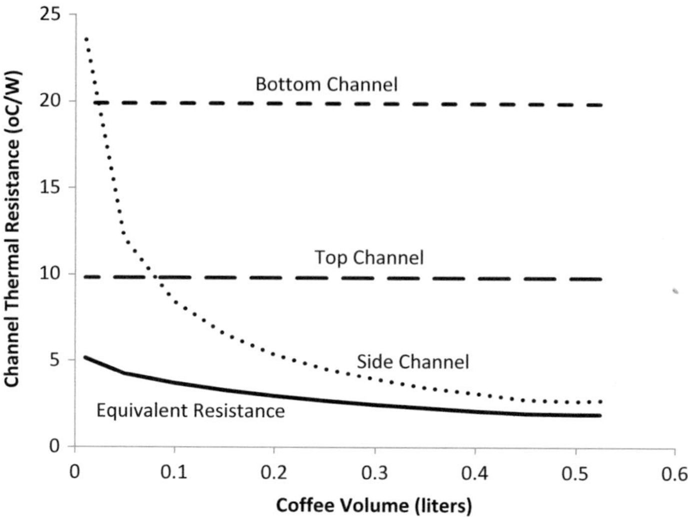

Fig. 7.15 Equivalent resistance of the three main thermal channels. The side channel includes the rim as a fin in series with the channel across the wall

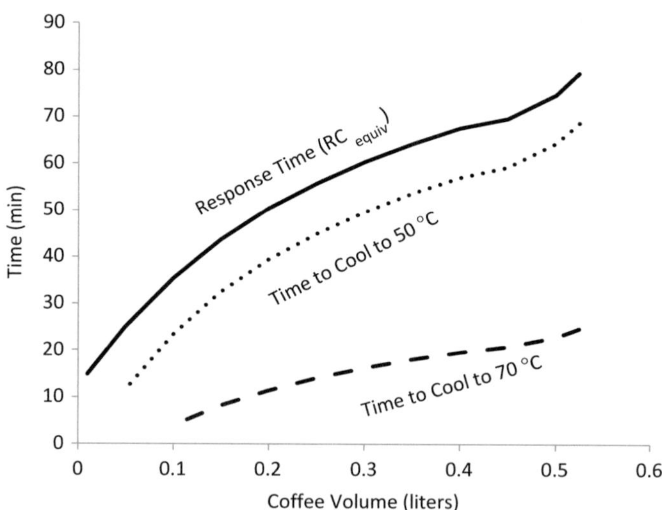

Fig. 7.16 Thermal response time for 1-node partially filled mug, and time to cool to drinking and tepid temperatures

7.4   3-Node Fin Analysis

Improved accuracy of the numerical method is obtained by discretizing space more finely. To demonstrate the basic method, the fin is broken into three equal sized elements, and nodes are placed in their centers (Fig. 7.17, top schematic). This grid is defined by first defining the nodal volumes and then placing the nodes at their centroids. Thermal resistances are shown between nodes and ambient as needed (bottom schematic). Each nodal element has a thermal capacitance associated with it, but since only the steady-state solution is sought, there is no change in energy stored, and the capacitances are not shown. A transient analysis using techniques developed previously would include appropriate capacitances for each node.

An alternative approach to a spatial grid would be to first place three unknown nodes, and then define the volumes that are associated with them, as in Fig. 7.18. In this case, a node is placed at the end, and that nodal element has half the volume of the other two. A good practice problem is to develop this approach and compare results. Conceptually, the finite volume approach is considered to be better, especially for transient problems, but the two methods yield comparable results. Care must always be taken when defining the resistances and capacitances to be consistent with whichever approach is taken.

There are three unknown temperatures, and three nodal equations (steady-state energy balances) are available. If all thermal resistances are treated as constant

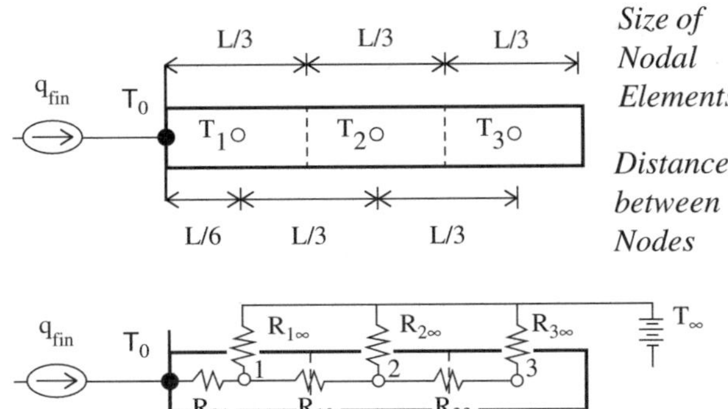

Fig. 7.17 Grid for 3-node network using the "define volumes, then place nodes" method (*top*) and resistance network (*bottom*)

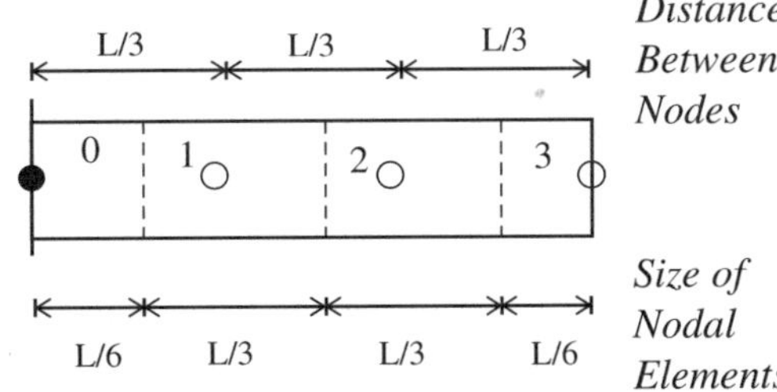

Fig. 7.18 3-Node grid using the "place nodes, then define volumes" method

(as is done here), these three equations constitute a closed set of linear algebraic equations that can be solved using a variety of methods (i.e., linear algebra). However, the objective here is to develop a numerical approach to solving steady-state problems, so this approach is not developed. Rather, the three nodal equations will be written directly into a form suitable for the Gauss–Seidel iteration method, a Method of Successive Substitution method, described in Appendix 3.

In general, as derived in Chap. 1, for a node "i" in thermal contact with neighbors "j", a nodal balance rearranged for node "i" in terms of its neighbors (some of which are also unknown) is:

$$T_i^{(k+1)} = \frac{\dot{Q}_i''' + \sum_j \dfrac{T_j^{(k,k+1)}}{R_{ij}}}{\sum_j \dfrac{1}{R_{ij}}}$$

In this general recursion equation, a superscript "k" is used to represent iteration number, and "k + 1" refers to the next iteration. For unknown neighbors, whether "k" or "k + 1" is used depends on the order in which the calculations are conducted. Superscript "k" represents a guess for an unknown temperature, while "k + 1" represents an improved guess. A general demonstration of a spreadsheet implementation of Gauss–Seidel is shown in Fig. 7.19. Input (and derived, not shown) parameters can be organized in a block of cells at the top. The iteration table placed below that (or in another worksheet) has a heading for iteration number and the three unknown temperatures. An initial guess to each unknown is made in the first row ("k" = 0). The three nodal recursion equations are entered into the "k" = 1 row. The arrows indicate which cells are referred to when the equations are inserted. The formulas also refer to the parameter block. The equations can be entered in any order, but the use of "k" or "k + 1" must be consistent with that order or a circular error will arise. Once these recursion relations are entered, the formulas can be copied down for as many rows are needed for convergence of the solution. Convergence is guaranteed if all the thermal resistances are constants. If there are variables, the method may fail, but often (usually for heat transfer problems) the method works just fine.

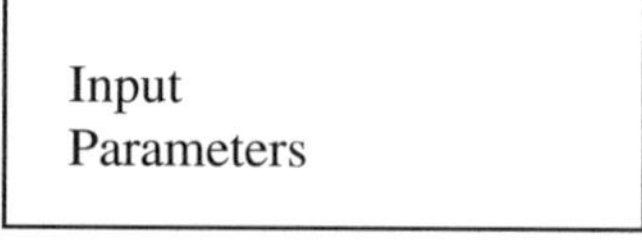

Fig. 7.19 Demonstration of a spreadsheet implementation of the Gauss–Seidel method

For this 3-node problem, the nodal equations will be implemented in the order 1, 2, 3. For the problem at hand, there are no heat sources. The base of the fin (fixed temperature T_0) receives a heat source, so a nodal balance there will yield the required value of the heat source (q_{fin}) once T_1 is solved for. The equation for node 1 is:

$$T_1^{(k+1)} = \frac{\dfrac{T_0}{R_{01}} + \dfrac{T_2^{(k)}}{R_{12}} + \dfrac{T_\infty}{R_{1\infty}}}{\dfrac{1}{R_{01}} + \dfrac{1}{R_{12}} + \dfrac{1}{R_{1\infty}}}$$

Notice that the superscripts on the right hand side of the first equation used are all "k", representing a "guess" for that temperature. There is only one of them in this case, but a more complicated network would have more. Nodal equation 2 is:

$$T_2^{(k+1)} = \frac{\dfrac{T_1^{(k+1)}}{R_{12}} + \dfrac{T_3^{(k)}}{R_{23}} + \dfrac{T_\infty}{R_{2\infty}}}{\dfrac{1}{R_{12}} + \dfrac{1}{R_{23}} + \dfrac{1}{R_{2\infty}}}$$

Here, an improved guess (k + 1) for T_1 is used in the recursion relation for T_2 since it is available. Nodal equation 3 is:

$$T_3^{(k+1)} = \frac{\dfrac{T_2^{(k+1)}}{R_{23}} + \dfrac{T_\infty}{R_{3\infty}}}{\dfrac{1}{R_{23}} + \dfrac{1}{R_{3\infty}}}$$

The improved guess for T_2 is used here. A spreadsheet implementation is shown in Fig. 7.20 for the original problem statement. The top block contains direct input parameters, the same as for the 1-node case, and the user of this program can simply insert different values for other problems that fall in the rectangular fin geometry. The derived parameters include primarily the resistance formulas, as follows.

The three conductive resistances are:

$$R_{01} = \frac{L/6}{kA} \quad \text{and} \quad R_{12} = R_{23} = \frac{L/3}{kA}$$

Notice that the distance from the wall to node 1 is half that between node 1 and 2, or between 2 and 3. The convective resistances are:

$$R_{1\infty} = R_{2\infty} = \frac{\bar{x}_{1\infty}}{k\bar{A}_{1\infty}} + \frac{1}{hP(L/3)} \quad \text{and} \quad R_{3\infty} = \frac{\bar{x}_{3\infty}}{k\bar{A}_{3\infty}} + \frac{1}{h[P(L/3) + A]}$$

The surface area associated with node 3 contains the sides, and the end of the fin. The first term represents the lateral conduction across the thickness of the fin, and is neglected here, but it could be easily included, approximated in a similar manner to the conduction across a cylinder or sphere in the 1-node transient analysis.

Fig. 7.20 Spreadsheet implementation of 3-node model showing 10 iterations and final converged solution after 100 iterations

INPUT PARAMETERS

t	0.001	m, fin thickness
w	0.25	m, fin width
L	0.1	m, fin length
k	200	W/m/K, fin thermal conductivity
h	50	W/m2/K, convection coef.
T0	100	oC, wall temperature
Tinf	0	oC, ambient temperature

DERIVED PARAMETERS

P	0.502	m, perimeter
A	0.0003	m2, fin area

thermal resistances

R01	0.3333	K/W, between base and 1
R12	0.6667	K/W, between 1 and 2
R23	0.6667	K/W, between 2 and 3
R1inf	1.1952	K/W, between 1 and ambient
R2inf	1.1952	K/W, between 2 and ambient
R3inf	1.1776	K/W, between 3 and ambient

GAUSS-SEIDEL ITERATION

iteration	T1	T2	T3
0	0	0	0
1	56.21487	21.97801	14.03349
2	62.39232	29.87977	19.07896
3	64.6133	32.72069	20.89296
4	65.41181	33.74209	21.54514
5	65.6989	34.10931	21.77963
6	65.80212	34.24134	21.86393
7	65.83923	34.28881	21.89424
8	65.85257	34.30588	21.90514
9	65.85737	34.31201	21.90905
10	65.85909	34.31422	21.91046

CHECK FOR CONVERGENCE

iteration	T1	T2	T3
99	65.86006	34.31546	21.91125
100	65.86006	34.31546	21.91125

The Gauss–Seidel iteration table contains a first guess (set to be equal to the fluid temperature, but anything could be placed here), and the second row contains the recursion formulas. Ten iterations are shown, and then the 99th and 100th iterations are shown as a check for convergence. In this case, convergence to within seven significant figures is demonstrated. Between the 10th and 99th iteration, there are

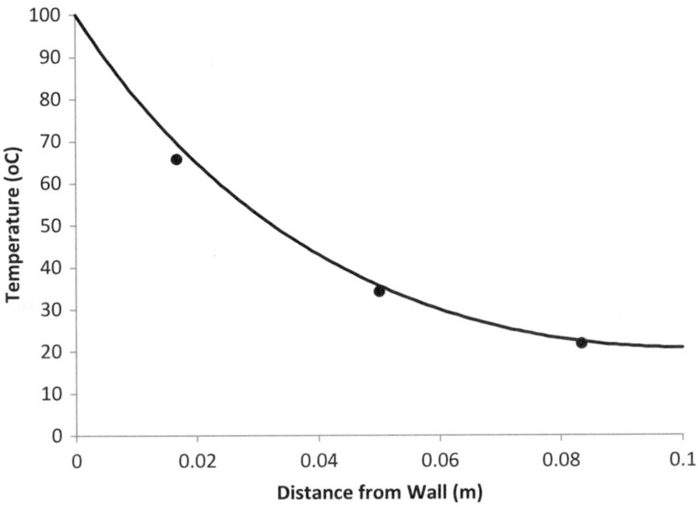

Fig. 7.21 Temperature vs. distance from the wall for the exact solution (*solid line*) and the 3-node model (*filled circles*)

Table 7.4 Comparison of the numerical models and the exact solution

Model	Heat transfer rate (W)	Percent error
1-Node	79.2	−27.7
3-Node	102.4	−6.5
Exact	109.6	–

changes to the fourth significant figure. In a computer program, a while loop test for convergence is usually a better practice than a "for loop," but spreadsheets really only allow the effect of a "for loop." When a test for convergence is made, it is important to test all nodes for convergence, not just one. It is common that a given node may not change for a particular iteration, but some of its neighbors will. At the next iteration, that given node will change, and a different one may not.

A comparison of the results of the 3-node model with the exact solution (using a corrected length) is made in Fig. 7.21 which shows the temperature and Table 7.4, which shows the heat transfer rate which is simply $q_{3\text{-}node} = (T_0 - T_1)/R_{01}$. The temperature predicted by the 3-node model is somewhat below the exact solution, but the agreement is really remarkable. The error in the heat transfer rate is 6.5 %, considered quite a good agreement for such a coarse grid. A more general "n-node" model would be straightforward to implement if improved accuracy were needed. Be ever mindful, however, that the underlying uncertainty in the inherent assumptions of the problem statement (especially the convection coefficient) completely dwarf those made here. The comparison of the numerical model with the "exact" model does not mean the "exact" model captures the real problem accurately. For example, the convection coefficient is highly uncertain, and may vary with position. The fluid temperature might not really be uniformly ambient, depending on the location of other fins. And so on...

Appendix 1. Constant-Area Fins: Differential Solution

Figure 7.2 shows an extended surface (fin) that has a base area A and perimeter p. The shape of the surface remains constant in the x direction, perpendicular to the base. The total length of the surface is L, and the material has a thermal conductivity k. The fin is exposed to a fluid at temperature T_∞ with an effective heat transfer coefficient, h. The fin is initially at the fluid temperature, and the base temperature is suddenly changed and maintained at a temperature T_0.

Model Development

A detailed derivation of the governing PDE for this case is developed for a general transient fin case, including lateral conduction. A cross-section of a differential slice of thickness dx at an arbitrary distance x placed at the center of the element is shown in Fig. 7.22. Heat enters the left face (at $x - dx/2$) by conduction, and splits into two channels, one leaving the surface into the fluid, and the other by conduction (at $x + dx/2$) into the adjacent slice.

A verbally stated energy balance applied to the slice, consistent with the assumed directions shown by the arrows of Fig. 7.22, is:

$$\begin{pmatrix} \text{Rate of Change of} \\ \text{Energy Stored in slice} \end{pmatrix} = \begin{pmatrix} \text{Rate of Conduction} \\ \text{into left face} \end{pmatrix} - \begin{pmatrix} \text{Rate of Conduction} \\ \text{out of right face} \end{pmatrix} - \begin{pmatrix} \text{Rate of Convection} \\ \text{out of sides} \end{pmatrix}$$

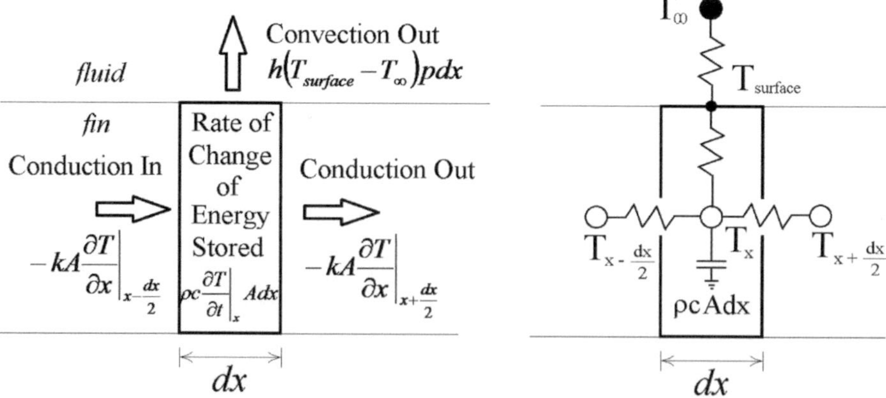

Fig. 7.22 Energy flows in a differential element in an extended surface. The right sketch shows the element in a resistance network, which allows conductive resistance between the center and the surface exposed to the fluid

Expressed mathematically:

$$\left(\rho c A dx \frac{\partial \overline{T}}{\partial t}\right) = \left[-kA \frac{\partial \overline{T}}{\partial x}\right]_{x-dx/2} - \left[-kA \frac{\partial \overline{T}}{\partial x}\right]_{x+dx/2} - hp dx\left(T_{surface} - T_\infty\right)$$

In the transient and conduction terms, the temperature, $\overline{T}$, represents the average temperature of the cross-section. In the convection term, the temperature is the surface temperature.

The conduction at $x + \Delta x$ (where $\Delta x = - dx/2$ or $+ dx/2$) can be related to that at the center (x) using a Taylor expansion:

$$\left[-kA \frac{\partial \overline{T}}{\partial x}\right]_{x+\Delta x} = \left[-kA \frac{\partial \overline{T}}{\partial x}\right]_{x} + \frac{\partial}{\partial x}\left[-kA \frac{\partial \overline{T}}{\partial x}\right]_{x} \Delta x + \cdots$$

The net conduction into the differential element is therefore:

$$\left[-kA \frac{\partial \overline{T}}{\partial x}\right]_{x-dx/2} - \left[-kA \frac{\partial \overline{T}}{\partial x}\right]_{x+dx/2} = \left(\left[-kA \frac{\partial \overline{T}}{\partial x}\right]_{x} + \frac{\partial}{\partial x}\left[-kA \frac{\partial \overline{T}}{\partial x}\right]\left(\frac{-dx}{2}\right)\right)$$

$$- \left(\left[-kA \frac{\partial \overline{T}}{\partial x}\right]_{x} + \frac{\partial}{\partial x}\left[-kA \frac{\partial \overline{T}}{\partial x}\right]\left(\frac{dx}{2}\right)\right)$$

$$\left[-kA \frac{\partial \overline{T}}{\partial x}\right]_{x-dx/2} - \left[-kA \frac{\partial \overline{T}}{\partial x}\right]_{x+dx/2} = \frac{\partial}{\partial x}\left[kA \frac{\partial T}{\partial x}\right]_{x} dx$$

The surface temperature can be related to the average temperature using a voltage divider branch between the center of the fin and its surface:

$$\frac{\overline{T} - T_\infty}{R_k + R_h} = \frac{T_{surface} - T_\infty}{R_h}$$

The resistances are in the lateral direction (y and z). Solving for the surface temperature:

$$T_{surface} - T_\infty = \left(\frac{R_h}{R_h + R_k}\right)\left(\overline{T} - T_\infty\right) = \frac{\overline{T} - T_\infty}{1 + {R_k}/{R_h}}$$

The conductive resistance is modeled as:

$$R_k = \frac{(A/p)/2}{k(p\Delta x/2)}$$

The numerator, $(A/p)/2$, is half of the characteristic half thickness of the section, which is the average conduction distance from the surface to the center. The area in the denominator is the average between the area at the center (zero) and at the surface. The convective resistance is straightforward:

$$R_h = \frac{1}{hp\Delta x}$$

The ratio is:

$$\frac{R_k}{R_h} = \frac{h(A/p)}{k} \equiv B$$

The governing partial differential equation (energy balance) is therefore:

$$\rho c A \frac{\partial \overline{T}}{\partial t} = \frac{\partial}{\partial x}\left[kA \frac{\partial \overline{T}}{\partial x} \right]_x - \frac{hp}{1+B}(\overline{T} - T_\infty)$$

The factor "B", a Biot number, accounts for temperature variation in the lateral direction, an effect that is usually neglected (and justifiably so) for fins.

For a rectangular fin, with $w \gg t$:

$$\frac{A}{p} = \frac{tw}{2(t+w)} = \frac{t}{2(1+t/w)} \approx \frac{t}{2}$$

Introducing a dimensionless temperature that has a value of zero at the base, and approaches unity as it approaches ambient:

$$\theta = \frac{\overline{T} - T_0}{T_\infty - T_0}$$

Introducing a dimensionless time and distance with normalization constants to be determined shortly:

$$\hat{t} = \frac{t}{\tau} \qquad \hat{x} = \frac{x}{L_t}$$

Assuming constant thermal conductivity and area and putting it together:

$$\frac{\rho c A(T_\infty - T_0)}{\tau} \frac{\partial \theta}{\partial \hat{t}} = \frac{kA(T_\infty - T_0)}{L_t^2} \frac{\partial^2 \theta}{\partial \hat{x}^2} - \frac{hp}{1+B}(T_0 + \theta(T_\infty - T_0) - T_\infty)$$

Simplifying and rearranging:

$$\frac{\rho c L_t^2}{k\tau} \frac{\partial \theta}{\partial \hat{t}} = \frac{\partial^2 \theta}{\partial \hat{x}^2} + \frac{hpL_t^2}{(1+B)kA}(1 - \theta)$$

The time constant and characteristic dimension can be defined by setting the coefficients to unity. For the convection term:

$$L_t = \sqrt{\frac{(1+B)kA}{hp}}$$

For the transient term:

$$\tau = \frac{\rho c L_t^2}{k}$$

The governing partial differential equation becomes:

$$\frac{\partial \theta}{\partial \hat{t}} = \frac{\partial^2 \theta}{\partial \hat{x}^2} - \theta + 1$$

Since the governing equation is first order with respect to time, a single initial condition must be specified, for problems that are in fact transient. For example, in the problem statement, the fin is initially at the fluid temperature (T_∞), and the base temperature is suddenly changed and maintained at T_0. Therefore, the initial condition (in dimensions) is:

$$T(x,0) = T_\infty$$

In dimensionless variables:

$$\theta(\hat{x},0) = 1$$

The governing equation is second order with respect to x, and therefore two conditions are required. One condition is that the base temperature is specified:

$$T(0,t) = T_0$$

In dimensionless variables:

$$\theta(0,\hat{t}) = 0$$

The second condition on x is trickier. The precise statement of it is that the rate of heat conduction to the end (at $x = L$) must equal the rate of convection away from the end:

$$-kA\frac{\partial T}{\partial x}\bigg|_{x=L} = hA\left(T_{(L,t)} - T_\infty\right)$$

In terms of dimensionless variables, with a dimensionless fin length $\hat{L} \equiv L/L_t$:

$$\frac{-k(T_\infty - T_0)}{L_t} \frac{\partial \theta}{\partial \hat{x}}\bigg|_{\hat{x}=\hat{L}} = h(T_\infty - T_0)\theta_{(\widetilde{L,t},\hat{t})}$$

$$\frac{\partial \theta}{\partial \hat{x}}\bigg|_{\hat{x}=\hat{L}} = -\left(\frac{hL_t}{k}\right)\theta_{(\hat{L},\hat{t})}$$

This type of boundary condition, a convection boundary condition, involves a gradient of the variable and the variable itself. It dramatically increases the algebra required to solve the problem and results in a complicated solution.

There is an alternative approach to the second boundary condition which will be developed here, namely the corrected length method. In it, as sketched in Fig. 7.3, the heat transfer out the end of the fin is considered by making the fin slightly longer than the actual fin, with an insulated end.

The length of the fin used is called a "corrected" length, L_c, expressed as the original length plus the added length:

$$L_c = L + \Delta L$$

The area added onto the sides by the added length (the perimeter times the added length ΔL) is the same as the end of the fin. That is:

$$p\Delta L = A$$

The corrected length is therefore:

$$L_c = L + \frac{A}{p}$$

The boundary condition on the end of the extended fin is determined by an energy balance at the end, in which heat conducted to the end from the left is set to the heat convected away, which is set to zero:

$$-kA\frac{\partial T}{\partial x}\bigg|_{x=L_c} = 0$$

Introducing dimensionless variables and cancelling factors results in a simpler boundary condition:

$$\frac{\partial \theta}{\partial \hat{x}}\bigg|_{\hat{x}=L_c/L_t} = 0$$

To restate the result of the model development, the governing equation and initial and boundary conditions are:

$$\text{Governing Equation}: \frac{\partial \theta}{\partial \hat{t}} = \frac{\partial^2 \theta}{\partial \hat{x}^2} - \theta + 1$$

$$\text{Initial Condition}: \theta(\hat{x}, 0) = 1$$

$$\text{Boundary Condition at base}: \theta(0, \hat{t}) = 0$$

$$\text{Boundary Condition at end}: \left.\frac{\partial \theta}{\partial \hat{x}}\right|_{\hat{x}=L_c/L_t} = 0$$

The governing equation is a linear partial differential equation with three terms (transient, conduction, convection) and no parameters (they are "buried" in the dimensionless variables). The second boundary condition does contain a parameter (its location, namely the dimensionless thermal length of the fin). The use of dimensionless variables has reduced the number of parameters in the system from seven (A, p, L, k, h, T_0, T_∞) to one.

As is often the case in heat transfer applications, it is not so much the temperature that is sought, but the heat transfer rate. For a fin, the rate at which heat transfer enters the base of the fin is:

$$q_{fin} = -kA \left.\frac{\partial T}{\partial x}\right|_{x=0}$$

In dimensionless variables:

$$q_{fin} = -\frac{kA(T_\infty - T_0)}{L_t} \left.\frac{\partial \theta}{\partial \hat{x}}\right|_{x=0}$$

Introducing the thermal length:

$$q_{fin} = \left.\frac{\partial \theta}{\partial \hat{x}}\right|_{x=0} \sqrt{\frac{hpkA}{1+B}}(T_0 - T_\infty)$$

The derivative of dimensionless temperature with respect to dimensionless distance is all that is required to determine the heat transfer rate.

Model Implementation: Analytical Solution

The general solution can be expressed as the sum of a transient and steady-state solutions:

$$\theta = \theta_{transient}(\hat{x}, \hat{t}) + \theta_{ss}(\hat{x})$$

The transient solution involves all three terms and requires a lot of work to develop a complete solution that satisfies the initial and boundary conditions. The full solution is developed here, mainly to demonstrate the basic method of solving linear partial differential equations. The steady-state solution is much easier to obtain, and if the problem can be effectively modeled as being at steady-state, it can be used without bothering with the transient case.

In practice, however, the transient response of a heat fin may be important, for example, for electronic equipment that is cycled on and off. During the initial start-up period, the rate of heat transfer into a fin can start high, as heat is going to change the energy stored, and gradually decay to its steady-state response and dissipate heat to ambient. A good design might take advantage of that behavior.

Transient Solution

The governing equation for the transient solution is:

$$\frac{\partial \theta_{transient}}{\partial \hat{t}} = \frac{\partial^2 \theta_{transient}}{\partial \hat{x}^2} - \theta_{transient}$$

Note that the term "1" in the governing equation will be grouped into the steady state solution. Using the separation of variables technique, the solution is assumed to be expressible as two separable functions: one depending on position, and the other on time:

$$\theta_{transient} = \mathrm{T}(\hat{t})X(\hat{x})$$

Insertion into the governing equation:

$$X\frac{\mathrm{d}\mathrm{T}}{d\hat{t}} = \mathrm{T}\left(\frac{d^2X}{d\hat{x}^2} - X\right)$$

Dividing through by XT:

$$\frac{1}{\mathrm{T}}\frac{\mathrm{d}\mathrm{T}}{d\hat{t}} = \left(\frac{1}{X}\frac{d^2X}{d\hat{x}^2} - 1\right) = -\lambda^2$$

In this expression, the term on the left is a function of time only, the term in the middle is a function of position only, and it is recognized that that can only be possible if they are both equal to at most a constant, called a separation constant, and given the symbol λ, whose value will be determined by application of the initial and boundary conditions. The negative sign avoids complex numbers and squaring makes for a cleaner final solution (not obvious why at this point).

The time-dependent equation is:

$$\frac{\mathrm{d}\mathrm{T}}{d\hat{t}} + \lambda^2\mathrm{T} = 0$$

The general solution is:

$$T = c_0 e^{-\lambda^2 \hat{t}}$$

The space-dependent equation is:

$$\frac{d^2 X}{d\hat{x}^2} + \lambda^2 X = 0$$

The general solution is:

$$X = c_1 \sin \lambda \hat{x} + c_2 \cos \lambda \hat{x}$$

The full general solution, in terms of undetermined coefficients $A = c_0 c_1$ and $B = c_0 c_2$ is:

$$\theta_{transient} = e^{-\lambda^2 \hat{t}} \left(A \sin \lambda \hat{x} + B \cos \lambda \hat{x} \right)$$

The values of A, B, and λ are determined by application of the initial and boundary conditions. First, applying the boundary condition at the base of the fin, $\theta(0, \hat{t}) = 0$

$$0 = e^{-\lambda^2 \hat{t}} \left(A \sin \lambda 0 + B \cos \lambda 0 \right) = B e^{-\lambda^2 \hat{t}}$$

The only way this condition can be satisfied is if the coefficient B is set to zero. The general solution, after application of boundary condition at the wall, has been reduced to:

$$\theta_{transient} = A e^{-\lambda^2 \hat{t}} \sin \lambda \hat{x}$$

Application of the second boundary condition:

$$\left. \frac{\partial \theta}{\partial \hat{x}} \right|_{\hat{x} = \hat{L} = L_c / L_t} = 0 = A\lambda e^{-\lambda^2 \hat{t}} \cos \lambda \hat{L}$$

Where a dimensionless length (the thermal length of the fin) has been defined as:

$$\hat{L} = \frac{L_c}{L_t}$$

The only way this boundary condition can be met (without setting A or λ to zero, which would make satisfying the subsequent initial condition impossible) is to set the separation constant to specific values, namely:

$$\cos \lambda \hat{L} = 0$$

This constraint is true for the following infinite number of discrete values of the separation constant, namely:

$$\lambda\hat{L} = \frac{\pi}{2}, \frac{3\pi}{2}, \frac{5\pi}{2}, \ldots, \frac{(2n+1)\pi}{2}, \ldots$$

These "quantized" values of l are called eigenvalues.

The solution can now be expressed as the sum of all possible solutions (prior to satisfying the initial condition) for these discrete values of λ, with a different coefficient (A) possible for each term:

$$\theta_{transient} = A_0 e^{-\left(\frac{\pi}{2L}\right)^2 \hat{t}} \sin\frac{\pi\hat{x}}{2\hat{L}} + A_1 e^{-\left(\frac{3\pi}{2L}\right)^2 \hat{t}} \sin\frac{3\pi\hat{x}}{2\hat{L}} + \cdots$$

$$+ A_n e^{-\left(\frac{(2n+1)}{2L}\right)^2 \hat{t}} \sin\frac{(2n+1)\pi\hat{x}}{2\hat{L}} + \cdots$$

Expressed in terms of an infinite series:

$$\theta_{transient} = \sum_{n=0}^{\infty} A_n e^{-\left(\frac{2n+1}{2L^2}\right)\hat{t}} \sin\frac{(2n+1)\pi\hat{x}}{2\hat{L}}$$

Notice that the transient solution decays to zero as time goes to infinity. Last, but not least, the initial condition must be satisfied:

$$\theta(\hat{x}, 0) = 1$$
$$1 = \sum_{n=1}^{\infty} A_n \sin\frac{(2n+1)\pi\hat{x}}{\hat{L}}$$

To determine the coefficients A_n, the final step in this tortured affair, multiply both sides by a sin term with an integer "m" as a frequency, and integrate over the range of x:

$$\int_{\hat{x}=0}^{\hat{L}} \sin\frac{(2m+1)\pi\hat{x}}{2\hat{L}} d\hat{x} = \sum_{n=1}^{\infty} \left[\int_{\hat{x}=0}^{\hat{L}} A_n \sin\frac{(2m+1)\pi\hat{x}}{2\hat{L}} \sin\frac{(2n+1)\pi\hat{x}}{2\hat{L}} d\hat{x} \right]$$

The integral on the right is zero (sketch two sinusoids of different frequencies, and look at their areas) when m is not equal to n. That is, for every value of n, all but one of the terms in the infinite series are identically zero. Therefore:

$$\int_{\hat{x}=0}^{\hat{L}} \sin\frac{(2n+1)\pi\hat{x}}{2\hat{L}} d\hat{x} = \int_{\hat{x}=0}^{\hat{L}} A_n \left(\sin\frac{(2n+1)\pi\hat{x}}{2\hat{L}} \right)^2 d\hat{x}$$

Performing the integration on the left hand side:

$$\int_{\hat{x}=0}^{\hat{L}} \sin \frac{(2n+1)\pi\hat{x}}{2\hat{L}}\, d\hat{x} = \frac{2\hat{L}}{(2n+1)\pi}\left| -\cos \frac{(2n+1)\pi\hat{x}}{2\hat{L}}\right|_0^{\hat{L}}$$

$$= \frac{-2\hat{L}}{(2n+1)\pi}\left(\cos \frac{(2n+1)\pi}{2} - 1\right) = \frac{2\hat{L}}{(2n+1)\pi}$$

The right hand side, with the help of a trigonometric identity for $\sin^2 x = (1 - \cos x)/2$, evaluates to a single number (noting that the second term involves a cosine integrated over a full cycle, and its integral is thus zero):

$$\int_{\hat{x}=0}^{\hat{L}} A_n\left(\sin \frac{(2n+1)\pi\hat{x}}{2\hat{L}}\right)^2 d\hat{x} = A_n \int_{\hat{x}=0}^{\hat{L}} \frac{1}{2}\left(1 - \cos \frac{2(2n+1)\pi\hat{x}}{2\hat{L}}\right)dx = \frac{A_n\hat{L}}{2}$$

The coefficient is therefore determined:

$$A_n = \frac{4}{(2n+1)\pi}$$

The complete transient solution is thereby attained and expressed as an infinite series:

$$\theta_{transient} = \sum_{n=0}^{\infty} \frac{4}{(2n+1)\pi}e^{-\left(\frac{(2n+1)\pi}{2\hat{L}}\right)^2 \hat{t}} \sin \frac{(2n+1)\pi\hat{x}}{2\hat{L}}$$

Expressing this result for the first few terms:

$$\theta_{transient} = \frac{4}{\pi}\left[\frac{1}{1}e^{-\left(\frac{\pi}{2\hat{L}}\right)^2 \hat{t}} \sin \frac{\pi x}{2\hat{L}} + \frac{1}{3}e^{-\left(\frac{3\pi}{2\hat{L}}\right)^2 \hat{t}} \sin \frac{3\pi x}{2\hat{L}} + \frac{1}{5}e^{-\left(\frac{5\pi}{2\hat{L}}\right)^2 \hat{t}} \sin \frac{5\pi x}{2\hat{L}} + \cdots\right]$$

Notice how each term is multiplied by an exponential factor with a negative exponent that grows in absolute value with the number of the term. Except for very early time, where $t \ll 1$, the first term dominates the solution.

To determine the heat transfer rate associated with the transient solution, the derivative with respect to x is taken and evaluated at $x = 0$:

$$\left.\frac{\partial \theta_{transient}}{\partial \hat{x}}\right|_{\hat{x}=0} = \sum_{n=0}^{\infty} 2e^{-\left(\frac{(2n+1)\pi}{2\hat{L}}\right)^2 \hat{t}} = 2e^{-\left(\frac{\pi}{2\hat{L}}\right)^2 \hat{t}} + 2e^{-\left(\frac{3\pi}{2\hat{L}}\right)^2 \hat{t}} + 2e^{-\left(\frac{5\pi}{2\hat{L}}\right)^2 \hat{t}} + \cdots$$

The heat transfer rate associated with the transient solution is therefore:

$$q_{fin,transient} = \sqrt{\frac{hpkA}{1+B}}(T_0 - T_\infty)\sum_{n=0}^{\infty} 2e^{-\left(\frac{(2n+1)\pi}{2\hat{L}}\right)^2 \hat{t}}$$

Steady-State Solution

The steady-state solution is much easier, especially with the corrected length boundary condition. The governing equation (a linear ordinary differential equation with constant coefficients) and boundary conditions are:

$$\frac{d^2\theta_{ss}}{d\hat{x}^2} - \theta_{ss} = -1$$

$$\left.\frac{\partial\theta}{\partial\hat{x}}\right|_{\hat{x}=\hat{L}} = 0$$

The general solution comprises a homogeneous and particular solution:

$$\theta_{ss} = \theta_{ss,H} + \theta_{ss,P}$$

The homogeneous solution is that with the right hand side set to zero, and the solution is:

$$\theta_{ss,H} = c_1 e^{-\hat{x}} + c_2 e^{\hat{x}}$$

The particular solution is:

$$\theta_{ss,P} = K$$

Substitution of these expressions into the governing equation yields $K = 1$, and the general solution is therefore:

$$\theta_{ss} = c_1 e^{-\hat{x}} + c_2 e^{\hat{x}} + 1$$

Application of the boundary condition at the base (at $x = 0$):

$$0 = c_1 + c_2 + 1$$

Application of the second boundary condition (at $x = L_c$):

$$0 = -c_1 e^{-\hat{L}} + c_2 e^{\hat{L}}$$

Combining them:

$$0 = (c_2 + 1)e^{-\hat{L}} + c_2 e^{\hat{L}}$$

Solving for c_2:

$$c_2 = \frac{-e^{-\hat{L}}}{e^{-\hat{L}} + e^{\hat{L}}}$$

And

$$c_1 = \frac{e^{-\hat{L}}}{e^{-\hat{L}} + e^{\hat{L}}} - 1 = \frac{e^{-\hat{L}} - \left(e^{-\hat{L}} + e^{\hat{L}}\right)}{e^{-\hat{L}} + e^{\hat{L}}} = \frac{-e^{\hat{L}}}{e^{-\hat{L}} + e^{\hat{L}}}$$

The steady-state solution is therefore:

$$\theta_{ss} = 1 - \left(\frac{e^{\hat{L}} e^{-\hat{x}} + e^{-\hat{L}} e^{\hat{x}}}{e^{-\hat{L}} + e^{\hat{L}}}\right) = 1 - \left(\frac{e^{(\hat{L}-\hat{x})} + e^{-(\hat{L}-\hat{x})}}{e^{-\hat{L}} + e^{\hat{L}}}\right)$$

This equation can be expressed in a hyperbolic cosine form:

$$\cosh(y) = \frac{e^y + e^{-y}}{2}$$

$$\sinh(y) = \frac{e^y - e^{-y}}{2}$$

$$\tanh(y) = \frac{\sinh(y)}{\cosh(y)} = \frac{e^y - e^{-y}}{e^y + e^{-y}}$$

$$\theta_{ss} = 1 - \frac{\cosh(\hat{L} - \hat{x})}{\cosh(\hat{L})}$$

Returning to dimensions:

$$\frac{T_{ss} - T_0}{T_\infty - T_0} = 1 - \frac{\cosh\left(\frac{L_c - x}{L_t}\right)}{\cosh\left(\frac{L_c}{L_t}\right)}$$

The derivative evaluated at $x = 0$ is:

$$\left.\frac{\partial \theta_{ss}}{\partial \hat{x}}\right|_{\hat{x}=0} = -c_1 + c_2 = \frac{e^{\hat{L}} - e^{-\hat{L}}}{e^{\hat{L}} + e^{-\hat{L}}} = \frac{\sinh(\hat{L})}{\cosh(\hat{L})} = \tanh(\hat{L})$$

The steady-state heat transfer rate, entering the definition of $\hat{L}$, is therefore:

$$q_{fin,ss} = \tanh\left(\frac{L_c}{L_t}\right) \sqrt{\frac{hpkA}{1+B}}(T_0 - T_\infty)$$

The full solution is therefore mercifully upon us:

$$\theta = \left[\sum_{n=0}^{\infty} \frac{4}{(2n+1)\pi} e^{-\left(\frac{(2n+1)\pi}{2\hat{L}}\right)^2 \hat{t}} \sin\frac{(2n+1)\pi\hat{x}}{2\hat{L}}\right] + \left[1 - \frac{\cosh(\hat{L} - \hat{x})}{\cosh(\hat{L})}\right]$$

where the first bracketed term is the transient solution, and the second is the steady-state solution. The heat transfer rate is:

$$q_{fin} = \sqrt{\frac{hpkA}{1+B}}(T_0 - T_\infty)\left[\tanh\left(\frac{L_c}{L_t}\right) + \sum_{n=0}^{\infty} 2e^{-\left(\frac{(2n+1)\pi}{2\tilde{L}}\right)^2 \hat{t}}\right]$$

where a transverse Biot number is defined as $B = \dfrac{h(A/p)}{k}$, the corrected length is $L_c = L + \frac{A}{p}$, and the thermal length is $L_t = \sqrt{\dfrac{(1+B)kA}{hp}}$.

Personal comments: I consider this really elegant mathematical solution to the transient rim problem to be "Good Mathematics" and "Bad Engineering." Engineering is about addressing a human need. To make this value judgment about whether the method is good or bad engineering requires me to revisit, and perhaps modify, the problem statement, an iteration of the modeling process. If the objective is to merely find the heat transfer rate, then the 1-node solution is really simple mathematically, and yields the same physics as the PDE solution.

The big problem I have with the PDE solution is not that it requires a sophisticated mathematical excursion. If that's what it takes, then so be it. My problem is that the solution ties me to some tight restrictions. The base wall temperature must be fixed in time. The initial condition must be that the fin is entirely at ambient, etc. The 1-node model gives all that. An extension to an n-node solution requires some more work, but that numerical method can be applied to much more general situations quite easily, once the technique is mastered. For example, suppose a fin is designed to dissipate heat from a motor that turns on and off intermittently, but has a clear maximum time it is on. A fin might be designed that takes advantage of high initial heat rates (where heat goes in part to changing the energy stored, and the rest transferred to the fluid).

In short, the 1-node model teaches the essential physics and yields key design parameters. The multi-node numerical method yields more detail and is arguably more quantitatively accurate than the 1-node model. On the other hand, there is so much uncertainty in convection coefficient values that accuracy can be misleading.

Heat transfer modeling is indeed an art.

Appendix 2. Variable Area Fins: Finite Difference Numerical Solution

Figure 7.23 shows a circumferential fin attached to a pipe, an example of an extended surface (fin) that has a base area A_{base} and base perimeter p_{base} and a cross-sectional area $A(x)$ and perimeter $p(x)$ that vary in the distance perpendicular to the base (x, radial position). The total length of the fin is L, and the material has a thermal conductivity k. The fin is exposed to a fluid at temperature T_∞ with an

Fig. 7.23 Circumferential fin of length L, thickness t place on a pipe of outer diameter D. The *left* sketch is an end view of the pipe; the *right* view is a cross-section from a side view of the pipe

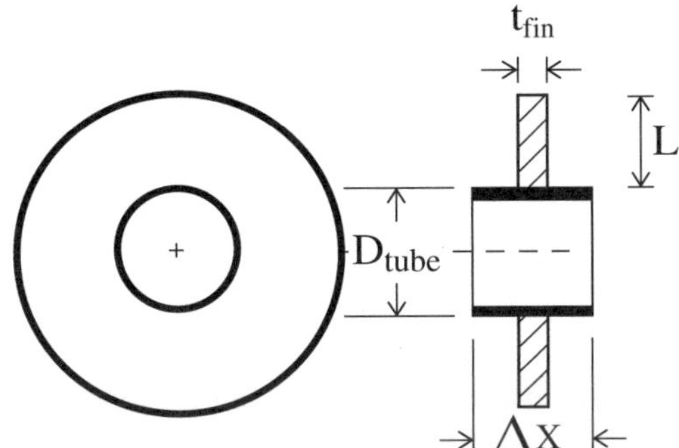

effective heat transfer coefficient, h. The fin is initially at the fluid temperature, and the base temperature is suddenly changed and maintained at a temperature T_0.

Model Development

The governing differential equation was derived in the previous appendix by application of the energy balance to a differential element:

$$\rho c A \frac{\partial \overline{T}}{\partial t} = \frac{\partial}{\partial x}\left[kA\frac{\partial \overline{T}}{\partial x}\right]_x - \frac{hp}{1+B}(\overline{T} - T_\infty)$$

With an initial condition and two boundary conditions (at the base and end of the fin). The difference between this case and the constant-area is that the area cannot be taken out of the conduction derivative term. Assuming constant thermal conductivity, using the chain rule for the conduction term and rearranging:

$$\frac{\partial T}{\partial t} = \alpha \frac{\partial^2 T}{\partial x^2} + \alpha \frac{1}{A}\frac{\partial A}{\partial x}\frac{\partial T}{\partial x} - \frac{hp}{\rho c A(1+B)}(\overline{T} - T_\infty)$$

In a finite difference explicit numerical solution, the time derivative is expressed as a forward finite difference approximation, and the spatial derivatives as central differences. The boundary conditions are expressed as forward for $x = 0$, and backward for $x = L$ second-order finite differences. The details are not developed here.

Appendix 3. Iterative Solutions of Nonlinear Algebraic Equations

Nonlinear algebraic equations arise in many problems of engineering interest. They cannot be solved analytically, in general. Rather, some form of iterative procedure is required. In an iterative procedure, a guess to the solution is first made. Then a recursion formula based on the governing equations is used to make another guess. Then another guess is made, then another and another... Once the initial guess is made, the algorithm takes over for all subsequent guesses (called an iteration). If the method is successful, successive iterations eventually "converge" to the exact solution. It never reaches it, but a CONVERGENCE CRITERION can be defined that means "Close Enough!" The method is said to DIVERGE if the absolute value grows to infinity. Some systems exhibit CHAOS.

Some general points...

- Many different algorithms have been "invented." Heat transfer problems are generally solvable using the Method of Successive Substitution. It is easy to implement and usually works. It can easily be extended to systems of equations. Newton's method is another, and will be introduced for a single nonlinear algebraic equation and outlined for systems of nonlinear equations. It is much harder to implement with systems (with more than one unknown) and is occasionally necessary in heat transfer problems.
- There is no guarantee that a given method will work, that is, find the desired solution to the problem.
- If an iterative procedure converges, it does so to a correct solution of the system of equations. However, that solution may make no physical sense (i.e., a negative absolute temperature).
- If an implementation diverges, it might converge for another initial guess.

Single Nonlinear Algebraic Equations

Solution of a single nonlinear algebraic equation is the natural place to start. The methods can be extended to systems of equations (with more than one unknown). The quadratic equation will be used to demonstrate how to implement the methods and to compare the response on a system that has an analytical solution. In addition, if the goal is not to solve the equation, but to study CHAOS, iterative techniques on the quadratic equation are a great place to start.

Quadratic equation:

$$ax^2 + bx + c = 0$$

The quadratic equation is about the simplest nonlinear algebraic equation there is. As a nonlinear equation, it has the rare distinction of having an exact solution. It has two solutions, its two roots x_1 and x_2:

$$x_1 = \frac{-b + \sqrt{b^2 - 4ac}}{2a} \qquad x_2 = \frac{-b - \sqrt{b^2 - 4ac}}{2a}$$

Keep in mind that in what follows, the objective is to learn new techniques that can be implemented to solve more difficult nonlinear equations, not to solve the quadratic equation...

Method of Successive Substitution: x = g(x)

The method of successive substitution is easy to implement, and it works often. It can be used to solve for many systems of linear or nonlinear equations as well.

STEP 1: Rearrange equation:

The governing nonlinear equation is rearranged into the recursion form:

$$x_{k+1} = g(x_k)$$

where the subscript "k" refers to the iteration number. There are an infinite number of ways this rearrangement can be done. For example, the quadratic equation written in its conventional form is:

$$ax^2 + bx + c = 0$$

Three equivalent forms of the quadratic equation (with subscripts showing the iteration number) are:

$$\text{Form 1}: \quad x_{k+1} = \frac{-1}{b}\left[c + ax_k^2\right] \qquad \text{(isolating the 2}^{\text{nd}}\text{ term)}$$

$$\text{Form 2}: \quad x_{k+1} = \pm\sqrt{\frac{-1}{a}[c + bx_k]} \qquad \text{(isolating the 1}^{\text{st}}\text{ term)}$$

$$\text{Form 3}: \quad x_{k+1} = ax_k^2 + (b+1)x_k + c \qquad \text{(adding x to both sides)}$$

STEP 2: Make an initial guess, x_0, and ITERATE:

Examples:

A spreadsheet was used to investigate the how the method works. Three cases are shown in Figs. 7.24, 7.25, 7.26, 7.27, 7.28:

Case 7.1 shows a situation in which both roots are found, one of them by two forms. A plot of the "trajectory" is shown.

Case 7.2 shows one form converging, one form "crashing" immediately (because the square root of a negative number was asked for), and one form diverging.

Case 7.3 shows CHAOS. The iteration trajectory neither converges nor diverges. Sensitivity to initial guess is shown by plotting two trajectories with nearly identical initial guesses.

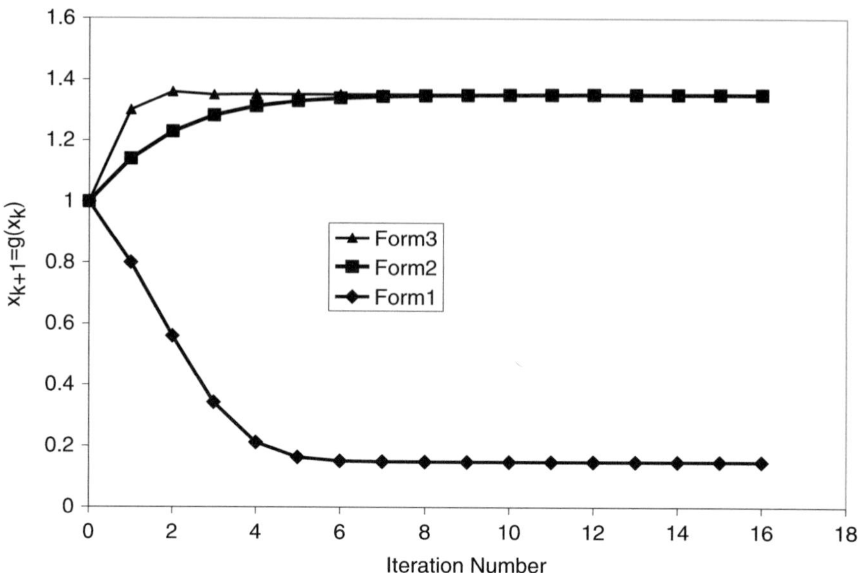

Fig. 7.24 Iteration trajectory for Case 7.1

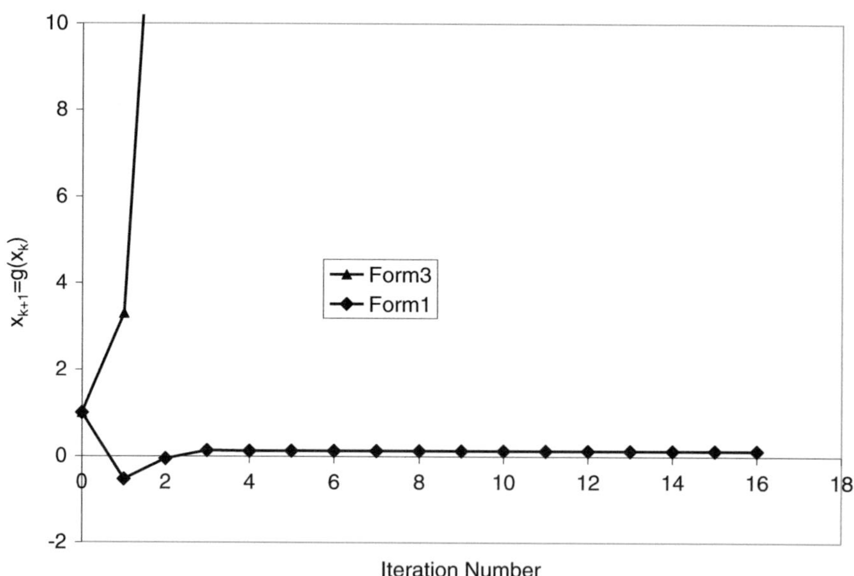

Fig. 7.25 Iteration trajectory for Case 7.2

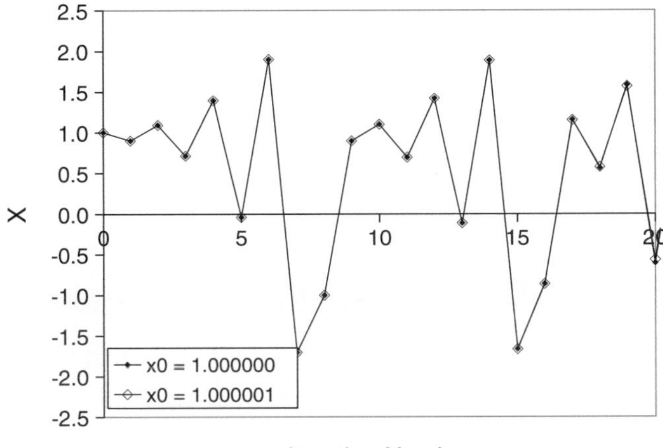

Fig. 7.26 First 20 iterations for Case 7.3

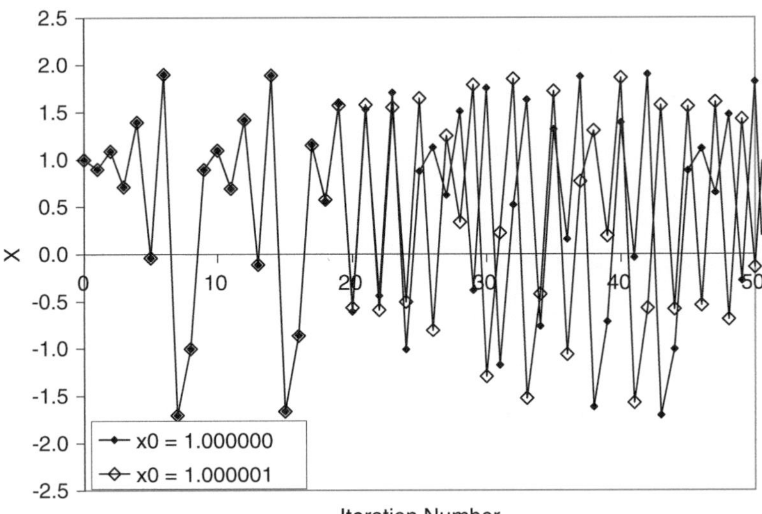

Fig. 7.27 First 50 iterations for Case 7.3

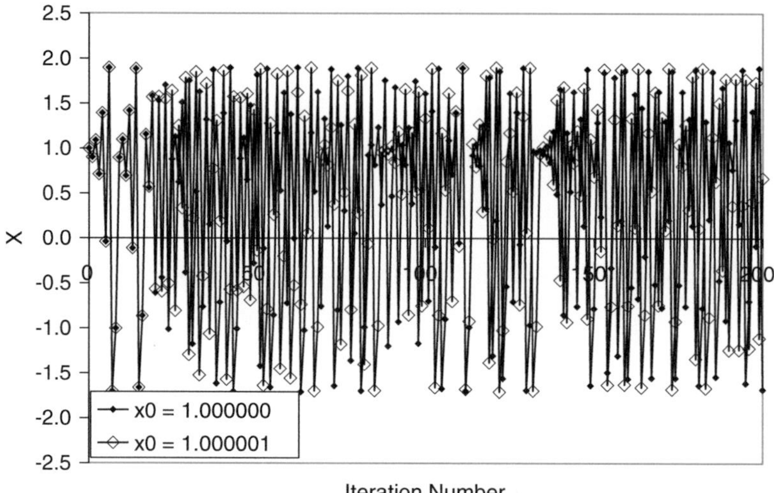

Fig. 7.28 First 200 iterations for Case 7.3

Method of Successive Substitution [x = g(x)] Applied to Quadratic Equation

a=   -1
b=   1.5
c=   -0.2
Exact Solution
  x1=   0.14792
  x2=   1.35208

$$ax^2 + bx + c = 0$$

ITERATION TABLE

Iteration	Form1	Form2	Form3
0	1	1	1
1	0.8	1.140175	1.3
2	0.56	1.228928	1.36
3	0.3424	1.281948	1.3504
4	0.211492	1.312601	1.35242
5	0.163153	1.330001	1.35201
6	0.151079	1.339777	1.352094
7	0.14855	1.345238	1.352077
8	0.148045	1.348279	1.35208
9	0.147945	1.34997	1.35208
10	0.147925	1.350909	1.35208
11	0.147921	1.35143	1.35208
12	0.14792	1.351719	1.35208
13	0.14792	1.35188	1.35208
14	0.14792	1.351969	1.35208
15	0.14792	1.352018	1.35208
16	0.14792	1.352046	1.35208

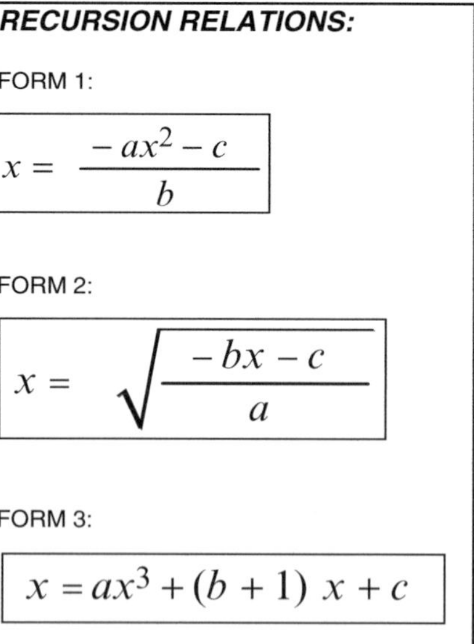

RECURSION RELATIONS:

FORM 1:

$$x = \frac{-ax^2 - c}{b}$$

FORM 2:

$$x = \sqrt{\frac{-bx - c}{a}}$$

FORM 3:

$$x = ax^3 + (b + 1)\, x + c$$

Case 7.1 Three forms converge to different solutions

Method of Successive Substitution [x = g(x)] Applied to Quadratic Equation

```
a=        1
b=        1.5
c=        -0.2
Exact Solution
     x1=  0.123212
     x2=  -1.623212
```

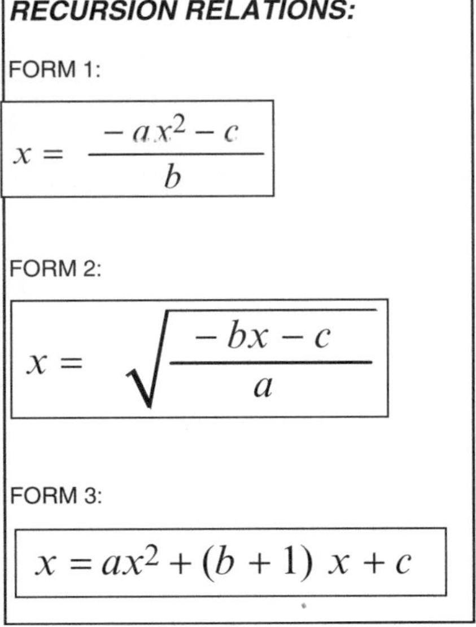

ITERATION TABLE

Iteration	Form1	Form2	Form3
0	1	1	1
1	-0.533333	#NUM!	3.3
2	-0.056296	#NUM!	18.94
3	0.13122	#NUM!	405.8736
4	0.121854	#NUM!	165747.9
5	0.123434	#NUM!	2.75E+10
6	0.123176	#NUM!	7.55E+20
7	0.123218	#NUM!	5.7E+41
8	0.123211	#NUM!	3.25E+83
9	0.123213	#NUM!	1.1E+167
10	0.123212	#NUM!	#NUM!
11	0.123212	#NUM!	#NUM!
12	0.123212	#NUM!	#NUM!
13	0.123212	#NUM!	#NUM!
14	0.123212	#NUM!	#NUM!
15	0.123212	#NUM!	#NUM!
16	0.123212	#NUM!	#NUM!

Case 7.2 Two forms diverge

	a=	-1.0
	b=	-1.0
	c=	1.9
epsilon=	0.000001	

Exact Solution

x1= -1.9662881
x2= 0.966288

Iteration	Form1	Form1
0	1.000000	1.000001
1	0.900000	0.899998
2	1.090000	1.090004
3	0.711900	0.711892
4	1.393198	1.393210
5	-0.041002	-0.041033
6	1.898319	1.898316
7	-1.703614	-1.703605
8	-1.002302	-1.002269
9	0.895390	0.895456
10	1.098277	1.098158
11	0.693789	0.694049
12	1.418657	1.418296
13	-0.112589	-0.111563
14	1.887324	1.887554
15	-1.661991	-1.662859
16	-0.862214	-0.865100
17	1.156587	1.151602
18	0.562308	0.573813
19	1.583810	1.570739
20	-0.608455	-0.567221

$$x = \frac{-ax^2 - c}{b}$$

Case 7.3 CHAOS

Newton's Method: f(x) = 0

Newton's method is an iterative method that uses the rate of change of the function
to help make an "intelligent" choice for the next iteration. The recursion relation
can be derived with the help of a graphical analysis. The following sketch shows a
general function f(x) in the vicinity of its solution, where it equals 0.

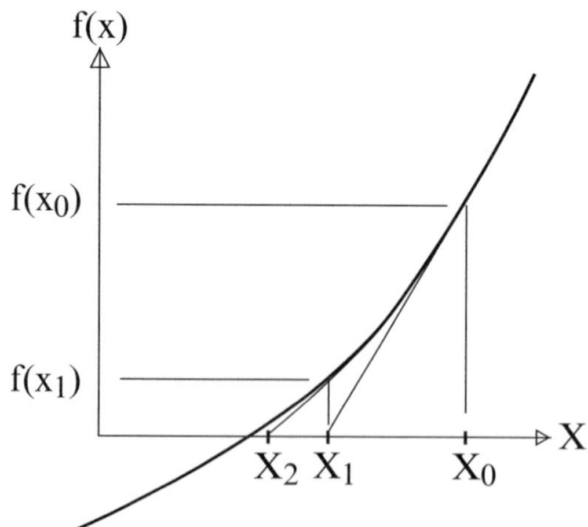

The first "guess" is x_0 and the function value is $f(x_0)$ at that point. The slope of the function: $\left.\dfrac{df}{dx}\right|_{x_0}$ is drawn at that point. The next iteration value chosen, x_1, is taken to be where that slope intersects the x-axis.

$$\left.\frac{df}{dx}\right|_{x_0} = \frac{f(x_0) - 0}{x_0 - x_1}$$

Solving this equation for x_1:

$$x_1 = x_0 - \frac{f(x_0)}{\left.\dfrac{df}{dx}\right|_{x_0}}$$

In general, the recursion relation is:

$$x_{k+1} = x_k - \frac{f(x_k)}{\left.\dfrac{df}{dx}\right|_{x_k}}$$

To implement the method, an initial "guess" is made, and the recursion relation is used in an iterative fashion.

Newton's Method for Systems of Nonlinear Algebraic Equations

In a resistance network, if the thermal resistances are expressed as functions of temperature (which are unknown), for example, with radiation coefficients, then the system of algebraic equations that results from discretizing space will be nonlinear. In most heat transfer applications, the relatively easy to implement Method of Successive Substitutions (i.e., Gauss–Seidel method applied to 3-node fin) works well. When it fails, Newton's method (which is harder to implement, but more likely to succeed) can be implemented. The system of n equations in n unknown temperatures can be written in the form (one for each node of a space-discretized network):

$$f_1(T_1, T_2, \ldots, T_n) = 0$$
$$f_2(T_1, T_2, \ldots, T_n) = 0$$
$$\ldots$$
$$f_n(T_1, T_2, \ldots, T_n) = 0$$

The method involves making an initial guess for all unknown temperatures, and then making an improved guess based on how the function is changing. The initial guess is given a superscript "k" and the improved guess the superscript "k + 1". The total derivative of each nodal equation consists of the contribution from each unknown variable:

$$df_1 = \left.\frac{\partial f_1}{\partial T_1}\right|^{(k)}\left(T_1^{(k+1)} - T_1^{(k)}\right) + \left.\frac{\partial f_1}{\partial T_2}\right|^{(k)}\left(T_2^{(k+1)} - T_2^{(k)}\right) + \cdots + \left.\frac{\partial f_1}{\partial T_n}\right|^{(k)}\left(T_n^{(k+1)} - T_n^{(k)}\right)$$
$$= f_1^{(k+1)} - f_1^{(k)} \approx 0 - f_1^{(k)}$$
$$df_2 = \left.\frac{\partial f_2}{\partial T_1}\right|^{(k)}\left(T_1^{(k+1)} - T_1^{(k)}\right) + \left.\frac{\partial f_2}{\partial T_2}\right|^{(k)}\left(T_2^{(k+1)} - T_2^{(k)}\right) + \cdots + \left.\frac{\partial f_2}{\partial T_n}\right|^{(k)}\left(T_n^{(k+1)} - T_n^{(k)}\right)$$
$$= f_2^{(k+1)} - f_2^{(k)} \approx 0 - f_2^{(k)}$$
$$\cdots$$
$$df_n = \left.\frac{\partial f_n}{\partial T_1}\right|^{(k)}\left(T_1^{(k+1)} - T_1^{(k)}\right) + \left.\frac{\partial f_n}{\partial T_2}\right|^{(k)}\left(T_2^{(k+1)} - T_2^{(k)}\right) + \cdots + \left.\frac{\partial f_n}{\partial T_n}\right|^{(k)}\left(T_n^{(k+1)} - T_n^{(k)}\right)$$
$$= f_n^{(k+1)} - f_n^{(k)} \approx 0 - f_n^{(k)}$$

The right hand side represents an approximation to the total change. The zero reflects that the function evaluated at the next iteration is supposed to be zero, but it should be closer to zero than the previous iteration. Since all the functions and their derivatives (possibly by numerical approximation) can be evaluated at the "k" iteration, this represents a system of linear algebraic equations that can be solved for the temperatures at the next iteration.

Workshop 7.1: Triangular Fins

Compare the steady-state performance of triangular fins with rectangular fins with the same base thickness (t) and the same total mass of material. Assume that the depth of the fins (not shown) are large compared to the thickness. Think about it before analyzing it.

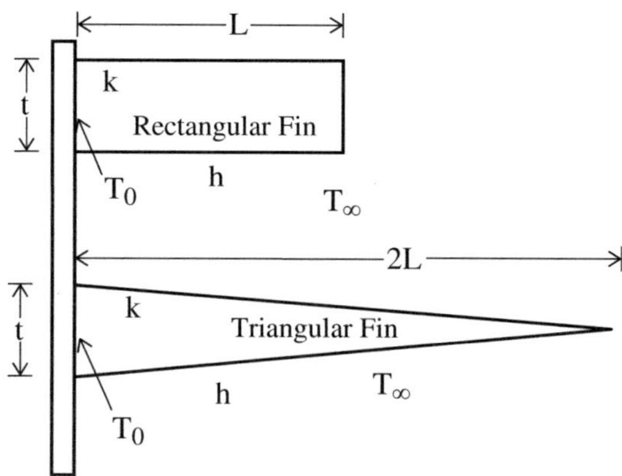

Level 1: Develop a 1-node analysis for the triangular fin (place the node at the centroid), and develop an expression for the ratio of the heat flow from the rectangular fin to that of the corresponding triangular fin. Is there a key dimensionless parameter to plot against?

Level 2: Develop a few-node model for both in spreadsheets, and compare the results with the 1-node model.

Level 3: Develop an "n" node model for both using a program code, and compare with the 1-node model.

Workshop 7.2: Circumferential Rectangular Fin

Analyze the steady-state performance of a fin constructed of a rectangular plate placed on the outer surface of a pipe, a common design. The fin is exposed to an ambient fluid (T_∞) with an effective heat transfer coefficient h. The spacing between fins (Δx) is shown but not used in this workshop in which a single fin is isolated.

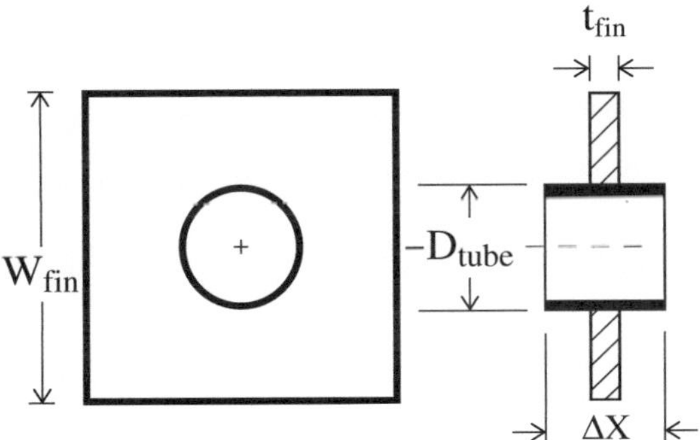

- Conduct a 1-node analysis for a thermally short model.
- Develop a 1-node thermally long model.
- Develop a few-node model (identify and take advantage of symmetry).

Steady-State Conduction 8

After the initial mug heating phase when hot coffee is poured into a ceramic mug, the coffee and mug gradually cool together. During this time, the rate of heat transfer from the coffee to the inner wall of the mug is approximately equal to the rate of heat transfer from the outer wall to ambient. That near balance defines a quasi-steady state situation for the mug itself, and is why the lumping of its capacity with the coffee is reasonable. For problems that are steady-state in nature (or quasi-steady state), a transient model can be developed and then run to equilibrium to obtain the steady-state solution. However, it is possible to solve directly for the steady-state solution, which involves somewhat different mathematics to execute. Methods for doing so were initiated in the previous chapter on fins, and elaborated on in this chapter. Furthermore, the mathematics involved is similar to that required to execute a step forward using an implicit transient. The method is first developed and applied to the mug of coffee, and then to a classic 2D problem. Extension to 3D is straightforward.

In general, heat transfer problems are transient and three dimensional. That is, the temperature varies with time and position. Many heat transfer applications are inherently constant in time, or are best modeled as such. The objective of this chapter is to develop a numerical method for problems that are inherently steady-state. The method is developed in a simple 2D (x,y) space, and the extension to 3D is straightforward, conceptually.

In classic heat transfer texts, conduction shape factors (S) are introduced for conduction in purely 2D geometries of homogeneous materials. The heat transfer rate per unit depth (q/P, in W/m, where P is the depth, or the dimension into the page) between two surfaces separated by a homogeneous material of thermal conductivity k and with temperatures maintained at T_1 and T_2 can be expressed as:

$$q/P = kS(T_1 - T_2)$$

The conduction shape factor is a measure of the conductive thermal resistance between two surfaces with a solid between them and depends only on the geometry

© Springer International Publishing Switzerland 2015

G. Sidebotham, *Heat Transfer Modeling: An Inductive Approach*,

DOI 10.1007/978-3-319-14514-3_8

of the 2D object. Note that the conductive thermal resistance can be expressed in terms of the shape factor, namely,

$$R_k = \frac{1}{kS}$$

Tables of conduction shape factors are available in classic heat transfer texts and on the Internet. A conduction shape factor workshop is included at the end of this chapter.

The treatment of the rim as a fin with a 1-node analysis in the previous chapter allowed for a good prediction of the overall behavior of the cooling mug of coffee. The focus of this chapter will be on inherently 2D detail that the 1-node model cannot reveal. Focusing on the mug wall, there is a discontinuity in the type of fluid at the free surface of the coffee. There is hot coffee on the inside surface up to the coffee level, and warm, humid air above that. However, the temperature cannot be discontinuous in either a solid or fluid.

Outside of the thermal boundary layer, the temperature of the fluid adjacent to the mug surface is different than the temperature of the mug surface temperature. Since the convection coefficient between coffee and surface is much higher than that between air and the surface, the temperature of the mug inner surface exposed to coffee is close to the coffee temperature, while a big drop in temperature occurs between the outer surface and air. To make an extreme case, in this chapter one hypothetical scenario is analyzed in which the convection on both air and coffee contact surfaces are so large that the surface temperature is fixed by the fluid temperature. A discontinuity in surface temperature is imposed, even though that would be impossible to achieve experimentally. In reality, the temperature gradient can be very steep, but never discontinuous. Mathematically, a discontinuous surface temperature is frequently prescribed in the literature.

8.1   Graphical Method

The following "warm-up" exercise outlines a procedure for solving for the temperature field and heat transfer rates in a 2D steady-state conduction situation, where the temperature is fixed along key surfaces. The application is modeled after the mug of coffee problem. There are two important ways this model differs from the actual experiment. First, the boundary conditions are of a fixed temperature type, not convection/radiation. That is, the mug surfaces are fixed at the fluid temperatures. Second, these "fluid" temperatures are fixed at values of 0 and 100 °C. This method can be used to estimate the conductive resistance between two surfaces that are NOT maintained at fixed temperature, to be used in a 1D approximation to a 2D problem.

The general steps of the graphical method are:

- Draw geometry to scale. It MUST be to scale.
- Sketch isotherms.
- Sketch heat flux lines.

The following warm-up exercise breaks these steps into a detailed procedure.

Warm-up
1. On a piece of paper, draw two vertical lines across the entire page (Fig. 8.1 shows steps 1–6). These lines represent the walls of a solid material, with a dimension W (=πD) into the page.
2. Draw a thicker line on the left line from somewhere in the middle to the bottom. The surface represented by the thick line is the "coffee" and has been maintained at 100 °C for a long time. The other surfaces (both walls) have been maintained at 0 °C.
3. Near the bottom of the page, place a dot where you think the temperature would be 50 °C. Place another dot about a centimcter higher up.
4. Place a dot near the coffee surface where you think it would be 50 °C. Place another dot about a centimeter away.
5. Alternate between the two ends until the entire curve is traced out. If a dot is placed where you wished you hadn't, just leave it, and place other dots where you think they should be.
6. Define the 50 °C isotherm by adding more dots in between others.
7. Near the bottom of the page, place a dot where you think the temperature would be 25 and 75 °C (Fig. 8.2 shows steps 7–9). Add two more dots, about a centimeter away.
8. Near the coffee surface, place dots at 25 and 75 °C. Add two more dots, about a centimeter away.
9. Add more dots to define the 25 and 75 °C isotherms.
10. Draw a horizontal line near the bottom of the page from the hot wall to the cold wall (Fig. 8.3 shows steps 10–12). The line just drawn is a heat flux line. As additional heat flux lines are drawn, observe the following three rules: each line starts at the 100 °C isotherm and ends at the 0 °C isotherm, each "pseudo box" (called an element) created should be nominally square (Fig. 8.4), and each line must be perpendicular to any isotherm it crosses.
11. Draw a heat flux line around the coffee surface, keeping each element approximately square.
12. Return to the bottom and add another heat flux line. Alternate between top and bottom until the web is complete. It is possible that the final heat flux lane will not have square elements.

Heat Transfer Analysis of Graphical Sketch
Consider any mesh element created, for example in Fig. 8.5, with thermal resistances between the isothermal lines. Two of its sides are isotherms, and the two other lines are heat flux lines. Heat does not cross a heat flux line and are drawn as insulated surfaces (hatched lines). There is solid material between them, so heat flows from the hot side to the cold side by conduction. The rate of heat transfer is:

$$q_{element} = \frac{\Delta T_{element}}{R}$$

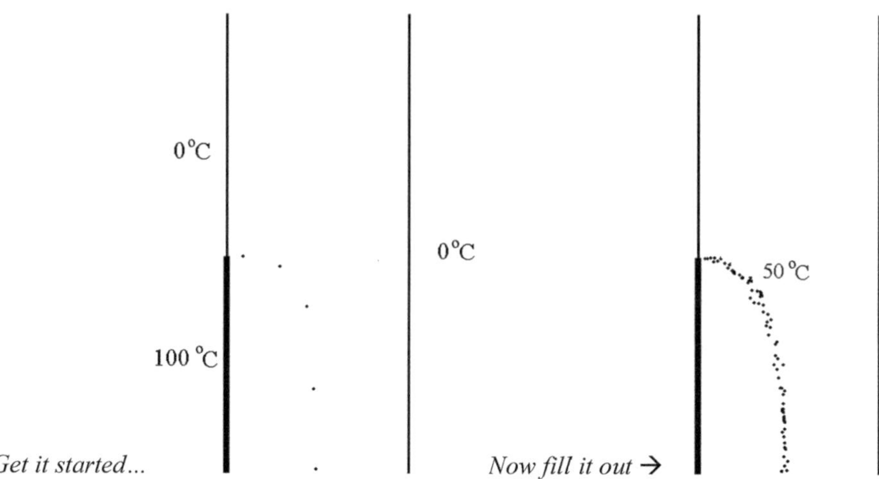

Fig. 8.1 Placement of 50 °C isotherm. Steps 1–6 of the warm-up exercise. The two *vertical lines* represent the surfaces of a solid in between. The temperature of the *right* face is maintained at 0 °C. The *left* face is maintained at 0 °C on the *top*, and the *bottom* (the *thicker line*) is maintained at 100 °C. The *dots* on the *left* sketch are best guesses as to where the temperature is 50 °C. Additional *dots* are added to the sketch on the *right* in order to define the 50 °C isotherm

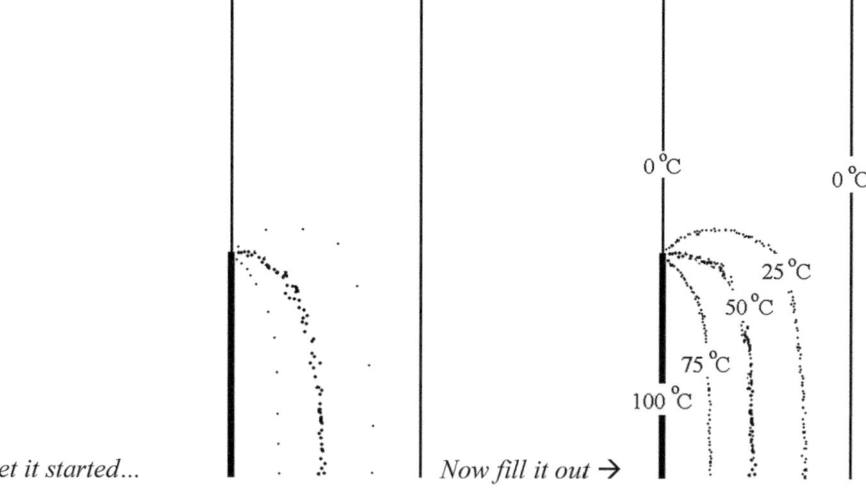

Fig. 8.2 Placement of 25 and 75 °C isotherms. Steps 7–9 of the warm-up exercise

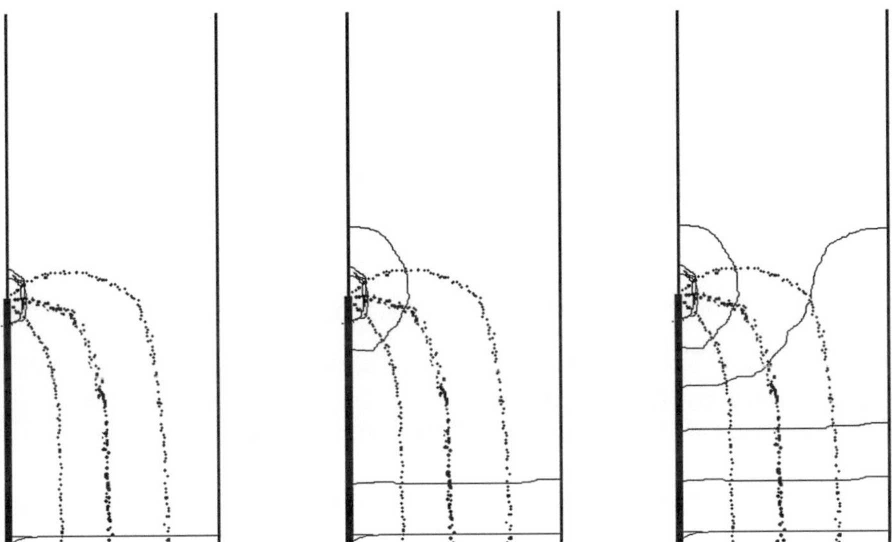

Fig. 8.3 Placement of heat flux lines. Steps 10–12 of the warm-up exercise

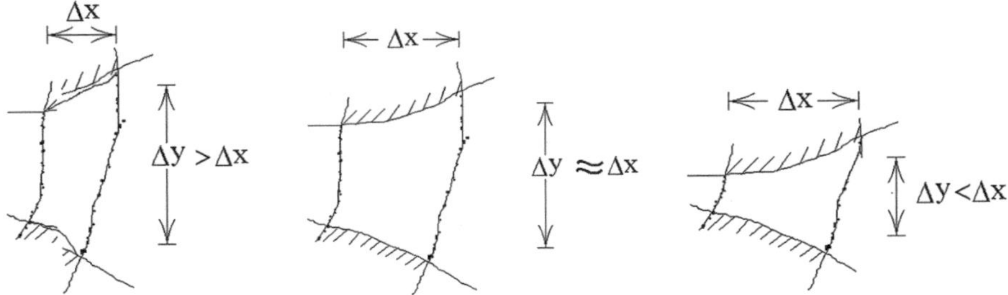

Fig. 8.4 Demonstration of "skewed" and "squarish" mesh elements, which can take some practice to achieve

Fig. 8.5 "Squarish" element with a conductive resistance between adjacent isotherms. The dimension "P" is the depth into the page

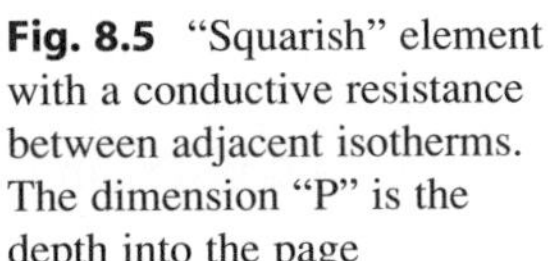

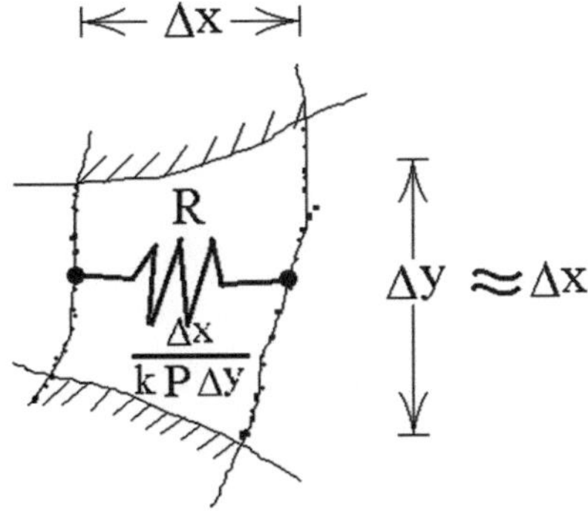

Here $\Delta T_{element}$ is the temperature difference between the two isotherms (25 °C in this example, which is the overall temperature difference divided by the number of temperature increments). Expressing the conductive resistances in terms of the dimensions of the cell, where Δx is the average distance between the isotherms, Δy is the average distance between the heat flow lines, and P is the depth of the wall (into the page) so that $\Delta y P$ is the area of an isothermal surface:

$$q_{element} = \frac{\Delta T_{element}}{\left(\dfrac{\Delta x}{k \Delta y P}\right)}$$

If the intent to keep each cell square was successfully followed, then the distance between isotherms (Δx) is nominally the same as the distance between heat flux lines (Δy), and they cancel each other. Therefore:

$$q_{element} = kP\Delta T_{element}$$

At steady-state, the rate at which heat enters an element equals the rate it leaves. Adjacent elements along the same heat flow path (called a heat flow "lane") experience the same rate of heat flux. That is:

$$q_{element} = q_{lane}$$

The temperature drop across an element is simply related to the over-all temperature drop by the number of temperature increments, $N_{increments}$:

$$N_{increment}\, \Delta T_{element} = \Delta T_{overall}$$

$$q_{lane} = kP \frac{\Delta T_{overall}}{N_{increments}}$$

Each lane acts in parallel with the others in transferring heat from the hot surface to cold. There is nothing to distinguish one lane from another, so each lane contributes the same rate of heat transfer. That may seem surprising. However, where isotherms are close together (steep gradients), the heat flux lines will also be close together, so the same rate of heat transfer occurs in a narrow lane. The total heat transfer rate is simply the sum of all the lanes created in the drawing, or the number of lanes, M_{lane}, which can be a fraction for the last lane drawn and the web closes on itself. The total heat transfer rate is therefore:

$$q_{total} = k\left(\frac{PM_{lanes}}{N_{increments}}\right)(T_H - T_C) \equiv kS(T_H - T_C) = \frac{(T_H - T_C)}{R_{HC}}$$

The shape factor is therefore:

$$S = \left(\frac{PM_{lanes}}{N_{increments}}\right)$$

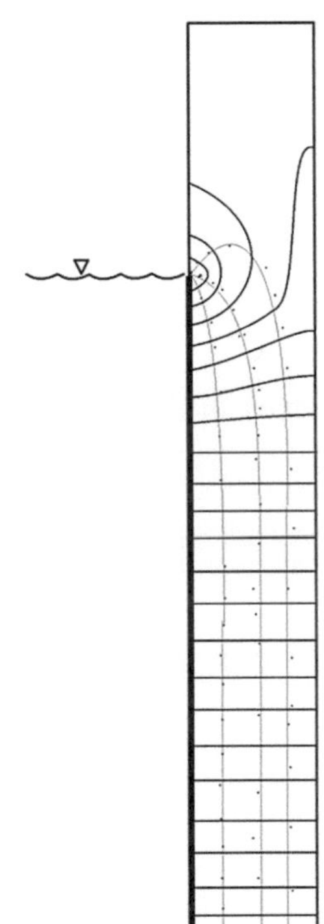

Fig. 8.6 Graphical analysis of side wall of the mug. The ratio of the wall thickness to the length of the air column on the inside (*left* surface above the coffee level) is to scale (0.3 × 2.2) with the problem that will be solved numerically

The conductive thermal resistance between the two surfaces is:

$$R_{HC} = \frac{N_{increments}}{kPM_{lanes}}$$

The values for M and N will depend on the specific drawing. For the drawing of Fig. 8.3, there are four temperature increments (N = 4), and a little more than six heat flow lanes (M = 6.2, say).

Figure 8.6 shows another sketch of a similar problem. It is a side view, to scale, but considered infinite in extent below the coffee level, and this problem will be solved with a numerical method in the next section. The isotherms and heat flux lines were constructed according to the procedure in the warm-up exercise. There are four temperature increments ($N_{increments}$) and 23 heat flow lanes generated (M_{lanes}). Therefore, the heat transfer rate is:

$$q_{total} = k\frac{PM_{lanes}}{N_{increments}}\Delta T_{overall}$$

The shape factor is:

$$S = \frac{PM_{lanes}}{N_{increments}} = \frac{P(23)}{4} = 5.75P$$

The heat transfer rate for the coffee mug problem ($P = \pi D$, $k = 1.0$ W/m/K, when $T_H = 70$ °C and $T_C = 23$ °C) is: $q_{total} = k \left(\dfrac{PM_{lanes}}{N_{increments}} \right) (T_H - T_C)$

$$q_{total} = \left(1.0 \frac{W}{m\,°C} \right) \left(\frac{(0.25\,m)(23)}{(4)} \right) (70 - 23)°C \; q_{total} = 67.6\,W$$

The method reveals a region of high activity near the coffee surface, where heat enters the mug wall from the left (coffee side) and exits to the right (to the air just above the coffee surface). Above a distance of a couple wall widths from the coffee surface, the temperature is nearly ambient, and below a couple wall widths, the heat transfer is 1D (straight isotherms and heat flow lanes). The graphical method quickly provides an excellent feel for this problem, and it can be amazingly accurate. The solution from the up and coming finite element solution is 69.84 W when the mug surface temperatures are artificially set equal to the adjacent fluid temperatures.

This "graphical method" of estimating heat transfer rates in a 2D solid can be a valuable-modeling tool in helping to understand the temperature field in a complicated shape. It can often be enough to help an Engineer make a design decision. If a more detailed analysis is considered necessary, the graphical method can help pinpoint regions of high action (where a fine grid would help) and regions of low action (where a coarse grid is sufficient).

The method can also be used to turn an inherently 2D problem into a 1D approximation when two surfaces can be identified. Heat flows from one surface to the other, without branching that gives rise to more complicated 2D mathematics.

To reiterate, a caveat on the method is that the drawing on which isotherms and then heat flow lanes are drawn must be to scale. Try a quick analysis on another drawing with a different spacing between the walls. If the walls are farther apart, the elements will be bigger, making fewer lanes, increasing the thermal resistance.

8.2   Numerical Method for 2D Steady-State Conduction

The finite element method provides a more detailed analysis than the graphical method. In it, the physical space is broken into discrete elements, each represented by a node. Thermal resistances between nodes and boundaries are defined. Each node has a capacitance associated with it, but can be omitted from drawings with the understanding that there is no change in energy stored in a steady-state problem. The elements can be any shape, with care being made to account for all thermal resistance. Finer grids near regions of high gradients can be considered.

The method will be developed with a specific problem statement as a base case defined to focus on the wall of the mug of coffee problem, and several scenarios will be run to explore the response. In the following problem statement, the side wall of the mug is treated as a rectangular prism (making a thin-wall approximation), with one side exposed to ambient air, and the other side exposed to hot coffee at a fixed temperature (the quasi-steady approximation) below the free surface and ambient above it. The top is exposed to ambient, and the bottom is exposed to a floor with an effective heat transfer coefficient (U_b). The height of the wall (2.2 cm) is much shorter than the wall of the mug investigated in earlier chapters in order to focus on the thermal response in the vicinity of the coffee surface. The treatment of the bottom surface is included for the purpose of introducing a variety of 2D responses.

Engineering Problem Statement

The outer surface and top of a 0.3 cm thick (w), 2.2 cm tall (H), 25 cm deep (P) solid wall of a material with a thermal conductivity $k = 1.0$ W/m/K is exposed to a fluid (air) at a temperature of $T_C = 23$ °C with a convection coefficient $h_C = 13.9$ W/m^2/°C. The other surface (inner surface) is exposed to air at T_C with h_C from the top to the free surface (a distance $L = 0.6$ cm below the top), and to coffee at a temperature $T_H = 70$ °C with a convection coefficient of $h_H = 420$ W/m^2/°C from the free surface to the bottom. The bottom surface is exposed to ambient temperature through an effective heat transfer coefficient U_b. Setting U_b to a "low value" like $1(10)^{-6}$ W/m^2/°C simulates an adiabatic bottom. Setting it to a "large value" like $1(10)^{6}$ W/m^2/°C simulates placement on a highly conductive floor (like copper). Using a finite element numerical method on a coarse numerical grid, calculate the steady-state temperature distribution in the wall, and the rate of heat transfer between the hot (@ T_H) and cold (@ T_C) surfaces (with a temperature difference $\Delta T_{overall}$).

Run the model for other hypothetical cases:

- *Fixed boundaries*: set the two convection coefficients to sufficiently high values that the surface temperatures are approximately equal to the fluid temperatures. Compare the graphical method with the numerical one for this case. Define a "Shape Factor" and overall thermal resistance for this geometry defined by:

$$q_{total} = kS\Delta T_{overall} = \frac{T_H - T_C}{R_{HC}}$$

- *Styrofoam*: set the thermal conductivity of the mug to that of Styrofoam ($k = 0.033$ W/m/K).
- *Fan*: set the cold side heat transfer coefficient to higher values (say, up to 50 W/m^2/°C) to simulate a fan blowing on the mug.
- *Conductive floor*: set the bottom heat transfer coefficient (U_b) to a value of 333 W/m^2/°C to simulate the base of the mug being fixed at ambient temperature by placement on a highly conductive floor.

8.2.1 Finite Element Numerical Analysis

A finite element approach to modeling this problem with a relatively coarse grid will be developed, and implemented in an Excel Workbook. One version contains a MACRO that allows easier implementation of different scenarios.

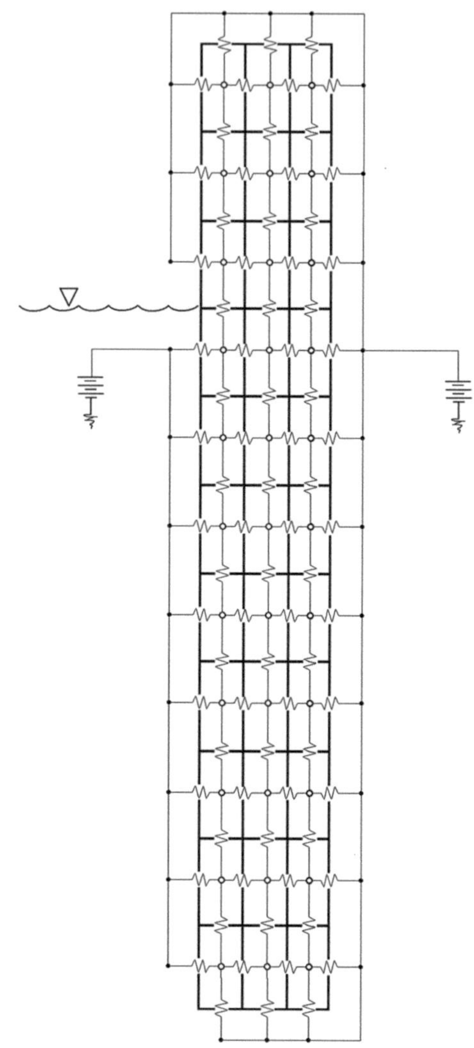

Fig. 8.7 2D grid used to model the quasi-steady state temperature distribution in the wall of the mug

8.2.1.1 Grid

The wall is broken up into the rectangular grid shown in Fig. 8.7. The first step is to break the region into elements. In this case, three "columns" are chosen, three "rows" are chosen for above the coffee level and eight "rows" below the coffee level. These two choices define the size of the elements, $\Delta x = w/3$ (=1 mm for the base case) and $\Delta y = L/3$ (=2 mm for the base case). The location of the free surface coincides with the boundary between two elements. The total height of the wall is $11*\Delta y$ (=22 mm). Nodes represented by open circles are placed in the center of each element. Filled circles show known fluid temperatures. The nodal temperatures represent average temperatures in a finite region. Surface temperatures can be determined from voltage divider calculations once the nodal temperatures are determined.

It would be conceptually straightforward, and require more detail to implement, to consider the thick wall case, where the surface area of the horizontal channels varied with x. Either a constant grid spacing ($\Delta x =$ constant) or a constant elemental volume scheme could be used.

8.2.1.2 General Nodal Equation

Figure 8.8 shows a random element, i, with neighbors "j" in contact through thermal resistances R_{ij}. A transient energy balance for this node, including a thermal energy production term is:

$$C_i \frac{dT_i}{dt} = q_{gen,i} + \sum_j \frac{T_j - T_i}{R_{ij}}$$

Rearranging and recognizing that the temperature T_i is common to each term in the summation:

$$C_i \frac{dT_i}{dt} = q_{gen,i} + \sum_j \frac{T_j}{R_{ij}} - T_i \sum_j \frac{1}{R_{ij}}$$

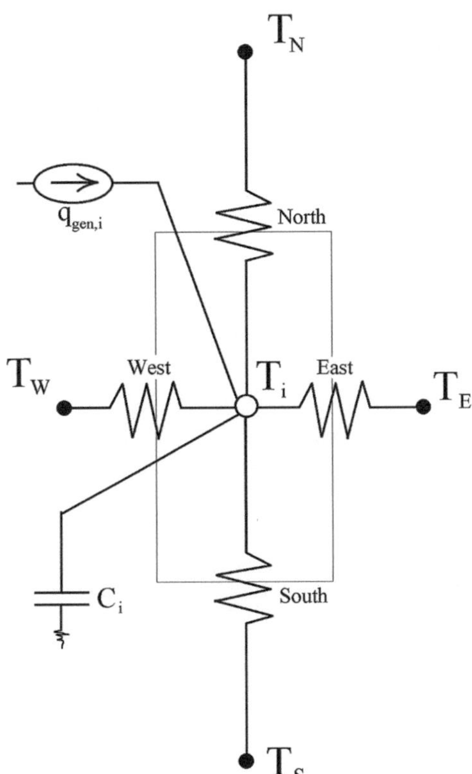

Fig. 8.8 General nodal RC network for node "i"

"Solving" for T_i:

$$T_i = \frac{-C_i \dfrac{dT_i}{dt} + q_{gen,i} + \sum_j \dfrac{T_j}{R_{ij}}}{\sum_j \dfrac{1}{R_{ij}}}$$

At steady-state, the transient term is identically zero, and the steady-state temperature of node i in terms and its neighbors is:

$$T_{i,SS} = \frac{q_{gen,i} + \sum_j \dfrac{T_{j,SS}}{R_{ij}}}{\sum_j \dfrac{1}{R_{ij}}}$$

If the physical space is broken up into a total of N elements, each represented by a node, there are a total of N equations of this type. The equations are algebraic, so a system of N algebraic equations in N unknown temperatures is represented. The system is linear if the thermal energy production and thermal resistances are constant. If they vary with temperature (i.e., thermal conductivity depends on temperature), the system is nonlinear.

For interior nodes, that is, those whose four faces are shared with another element can be written in the general form:

$$T_{i,SS} = \frac{q_{gen,i} + \dfrac{T_{North}}{R_{North}} + \dfrac{T_{East}}{R_{East}} + \dfrac{T_{South}}{R_{South}} + \dfrac{T_{West}}{R_{West}}}{\dfrac{1}{R_{North}} + \dfrac{1}{R_{East}} + \dfrac{1}{R_{South}} + \dfrac{1}{R_{West}}}$$

Here, the neighbors are given N, S, E, W designations. The thermal resistances are all conductive resistances, and since the x and y spacing is constant, there are only two variations:

$$R_{North} = R_{South} = \frac{\Delta y}{k \Delta x P}$$

$$R_{East} = R_{West} = \frac{\Delta x}{k \Delta y P}$$

If the thin-wall approximation were relaxed, the East and West conductive resistances would vary with radial position in the wall.

8.2.2 Boundary Resistances

Boundary elements, those that have at least one face exposed to a fluid, have a similar form, but one (or two, for corners) of the temperatures is the fluid, and the thermal resistances between the fluid and boundary node consist of a convective resistance in series with a conductive resistance (over half the element).

Top boundary: For the nodal element on the top row in the middle (top of the mug), the resistance channel to the north is modified:

$$T_{Top} = \frac{q_{gen,i} + \dfrac{T_C}{R_{Top}} + \dfrac{T_{East}}{R_{East}} + \dfrac{T_{South}}{R_{South}} + \dfrac{T_{West}}{R_{West}}}{\dfrac{1}{R_{Top}} + \dfrac{1}{R_{East}} + \dfrac{1}{R_{South}} + \dfrac{1}{R_{West}}}$$

The thermal resistance for elements exposed to the top surface is a convective resistance in series with a conductive resistance between the node and the surface:

$$R_{Top} = \frac{(\Delta y)/2}{k\Delta x P} + \frac{1}{h_C \Delta x P}$$

Rearranging:

$$R_{Top} = \frac{1}{h_C \Delta x P}\left(\frac{h_C \Delta y}{2k} + 1\right) = \frac{B_{top} + 1}{h_C \Delta x P}$$

This form reveals a Biot number for the element, $B_{top} = h_C \Delta y/2/k$, whose value yields insight into the temperature of the surface. For the base case, in SI units,

$$B_{top} = \frac{(13.9)(0.002)}{(2)(1)} = 0.014$$

The Biot number for this node being much less than 1 means that the temperature in the element will be effectively uniform. In other words, the surface temperature of the nodal element will be closer to the nodal temperature than to the fluid temperature. Yet another way to say it, it is easier for heat to flow by conduction inside the element than it is for heat to leave the surface into the fluid.

Right boundary: Elements that are exposed to the fluid on the right (ambient air) experience a different resistance channel to the east:

$$T_{Right} = \frac{q_{gen,i} + \dfrac{T_{North}}{R_{North}} + \dfrac{T_C}{R_{Right}} + \dfrac{T_{South}}{R_{South}} + \dfrac{T_{West}}{R_{West}}}{\dfrac{1}{R_{North}} + \dfrac{1}{R_{Right}} + \dfrac{1}{R_{South}} + \dfrac{1}{R_{West}}}$$

The resistance between the element and the fluid is a conductive resistance over half the element in series with a convective resistance:

$$R_{Right} = \frac{(\Delta x)/_2}{k\Delta yP} + \frac{1}{h_C \Delta yP}$$

Rearranging:

$$R_{Right} = \frac{1}{h_C \Delta yP}\left(\frac{h_C \Delta x}{2k} + 1\right) = \frac{B_{right} + 1}{h_C \Delta yP}$$

The Biot number here is different from the top one because the distance from the center of the node to the surface is the horizontal distance, whereas it was the vertical distance for the top boundary. For the base case:

$$B_{Right} = \frac{(13.9)(0.001)}{(2)(1)} = 0.007$$

This value is half that for the top boundary (because the horizontal grid spacing Δx is half that of the vertical spacing Δy).

Bottom boundary: Elements that are exposed to the "fluid" on the bottom (simulated floor) experience a different resistance channel to the south:

$$T_{Bottom} = \frac{q_{gen,i} + \dfrac{T_{North}}{R_{North}} + \dfrac{T_{East}}{R_{East}} + \dfrac{T_C}{R_{Bottom}} + \dfrac{T_{West}}{R_{West}}}{\dfrac{1}{R_{North}} + \dfrac{1}{R_{East}} + \dfrac{1}{R_{Bottom}} + \dfrac{1}{R_{West}}}$$

The resistance between the element and the fluid is a conductive resistance over half the element in series with a convective resistance:

$$R_{Bottom} = \frac{(\Delta y)/_2}{k\Delta xP} + \frac{1}{U_b \Delta xP}$$

Rearranging:

$$R_{Bottom} = \frac{1}{U_B \Delta xP}\left(\frac{U_B \Delta y}{2k} + 1\right) = \frac{B_{right} + 1}{U_B \Delta xP}$$

For the base case, with a simulated adiabatic bottom (low U_B):

$$B_{Bottom} = \frac{(1E - 6)(0.002)}{(2)(1)} = 1E - 9$$

Left boundary, below coffee level: Elements that are exposed to the fluid on the left below the coffee line (coffee) experience a different resistance channel to the west:

$$
T_{Left,H} = \frac{q_{gen,i} + \dfrac{T_{North}}{R_{North}} + \dfrac{T_{East}}{R_{East}} + \dfrac{T_{South}}{R_{South}} + \dfrac{T_H}{R_{Left,H}}}{\dfrac{1}{R_{North}} + \dfrac{1}{R_{East}} + \dfrac{1}{R_{South}} + \dfrac{1}{R_{Left,H}}}
$$

The resistance between the element and the fluid is a conductive resistance over half the element in series with a convective resistance:

$$
R_{Left,H} = \frac{(\Delta x)/2}{k\Delta yP} + \frac{1}{h_H \Delta yP}
$$

Rearranging:

$$
R_{Left,H} = \frac{1}{h_H \Delta yP}\left(\frac{h_H \Delta x}{2k} + 1\right) = \frac{B_{left,H} + 1}{h_H \Delta yP}
$$

For the base case:

$$
B_{Left,H} = \frac{(420)(0.001)}{(2)(1)} = 0.21
$$

Due to the much higher convection coefficient between the coffee and surface, this value is larger than the air-based Biot numbers. It is still less than 1, however.

Left boundary, above coffee level: Elements that are exposed to the fluid on the left above the coffee level (humid air inside rim) experience a different resistance channel to the west:

$$
T_{Left,C} = \frac{q_{gen,i} + \dfrac{T_{North}}{R_{North}} + \dfrac{T_{East}}{R_{East}} + \dfrac{T_{South}}{R_{South}} + \dfrac{T_C}{R_{Left,C}}}{\dfrac{1}{R_{North}} + \dfrac{1}{R_{East}} + \dfrac{1}{R_{South}} + \dfrac{1}{R_{Left,C}}}
$$

The resistance between the element and the fluid is a conductive resistance over half the element in series with a convective resistance:

$$
R_{Left,C} = \frac{(\Delta x)/2}{k\Delta yP} + \frac{1}{h_C \Delta yP}
$$

Rearranging:

$$
R_{Left,C} = \frac{1}{h_C \Delta yP}\left(\frac{h_C \Delta x}{2k} + 1\right) = \frac{B_{left,C} + 1}{h_C \Delta yP}
$$

For the base case:

$$B_{Left,C} = \frac{(13.9)(0.001)}{(2)(1)} = 0.007$$

This Biot number is the same as for the right boundary.

Corner Nodes

The four corner nodes experience two modifications from the general NESW form, but all resistances have been defined already. The energy balance equations are, for the top left:

$$T_{TopLeft} = \frac{q_{gen,i} + \dfrac{T_C}{R_{Top}} + \dfrac{T_{East}}{R_{East}} + \dfrac{T_{South}}{R_{South}} + \dfrac{T_C}{R_{Left,C}}}{\dfrac{1}{R_{Top}} + \dfrac{1}{R_{East}} + \dfrac{1}{R_{South}} + \dfrac{1}{R_{Left,C}}}$$

For the top right:

$$T_{TopRight} = \frac{q_{gen,i} + \dfrac{T_C}{R_{Top}} + \dfrac{T_C}{R_{Right}} + \dfrac{T_{South}}{R_{South}} + \dfrac{T_{West}}{R_{West}}}{\dfrac{1}{R_{Top}} + \dfrac{1}{R_{Right}} + \dfrac{1}{R_{South}} + \dfrac{1}{R_{West}}}$$

For the bottom right:

$$T_{BottomRight} = \frac{q_{gen,i} + \dfrac{T_{North}}{R_{North}} + \dfrac{T_C}{R_{Right}} + \dfrac{T_C}{R_{Bottom}} + \dfrac{T_{West}}{R_{West}}}{\dfrac{1}{R_{North}} + \dfrac{1}{R_{Right}} + \dfrac{1}{R_{Bottom}} + \dfrac{1}{R_{West}}}$$

For the bottom left:

$$T_{BottomLeft} = \frac{q_{gen,i} + \dfrac{T_{North}}{R_{North}} + \dfrac{T_{East}}{R_{East}} + \dfrac{T_C}{R_{Bottom}} + \dfrac{T_H}{R_{Left,H}}}{\dfrac{1}{R_{North}} + \dfrac{1}{R_{East}} + \dfrac{1}{R_{Bottom}} + \dfrac{1}{R_{Left,H}}}$$

Surface Temperatures

Once the nodal temperatures are determined, the surface temperatures can be inferred from them using a voltage divider analysis. In general, for a boundary element (at temperature T_i) in thermal contact with a fluid (at temperature T_∞), the voltage divider equation can be expressed as:

$$q_{i\infty} = \frac{(T_i - T_\infty)}{R_k + R_h} = \frac{(T_{S,i} - T_\infty)}{R_h}$$

In this expression, $T_{S,i}$ is the surface temperature of node i, and the conductive (R_k) and convective (R_h) thermal resistances are defined appropriately for the element in question. Rearranging for the surface temperature:

$$T_{S,i} = T_\infty + \left(\frac{R_h}{R_k + R_h}\right)(T_i - T_\infty)$$

Expressed as the ratio of thermal resistances:

$$T_{S,i} = T_\infty + \left(\frac{1}{R_k/R_h + 1}\right)(T_i - T_\infty)$$

The ratio of the conductive to convective resistance is a Biot number. For example, for the top middle node:

$$\frac{R_k}{R_h} = \frac{\dfrac{(\Delta y/2)}{k\Delta x P}}{\dfrac{1}{h_C \Delta x P}} = \frac{h_C \Delta y}{2k} = B_{Top}$$

Corner Surface Temperatures

The corner surface temperatures are a little different, and can be handled by an energy balance on a resistance network defined at the corner of the corner elements. Figure 8.9 shows the bottom left corner element. The corner is in thermal contact with the surface "node" to the north (a distance $\Delta y/2$ away, across a conductive area $P\Delta x/4$), to the east with the surface "node" (a distance $\Delta x/2$ away, across a

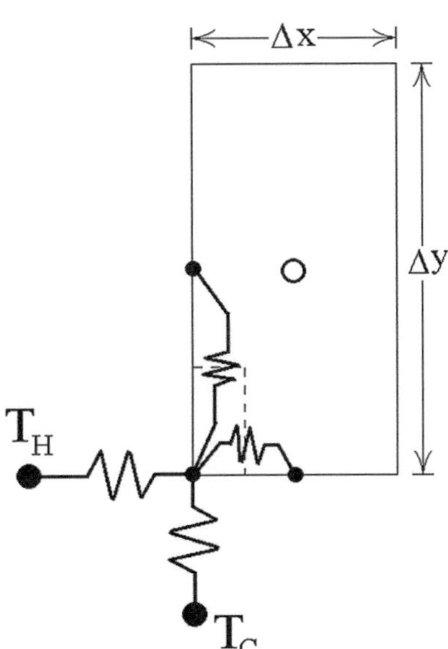

Fig. 8.9 *Bottom left corner* subelement used to calculate the surface temperature at the corner

conductive area PΔy/4), to the south with the cold fluid (across a convective area PΔx/4), and to the east with the hot fluid (across a convective area PΔy/4). The bottom left corner surface temperature, paying close attention to how the resistances are defined, is related to its four neighbors by:

$$T_{BottomLeftSurface} = \frac{\dfrac{T_{S,above}}{\frac{(\Delta y/2)}{k(\Delta x/4)P}} + \dfrac{T_{S,right}}{\frac{(\Delta x/2)}{k(\Delta y/4)P}} + \dfrac{T_C}{\frac{1}{U_B(\Delta x/4)P}} + \dfrac{T_H}{\frac{1}{h_H(\Delta y/4)P}}}{\dfrac{1}{\frac{(\Delta y/2)}{k(\Delta x/4)P}} + \dfrac{1}{\frac{(\Delta x/2)}{k(\Delta y/4)P}} + \dfrac{1}{\frac{1}{U_B(\Delta x/4)P}} + \dfrac{1}{\frac{1}{h_H(\Delta y/4)P}}}$$

For the top left corner surface:

$$T_{TopLeftSurface} = \frac{\dfrac{T_C}{\frac{1}{h_C(\Delta x/4)P}} + \dfrac{T_{S,right}}{\frac{(\Delta x/2)}{k(\Delta y/4)P}} + \dfrac{T_{S,below}}{\frac{(\Delta y/2)}{k(\Delta x/4)P}} + \dfrac{T_C}{\frac{1}{h_C(\Delta y/4)P}}}{\dfrac{1}{\frac{1}{h_C(\Delta x/4)P}} + \dfrac{1}{\frac{(\Delta x/2)}{k(\Delta y/4)P}} + \dfrac{1}{\frac{(\Delta y/2)}{k(\Delta x/4)P}} + \dfrac{1}{\frac{1}{h_C(\Delta y/4)P}}}$$

For the top right corner:

$$T_{TopRightSurface} = \frac{\dfrac{T_C}{\frac{1}{h_C(\Delta x/4)P}} + \dfrac{T_C}{\frac{1}{h_C(\Delta y/4)P}} + \dfrac{T_{S,below}}{\frac{(\Delta y/2)}{k(\Delta x/4)P}} + \dfrac{T_{S,left}}{\frac{(\Delta x/2)}{k(\Delta y/4)P}}}{\dfrac{1}{\frac{1}{h_C(\Delta x/4)P}} + \dfrac{1}{\frac{1}{h_C(\Delta y/4)P}} + \dfrac{1}{\frac{(\Delta y/2)}{k(\Delta x/4)P}} + \dfrac{1}{\frac{(\Delta x/2)}{k(\Delta y/4)P}}}$$

For the bottom right corner surface:

$$T_{BottomRightSurface} = \frac{\dfrac{T_{S,above}}{\frac{(\Delta y/2)}{k(\Delta x/4)P}} + \dfrac{T_C}{\frac{1}{h_C(\Delta y/4)P}} + \dfrac{T_C}{\frac{1}{U_B(\Delta x/4)P}} + \dfrac{T_{S,left}}{\frac{(\Delta x/2)}{k(\Delta y/4)P}}}{\dfrac{1}{\frac{(\Delta y/2)}{k(\Delta x/4)P}} + \dfrac{1}{\frac{1}{h_C(\Delta y/4)P}} + \dfrac{1}{\frac{1}{U_B(\Delta x/4)P}} + \dfrac{1}{\frac{(\Delta x/2)}{k(\Delta y/4)P}}}$$

Once all the temperatures are known, the total heat transfer rate from the coffee to the side wall can be calculated by adding the contributions at each node exposed to coffee. A good debugging tool is to also calculate the total heat transfer rate out of all other faces, which should be equal and opposite at steady-state. If not, there is a bug somewhere.

All the pieces are now in place, and it is time to put them all together.

Model Implementation

An iterative procedure can be applied to a system of N algebraic equations in N unknowns by guessing a solution, followed by repeated application of a recursion relation formed from the governing equations. The nodal equations developed in the previous section define a particular set of recursion relations. This form, where for node i, T_i is expressed in terms of its neighbors, that is, the Method of Successive Substitution (MOSS). Depending on the system, this iteration procedure will result in either convergence to a correct mathematical solution (which might be an unphysical solution), or divergence (chaos or numerical instability), which means the particular method used has failed. It is possible that the same method would work with a different set of initial guesses, or that the recursion relations can be expressed differently.

A strongly diagonal linear system of equations is one in which all the diagonal elements of each row are greater than or equal to the sum of the absolute values of off-diagonal elements. For strongly diagonal linear systems, the MOSS will converge to the correct solution. Heat conduction problems with constant thermal energy production and resistances are strongly diagonal and therefore the MOSS will work. However, if the thermal energy production and thermal resistances are variable (for example, with radiation coefficients), making it a nonlinear system, the method may work, and is generally worth trying. If not, the recursion relations for nodes that have a high degree of nonlinearity (radiation coefficients are likely culprits) can be written in another form. If that fails, then other iterative methods, particularly Newton's method, are possible. However, setting up a transient problem and allowing it to reach equilibrium is a legitimate approach if the MOSS fails after a little tweaking.

A word of caution is in order. Implementation of the MOSS appears to give a "transient solution" if guessed temperatures are plotted against iteration number. It is not a transient solution, however, especially if the response time of nodal elements varies significantly. If a transient solution is desired, set up and solve it as a transient problem.

Superscripts "k" and "k + 1" are used to identify iteration numbers. In the Jacobi form, the general equation for node i with superscripts is:

$$T_{i,SS}^{(k+1)} = \frac{q_{gen,i}^{(k)} + \sum_j \dfrac{T_{j,SS}^{(k)}}{R_{ij}^{(k)}}}{\sum_j \dfrac{1}{R_{ij}^{(k)}}}$$

Superscripts have to be assigned to all factors on the right hand side, which indicates thermal energy production and resistances are evaluated at the local "guessed" temperature, at iteration "k". The Gauss–Seidel method could alternatively be used in which each nodal temperature is updated to "k + 1" as soon as available. The order in which the calculations are performed is important in this method. Systems converge with fewer iterations, generally, but the computational savings is not dramatic, so the Jacobi method is implemented here.

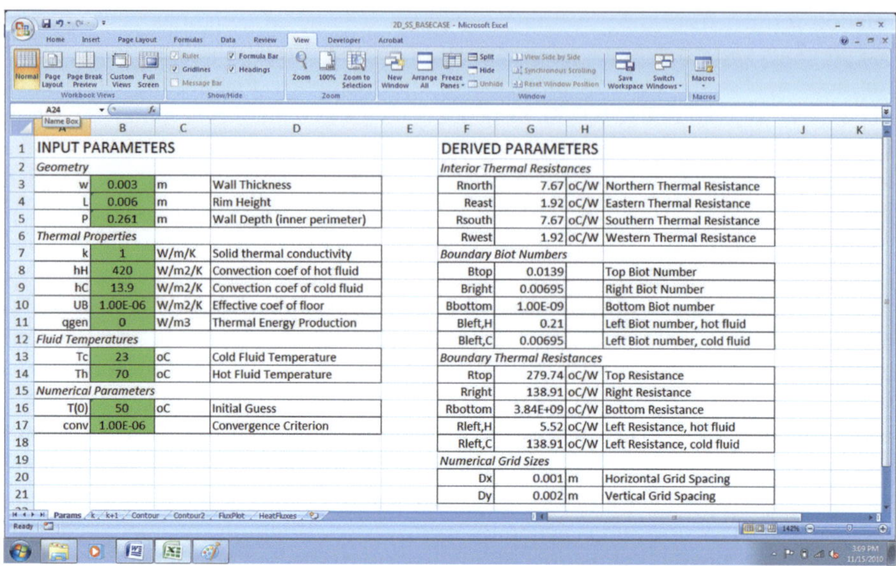

Fig. 8.10 Screenshot of the "Params" worksheet of the excel workbook used to solve the quasi-steady state mug side wall problem

This problem was solved using a macro-enabled Excel workbook, which is described in detail here, followed by a presentation of results from running the program for different scenarios. Figure 8.10 shows a screenshot of the workbook with the worksheet entitled "Params" accessed. The other worksheets will be described from left to right. The input parameters (grouped into geometry, thermal properties, and fluid temperatures) on the left can all be changed by the user. The derived parameters incorporate the equations developed for the model (grouped into interior resistances, boundary Biot numbers and resistances, and numerical grid sizes). The workbook can be protected so only the shaded cells can be changed, preventing inadvertent changes to the program.

The next sheet, named "k", shown in Fig. 8.11, shows the "guessed" temperatures. The worksheet is shown at the very first iteration ($k = 0$) in which the initial guess is made. The x position is along the top row, and the y position is in the first column (in millimeters). The cells around the perimeter represent the fluid temperatures. The hot fluid (coffee) is shaded in red; the cold fluid (ambient air) is shaded in blue. Inside those, the cells shaded in gray are the surface temperatures calculated from the guessed temperatures and the adjacent fluids according the model. Finally, the 33 nodal temperatures are shown in the heavy boxed region in bold. At this iteration ($k = 0$), all the temperatures are set to 50.00, somewhat arbitrarily.

To the right of the temperatures are two system checks. The "Test for Convergence" references the next sheet, to be discussed shortly, and this cell finds the maximum value of any change that takes place between iteration k and $k + 1$. At this point, the maximum change is 0.0869. Underneath, an if-then-else statement

x (mm)=	0	0.5	1.5	2.5	3			Test for Convergence	
y(mm)	23	23	23	23	23	23	23	MAXIMUM FRACTIONAL CHANGE	
6	23	49.4	49.6	49.6	49.6	49.4	23	0.086934023	
5	23	49.8	**50.00**	**50.00**	**50.00**	49.8	23	*Not Converged*	
3	23	49.8	**50.00**	**50.00**	**50.00**	49.8	23	**Author:** Run MACRO shortcut key CTRL-i to execute an iteration	
1	23	49.8	**50.00**	**50.00**	**50.00**	49.8	23		
-1	70	53.5	**50.00**	**50.00**	**50.00**	49.8	23	*Overall Heat Flux Test*	
-3	70	53.5	**50.00**	**50.00**	**50.00**	49.8	23	HEAT FLUX IN (Watts)=	28.96289
-5	70	53.5	**50.00**	**50.00**	**50.00**	49.8	23	HEAT FLUX OUT (Watts)=	-3.01073
-7	70	53.5	**50.00**	**50.00**	**50.00**	49.8	23	FRACTIONAL DIFFERENCE=	0.896049
-9	70	53.5	**50.00**	**50.00**	**50.00**	49.8	23		
-11	70	53.5	**50.00**	**50.00**	**50.00**	49.8	23		
-13	70	53.5	**50.00**	**50.00**	**50.00**	49.8	23		
-15	70	53.5	**50.00**	**50.00**	**50.00**	49.8	23		
-16	70	53.5	50.0	50.0	50.0	49.8	23		
	70	23	23	23	23	23	23		

Fig. 8.11 Worksheet showing the "k" iteration number. The cells are arranged in a 2D grid like the resistance network

compares that change with the user-defined convergence criterion. If the change is greater than the criterion, the cell returns "Not Converged" as shown here. Otherwise, it returns "Converged." The comment by the author will be discussed in the next sheet.

Below that comment is a test for overall heat balance. The total heat flux from the hot fluid into the wall is calculated by adding up all the individual heat fluxes. The total heat flux out is the heat flux from the cold fluid into the wall (with the bottom considered part of the cold fluid). This value is negative, meaning that heat is actually leaving the wall. The sum of the two is the net heat flux into the wall, which when properly balanced, should equal zero. This overall balance is more of a check on the overall implementation of the model. A system can converge to a set of temperatures, but if there are errors in incorporating the model, an imbalance in the heat flux may be revealed, without necessarily pinpointing precisely where it is.

The next worksheet, "k + 1", shown in Fig. 8.12, is where the recursion relations are enacted. The structure of the temperature table is the same as for worksheet "k" except that the recursion relations are implemented in all 33 nodal cells, which refer to the temperatures "guessed" in worksheet "k". At the point shown, after one iteration, the boundary cells have changed. The table on the right calculates the difference between k and k + 1 for each node. Notice that the temperatures in the middle of the wall have not changed (because all the neighbors were at the same temperature), but the difference between "k" and "k + 1" iterations is $1.42(10)^{-16}$, an indication of the precision of the particular computer being used. At the bottom, the maximum from each column is determined, and the maximum of these is what is determined in worksheet "k".

x (mm)=	0	0.5	1.5	2.5	3			FRACTIONAL CHANGES		
y(mm) 23	23	23	23	23	23	23				
6	23					23				
5	23	49.561	49.918	49.561		23		8.78E-03	1.64E-03	8.78E-03
3	23	49.7538	50	49.7538		23		4.92E-03	1.42E-16	4.92E-03
1	23	49.7538	50	49.7538		23		4.92E-03	1.42E-16	4.92E-03
-1	70	53.7584	50	49.7538		23		7.52E-02	1.42E-16	4.92E-03
-3	70	53.7584	50	49.7538		23		7.52E-02	1.42E-16	4.92E-03
-5	70	53.7584	50	49.7538		23		7.52E-02	1.42E-16	4.92E-03
-7	70	53.7584	50	49.7538		23		7.52E-02	1.42E-16	4.92E-03
-9	70	53.7584	50	49.7538		23		7.52E-02	1.42E-16	4.92E-03
-11	70	53.7584	50	49.7538		23		7.52E-02	1.42E-16	4.92E-03
-13	70	53.7584	50	49.7538		23		7.52E-02	1.42E-16	4.92E-03
-15	70	54.3467	50	49.7051		23		8.69E-02	1.20E-10	5.90E-03
-16	70					23				
	70	23	23	23	23	23				
							MAXIMUM	8.69E-02	1.64E-03	8.78E-03

Fig. 8.12 Worksheet showing the "k + 1" iteration

Execution of another iteration involves copying the 33 nodal temperatures in worksheet "k + 1" and pasting their values into worksheet "k". To emphasize, worksheet "k" has numerical values that represent a guess to the solution, while worksheet "k + 1" has formulas that look to those values, based on the recursion relation. When another iteration is to be made, the values at "k + 1" are used to update. In excel, the cells in "k + 1" are copied, and then "paste special as values." A MACRO program was written to do all these steps at once by entering the shortcut key "CTRL-i". Pressing and holding CTRL-i while in the worksheet "k" will cause repeated executions, as fast as the computer will allow, and the state of the convergence can be monitored. The program does not count how many iterations are made (although you could modify it). For the problem at hand, it is the final solution that is sought, not the route taken to get there.

Figure 8.13 shows the worksheet "k" after some several hundred (not counted) iterations required to reach convergence to the user-defined criterion. Figure 8.14 shows a contour plot of just the 33 nodal elements, while Fig. 8.15 shows a contour plot that includes the surface temperatures and the fluid temperatures. The problem with this last plot is related to a limitation of making contour plots with the version of Excel used which restricts the plot to uniformly spaced x and y series. If the x and y spacing are not uniform, there is a distortion in the plot. This distortion is not readily apparent in the base case, but it will be in some of the other cases considered.

Away from the coffee-free surface, the converged solution confirms what was learned from the transient analysis, namely, after the mug heating phase, there is a relatively small drop in temperature from the hot coffee to the mug inner wall, and then across the inner wall. A large temperature drop occurs between the outer wall and the ambient air. Also, there is little variation in the vertical direction from the floor to about 3 mm below the coffee surface. The heat transfer is effectively 1D

x (mm)=	0	0.5	1.5	2.5	3	
y(mm)	23	23	23	23	23	23	23
6	23	52.6	52.8	52.9	52.8	52.6	23
5	23	53.0	53.20	53.33	53.18	53.0	23
3	23	54.9	55.16	55.27	55.10	54.9	23
1	23	58.2	58.44	58.39	58.07	57.8	23
-1	70	64.9	63.88	62.57	61.72	61.5	23
-3	70	66.7	66.03	64.93	64.10	63.8	23
-5	70	67.6	67.08	66.19	65.45	65.2	23
-7	70	68.0	67.63	66.87	66.18	65.9	23
-9	70	68.3	67.91	67.22	66.57	66.3	23
-11	70	68.4	68.06	67.41	66.78	66.5	23
-13	70	68.5	68.14	67.50	66.88	66.6	23
-15	70	68.5	68.17	67.54	66.93	66.6	23
-16	70	68.5	68.2	67.5	66.9	66.6	23
	70	23	23	23	23	23	23

Test for Convergence	
MAXIMUM FRACTIONAL CHANGE	
9.97343E-07	
Converged	
Author: Run MACRO shortcut key CTRL-i to execute an iteration	

Overall Heat Flux Test	
HEAT FLUX IN (Watts)=	4.181607
HEAT FLUX OUT (Watts)=	-4.18125
FRACTIONAL DIFFERENCE=	8.58E-05

Fig. 8.13 Worksheet "k" after convergence to the final result

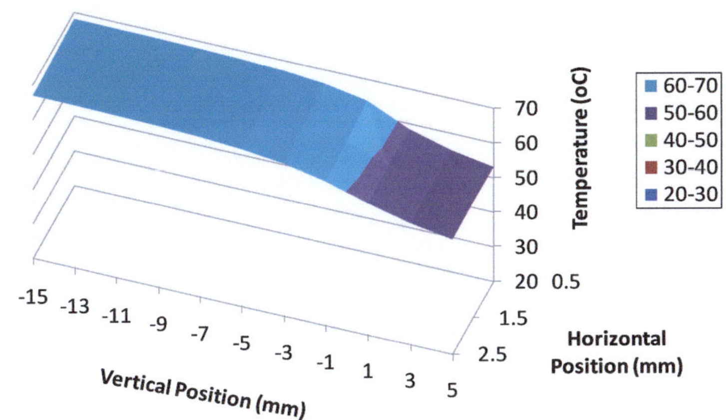

Fig. 8.14 Contour plot of converged solution showing just the 33 nodal elements

across the mug over most of the surface. This 1D behavior was inherently assumed in the lumped model, and also was revealed by the graphical analysis (where isotherms were vertical and heat flux lanes horizontal).

On the other hand, the 2D analysis reveals details near the coffee-free surface. Figure 8.16 shows the boundary heat fluxes along the inner and outer walls (the top and bottom not shown). Along the inner surface, the heat flux from the coffee to the mug increases dramatically as the coffee surface is approached from below, and then suddenly switches to a heat flux from the mug to the air immediately above the surface. This effect was predicted by the graphical solution which shows heat flux line spacing decreasing as the coffee surface is approached and the flux line turning

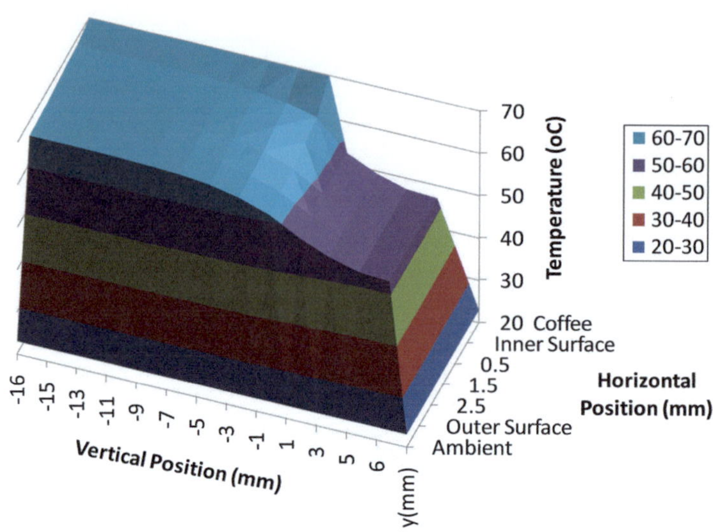

Fig. 8.15 Converged solution of the 33 nodal elements plus the surface nodes plus the fluid nodes

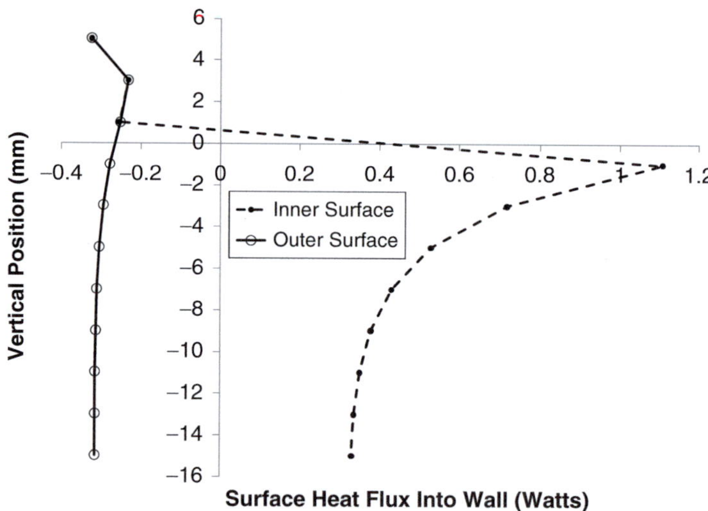

Fig. 8.16 Surface heat fluxes into the wall (x axis) as a function of vertical position. The total heat transfer rate between the submerged surface and that exposed to air (inside and out) is 4.18 W

back on itself. Heat that enters near the coffee-free surface must transfer across the mug wall (as everywhere below it) but also must transfer vertically into the rim, resulting in a high heat flux near the surface. Also, the heat loss from the top corners is higher than from lower nodes due to the heat transferred vertically to the top surface (in parallel with the horizontal heat flux). The heat flux on the inner and outer surfaces of the rim are essentially equal and closer to the mug temperature (at the coffee surface) than ambient, indicating that the rim behaves like a thermally short heat transfer fin.

8.2.3   Case Studies

Each case study from the original problem statement is introduced, followed by its converged solution, then the contour plot and heat flux plot.

8.2.3.1 Fixed Boundary Temperatures

Figure 8.17 shows the converged results and Fig. 8.18 the contour plot and surface heat flux when the convection coefficients are artificially set to sufficiently high values to force the surface temperatures equal to the adjacent fluid temperatures. This case corresponds to that of the graphical method.

There are steep gradients in temperature across the wall, as expected since the overall temperature drop across the wall is forced to be that way. Also, the gradients in the vertical direction are quite steep in the region just above the coffee level. The rim acts more like a "handle" than a "fin" because three sides are forced to ambient temperature. This case is considered mainly as a direct comparison to the graphical method example, and the heat flux of both methods agrees very well.

y(mm)	x (mm)=	0	0.5	1.5	2.5	3	
	23	23	23	23	23	23	23
6	23	23.0	23.0	23.0	23.0	23.0	23
5	23	23.0	23.06	23.11	23.05	23.0	23
3	23	23.0	23.49	23.76	23.34	23.0	23
1	23	23.0	26.76	27.18	24.66	23.0	23
-1	70	70.0	58.39	42.32	29.18	23.0	23
-3	70	70.0	61.66	45.73	30.50	23.0	23
-5	70	70.0	62.08	46.37	30.78	23.0	23
-7	70	70.0	62.14	46.48	30.83	23.0	23
-9	70	70.0	62.15	46.50	30.84	23.0	23
-11	70	70.0	62.16	46.50	30.84	23.0	23
-13	70	70.0	62.16	46.50	30.84	23.0	23
-15	70	70.0	62.16	46.50	30.84	23.0	23
-16	70	70.0	62.2	46.5	30.8	23.0	23
	70	23	23	23	23	23	23

Test for Convergence

MAXIMUM FRACTIONAL CHANGE
4.37626E-07

Converged

| Author: |
| Run MACRO shortcut key |
| CTRL-i |
| to execute an iteration |

Overall Heat Flux Test

HEAT FLUX IN (Watts)=	69.83995
HEAT FLUX OUT (Watts)=	-69.8399
FRACTIONAL DIFFERENCE=	5.12E-07

Fig. 8.17 Converged results for the case of fixed surface temperatures (achieved by setting the convection coefficients to sufficiently high values)

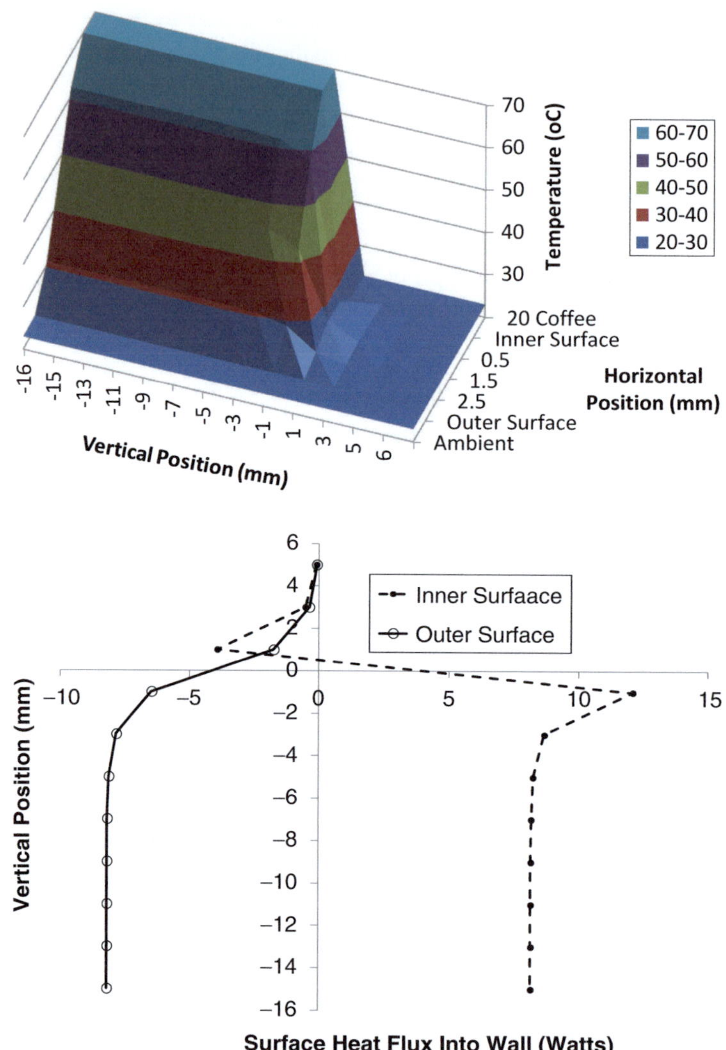

Fig. 8.18 Contour and heat flux plot for the case of fixed boundary temperatures. The x-scale was increased from 1.2 to 15 compared to the base case. The total heat transfer rate between the submerged surface and ambient air is 69.84 W

8.2.3.2 Styrofoam Mug

Figure 8.19 shows the converged results and Fig. 8.20 the contour plot and surface heat flux when the thermal conductivity is changed from a value of 1.0 W/m/K (typical of "ceramic") to a value of 0.033, typical of Styrofoam. There is a large temperature drop across the mug wall, and the surface heat flux is much lower than the base case (ceramic mug). The rim acts like a thermally long fin in this case, with the temperature at the top much closer to ambient than to the base of the rim (at the coffee surface).

y(mm)	x (mm)=	0	0.5	1.5	2.5	3	
	23	23	23	23	23	23	23
6	23	24.0	24.2	24.4	24.2	24.0	23
5	23	24.5	24.76	24.95	24.71	24.4	23
3	23	26.7	27.49	27.83	27.14	26.4	23
1	23	33.0	35.11	34.85	32.54	30.9	23
-1	70	68.7	60.65	50.01	42.02	38.7	23
-3	70	69.2	64.11	54.62	45.96	42.0	23
-5	70	69.3	64.78	55.90	47.24	43.0	23
-7	70	69.3	64.95	56.25	47.62	43.3	23
-9	70	69.3	65.00	56.35	47.73	43.4	23
-11	70	69.3	65.01	56.38	47.76	43.5	23
-13	70	69.3	65.01	56.39	47.77	43.5	23
-15	70	69.3	65.01	56.39	47.77	43.5	23
-16	70	69.3	65.0	56.4	47.8	43.5	23
	70	23	23	23	23	23	23

Test for Convergence	
MAXIMUM FRACTIONAL CHANGE	
7.91353E-07	
Converged	

Author:
Run MACRO shortcut key
CTRL-i
to execute an iteration

Overall Heat Flux Test	
HEAT FLUX IN (Watts)=	1.352793
HEAT FLUX OUT (Watts)=	-1.3528
FRACTIONAL DIFFERENCE=	-5.4E-06

Fig. 8.19 Converged solution for a Styrofoam mug

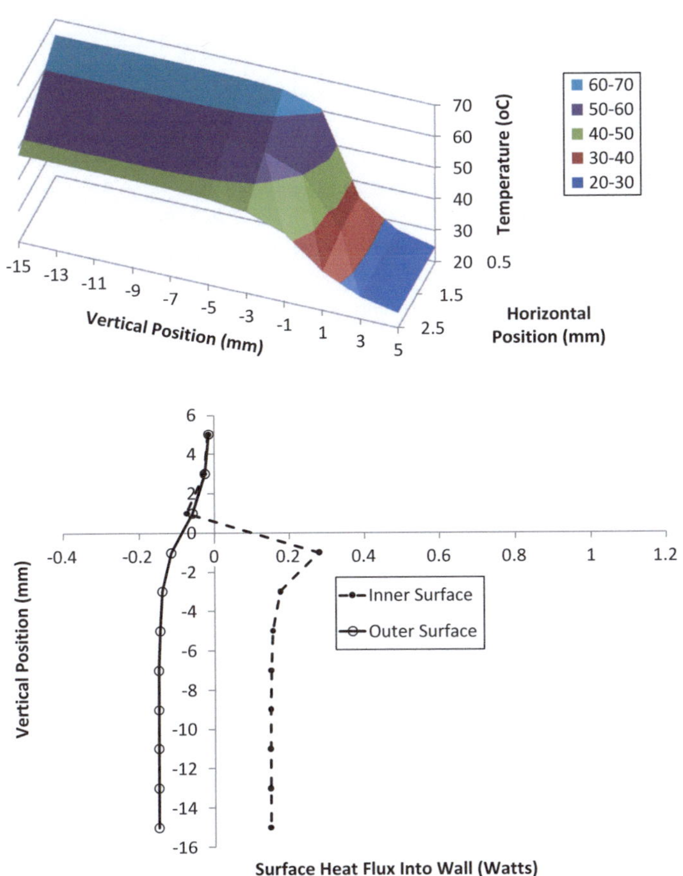

Fig. 8.20 Contour plot and surface heat flux for a Styrofoam mug. The scale on the surface flux is that same as the base case, revealing the lower heat transfer rate across the mug wall

8.2.3.3 Copper Floor

The base case assumed an adiabatic floor, that is, it prevented heat from flowing through the bottom surface by using a sufficiently low value of the effective heat transfer coefficient to the floor (U_B). In reality, heat will flow from the mug to the floor. To simulate the other extreme, namely, a highly conductive floor capable of providing an effective means of pulling heat from the base of the wall, the floor heat transfer coefficient (U_b) is set to 333 W/m^2/°C (which considers the conduction across a mug thickness). With this value of U_B, the converged solution is shown in Fig. 8.21 and the contour plot and heat flux plots in Fig. 8.22. It is apparent that the floor effects increase the heat flux (by more than a factor of 2). However, the effects are confined to a few millimeters, so for the much taller mug wall than analyzed here, the majority of the wall will experience 1D heat transfer effects (except near the rim base and the floor).

y(mm)	x (mm)=	0	0.5	1.5	2.5	3	
	23	23	23	23	23	23	23
6	23	52.4	52.6	52.7	52.6	52.4	23
5	23	52.8	52.99	53.12	52.98	52.8	23
3	23	54.7	54.94	55.06	54.88	54.7	23
1	23	58.0	58.21	58.15	57.83	57.6	23
-1	70	64.8	63.65	62.29	61.42	61.2	23
-3	70	66.4	65.69	64.51	63.64	63.4	23
-5	70	67.1	66.48	65.45	64.64	64.4	23
-7	70	67.1	66.53	65.50	64.69	64.4	23
-9	70	66.6	65.88	64.69	63.80	63.5	23
-11	70	65.3	64.26	62.68	61.61	61.3	23
-13	70	62.5	60.95	58.66	57.28	57.0	23
-15	70	57.0	54.24	50.59	49.32	49.1	23
-16	70	50.1	46.4	44.0	42.7	42.6	23
	70	23	23	23	23	23	23

Test for Convergence

MAXIMUM FRACTIONAL CHANGE
9.88294E-07

Converged

> **Author:**
> Run MACRO shortcut key CTRL-i
> to execute an iteration

Overall Heat Flux Test

HEAT FLUX IN (Watts)=	9.471576
HEAT FLUX OUT (Watts)=	-9.47123
FRACTIONAL DIFFERENCE=	3.67E-05

Fig. 8.21 Converged solution for the case of a highly conductive (i.e., copper) floor

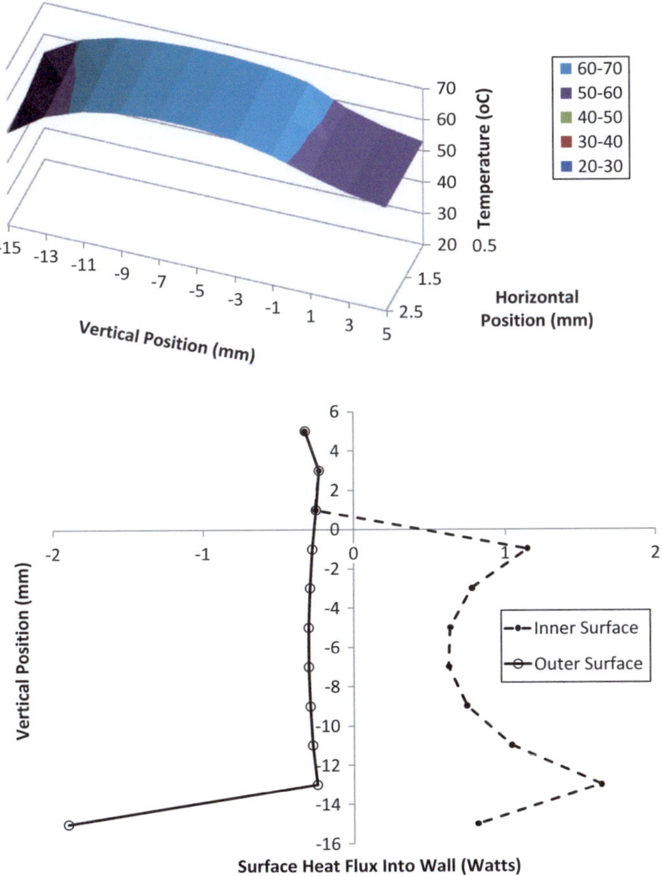

Fig. 8.22 Contour plot and heat flux plots when the floor is highly conductive (i.e., copper floor). The x-scale was increased from 1.2 to 2 in order to observe the floor effects

8.2.3.4 Fan

The effectiveness of the heat transfer from the outer surface to the ambient can be enhanced by blowing across it (Fig. 8.23). This effect can be simulated by increasing the value of the convection coefficient on the outside. Changing it to a value of 50 W/m^2/°C (a factor of 3.5 greater than the base case) results in the contour and heat flux plots of Fig. 8.24. The heat flux increases by a factor of 2.4, and there is a greater drop in temperature across the mug wall. Recall that the outer convective resistance dominates the total resistance from coffee to ambient in the lumped model, and therefore lowering that resistance has a large effect. Also, the rim as a fin is thermally longer than the base case, as the temperature at the top of the rim is closer to ambient than the rim base.

y(mm)	x (mm)=	0	0.5	1.5	2.5	3	
	23	23	23	23	23	23	23
6	23	36.1	36.5	36.7	36.4	36.1	23
5	23	36.8	37.14	37.36	37.12	36.8	23
3	23	40.0	40.38	40.59	40.24	39.8	23
1	23	45.6	46.15	46.05	45.33	44.8	23
-1	70	59.1	56.85	53.95	51.94	51.2	23
-3	70	62.5	60.89	58.26	56.17	55.4	23
-5	70	64.0	62.76	60.48	58.48	57.6	23
-7	70	64.8	63.69	61.62	59.68	58.8	23
-9	70	65.2	64.16	62.20	60.30	59.4	23
-11	70	65.4	64.40	62.49	60.61	59.7	23
-13	70	65.5	64.52	62.63	60.77	59.8	23
-15	70	65.5	64.56	62.69	60.83	59.9	23
-16	70	65.5	64.6	62.7	60.8	59.9	23
	70	23	23	23	23	23	23

Test for Convergence	
MAXIMUM FRACTIONAL CHANGE	
9.53685E-07	

Converged

Author:
Run MACRO shortcut key
CTRL-i
to execute an iteration

Overall Heat Flux Test

HEAT FLUX IN (Watts)=	10.52989
HEAT FLUX OUT (Watts)=	-10.5303
FRACTIONAL DIFFERENCE=	-3.6E-05

Fig. 8.23 Converged solution for the case of a fan blowing across the side wall

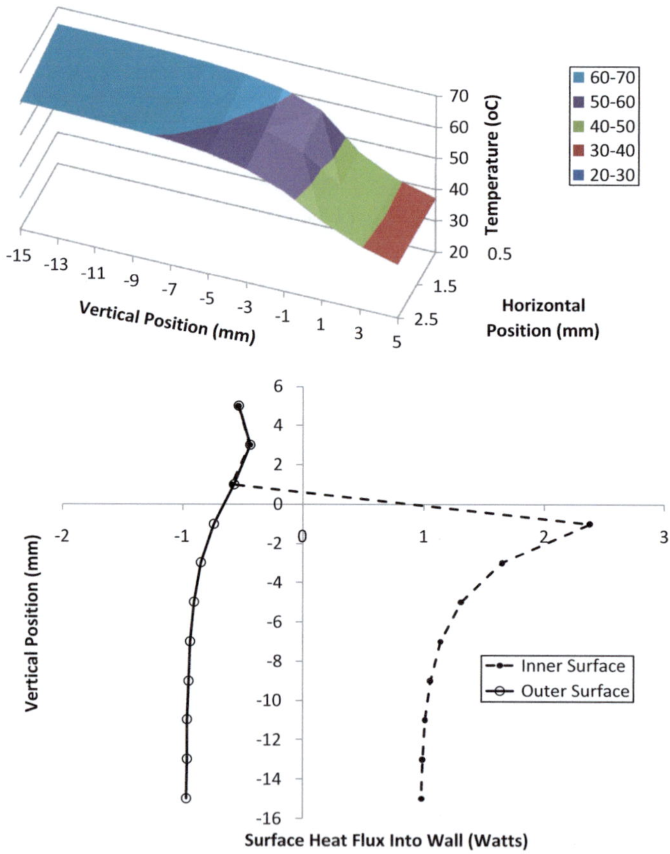

Fig. 8.24 Contour and surface heat flux plots for the case of an enhanced convection on the outer surface of the mug (i.e., a fan blowing on it). Again, the x-scale is changed compared to the base case

It is intriguing that the natural response of an impatient coffee drinker is to blow on the surface of coffee to cool it down, and that evolution has not progressed to the point that she blows on the side.

8.3 Classic 2D Problem Statement

The bottom surface (called south) of a long $w \times H$ m^2 rectangular structural member (Fig. 8.25, with length L into the page, width $w = 4.2$ m, height $H = 1.3$ m) is maintained at temperature T_H ($=100$ °C). The other three surfaces (north, east, and west) are maintained at a temperature $T_C = 0$ °C. The member has a thermal conductivity $k = 20$ W/m/K (typical of steel). A uniform internal generation of thermal energy is possible (i.e., conversion of electrical energy into thermal energy). For the case of no internal heat generation, determine the steady-state temperature distribution within the member and heat transfer rate (per unit length) from the bottom surface to the three other sides (q_{HC}/L) and the conduction shape factor. Determine the internal heat generation rate (in W/m^3) that will yield a maximum temperature of 200 °C.

8.3.1 Numerical Analysis

As with transient problems, the region of interest can be divided into discrete elements. Each element is represented by a node. Rather than think of each element as having a uniform temperature, the nodal temperature represents an average temperature of the element. That way, boundary temperatures between nodes can be approximated through voltage divider relations, giving more information. The accuracy of the model improves as the region is divided up into smaller and smaller elemental sizes, at the cost of initial setup work, and computational effort.

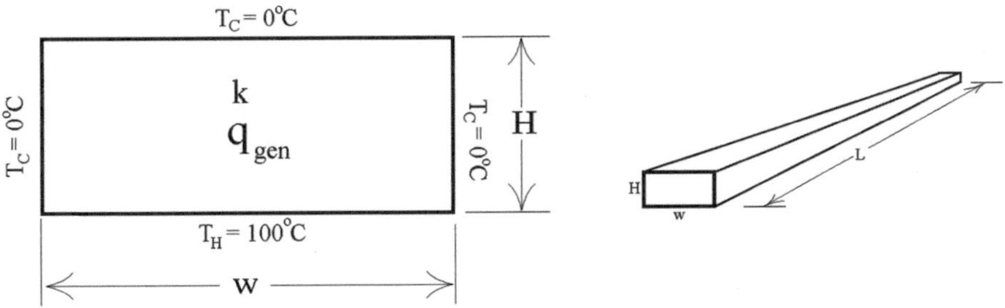

Fig. 8.25 Cross-section (*left*) and oblique projection (*right*) of 2D rectangular member

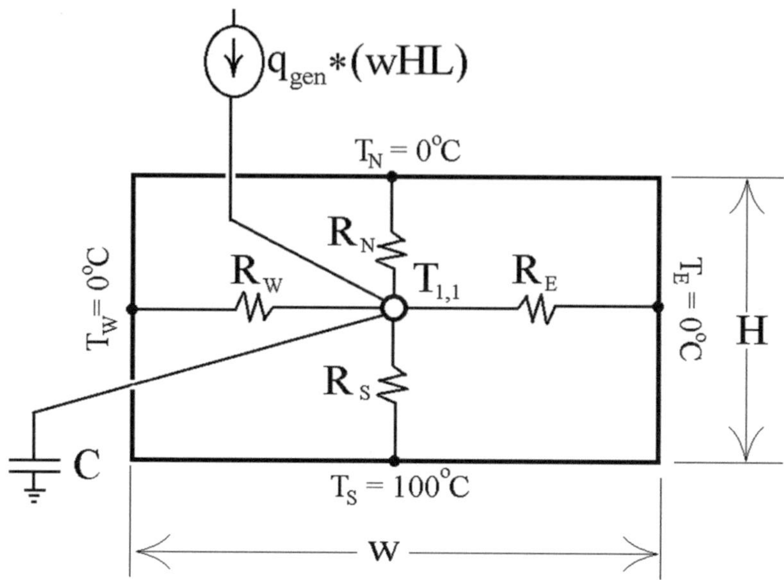

Fig. 8.26 Thermal resistance network for a 1 × 1 model of the classic 2D problem. The surface temperatures specified are for the classic 2D problem

8.3.2   1 × 1 Model

A lumped capacity thermal resistance network is shown in Fig. 8.26 for a more general case, where the temperatures of all four faces can be set to different values and there is internal heat generation generated. The thermal capacitance (C = ρc*(wHL)) of the system is shown, along with the internal heat generation (q_{gen} being the volumetric heat addition, that is, heat added per unit volume). The nodal temperature is given a subscript 1,1, in anticipation of generalizing to a finer grid, and represents the average temperature in the member. The node is in thermal contact with the four sides through conductive resistances (north, east, south, and west). The three sides can be at different temperatures, but for the classic problem are set to be the same temperature. It is relatively easy to extend the problem further, by specifying convective boundary conditions, a good workshop problem.

An energy balance applied to the single node is:

$$C\frac{dT_{1,1}}{dt} = q_{gen}wHL + \frac{T_N - T_{1,1}}{R_N} + \frac{T_E - T_{1,1}}{R_E} + \frac{T_S - T_{1,1}}{R_S} + \frac{T_W - T_{1,1}}{R_W} = 0 \text{ @ } SS$$

Recognizing that the four conductive resistance terms have the nodal temperature in common, this expression can be written (now at steady-state) as:

$$0 = q_{gen}wHL + \left(\frac{T_N}{R_N} + \frac{T_E}{R_E} + \frac{T_S}{R_S} + \frac{T_W}{R_W}\right) - T_{1,1}\left(\frac{1}{R_N} + \frac{1}{R_E} + \frac{1}{R_S} + \frac{1}{R_W}\right)$$

Solving for the nodal temperature:

$$T_{1,1} = \frac{q_{gen}wHL + \left(\dfrac{T_N}{R_N} + \dfrac{T_E}{R_E} + \dfrac{T_S}{R_S} + \dfrac{T_W}{R_W}\right)}{\left(\dfrac{1}{R_N} + \dfrac{1}{R_E} + \dfrac{1}{R_S} + \dfrac{1}{R_W}\right)}$$

Alternatively, expressing the heat generation rate as a function of all temperatures:

$$q_{gen} = \frac{T_{1,1}\left(\dfrac{1}{R_N} + \dfrac{1}{R_E} + \dfrac{1}{R_S} + \dfrac{1}{R_W}\right) - \left(\dfrac{T_N}{R_N} + \dfrac{T_E}{R_E} + \dfrac{T_S}{R_S} + \dfrac{T_W}{R_W}\right)}{wHL}$$

The four thermal resistances are conductive, and expressed as the distance between the nodes divided by the thermal conductivity and area. The north and south resistances are the same (and renamed "v" for vertical). Similarly for the east and west (named "h" for horizontal):

$$R_N = R_S = R_{vert} = \frac{(H/2)}{kwL}$$

$$R_E = R_W = R_{horiz} = \frac{(w/2)}{kHL}$$

$$T_{1,1} = \frac{q_{gen}wHL + \left(\dfrac{T_N}{R_v} + \dfrac{T_E}{R_h} + \dfrac{T_S}{R_v} + \dfrac{T_W}{R_h}\right)}{\left(\dfrac{2}{R_v} + \dfrac{2}{R_h}\right)}$$

$$T_{1,1} = \frac{q_{gen}wHL\left(\dfrac{H}{2kwL}\right) + (T_N + T_S + {}^{H}\!/_{w}(T_E + T_W))}{2(1 + {}^{H}\!/_{w})}$$

$$T_{1,1} = \frac{q_{gen}H^2}{4k(1 + H/w)} + \frac{T_N + T_S + (H/w)(T_E + T_W)}{2(1 + H/w)}$$

The rate of heat transfer from the bottom (south) surface is:

$$q = \frac{T_s - T_{1,1}}{R_s}$$

This channel splits into three parallel channels to the other three surfaces. In terms of a conduction shape factor (which does not apply if there is heat generation):

Example 1   Square member (H = w), $q_{gen} = 0$

$$T_{1,1} = \frac{q_{gen}H^2}{8k} + \frac{1}{4}(T_N + T_S + T_E + T_W)$$

For the classic problem stated (no heat generation, three sides fixed at equal temperatures), the temperature of the node is the geometric average of its four neighbors:

$$T_{1,1} = 0 + \frac{1}{4}(0 + 100 + 0 + 0) = 25\,°C$$

The heat transfer rate is:

$$q = \frac{T_s - T_{1,1}}{R_s} = \frac{100 - 25}{\dfrac{H/2}{kwL}} = 150kL = 1.5kL\Delta T_{overall}$$

The last expression defines the shape factor to be 1.5, by inserting the overall temperature drop between surface 1 (the floor at 100 °C) and surface 2 (the other three sides at 0 °C).

Example 2   For the problem statement (H = 1.3 m, W = 4.2 m, $q_{gen} = 0$)

$$T_{1,1} = 0 + \frac{(0 + 100 + (1 + 1.3/4.2)(0 + 0))}{2(1 + 1.3/4.2)} = 38.2\,°C$$

The heat transfer rate is:

$$q = \frac{T_s - T_{1,1}}{R_s} = \frac{2wkL}{H}\frac{(T_s - T_{1,1})}{\Delta T_{overall}}\Delta T_{overall} = \frac{2(4.2\,\mathrm{m})kL}{1.3}\frac{(100 - 38.2)}{(100 - 0)}\Delta T_{overall}$$

$$= 3.99kL\Delta T_{overall}$$

The shape factor for this 4.2 by 1.3 shape is 3.99.

Example 3   For the problem statement (H = 1.3 m, W = 4.2 m, and $T_{max} = 200$ °C)

$$q_{gen} = \frac{4k(1 + H/w)}{H^2}\left(T_{1,1} - \frac{T_N + T_S + H/w(T_E + T_W)}{2(1 + H/w)}\right)$$

$$q_{gen} = 10{,}031\,\mathrm{W/m^3}$$

A good check on the execution is to see whether the net heat flux out of all four sides equals the total heat generation rate.

Improved 1-Node Model

The problem statement for this classic case is symmetric about the center vertical plane, and an improved 1-node model can be made by considering that plane to be an insulated surface, and placing the node at the center of one half of the rectangle.

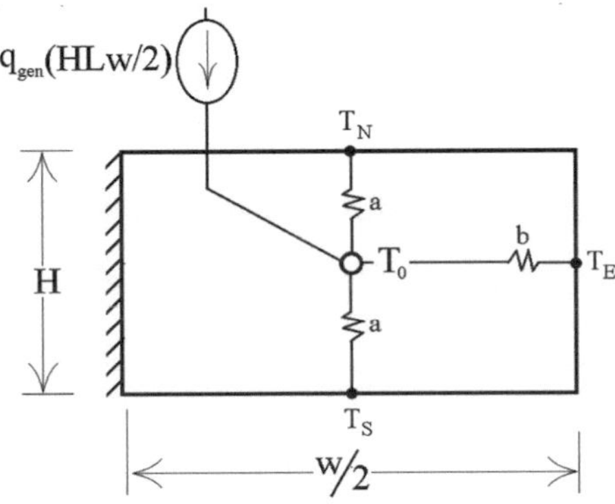

$$T_0 = \frac{\dfrac{q_{gen}wHL}{2} + \left(\dfrac{T_N}{R_a} + \dfrac{T_E}{R_b} + \dfrac{T_S}{R_a}\right)}{\left(\dfrac{2}{R_a} + \dfrac{1}{R_b}\right)}$$

Note that these resistances differ from the previous model (distances and area) and are given different subscripts (a and b) to distinguish. A good practice exercise is to compare the results of this model with the previous, and with the more detailed models that follow. It is emphasized that the 1-node model yields closed-form solutions, while the detailed models require iterative calculations.

8.3.3   Few-Node Model

It is hopefully obvious that breaking the region into a finer grid will yield more detail. A "few-node" model is defined here as one which is readily solved in a spreadsheet, where each node can be listed as its own column. A fun classroom exercise (Workshop 8.3) that clearly demonstrates the nature of the iterative calculation is to assemble the class into a grid pattern, and treat each person as a node. Figure 8.27 shows the right half broken into four equal sized nodes, with the centerline as an insulated boundary, and each node labeled as both a person, and in a Cartesian (xy) indexing scheme (discussed further in the next section). Horizontal resistances are not shown, as heat does not flow across the centerline. Also, the heat generation sources are shown, but each node receives ¼ of the total heat generated in the right half (which is half of the total heat generated). It is a common error to

Fig. 8.27 A 4-node model. Using symmetry, the *right* side of the member is considered, broken into four equal sized elements. The *left* surface is an insulated surface

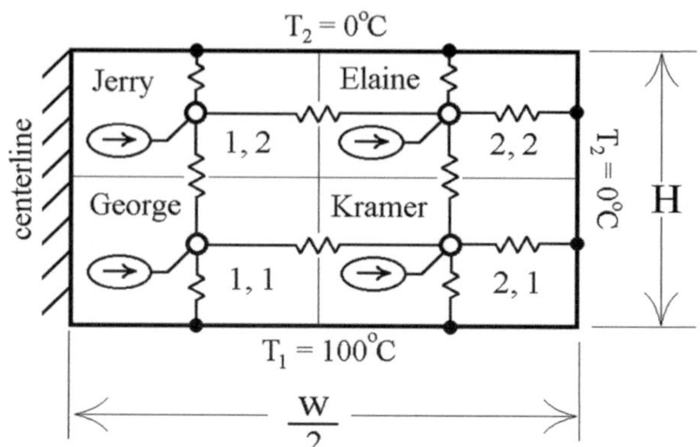

forget that symmetry has been used. For example, the total heat transfer from the south to the three other sides equals twice the heat leaving the south side of the right half, an easy calculation, but one which might slip by with being done.

A spreadsheet implementation of this 4-node model is shown in Fig. 8.28. In the derived parameters block, the horizontal and vertical resistances are those between nodes. The resistances between the nodes and their associated boundaries are half since the distance is half, and accounted for when the formulas are entered into the iteration table, which uses the Gauss–Seidel method. The table to the right of the iteration table shows the absolute value of the change between iterations. Only 10 iterations are shown, but 100 iterations are conducted to produce the final results (but as shown here, 10 iterations are sufficient for convergence to 5 decimal places in temperature change).

The converged temperature results reported in Fig. 8.29 are oriented in the original grid pattern, and both sides of the rectangle are shown. The heat balances show that the heat entering the two bottom nodes is equal and opposite to the heat leaving the top and right faces, a good check for consistency. The net heat flux in is 8 orders of magnitude lower than the heat flux from the bottom surface, and is considered to be a demonstration of computer precision.

The same spreadsheet is used for the case of a non-zero internal heat generation, specifically one that yields a maximum temperature of 200 °C, and is shown in Fig. 8.30. The heat generation (q_{gen}) is an input parameter to this problem, but its value is not specified beforehand. Rather, an iterative solution (on top of the Gauss–Seidel iteration) is needed. A guess to the heat generation is made, and the Gauss–Seidel iteration is executed until it converges. Then another guess for q_{gen} is made. In this case, the "goal seek" feature of MS-Excel was used rather than a brute force entry of q_{gen} values. The final result is that a heat generation rate of 12,555.32 W/m³ yields a maximum temperature of 200 °C.

The converged temperature results are reported in Fig. 8.31. The heat balances show that heat leaves all surfaces, and that the total rate of heat loss equals the total internal heat generation rate.

INPUT PARAMETERS

k	20	W/m/K	thermal conductivity
W	4.2	m	full width
H	1.3	m	Height
Tsouth	100	oC	Base temperature
Teast	0	oC	side temperature
Tnorth	0	oC	top temperature
qgen	0	W/m^3	volumetric heat generation rate

DERIVED PARAMETERS

Dx	1.05	m	horizontal distance between nodes
Dy	0.65	m	vertical distance between nodes
Rh	0.080769		horizontal resistance between nodes
Rv	0.030952		vertical distance between nodes
qgen,node	0		heat generation per node

CARTESIAN GRID INDEX

1, 2	2, 2
1, 1	2, 1

ITERATION TABLE (GAUSS-SEIDEL METHOD)

k	T11	T21	T12	T22
0	0.000	0.000	0.000	0.000
1	59.115	53.656	17.473	14.544
2	70.358	58.199	22.443	16.098
3	72.341	58.757	23.206	16.302
4	72.630	58.833	23.314	16.331
5	72.670	58.843	23.329	16.335
6	72.676	58.845	23.332	16.335
7	72.677	58.845	23.332	16.335
8	72.677	58.845	23.332	16.335
9	72.677	58.845	23.332	16.335
10	72.677	58.845	23.332	16.335

CHANGE FROM 'k' to 'k+1'

T11	T21	T12	T22
-	-	-	-
59.11528	53.65600	17.47308	14.54385
11.24229	4.54305	4.97035	1.55381
1.98371	0.55764	0.76234	0.20478
0.28849	0.07599	0.10847	0.02833
0.04067	0.01058	0.01523	0.00396
0.00570	0.00148	0.00213	0.00055
0.00080	0.00021	0.00030	0.00008
0.00011	0.00003	0.00004	0.00001
0.00002	0.00000	0.00001	0.00000
0.00000	0.00000	0.00000	0.00000

Fig. 8.28 Spreadsheet implementation of 4-node 2D classic problem, with heat generation set to zero

8.3.4 n × m Model

An "n × m" model (or n × m × p for a 3D space) is defined to be one in which the number of nodes is specified as a numerical input parameter that can be readily changed in a computer program. The few-node model is considered to be one where a specific breakdown of space is decided on beforehand. The few-node model can be readily solved in a spreadsheet or a computer program. The n × m model cannot readily be solved with a spreadsheet. This section demonstrates the method and considers a case study of the computation requirements needed to solve these problems as a function of the number of nodes.

Fig. 8.29 Results from the spreadsheet solution for a 4-node model. Temperature results are shown on both sides of the *centerline*, but calculations were conducted on the *right* side, with the *centerline* as an insulated surface. The heat balances (in Watts) are conducted as a check of the results

TEMPERATURE RESULTS (after 100 iterations)

	0	0	0	0	
0	**16.3**	**23.3**	**23.3**	**16.3**	0
0	**58.8**	**72.7**	**72.7**	**58.8**	0
	100	100	100	100	

HEAT BALANCES (Right half)

INTO BOTTOM	4424.7
INTO RIGHT	-1861.6
INTO TOP	-2563.1
TOTAL IN	4.3E-05
TOTAL GEN	0.0E+00

SHAPE FACTOR
4.42

INPUT PARAMETERS

k	20	W/m/K	thermal conductivity
W	4.2	m	full width
H	1.3	m	Height
Tsouth	100	oC	Base temperature
Teast	0	oC	side temperature
Tnorth	0	oC	top temperature
qgen	12555.32	W/m^3	volumetric heat generation rate

DERIVED PARAMETERS

Dx	1.05	m	horizontal distance between nodes
Dy	0.65	m	vertical distance between nodes
Rh	0.080769		horizontal resistance between nodes
Rv	0.030952		vertical distance between nodes
qgen,node	8569.007		heat generation per node

CARTESIAN GRID INDEX

1, 2	2, 2
1, 1	2, 1

ITERATION TABLE (GAUSS-SEIDEL METHOD)

k	T11	T21	T12	T22
0	0.000	0.000	0.000	0.000
1	137.511	124.812	119.041	104.987
2	186.835	154.667	145.512	114.627
3	198.041	158.025	149.916	115.843
4	199.723	158.474	150.551	116.009
5	199.961	158.536	150.640	116.032
6	199.995	158.544	150.653	116.036
7	199.999	158.546	150.655	116.036
8	200.000	158.546	150.655	116.036
9	200.000	158.546	150.655	116.036
10	200.000	158.546	150.655	116.036

CHANGE FROM 'k' to 'k+1'

T11	T21	T12	T22
-	-	-	-
137.51136	124.81222	119.04119	104.98749
49.32335	29.85526	26.47083	9.63920
11.20588	3.35775	4.40403	1.21587
1.68206	0.44834	0.63490	0.16668
0.23845	0.06219	0.08936	0.02324
0.03346	0.00869	0.01252	0.00325
0.00469	0.00122	0.00175	0.00045
0.00066	0.00017	0.00025	0.00006
0.00009	0.00002	0.00003	0.00001
0.00001	0.00000	0.00000	0.00000

Fig. 8.30 Spreadsheet implementation of 4-node 2D classic problem, with heat generation set to a value that yields a maximum nodal temperature of 200 °C. The "goal seek" feature of MS-Excel was used for this purpose

Fig. 8.31 Results from the spreadsheet solution for the 4 node model with internal heat generation set to a value that yields a maximum temperature of 200 °C. Temperature results are shown on both sides of the *centerline*, but calculations were conducted on the *right* side, with the *centerline* as an insulated surface. The heat balances (in Watts) are conducted as a check of the results. The shape factor is not applicable to a case where internal heat generation occurs

TEMPERATURE RESULTS (after 100 iterations)

	0	0	0	0	
0	116.0	150.7	150.7	116.0	0
0	158.5	200.0	200.0	158.5	0
	100	100	100	100	

HEAT BALANCES (Right half)

INTO BOTTOM	-10244.5
INTO RIGHT	-6799.2
INTO TOP	-17232.4
TOTAL IN	-3.4E+04
TOTAL GEN	3.4E+04

SHAPE FACTOR
Not Applicable

Indexing Confusion

There are two common ways to index a 2D variable, and it is easy to introduce bugs in codes by confusing them, as demonstrated in Fig. 8.32. In matrix indexing (the natural mode in MATLAB), the grid is considered to be row/column, that is, the first number is the row, the second the column. Cell 1,1 is in the top left corner and cell n,m is in the bottom right of a grid. In Cartesian indexing (x/y), the first number is the x position and the second is the y position. Cell 1,1 is at the origin of a Cartesian grid (bottom left corner) and cell n,m is at the top right. For the general node i,j, the relationship between the immediate neighbors is different, as shown in the isolated i,j nodes with their neighbors.

MATRIX (row by column) n X m INDEXING

1,1	1,2	1,3	1,4	...	1,m
2,1	2,2	2,3	2,4	...	2,m
3,1	3,2	3,3	3,4	...	3,m
...	...	...	...	...	...
n,1	n,2	n,3	n,4	...	n,m

CARTESIAN (x by y) n X m INDEXING

1,m	2,m	3,m	4,m	...	n,m
...	...	...	...	...	...
1,3	2,3	3,3	3,4	...	n,3
1,2	2,2	3,2	4,2	...	n,2
1,1	2,1	3,1	4,1	...	n,1

NODE i, j and Neighbors

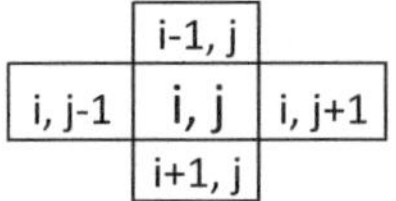

NODE i, j and Neighbors

Fig. 8.32 Matrix and Cartesian indexing scheme

The nodal equation for node i,j in the form suitable for the Jacobi iteration scheme (with superscripts "k" or "k + 1" for iteration number) is:

$$T_{i,j}^{(k+1)} = \frac{q_{gen,i,j}^{(k)} + \dfrac{T_{i-1,j}^{(k)}}{R_{north,i,j}} + \dfrac{T_{i,j+1}^{(k)}}{R_{east,i,j}} + \dfrac{T_{i+1,j}^{(k)}}{R_{south,i,j}} + \dfrac{T_{i,j-1}^{(k)}}{R_{west,i,j}}}{\dfrac{1}{R_{north,i,j}} + \dfrac{1}{R_{east,i,j}} + \dfrac{1}{R_{south,i,j}} + \dfrac{1}{R_{west,i,j}}} \quad \text{for MATRIX indexing}$$

$$T_{i,j}^{(k+1)} = \frac{q_{gen,i,j}^{(k)} + \dfrac{T_{i,j+1}^{(k)}}{R_{north,i,j}} + \dfrac{T_{i+1,j}^{(k)}}{R_{east,i,j}} + \dfrac{T_{i,j-1}^{(k)}}{R_{south,i,j}} + \dfrac{T_{i-1,j}^{(k)}}{R_{west,i,j}}}{\dfrac{1}{R_{north,i,j}} + \dfrac{1}{R_{east,i,j}} + \dfrac{1}{R_{south,i,j}} + \dfrac{1}{R_{west,i,j}}} \quad \text{for CARTESIAN indexing}$$

The heat generation rate is given a superscript "k" to indicate that it may depend on temperature. It is also possible for some or all of the thermal resistances to be temperature dependent, and these therefore would be recalculated at each iteration, but the superscripts are left off for brevity. Notice that 16 elementary operations $(+, -, *, /)$ are needed to execute one of these expressions. It is possible to reduce that number for the case of equal spacings (when $R_{east} = R_{west}$, for example).

A pseudo-code (MATLAB based) for the n × m problem that solves for the heat generation rate required for the maximum temperature to be 200 °C using the Gauss–Seidel iteration with Cartesian indexing is presented in the Appendix to this chapter. The Gauss–Seidel iteration method will yield faster convergence than the Jacobi method and requires careful consideration of the order in which the computer code calculates the next iteration values. Results from modifications of this program are presented in this section. Figure 8.33 shows the temperature profile for the base case of a 4.2 m wide × 1.3 m high rectangle, with the bottom surface set to 100 °C and the three sides set to 0 °C. The centerline (at x = 0) is an insulated

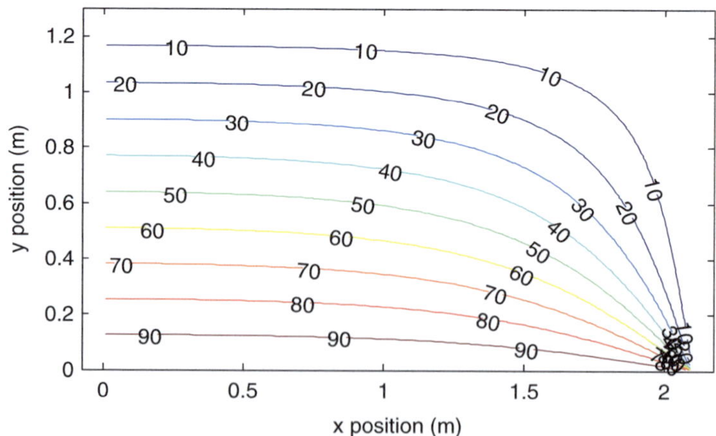

Fig. 8.33 Contour plot of the right half of the 4.2 m × 1.3 m rectangle (using a 100 × 100 grid)

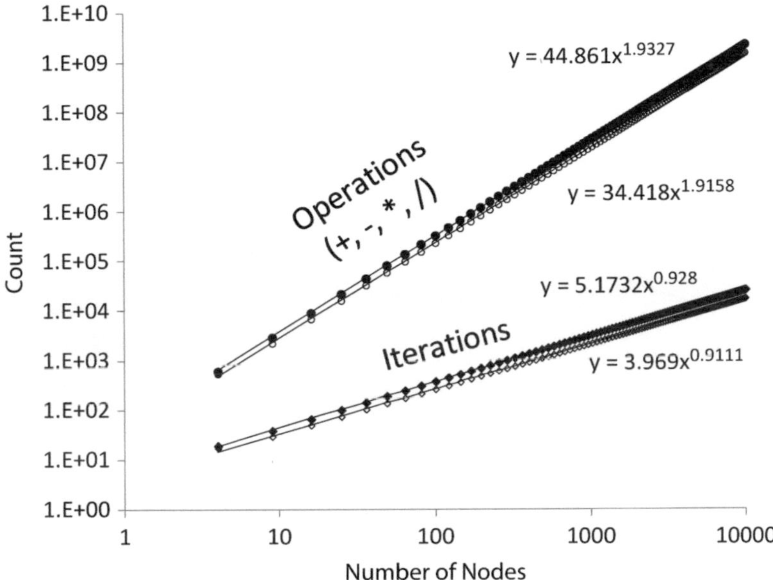

Fig. 8.34 Computational requirements for the 4.2×1.3 rectangle. The convergence criterion (defined as the maximum change of any node from one iteration to the next) is set to be $1(10)^{-6}\,^{\circ}\text{C}$. The *solid symbols* are for the Jacobi method and the *open symbols* are for the Gauss–Seidel method

surface. This figure shows what a well-constructed graphical analysis would show. The heat transfer is effectively 1D over the left half of the space.

The computational requirements for this case are shown in Fig. 8.34, on a log–log scale for a 2×2 up to a 100×100 grid. The number of iterations required for the solution to converge to the defined convergence criterion and the total number of elementary operations required are shown for both the Jacobi (filled symbols) and Gauss–Seidel (open symbols) as a function of number of nodes ($n = m$ for all cases). These values are determined by adding simple counter statements (i.e., $iter = iter + 1$) at appropriate places in the code (not shown in the Appendix code for brevity). Power law fits to each curve are shown. The results show that the number of iterations required to converge increases almost linearly (to the 0.9 power approximately) with the number of nodes. That is, the more nodes, the more iterations are required for the mathematical system to converge to a solution. Then, since one nodal equation is needed for each node, the number of elementary operations is proportional to the total number of nodes. The elementary operations are not exactly a power of 1 higher than the iterations because the centerline nodes have fewer elementary operations required than interior nodes, and the percentage of these changes with the number of nodes. The main point is that the computational requirements increase dramatically with the number of nodes taken, and is a fundamental challenge of computational methods in general. A good understanding and experience aimed at addressing specific questions using a minimum number of nodes is an important and valuable asset.

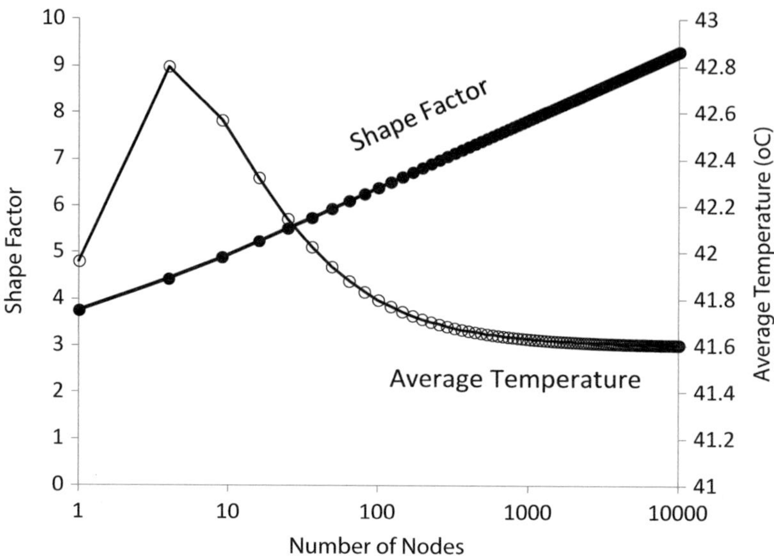

Fig. 8.35 Calculated shape factor as a function of number of nodes for the base case (semi-logx axes)

Grid Insensitivity

A formal way to determine the minimum number of nodes needed to attain a certain degree of accuracy is to demonstrate grid insensitivity. The general concept is that the results of a numerical calculation should be independent of the size of the grid, which is a numerical parameter, not a physical one. That can only be obtained if the calculation is done at least twice with two grid sizes. If the results agree sufficiently, then the coarser grid is shown to be sufficient.

Demonstration of grid insensitivity requires that something specific be compared at different grid sizes. There is, however, some art required to appropriately define a specific outcome, such as a temperature at a specific location, or an average temperature, or a heat transfer rate (which is generally a good global measure). For example, Fig. 8.35 shows two different measures for this case study; the predicted shape factor (heat transfer rate from the bottom surface to the other three surfaces) and the average temperature as a function of the number of nodes (using semi-logx axes). This plot suggests that the shape factor does NOT show grid insensitivity. As the number of nodes increases, the heat transfer rate continues to increase, with no sign of leveling off. However, the temperature (which is plotted in a narrow 2 °C temperature band) shows a leveling off to a value of 41.6 °C. The 1-node model actually gives a better agreement than the 2×2 model.

8.3.5 Comment on Exact Solution

The exact solution to this problem involves solving a boundary-value type partial differential equation, resulting in an infinite series solution for the temperature field.

Evaluation of the heat transfer rate between the bottom and side surfaces yields a non-converging power series. That is, the heat transfer rate is infinite, due to the discontinuity in temperature that occurs at the corner. The mathematical statement of the classic problem is a physical impossibility, however elegant.

$$q = \sum_{n=0}^{\infty} \frac{\left[(-1)^{n+1} + 1\right] \cos{(n\pi - 1)}}{n \tanh(n\pi w/H)}$$

Even though the magnitude of successive terms decreases with n (becoming 1/n for large n), this series fails the formal testing for convergence. The culprit for this behavior is the physically unattainable discontinuous temperature jump at the corner, which results in locally infinite heat transfer rates. This result explains why the numerically predicted shape factor increased with the number of nodes, as smaller and smaller nodes were placed in the corner.

8.3.6 Results with Internal Heat Generation

The same system was solved with an internal heat generation rate (q_{gen}), specifically, that value that yields a maximum temperature of 200 °C. The code (reported in the Appendix) was written with a Regula-Falsi scheme in which a high and low value of heat generation was initially selected (assisted by 1-node results). Use of the low value (0 initially) results in a maximum temperature below the target, and use of the high value (twice the 1-node prediction was sufficient) results in a maximum temperature above the target. An iteration loop was implemented in which the next "guess" for heat generation was taken to be the average of the high and low values (a Golden Ratio algorithm would be a little faster, but requires more explanation). If the maximum temperature attained at this average value exceeded the target, then the high value is replaced by this new value, and vice versa if the maximum temperature is too low. A Gauss–Seidel iteration loop is used for each "guessed" value of q_{gen}.

The code for this problem has three nested loops. The outer loop simply runs the algorithm for increasing number of horizontal nodes (n, with vertical nodes, m, set equal to n). The next loop executes the Regula-Falsi method in which a value of heat generation is "guessed." The inner loop executes the Gauss–Seidel iteration scheme with this selected value.

Temperature contour results for a 100 × 100 grid are shown in Fig. 8.36. There are strong temperature gradients toward the three isothermal surfaces as heat generated in the interior must conduct to these surfaces. The peak temperature is located on the centerline, and somewhat closer to the hotter bottom surface.

The effect of grid size is shown in Fig. 8.37 in which the internal heat generation rate required to achieve a maximum temperature of 200 °C is plotted against the number of nodes. Grid insensitivity is demonstrated above approximately 100 nodes (a 10 × 10 grid). However, even the 2 × 2 model gives a predicted heat generation rate that is within 10 % of the grid insensitive value. The 1-node model

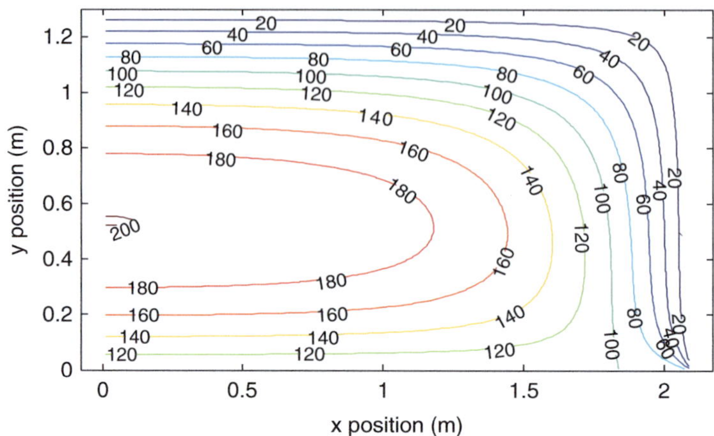

Fig. 8.36 Temperature contour for the case with internal heat generation that yields a maximum temperature of 200 °C. A 100×100 grid was used for this plot

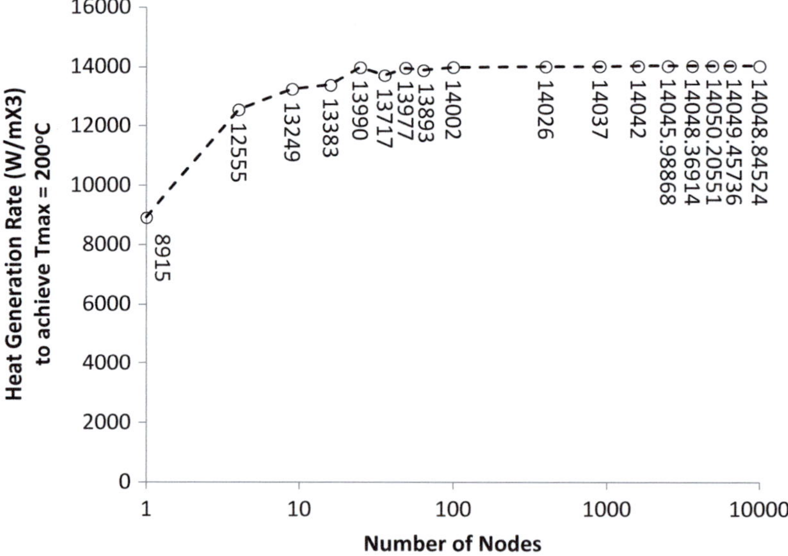

Fig. 8.37 Internal heat generation rate required to achieve a maximum temperature of 200 °C as a function of number of nodes. Values are shown to ten decimal places to demonstrate computer precision issues

exhibits a large deviation, but places the system in the right neighborhood. Any input parameter value can be changed to a vastly different problem, and the 1-node model will find an excellent first approximation.

The numerical values are shown (to ten significant figures) in Fig. 8.37 for the cases that appear from the plot to be equal. These values are not monotonically increasing (which would suggest the result was still sensitive to the grid size), but rather fluctuate about an average value. This behavior is attributed to perhaps both computer precision, and that a specific convergence criterion was used $(1(10)^{-6})$ for both the Regula-Falsi algorithm and the Gauss–Seidel algorithm. There is a curious bump for the case of 25 nodes (5×5) which is similarly related to computational issues, but in this case, it is the graininess of the spatial grid at fault. The basic shape of this curve would be the same, with differences in minor detail if a different convergence criterion were selected.

Workshop 8.1. Graphical Method on Classic 2D Problem

The graphical conduction method is a useful tool for helping visualize the flow of heat and estimating conductive resistances in a complicated 2D solid under quasi-steady state conditions (the temperature may be slowly changing with time). It involves first specifying two surfaces (1 and 2) that are set to fixed temperatures, then sketching isotherms and heat flow lines (analogous to streamlines in fluid flow) in a prescribed manner. The rate of heat transfer (for a length L into the page) between the two surfaces is obtained from the sketch. A conduction shape factor (S) can be defined as: $q_{12} = kLS(T_1 - T_2)$. Published exact solutions for specific shapes can be used to compare results of this approximate method.

The base of a long 2×1 m^2 rectangular structural member is maintained at temperature T_2 ($=100\ °C$) and the other three sides are maintained at temperature T_1 ($=0\ °C$). The member has a thermal conductivity k, and heat flows between the base and the three sides at a rate q_{12}. Use the graphical method to estimate the conduction shape factor, S.

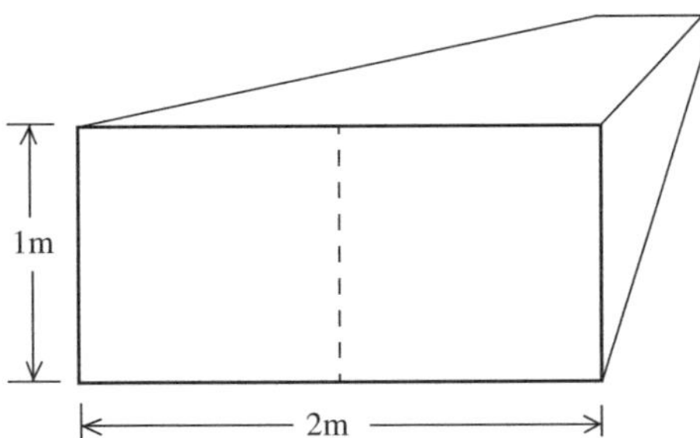

Graphical method
1. Draw object to scale (already done here).
2. Use symmetry to identify any insulated surfaces and "hatch" them (note: insulated surfaces are also heat flow lines).
3. Carefully sketch the average temperature isotherm, creating two temperature increments (note: isotherms are perpendicular to insulated surfaces).

4. Carefully sketch average isotherms to create four total temperature increments. Decide if it's worth going to 8, then 16, then 32. ...

5. Carefully sketch heat flow lines, keeping all boxes created approximately square (Note: heat flow lines are perpendicular to isotherms).

6. S = M/N, where M is the number of heat flow lanes created, and N is the number of temperature increments.

Workshop 8.2. Graphical and Thermal Resistance Practice

A steam pipe of outer diameter D = 0.1 m is encased in a concrete (k = 0.3 W/m/K) square channel of dimension w = 0.4 m. The steam pipe temperature is maintained at $T_1 = 120$ °C, while the outside of the concrete is maintained at $T_2 = 20$ °C. Estimate the rate of heat transfer through the concrete layer per unit length using three methods: conduction shape factor ("exact" solution taken from the literature), the graphical method, and by creating an approximate conductive resistance.

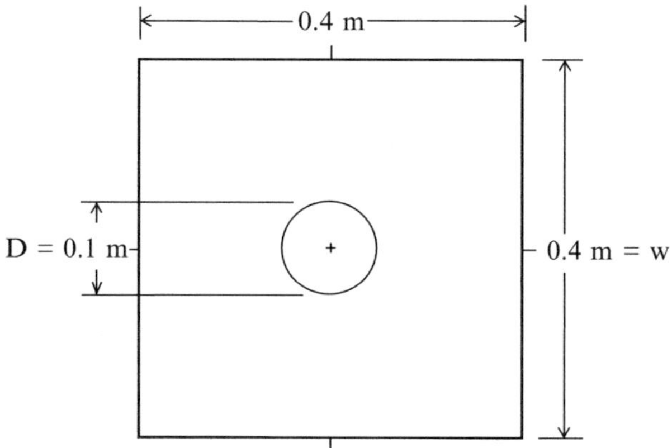

RESULTS:

Method	$q_{12}/L\,(W/m)$	% Differrence
Exact Conduction Shape Factor: $$S = \frac{2\pi}{\ln(1.08w/D)} = \frac{q_{12}/L}{k_{12}(T_2 - T_1)}$$		NA
Graphical Method		
Conductive Resistance		

Bonus: If the outside were exposed to air at 20 °C and a convective/radiative heat transfer coefficient of 15 W/m^2/K, would the heat transfer rate change substantially? (Justify with a resistance model calculation.)

Workshop 8.3. Graphical Analysis Practice

Conduct a graphical analysis on each of these five 2D shapes. Record the value of the shape factor next to each.

$$\text{Graphical Method} : \frac{q}{L} = k\left(\frac{M_{lanes}}{N_{increments}}\right)(T_1 - T_2) = kS(T_1 - T_2)$$

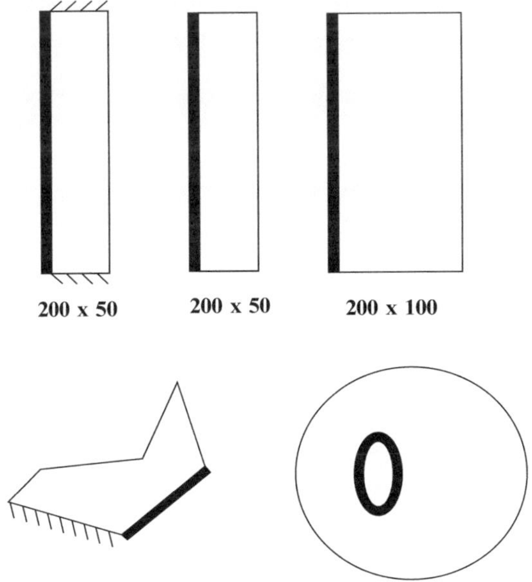

Heavy line is T1, light line is T2, hatched lines are insulated

Workshop 8.4. Classroom Fun!

For a class of 24 students, arrange them into a 6×4 system, where each student is assigned a node. The number of rows and columns can be chosen depending on the class size (for example, a class of 25 could use a 5×5 grid). The method is applied to a system where $\Delta x = \Delta y$, so that the iteration formula results in the next guess for each temperature to be equal to the geometric average of the four neighbors. For students on the perimeter, one of the neighbors is fixed. For students at the corners, two are fixed. Each student fills out a Worksheet, which requires verbally obtaining the current iteration neighbor's values (and a good ice-breaker, especially for shy students).

	0 °C	0 °C	0 °C	0 °C	0 °C	0 °C	
0 °C	•	•	•	•	•	•	0 °C
0 °C	•	•	•	•	•	•	0 °C
0 °C	•	•	•	•	•	•	0 °C
0 °C	•	•	•	•	•	•	0 °C
	100 °C	100 °C	100 °C	100 °C	100 °C	100 °C	

Worksheet instructions: use integer values only.

1. For the first iteration, use engineering judgment to make a guess as to your temperature based on your location. Enter in "MY GUESS" column, iteration 1 row.
2. Ask your four neighbors what their guesses are and enter them in the appropriate column for iteration 1.
3. Your next guess (iteration 2) is the average of your neighbors. Iterate until convergence (or you run out of time).
4. Enter your final results on a grid visible to the entire class.

Iteration	MY GUESS	North	East	South	West
1					
2					
3					
4					
5					
6					
7					
8					
9					
10					

An example of the final result of this exercise is shown in Fig. 8.38.

Workshop 8.5. Storage of Energetic Materials—The Chemically Reacting Pile

A pile of material that slowly reacts chemically (i.e., coal dust, fertilizer) is stored in a long $w \times H$ (use 2×1 m^2 as default) rectangular bin. The base is maintained at ambient temperature $T_\infty = 25$ °C and the other three sides are exposed to ambient air with an effective heat transfer coefficient of $U = 1.2$ W/m^2/K (across a thermal channel consisting of a conductive wall resistance and a convective/radiative resistance to ambient in series). The material has a thermal conductivity $k = 0.6$ W/m/K, and it generates heat internally at a constant volumetric rate (q_{gen} with units W/m^3, so that the rate of heat generation is q_{gen} times the volume). After storage for a sufficiently long time for a steady-state to be established, it is found

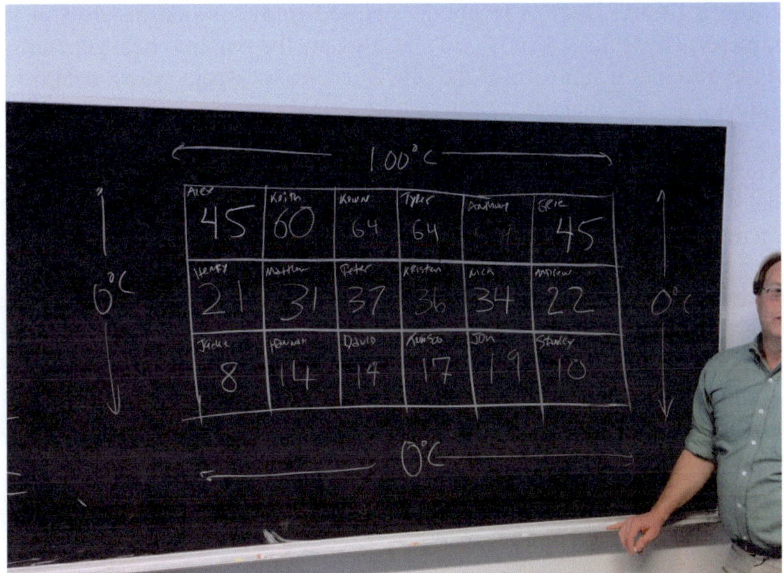

Fig. 8.38 Photo of results at the end of a "Class Fun" workshop. Approximately 30 min was allotted for this exercise. The results were tabulated before the system was fully converged as shown by the asymmetry. Notice that each node is the first name of the student, making this a "few-node" model

that the temperature at the center of the pile is $T_0 = 40$ °C. Determine the heat generation rate, q_{gen}, and the maximum observed temperature and its location. The material is known to spontaneous combust if it reaches a temperature $T_{ignition} = 60$ °C.

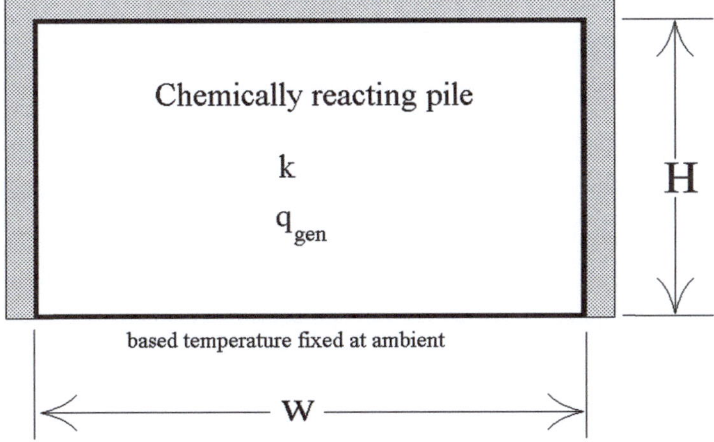

- Perform a 1-node analysis to obtain a closed-form solution in terms of system parameters (w, H, k, q_{gen}, h, T_∞, T_0). Calculate the required heat generation rate and the maximum safe size of the pile (for various aspect ratios, w/H).
- Conduct a 4-node analysis (see over for thermal resistance network) and use the MOSS (Gauss–Seidel method being one variation) to obtain a second approximation. Using symmetry, the model considers only the right side of the pile, with equal size nodal elements with the same aspect ratio (w/H) as the pile. Consider the temperature at the center to be the average temperature of two left side nodal elements. Set up a spreadsheet or program code with a block of input parameters, then an iteration table that implements the MOSS with sufficient iterations for convergence. Once the model is in place, use brute force, or "goal seek" or create a "while loop" to find the heat generation rate.
- Write program code to conduct an analysis with a finer "n × m" grid. Determine the maximum temperature in the pile and its location. Explore the effect of grid size on heat generation rate and on the computational requirements.
- Investigate the behavior of this material for different sizes and aspect ratios (i.e., peak temperature vs. height and/or width). Determine the maximum size pile that can be used without spontaneous combustion.

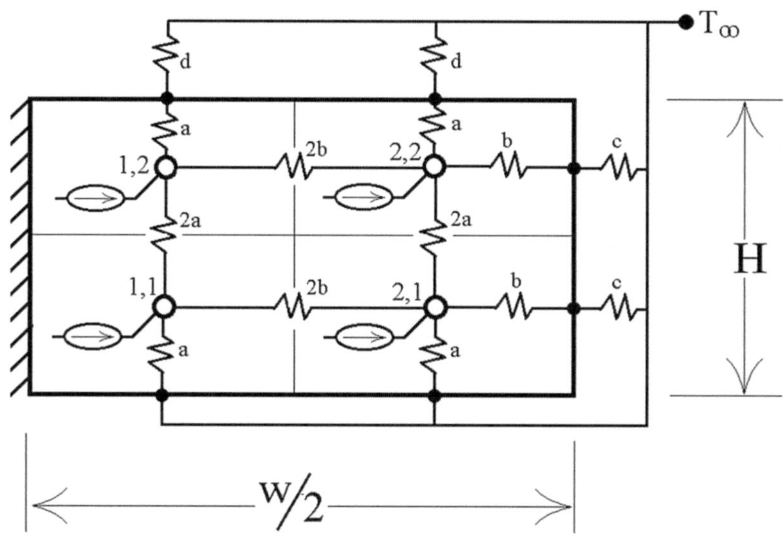

In general, for node i, with neighbors j, an energy balance is:

$$C_i \frac{dT_i}{dt} = q_{gen} * V_i + \sum_j \frac{T_j - T_i}{R_{ij}} \quad (= 0 \, @ \, Steady \, State)$$

where V_i is the volume associated with the nodal element, and R_{ij} is the thermal resistance between the node "i" and its neighbor "j". At steady-state, expanding the summation (noting that T_i is common to all summation terms):

$$0 = q_{gen} * V_i + \sum_j \frac{T_j}{R_{ij}} - T_i \sum_j \frac{1}{R_{ij}}$$

Note that either Celsius or Kelvin temperatures can be used (but consistently) because in the original energy balance temperature appears as a temperature difference everywhere. Rearranging for the nodal temperature:

$$T_i = \frac{q_{gen} * V_i + \sum_j \dfrac{T_j}{R_{ij}}}{\sum_j \dfrac{1}{R_{ij}}}$$

For example:

$$T_{1,2} = \frac{q_{gen} * (H/2) * (w/4) * L + \left(\dfrac{T_\infty}{R_a + R_d} + \dfrac{T_{2,2}}{2R_b} + \dfrac{T_{1,1}}{2R_a} \right)}{\left(\dfrac{1}{R_a + R_d} + \dfrac{1}{2R_b} + \dfrac{1}{2R_a} \right)}$$

Workshop 8.6. Cold Plate

In a refrigeration application, the lower left corner of an 11.2 cm by 34.4 cm block of steel (k = 73 W/m/K) is maintained at a temperature $T_0 = -20$ °C (by an evaporating refrigerant) and the top surface is exposed to a fluid at $T_\infty = 5$ °C with a convection coefficient of 35 W/m^2/K and exchanges radiation with walls at $T_w = 25$ °C. The other three sides are insulated. For the two nodal networks shown (one places the nodes, then defines the volumes, Fig. 8.39; the other defines volumes, then places the nodes, Fig. 8.40), solve for the temperatures of all nodes, and for the rates of heat transfer from the cold corner, to the fluid being cooled (by convection), and to the walls (by radiation).

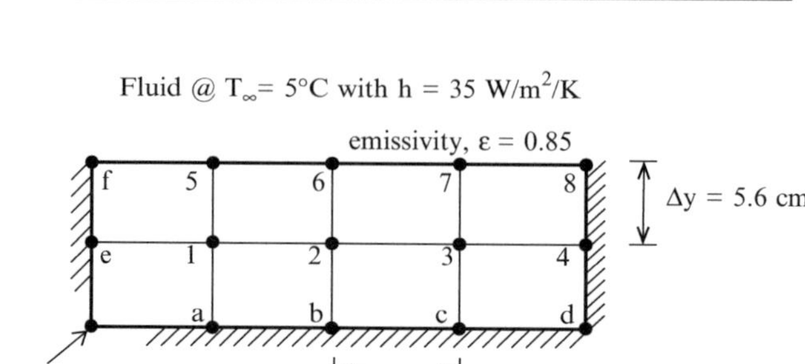

Fig. 8.39 Nodal network where the nodes are placed, then the volumes (not shown) are defined around them

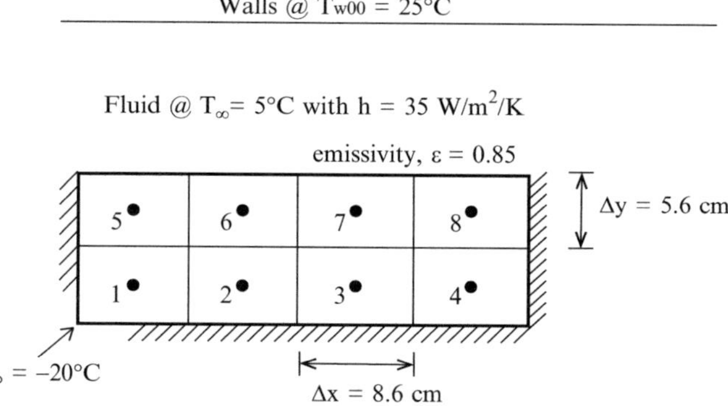

Fig. 8.40 Nodal network where the volumes are defined, then the nodes are placed in the center

Appendix. MATLAB Code for Classic 2D Problem

```matlab
%rectangle_GaussSeidel_TMAX_VARYn.m  George Sidebotham 2012, revised 2014
%Program to calculate temperatures and heat flux from a rectangle whose base
%is fixed at one temperature, and the other 3 sides are fixed at another,
%with a constant heat generation throughout.  This version finds the Qdot
%that yields a defined maximum temperature
clear all
clc
%INPUT PARAMS
k = 20;        %W/m/K, thermal conductivity
width = 2.1;   %m, halfwidth of rectangle
height = 1.3;  %m, height of rectangle
length = 1;    %m, depth "into the page"
Tsouth = 100;     %oC, bottom surface temperature (base)
Tc = 0;        %oC, temperature of other 3 sides
Teast = Tc;    %East wall temp
Tnorth = Tc;   %North wall temp
Tmax_set = 200;   %oC, set maximum temperature to find
converge = 1e-6;     %convergence criterion

%1-Node Model
Rnorth = height/2/(k*width*length);
Rsouth = Rnorth;
Reast = width/2/(k*height*length);
Qdot_1node = (-Tc/Rnorth - Tc/Reast - Tsouth/Rsouth + ...
    Tmax_set*(1/Rnorth+1/Reast+1/Rsouth))/(height*width*length);

%NUMERICAL GRID AND DEFINED PARAMS
CASES = 100;
Qset(1:CASES) = Qdot_1node;   %allocate final heat generation rate
%LOOP FOR NUMBER OF NODES
for n = 2:CASES;        %number of elements in x direction (columns)
    m = n;        %number of elements in y direction (rows)
    Qdot_low = 0;    %heat generation yields Tmax < Tmax_set
    Qdot_high = 2* Qdot_1node;%heat generation yields Tmax > Tmax_set
    dx = width/n;    %x grid size in x direction
    dy = height/m;        %y grid size in y direction
    Rh = dx/k/dy/length;        %oC/W, horizontal conductive resistance
    Rv = dy/k/dx/length;        %oC/W, vertical conductive resistance
    T(1:n,1:m) = Tc;    %"initialize" temperature to cold temperature
    Tnext(1:n,1:m) = Tc;   %"initialize" next guess
    for i = 1:n       %define x location vector
        x(i) = dx*(i-1/2);
    end
    for j = 1:m    %define y location vector
        y(j) = dy*(j-1/2);
    end
    %LOOP FOR REGULA-FALSI (High/Low) METHOD OF FINDING HEAT GENERATION
    while abs(Qdot_high - Qdot_low) > converge
        Qdot = 1/2*(Qdot_high + Qdot_low);   %Next guess for heat generation
        qgen = Qdot*(dx*dy*length);   %W, heat generation per cell
        change = converge+eps;       %needed to get into main iteration loop...
        %LOOP FOR GAUSS-SEIDEL ITERATION LOOP
        while change > converge
            change = 0;       %reset for this iteration
            %Bottom left corner
```

```matlab
        Tnext(1,1) = (qgen + T(2,1)/Rh + T(1,2)/Rv +…
                                Tsouth*2/Rv)/(1/Rh + 3/Rv);
        %Bottom row
        for i = 2:n-1
            Tnext(i,1) = (qgen + T(i+1,1)/Rh + Tnext(i-1,1)/Rh +…
                    T(i,2)/Rv + 2*Tsouth/Rv)/(2/Rh + 3/Rv);
        end
        %Bottom right corner
        Tnext(n,1) = (qgen + T(n,2)/Rv + Teast*2/Rh + Tsouth*2/Rv +...
            Tnext(n-1,1)/Rh)/(3/Rh+3/Rv);
        %Right wall
        for j = 2:m-1
            Tnext(n,j) = (qgen + T(n,j+1)/Rv + Teast*2/Rh +…
                Tnext(n,j-1)/Rv + T(n-1,j)/Rh)/(3/Rh +2/Rv);
        end
        %Top right corner
        Tnext(n,m) = (qgen + Tnorth*2/Rv + Teast*2/Rh +…
            Tnext(n,m-1)/Rv + T(n-1,m)/Rh)/(3/Rv + 3/Rh);
        %Top row
        for i = 2:n-1
            Tnext(i,m) = (qgen + Tnext(i+1,m)/Rh + T(i-1,m)/Rh +…
                T(i,m-1)/Rv + 2*Tnorth/Rv)/(2/Rh + 3/Rv);
        end
        %Top left corner
        Tnext(1,m) = (qgen + Tnext(2,m)/Rh + T(1,m-1)/Rv +…
            Tnorth*2/Rv)/(1/Rh + 3/Rv);
        %Left wall
        for j = 2:m-1
            Tnext(1,j) = (qgen + Tnext(1,j+1)/Rv + T(1,j-1)/Rv +…
                T(2,j)/Rh)/(1/Rh +2/Rv);
        end
        %Interior cells
        for i = 2:n-1
            for j = 2:m-1
                Tnext(i,j) = (qgen + T(i,j+1)/Rv + T(i+1,j)/Rh +...
                    Tnext(i,j-1)/Rv + Tnext(i-1,j)/Rh)/(2/Rh + 2/Rv);
            end
        end
        change = max(max(abs(T-Tnext)));  %Find the maximum change between
                                        iterations
        T = Tnext;    %update for next iteration
    end
    %RESET THE APPROPRIATE HIGH OR LOW VALUE
    if max(max(T)) > Tmax_set
        Qdot_high = Qdot;
    else
        Qdot_low = Qdot;
    end
    disp(sprintf('Qdot low, high n= %g %g %g',Qdot_low, Qdot_high, n))
end
%OUTPUT HEAT
Qset(n) = Qdot;
end
```

Part IV

Open Systems: Convection and Heat Exchangers

Nusselt Number Correlations 9

It can be argued that the exchange of energy between a solid and a fluid is important in most, if not virtually all, practical situations for which heat transfer plays a role. In the mug of coffee problem, the prediction of the cooling rate is most sensitive to the values of the convection coefficient and radiation coefficient used to model the heat transfer rate between the outer side wall and the surroundings. The side wall channel has the lowest thermal resistance of the three parallel channels (because it has the largest exposed area), but within this channel, the outer resistance is the largest. The effective Biot number of the combined system ($h''L''/k_f$) is low. The initial mug heating period is most sensitive to the value of the convection coefficient chosen between the coffee and the mug inner wall.

The goal of this chapter is to determine the convective heat transfer coefficients between the coffee and the mug and between the mug and ambient that have been used in the various thermal models. The general approach to using Nusselt number correlations is developed, and applied to the mug inner and outer walls. Rough order of magnitude "Rules of Thumb" presented in Chap. 3 are thereby tested for this case study. The next chapter explores more fundamentally the physics of the thermal boundary layer.

Nusselt number correlations can be obtained from a variety of printed and online sources, and only a few relevant examples are introduced here, as needed.

9.1 Flow Classification and Dimensionless Parameters

Convection flows can be classified as being external or internal. In external flows, the fluid surrounds the object. For example, a sailor testing for the direction of the wind wets her finger and holds it vertically. The fluid (air) surrounds the finger making this scenario one of external flow. On the other hand, a refrigerant that enters a condenser or evaporator does so inside a tube that enters a heat exchanger. The refrigerant is completely surrounded by the inner wall of the tube making this

© Springer International Publishing Switzerland 2015

G. Sidebotham, *Heat Transfer Modeling: An Inductive Approach*,

DOI 10.1007/978-3-319-14514-3_9

scenario one of internal flow. Internal flows and heat exchangers will be addressed in separate chapters.

There is also a distinction between local and average heat transfer coefficients. For the sailor, the human nervous system can discern that the rate of heat loss on the upwind side is greater than that on the downwind side, and thereby determine the direction of the wind. Therefore, the local convection coefficient on the upwind side is different than the downstream value. In fact, the local convection coefficient varies continuously across the entire surface. However, as is common in engineering analysis, if the total rate of heat loss is of interest, an averaged heat transfer coefficient would be used that yields the correct total heat transfer rate, but with no distinction with position.

Finally, flows can be classified as being forced, natural, or mixed. In forced convection, there is a discernible relative flow velocity between the solid object and the fluid that surrounds it (for external flows). Therefore, if the sailor can discern the wind direction by holding up a moist finger, then there is a significant wind speed, and the finger experiences a forced convection. Even a gentle breeze can be discerned this way. On the other hand, if there is a dead calm, the sailor will not experience a spatial difference and will conclude that there is no wind. However, "hot air rises." The sailor's finger is near body temperature and is warmer (and therefore less dense) than the surrounding air, so this layer of air will rise like a helium balloon. The word "convective" implies that there is fluid motion. In forced convection, the fluid motion is imposed. In natural convection, the fluid motion is induced by the heat transfer process itself.

Mixed convection flows are for those cases where there is a forced velocity, but buoyancy effects are strong enough to influence the flow field as well. A legitimate modeling strategy is to calculate both a forced and a natural convection coefficient, and use the higher of the two. This approach would underestimate the convection coefficient. Alternatively, a fourth-order polynomial fit can be made across a defined transition range that matches the slope and value at locations before and after the crossing point, where the correlations for natural and forced flow are equal.

Natural convective flows are driven by buoyancy and therefore require a gravitational field. One of the many engineering challenges of space travel is to design systems that are, for economic reasons, tested on the Earth where natural convective cooling effects (of electronic components, for example) occur. However, these systems are intended to be operational on a spacecraft which experiences microgravity. Fire safety is of particular concern because the high gas phase temperatures due to combustion cause intense upward flows on the Earth, but not on a spacecraft.

For engineering applications, convection coefficients for a variety of flow geometries have been correlated in terms of the dimensionless Nusselt number, defined as:

$$Nu_x = \frac{hx}{k_f}$$

where h is the convection coefficient, x is the appropriate linear dimension for the problem in question, and k_f is the thermal conductivity of the fluid. Special care must be taken to ensure that the correct dimension for the published correlation is used. For example, for a pipe of diameter D, length L, the dimension used differs depending on the flow situation (D for an external cross-flow, L for a vertical pipe in a still room). Also, the Nusselt number is not to be confused with the Biot number ($Bi = hx/k$, where k is that of the solid), which has an entirely different physical interpretation, namely, the ratio of a conductive to a convective thermal resistance associated with a solid. The Nusselt number will be shown in the next chapter to be the ratio of the conductive to convective thermal resistance across a thermal boundary layer.

Convection coefficients depend on the nature of the flow field (higher speeds mean better heat transfer) and the nature of the fluid in question (better conducting fluids with large thermal masses (ρc) convect better). Therefore, the Nusselt number depends on two other dimensionless numbers, one characterizing the flow field, the other characterizing the nature of the fluid.

For *forced convection*, the flow is characterized by the Reynolds number:

$$\mathrm{Re}_x = \frac{\rho V x}{\mu} = \frac{V x}{\nu}$$

where V is the forced flow velocity, ρ is the fluid density, μ the dynamic viscosity, $\nu = \mu/\rho$ is the kinematic viscosity. The other dimensionless number is a fluid property, the Prandtl number, which is the ratio of the kinematic viscosity to the thermal diffusivity of the fluid.

$$\mathrm{Pr} = \frac{\nu}{\alpha}$$

For *natural convection*, there is no forced velocity, but there is a flow. The temperature differences in the fluid give rise to density differences which, in a gravitational field, cause buoyant forces that induce fluid flows. The dimensionless number that characterizes this flow is the Grashof number:

$$Gr_x = \frac{g\beta |(T_w - T_\infty)| x^3}{\nu^2}$$

where g is the acceleration due to gravity, T_w is the wall surface temperature, T_∞ is the free stream fluid temperature, and β is the coefficient of expansion (how the density changes with temperature) given by:

$$\beta = -\frac{1}{\rho}\left(\frac{\partial \rho}{\partial T}\right)_P$$

For an ideal gas, $\beta = 1/T$, where T is the absolute temperature (R or K). This relation can be derived using the ideal gas law, $P = \rho R T$. The absolute value of the temperature difference between surface and fluid is required to prevent negative values of an inherently positive quantity.

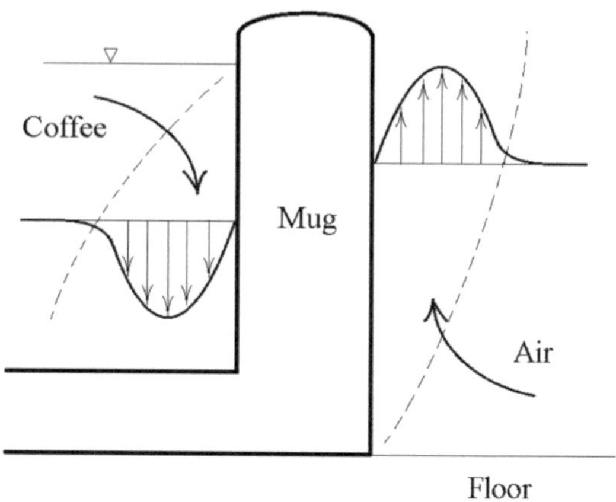

Fig. 9.1 Schematic of side wall of coffee mug

Many (if not most) of the natural convection correlations involve yet another combination dimensionless parameter, the Rayleigh number, defined by:

$$Ra_x = Gr_x \mathrm{Pr} = \frac{g\beta(T_w - T_\infty)x^3}{\nu\alpha}$$

A few forced convection scenarios could be imagined for the mug of coffee problem. During the initial pouring, it could be argued that there is a forced convection, and a characteristic velocity could be wildly stabbed at by estimating the time it takes to pour the coffee and distance travelled (like the height of the mug). Another variation would be to stir the coffee, in which case a velocity could be defined based on the stirring rate and diameter of the rotation. Yet another would have the impatient train commuter blowing on the top. Notice, however, that results from previous chapters suggest that blowing on the sides would actually be more effective. This result is counterintuitive as the vast majority of people would blow on the top, not the sides.

In the scenario detailed here, the mug of coffee is simply placed on a table and allowed to cool without intervention. This case is clearly one of natural convection. There is no imposed flow velocity, but there is fluid motion induced by gravity.

Another classification that can be made is whether the flow is internal or external. Internal flows are generally cases where fluid flows in some type of duct, and would therefore almost always be a forced convection case. In the present case, it would be easy to at first think that the inner convection environment is an internal flow, but it really is not a forced flow in a duct. Imagine, rather, that the mug has a very large diameter.

The schematic of Fig. 9.1 shows the side wall of a mug of coffee as a vertical wall with two different fluids on either side. On the left, the coffee is hotter than the wall, and a plunging flow in a thermal boundary layer next to the wall (indicated by the dashed line) is shown. The fluid near the wall is slightly colder, and therefore denser, than the bulk of the fluid far from the wall, and gravity induces a downward flow. On the right (air) side, the wall is hotter than the free stream air temperature, and the air in the thermal boundary layer is less dense than ambient air far from the mug, and gravity induces an upward flow. Note the no-slip condition, where the velocity of the fluid in contact with the solid is motionless.

So for modeling purposes, the coffee side (inner surface) can be classified as a natural convection on a vertical heated plate of height H, and width πD. The air side (outer surface) can be classified as a natural convection on a vertical cooled plate of height H_0 and width π(D + 2w).

A difficulty with this case is that the temperatures of the two fluids can be considered to be known (at an instant in time), but the temperatures of the two solid surfaces that they see are not known. The convection coefficients depend on the temperatures of the solid surfaces for these natural convection cases. The way around that circular argument is to develop an iterative procedure, a trial and error approach, developed later in this chapter. The key step in this iterative procedure is the calculation of the convection coefficients for an assumed value of surface and fluid temperatures. The problem will now be recast in a way that obviates, for now, the need for iteration.

9.2 Nusselt Number Example Calculation

Problem Statement
Calculate the convection coefficient between a vertical solid surface of height $H = 0.098$ m maintained at a temperature $T_{surface}$ and an adjacent stagnant fluid maintained at temperature T_{fluid}. Consider a range of temperatures for air and liquid water.

Analysis
A power law relationship ($y = ax^b$) between the Nusselt number and Rayleigh number has been shown to correlate a wide range of data for natural convection situations:

$$Nu_H = C\left(Gr_H Pr_f\right)^m$$

The constants C and m take on different values depending on the value of the Rayleigh number ($Ra = Gr_H Pr_f$). Table 9.1 lists published values for two ranges that occur in the case study under investigation. These ranges span nine orders of magnitude in the Rayleigh number, and thereby cover most practical problems of interest. For cases where the Rayleigh number is below the lowest range can be handled approximately by either treating the fluid as non-moving (a solid) or using the lowest value of Rayleigh number. This table will be modified shortly because it leads to a discontinuity at the point of transition (at $Ra = 10^9$).

Table 9.1 Published values of constants used in the Nusselt number correlation for natural convection on vertical plates

Rayleigh number range	Flow character	C	m
$(10)^4$–$(10)^9$	Laminar	0.59	0.25
$(10)^9$–$(10)^{13}$	Turbulent	0.021	0.40

Fig. 9.2 Properties of saturated liquid water at three temperatures

FROM ENGINEERING TOOLBOX:					
Property	*SI Unit*	*T = 60oC*	*T = 70oC*	*T = 80oC*
ρ	kg/m3	983	978	972
cp	J/kg/oC	4185	4191	4198
μ	kg/s/m	4.67E-04	4.04E-04	3.55E-04
Pr = ν/α	-	2.99	2.56	2.23
β	1/K	5.23E-04	5.85E-04	6.43E-04
DERIVED				
$\nu=\mu/\rho$	m2/s	4.75E-07	4.13E-07	3.65E-07
$\alpha=k/\rho/c$	m2/s	1.59E-07	1.61E-07	1.64E-07
k	W/m/oC	0.654	0.661	0.668

9.2.1 Property Values and Their Temperature Dependency

All the dimensionless parameters depend on fluid properties, whose values depend to some extent on the value of the temperature. There are a variety of sources of fluid properties, and care must be taken to properly interpret the units and exponents. Figure 9.2 lists fluid properties, for saturated liquid water at three temperature values, taken from an online source.[1] The compressed liquid approximation, namely that thermodynamic properties are generally functions of temperature only (not pressure), is invoked in using these parameter values (as long as the pressure is above the saturation pressure for that temperature).

Figure 9.3 lists fluid properties for atmospheric air at the same temperatures.[2] The only property value that depends significantly on pressure is the density, and the ideal gas law can be used when needed for problems at other than atmospheric pressure.

In both of these tables, the properties listed in the reference source are listed in the top section, and those properties derived from the others are listed in the "DERIVED" section. For air, properties were listed only at 60 and 80 °C, and the values at 70 °C are from linear interpolations. Different sources may list different properties.

To organize these property values and their combinations, it is possible to consider five of them as being primary property values (ρ, c_p, k, μ, β) and the others

---

[1] http://www.engineeringtoolbox.com/water-thermal-properties-d_162.html. Accessed 10/19/2010.

[2] http://www.engineeringtoolbox.com/air-properties-d_156.html. Accessed 10/19/2010.

Fig. 9.3 Property values for air at atmospheric pressure

FROM ENGINEERING TOOLBOX:					
Property	**SI Unit**	**T = 60oC**	**T = 70oC**	**T = 80oC**
ρ	kg/m3	1.067	1.034	1.000
cp	J/kg/oC	1009	1009	1009
k	W/m/oC	0.029	0.029	0.030
$\nu = \mu/\rho$	m2/s	1.89E-05	1.99E-05	2.09E-05
Pr = ν/α	-	0.709	0.7085	0.708
β	1/K	3.00E-03	2.92E-03	2.83E-03
DERIVED				
$\alpha = k/\rho/c$	m2/s	2.65E-05	2.80E-05	2.96E-05
μ	kg/s/m	2.02E-05	2.06E-05	2.09E-05

as derived from them (ν, α, Pr). These property values are shown for air and water over a temperature range of 0–100 °C in Fig. 9.4, designed to compare air against water values using a semi-logy plot, and Fig. 9.5, which plots air and water on separate linear y axes to observe the temperature dependence. All the property values for water exceed those of air, except for the coefficient of thermal expansion.

Notice that liquid water is not truly incompressible in heat transfer applications. Rather, if significant temperature variation occurs in water, the mild temperature dependency of density results in a gentle motion which profoundly affects its heat transfer behavior. A truly incompressible fluid would not experience a natural convection flow because that flow is due to horizontal density variation in a gravitational field. A rather interesting situation occurs when water is near its freezing point because the density of the liquid increases ever so slightly from 0 to 4 °C. The temperature dependency varies from property to property and between air and water. The specific role each property plays on the convection coefficient will be considered after they are calculated.

For most practical applications, property values are treated as suitably chosen constants, not functions of temperature, in the evaluation of convection coefficients. For natural convection problems, the convection coefficient depends directly on the temperature difference driving the heat because of the Grashof number. For forced convection, convection coefficients are generally considered to be independent of temperature in most cases. This practice will be further assessed after all the factors that contribute to convection coefficients are combined.

9.2.2 Rayleigh Number

Figure 9.6 shows the Rayleigh Number plotted (on semi-logy axes) against temperature difference for liquid water and air. All properties are evaluated at 70 °C. The Raleigh number is some four orders of magnitude larger in water than air. The main culprits for this difference are the density and thermal conductivity, with a minor specific heat effect. The coefficient of thermal expansion and dynamic viscosity

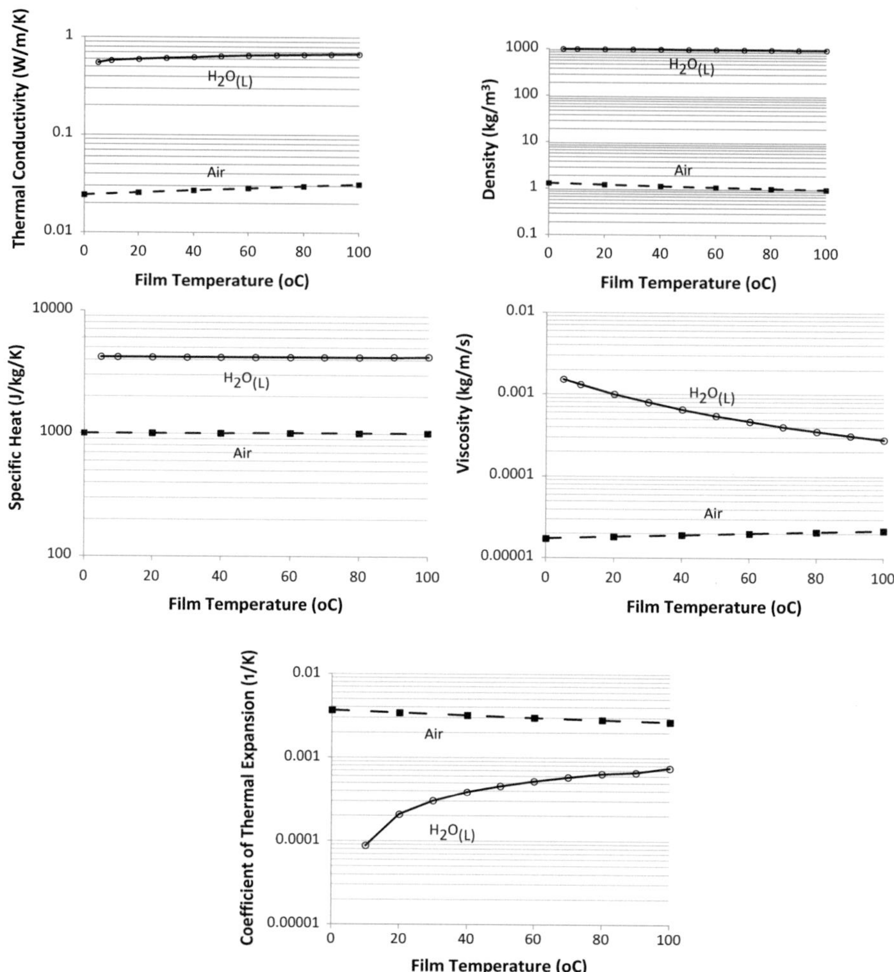

Fig. 9.4 Fluid properties for air and water between 0 and 100 °C. Semi-logy axes used to compare individual property values of liquid water and atmospheric pressure air

favor air. These comparisons will be revisited shortly. It is emphasized that the x-axis represents the difference between the fluid and surface temperature, and not on their values. A conceptual dilemma arises because the actual temperature of the fluid in the thermal boundary layer changes from one extreme to the other, and so do all the property values (some more than others). In practice, all property values are taken at the film temperature, the geometric average of the surface and fluid temperatures.

Technically, then, Fig. 9.6 applies to a specific temperature value. For example, for a temperature difference of 100 °C, the surface temperature would be 20 °C and the fluid temperature would be 120 °C, or vice versa. In practice however, this

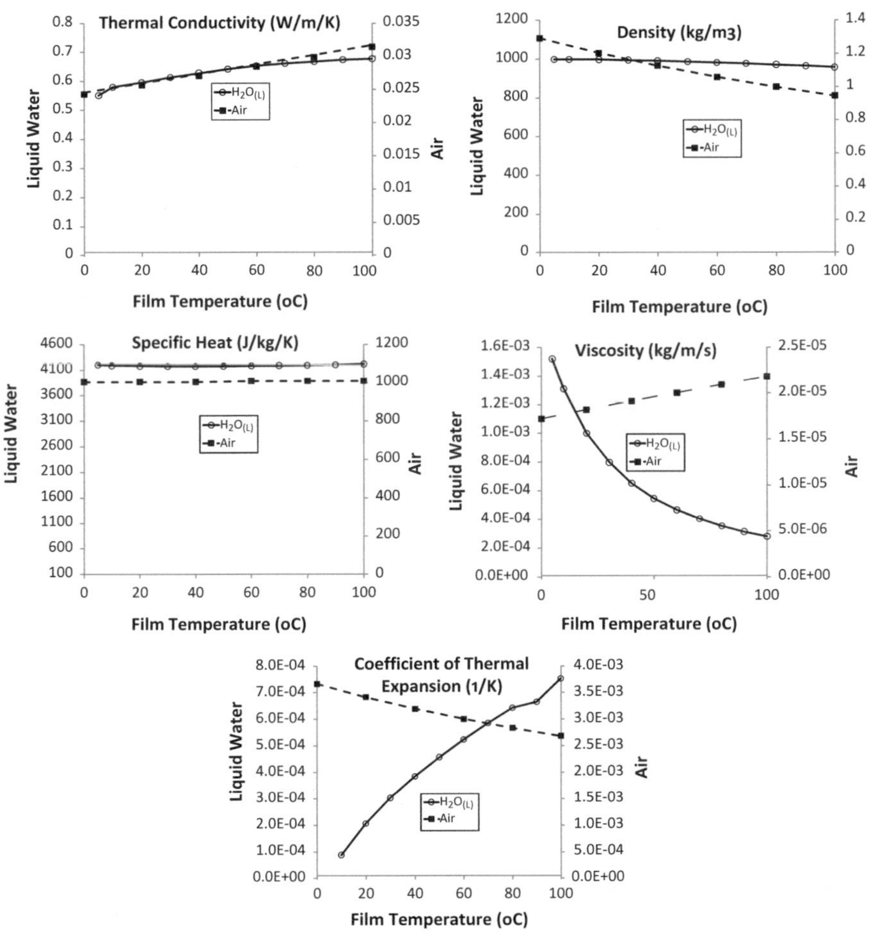

Fig. 9.5 Fluid properties of air and water between 0 and 100 °C. Linear axes with air as a secondary scale

distinction can usually be ignored, provided a reasonable film temperature for the problem at hand is selected, using sound engineering judgment.

9.2.3   Nusselt Number

Application of the semi-empirical power law expression ($Nu = C(Ra)^m$) results in the Nusselt number and is plotted against temperature difference in Fig. 9.7. The Nusselt number increases with temperature difference and is about eight times higher in water than in air. The downward jump in the curve for water at a Rayleigh number of approximately 12 is due to the change in power law coefficients, attributed loosely to a transition from laminar to turbulent flow within the thermal

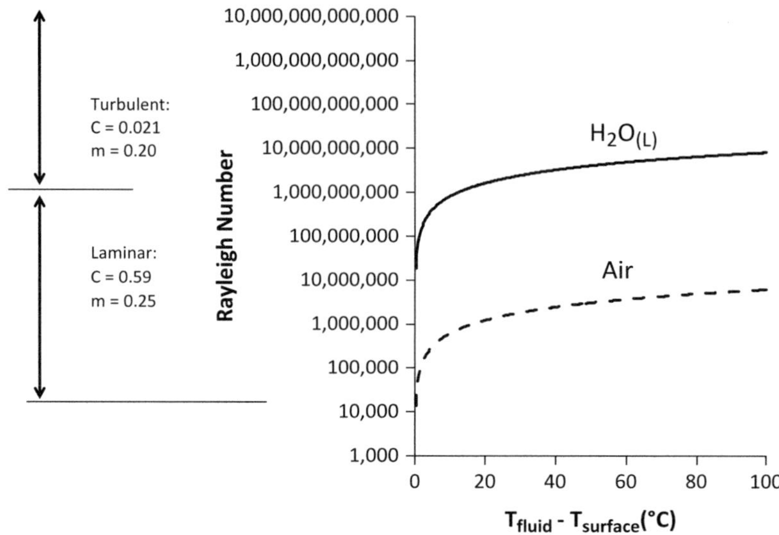

Fig. 9.6 Rayleigh number vs. temperature difference between fluid and surface. Property values evaluated at 70 °C

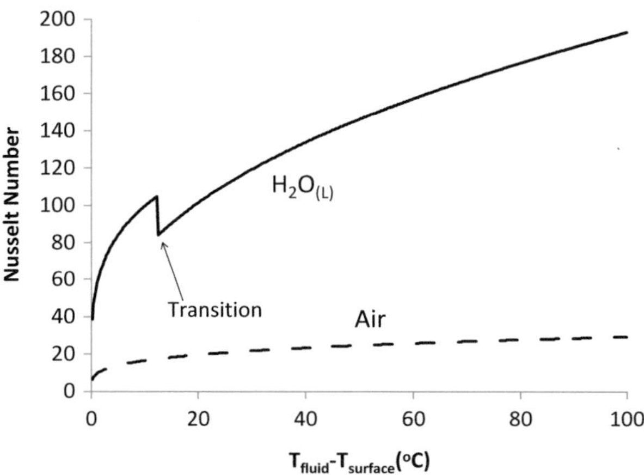

Fig. 9.7 Nusselt number vs. temperature difference between fluid and surface using published constants

boundary layer. Conceptually, the Nusselt number (and thus convection coefficient) should monotonically increase with temperature difference. It is conceivable that a discontinuous jump UP could occur at the onset of turbulence, owing to the enhanced mixing. A transition to turbulence, in any event, is very sensitive to specific geometries. The intent here is to merely generate a monotonically increasing function for modeling purposes. There is no such jump in air because the Rayleigh number is always below the transition value for the specific case considered here.

Table 9.2 Revised table of correlation parameters for natural convection on vertical plates

Rayleigh number range	Flow character	C	m
$(10)^4$–$4.545(10)^9$	Laminar	0.59	0.25
$4.545(10)^9$–$(10)^{13}$	Turbulent	0.021	0.40

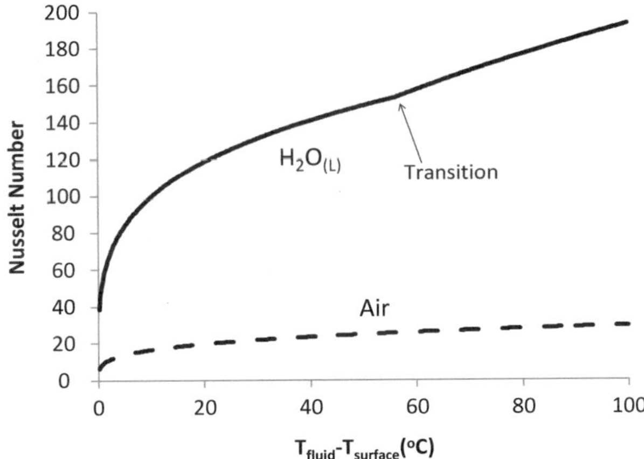

Fig. 9.8 Revised Nusselt number vs. temperature difference with modified transition point

To address that problem, the critical transition Rayleigh number can be redefined so that the transition occurs continuously (with a discontinuous slope). To do so formally, the critical Rayleigh number can be calculated by setting the Nusselt number equal using the two cases (1 and 2):

$$C_1(Ra_{crit})^{m_1} = C_2(Ra_{crit})^{m_2}$$

Rearranging and solving for the critical Rayleigh number:

$$Ra_{crit} = \left(\frac{C_2}{C_1}\right)^{\frac{1}{m_1-m_2}} = \left(\frac{0.021}{0.59}\right)^{\frac{1}{0.25-0.4}} = 4.545(10)^9$$

Table 9.2 and Fig. 9.8 show the revised table and plot of Nusselt number vs. temperature difference in which case a mild discontinuous slope is observed at the transition point, but not a discontinuous jump.

9.2.4 Convection Coefficients

Finally, the convection coefficients are plotted against temperature difference in Fig. 9.9, which is a linear axes, but different scales for air and water, and the same data plotted in Fig. 9.10, on semi-logy axes. This latter view more visually emphasizes the difference in magnitude between air and water. The convection

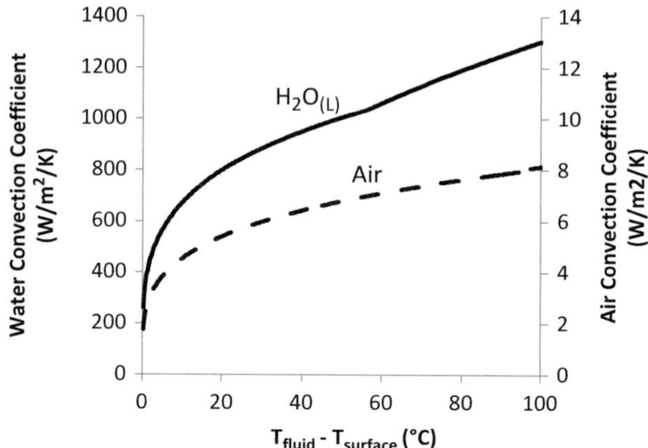

Fig. 9.9 Convection coefficients as a function of temperature difference. Film temperature is 70 °C

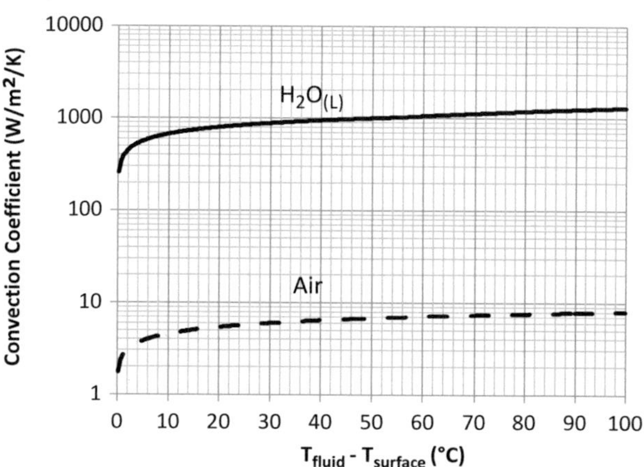

Fig. 9.10 Convection coefficient vs. temperature difference for water and air

coefficients in liquid water are more than 100 times those of air. The Nusselt number is eight times bigger, and the thermal conductivity, the last step in the calculation, is 20 times bigger. In a future chapter, the underlying physics behind this difference is explored further. This case study adds credence to the rule of thumb case, except the water goes a bit above the value of 1,000. It is pretty hard to maintain that temperature difference with water, however. Also, "Rules of Thumb" are made to be broken.

For both fluids, the value of the convection coefficients drops steeply as the temperature difference is reduced to zero. This effect is generally not observed for forced convection situations. For natural convection, it is the temperature difference which gives rise to a density difference which gives rise to a buoyant flow. The flow in a forced convection does not require a temperature difference to drive it. This effect is one reason that radiation (and evaporation, when relevant), becomes an important, if not dominant, mode of heat transfer as an object cools (or warms) to an ambient environment.

9.2.5   Natural Convection Film Property

A final step in this process is to develop a simple working formula that combines all the fluid property values into a single term, except for the vertical height H and the temperature difference. First, the definition of the Nusselt number is expressed in terms of the convection coefficient:

$$h = \frac{k_f Nu_H}{H} = \frac{k_f C}{H}(Ra)^m$$

Next, the Rayleigh number is expressed in terms of the primary property values:

$$Ra = \left[\frac{\beta}{\nu\alpha}\right]gH^3\Delta T = \frac{\beta}{\left(\frac{\mu}{\rho}\right)\left(\frac{k}{\rho c_p}\right)}gH^3\Delta T$$

Rearranging:

$$Ra = \left[\frac{\rho^2 c_p \beta}{\mu k}\right]gH^3\Delta T$$

Inserting this form of the Rayleigh number into the convection coefficient correlation and grouping property terms into brackets:

$$h = Cg^m\left[\frac{k^{1-m}\rho^{2m}c_p^m\beta^m}{\mu^m}\right]\frac{\Delta T^m}{H^{1-3m}}$$

The term in brackets is a fluid property (with very strange units) to be evaluated at the film temperature, and will be referred to as the "film property" for natural convection on a vertical plate. A similar approach can be taken for other correlations in other situations. Figure 9.11 shows the value of this property (with the factor Cg^m included) as a function of film temperature (not the difference between fluid and surface) for the laminar range (C = 0.59, m = 0.25), which covers the problem in question. The water film property is more sensitive to temperature than the air property.

Finally, a set of working formulas, strictly in SI units, that was used in the mug of coffee problem (n-node transient case) are given by:

$$h_{air} = \left(1.5646 - 0.0015T_{film}\right)\frac{\left|\left(T_{outerSurface} - T_{ambient}\right)\right|^{0.25}}{H^{0.25}}$$

$$h_{coffee} = \left(32.38T_{film}^{0.441}\right)\frac{\left|\left(T_{coffee} - T_{innerSurface}\right)\right|^{0.25}}{H^{0.25}}$$

To compare the relative effects of air and water further, it is useful to take the ratio of the convection coefficients, and look at all the individual factors that

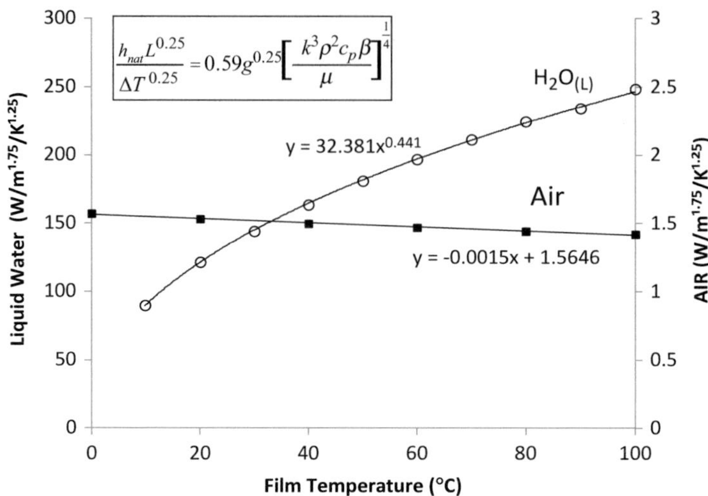

Fig. 9.11 Natural convection on a vertical plate film property vs. film temperature. Laminar flow restriction

make it up. Entering values at 70 °C from Figs. 9.2 and 9.3, keeping the individual effects separate:

$$\frac{h_{H2O}}{h_{air}} = \frac{\left(\frac{k_{water}}{k_{air}}\right)^{0.75}\left(\frac{\rho_{water}}{\rho_{air}}\right)^{0.5}\left(\frac{c_{p,water}}{c_{p,air}}\right)^{0.25}\left(\frac{\beta_{water}}{\beta_{air}}\right)^{0.25}}{\left(\frac{\mu_{water}}{\mu_{air}}\right)^{0.25}}$$

$$\frac{h_{H2O}}{h_{air}} = \frac{(22.8)^{0.75}(946)^{0.5}(4.15)^{0.25}(0.20)^{0.25}}{(19.6)^{0.25}}$$

$$\frac{h_{H2O}}{h_{air}} = \frac{(10.4)(30.7)(1.43)(0.67)}{(2.1)}$$

$$\frac{h_{H2O}}{h_{air}} = 145$$

The natural convection coefficient of water is 145 times as large as that in air, provided the same temperature difference is maintained (not so when the coffee mug is revisited) and the wall height is the same. This analysis reveals that the density difference has the largest effect (factor of 30), followed by the thermal conductivity (k). The specific heat has a mild effect (1.43), and the thermal expansion and viscosity favor the air. Further discussion of the underlying physics behind these effects is deferred until convection coefficients models based on governing laws are considered, rather than these empirical correlations.

The combined fluid property effects for natural convection cases (and a similar analysis could be conducted for forced convection cases) exhibits a strong temperature dependency for liquid water, and a mild temperature dependency for air. Therefore, the general practice of treating the property values as constants is shown to be well justified for air (and could be extended to most gases), but care needs to be taken when doing so with liquid water (or other liquids). For steady-state applications, it seems reasonable to select a suitable film temperature. However, for transient cases, recalculation of the convection coefficient as time proceeds may be needed to obtain precise simulations.

9.3 Application to the Mug of Coffee

It is apparent that in the coffee/mug problem, there really is no such thing as "A" convection coefficient because the temperature changes with time, and the natural convection coefficients depend on the changing temperature differences between the fluid and the corresponding wall. The convection coefficient rather is changing with time, as the temperature difference changes. In addition, there are spatial variations not addressed here. The Nusselt correlation used here give spatially averaged convection coefficients, that is, averaged over the length of the vertical plate. In reality, the convection coefficient is high near the leading edge of the flow (near the top surface on the coffee side, near the bottom on the air side), and decreases along the plate in the direction of induced flow as the boundary layer thickens, reducing the local temperature gradient driving the heat.

To further complicate matters, the Nusselt number calculations are based on steady-state experiments in which, for example, a vertical heated plate is MAINTAINED at a surface temperature and investigated. The thermal boundary layer has a thermal response time, and error is introduced if the fluid and/or surface temperature is changing with time rapidly compared to that response time. It is conceivable that this time constant can be estimated using methods developed in this text, once a way to estimate the thickness of the boundary layer is developed. A quasi-steady equilibrium is therefore implicitly assumed in the use of these correlations to the coffee/mug problem.

On the other hand, there is no problem thinking about "A" coffee temperature, even though it is now known that the coffee in the center is at a different temperature than the coffee right next to the mug. For modeling purposes, the choice of a single value for convection coefficient simplifies the model implementation.

9.3.1 Value Selection for Constant Convection Coefficient Models

In the 2-node model of Chap. 4, a single convection coefficient was chosen for both the coffee/inner mug surface and air/outer surface, even though it is clear that they are both changing with time. So what value should be chosen? It seems logical to select a temperature difference that exists during the period of time of most interest

Fig. 9.12 Thermal resistance network for the side channel of a 1-node lumped model of the mug of coffee problem

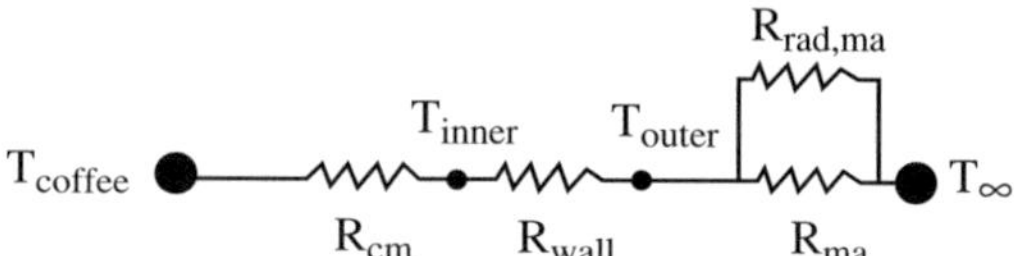

to the objective of the calculation. For example, if the objective is to calculate the time it takes for the coffee to cool from an initial temperature (say 82 °C) to a defined drinking temperature (say 70 °C), then it makes sense conceptually to determine convection coefficient values when the coffee is at 76 °C, the average coffee temperature during the event. That convection coefficient will underestimate its value when the coffee is above it, and overestimate it when the coffee is below, and a degree of cancellation of errors will occur. On the other hand, if the time to cool from a drinking temperature to a tepid temperature (say 50 °C), then 60 °C would be the average coffee temperature during the event.

Figure 9.12 shows a quasi-steady state resistance network for the side channel for the 1-node lumped capacity analysis (there is no capacitance attributed to the mug). The coffee temperature is treated as if it were being maintained at a fixed temperature, even though in the broader model, it gradually changes with time. At the outer wall, convection (R_{ma}) is modeled as acting in parallel with radiation ($R_{rad.ma}$).

At any time in the simulation, the difference between the average temperature of the coffee/mug system and ambient is well characterized (say 45 °C, when ambient is 25 °C and the coffee is 70 °C) . However, the Nusselt number correlation requires the temperature difference between the fluid and the solid surface adjacent to it, not the overall temperature drop from one fluid (coffee) to the other (ambient air). The 1-node lumped model allows the temperature of the two surfaces to be calculated, but there is a circular argument. The surface temperatures depend on the convection coefficients which depend on the surface temperatures. What can be done?

Develop an iterative scheme.

That is, make an intelligent guess (or a wild stab, if necessary) of both surface temperatures. Use the Nusselt number correlation to determine a tentative convection coefficient based on those guesses. Calculate new inner and outer surface temperatures based on these convection coefficients. Since these new temperatures are different than the first guess (i.e., you guessed wrong), the convection coefficients will be different from the first guess. But now the convection coefficients can be recalculated based on these updated values. An iterative procedure like this can often be done by hand, or it can be automated in a computer program.

While there are actually different iterative algorithms that could be implemented, the Method of Successive Substitution (MOSS), introduced in Chap. 7, works well with most 2- and 3-dimensional steady-state heat transfer problems, as already shown for the case where the resistance values are constant.

The two wall surface temperatures in Fig. 9.12, T_{inner} and T_{outer} are unknown. Expressing the nodal energy balances on the inner and outer surfaces in the form suitable for MOSS (using the Gauss–Seidel scheme where updates are used as soon as available) yields the two recursion relations:

$$T_{inner}^{(k+1)} = \frac{\dfrac{T_{coffee}}{R_{cm}^{(k)}} + \dfrac{T_{outer}^{(k)}}{R_{wall}}}{\dfrac{1}{R_{cm}^{(k)}} + \dfrac{1}{R_{wall}}} \qquad \text{NODE AT INNER WALL}$$

$$T_{outer}^{(k+1)} = \frac{\dfrac{T_{inner}^{(k+1)}}{R_{wall}} + \dfrac{T_\infty}{R_{ma,equiv}^{(k)}}}{\dfrac{1}{R_{wall}} + \dfrac{1}{R_{ma,equiv}^{(k)}}} \qquad \text{NODE AT OUTER WALL}$$

Superscripts $(k+1)$ and (k) have been applied to all unknowns that refer to the iteration number. A subscript (k) means the "guessed" values, and $(k+1)$ refers to the next guess. It is easy to confuse these iteration numbers with the "p" and "p + 1" that refer to present and future in a numerical simulation of a transient behavior, but they are fundamentally different things. A simulation step is a time prediction. An iteration is a guess.

The resistor values are evaluated using the following formulas. Across the mug wall:

$$R_{wall} = \frac{w}{k_{mug}(A_{inner} + A_{outer})/2}$$

where w is the mug thickness, k_{mug} its thermal conductivity, $A_{inner} = \pi DH$ is the inner surface area, and $A_{outer} = \pi(D + 2w)H_o$ is the outer surface area. This resistance is a constant.

The inner convective resistance is:

$$R_{cm} = \frac{1}{h_{coffee}A_{inner}}$$

where h_{coffee} depends on temperature as described previously.

The outer convective resistance is:

$$R_{ma} = \frac{1}{h_{air}A_{outer}}$$

where h_{air} depends on temperature as described previously.

The radiative coefficient is given by:

$$R_{ma,\,rad} = \frac{1}{h_{rad}A_{outer}}$$

where the radiative coefficient is given by Kirchhoff's Law (Chap. 3) as:

$$h_{rad} = \varepsilon\sigma\left[(T_{outer} + 273)^2 + (T_\infty + 273)^2\right](T_{outer} + 273 + T_\infty + 273)$$

where ε is the emissivity of the surface and $\sigma = 5.669(10)^{-8}$ W/m^2/K^4 is the Stefan–Boltzmann constant. The temperatures are input in Celsius, and converted to Kelvin in this equation, which must be in absolute units.

The equivalent resistance for the parallel convective/radiative channel is:

$$R_{ma,equiv} = \frac{1}{\dfrac{1}{R_{ma}} + \dfrac{1}{R_{ma,rad}}}$$

Since the area is the same for the convection and radiation, the equivalent resistance can be expressed as:

$$R_{ma,equiv} = \frac{1}{(h_{ma} + h_{ma,rad})A_{outer}}$$

Figure 9.13 shows a spreadsheet implementation of the MOSS for this problem. Final converged results are shown in the top section. Fixed parameters are defined below that. These cells refer to input parameters from a separate sheet (not shown). The wall resistance and surface areas are from the mug of coffee problem. The outer area is larger than the inner area because its radius is larger, and its height is higher. The emissivity is a typical value for ceramic. The effect of radiation can be simply tested by setting the emissivity to zero.

In the iteration table, the initial row (iteration 0) contains the initial guesses (set at 50 °C) for the inner and outer wall temperatures. On the same row, the film temperatures (average of fluid and adjacent wall surface) are calculated, followed by the inner convection, outer convection (from the Nusselt number correlation), and radiative coefficient, and finally the thermal resistances. These formulas are written once and copied down. On the iteration 2 row (k + 1), the recursion formulas are entered for T_{inner} and T_{outer}, which refer to the resistance values from the previous row (k) or from the same row (k + 1), depending on the order the calculations are done. These formulas are copied down. Ten iterations are shown for this case, which is sufficient for successive values to converge to within 0.01 °C. Formal convergence of the method can be tested since the numbers continue to change at a significant figure level less than shown.

The first "guess" of 50 °C for the surface temperatures yielded a convection coefficient of 744.5 W/m^2/K on the inner (coffee) surface, and 6.028 W/m^2/K on the outer (air) surface. The radiative coefficient on the outer is coincidentally very

RESULT

Rside=	2.63	oC/W
Uside=	12.56	oC/W

Fixed Parameters

emissivity =	0.9
Tcoffee (oC) =	70
Tamb (oC) =	25
Rwall (oC/W) =	0.094
Ainner (m2)=	0.0198
Aouter (m2)=	0.0303

iteration	Tinner (oC)	Touter (oC)	Tfilm_inner(oC)	Tfilm_outer(oC)	hcm	hma	hrad,ma	Rcm	Rma,equiv
0	50.00	50.00	60.00	37.50	744.5	6.028	6.119	0.0443	2.717
1	63.58	62.29	66.79	43.64	587.7	6.621	6.502	0.0562	2.515
2	67.11	65.60	68.55	45.30	486.9	6.752	6.608	0.0678	2.470
3	68.15	66.58	69.08	45.79	436.8	6.789	6.640	0.0755	2.457
4	68.47	66.88	69.24	45.94	417.0	6.800	6.650	0.0791	2.454
5	68.57	66.97	69.28	45.98	410.3	6.804	6.653	0.0804	2.452
6	68.60	67.00	69.30	46.00	408.2	6.805	6.653	0.0808	2.452
7	68.61	67.00	69.30	46.00	407.5	6.805	6.654	0.0810	2.452
8	68.61	67.01	69.31	46.00	407.3	6.805	6.654	0.0810	2.452
9	68.61	67.01	69.31	46.00	407.3	6.805	6.654	0.0810	2.452
10	68.61	67.01	69.31	46.00	407.3	6.805	6.654	0.0810	2.452

Fig. 9.13 Iteration table using the mug of coffee problem using the Gauss–Seidel method

close to the convection value. Based on these values, the recursion relations yield improved "guesses" of 63.58 and 62.29 °C for the inner and outer wall surface temperatures, respectively. The film temperature, convection and radiation coefficient, and thermal resistances are all recalculated, and the iteration scheme is repeated for as many rows as desired. It is best to confirm convergence rather than fixing the number of iterations and "hoping" it worked. This method is not guaranteed to converge, and chaotic or divergent iteration "trajectories" are conceivable. In that case, it is possible to isolate the unknown temperature in a different way (i.e., isolate the temperature in the radiation coefficient).

Workshop 9.1. Wind Chill Effect

The temperature today is 20 °C, but with the wind it feels like 5. *What's up with that?*

Physiologically, the sensation of cold is related more to the rate of heat loss than directly to the temperature of the fluid that the skin is exposed to. The wind chill effect (shown schematically) can be modeled by equating the actual rate of heat loss from a person's skin on a cold windy day (left sketch, with air temperature T_∞ and actual convection coefficient h_{actual}) to the heat loss on a hypothetical colder day with no wind (right sketch, with hypothetical air temperature $T_{\infty,WindChill} = (T_\infty - \Delta T_{WindChill})$ and a hypothetical convection coefficient $h_{hypothetical}$ that would exist under those conditions). Expressing the heat transfer rate with an overall heat transfer coefficient U:

$$q = U_{actual}A(T_{skin} - T_\infty) = U_{hypothetical}A(T_{skin} - T_{\infty,WindChill})$$

Note that while the convection coefficient is the primary variable that changes with the wind, the quantitative determination of wind chill effect involves several factors and depends on whether exposed skin or clothed skin is to be compared. Radiation acts in parallel with convection, there are conduction effects from the interior (Biot number effects). Finally, physiological thermoregulatory responses (i.e., constriction of blood vessels) are important. A good student project is to consider these effects one by one. The main goal of this workshop is to gain experience in estimating values of the convective heat transfer coefficient using Nusselt number correlations. The effect of wind speed, size of the object (extended finger vs. arm), and fluid type (air vs. water) will be considered.

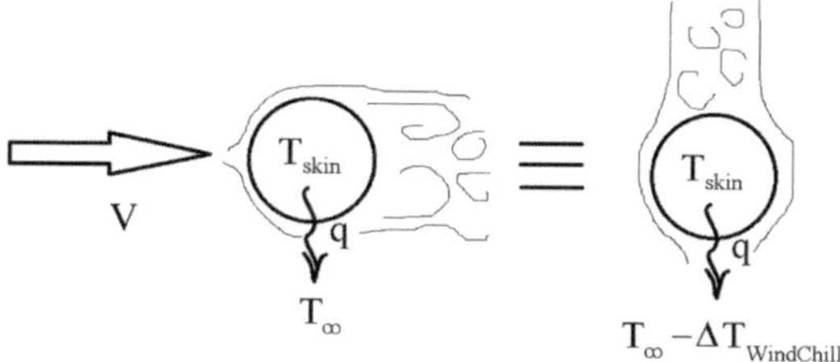

Problem Statement

- Write a Worksheet or program code that takes as input parameters (in SI units) D (outer diameter), wind speed V, and fluid properties (k, ρ, c_p, μ) and uses the Hilpert/Knudsen/Katz correlation to return the convection coefficient for forced convection across a cylinder.
- Write a Worksheet or program code that takes as input parameters (in SI units) D (outer diameter), surface temperature (T_{skin}), fluid temperature (T_∞), and fluid properties (β, k, ρ, c_p, μ) and uses the "Notsurewho" correlation to return the convection coefficient for natural convection across a horizontal cylinder.
- Use this code to complete the Results table, and notice both the values of convection coefficient, and the trends associated with size, wind speed, and fluid type.

Forced Convection, External Flow

Convection coefficient experimental data for heat exchange between a cylinder (outer diameter D) and a fluid (with properties k, ρ, c_p, μ and derived properties

$\nu = \mu/\rho, \ \alpha = k/(\rho c_p))$ with a cross-flow velocity (V) have been correlated using the Hilpert/Knudsen/Katz equation:

$$Nu_D = C Re_D^n Pr^{1/3}$$

$$\frac{hD}{k} = C\left(\frac{\rho V D}{\mu}\right)^n \left(\frac{\nu}{\alpha}\right)^{1/3}$$

In selecting fluid properties as constants evaluated at the film temperature. The constants C and n depend on the value of the Reynolds number (Re_D) given in the following table, modified from the published sources to give continuous functions between ranges:

Reynolds number range	C	n
0.4–4	0.989	0.330
4–35	0.911	0.385
35–4,083	0.683	0.466
4,083–40,045	0.193	0.618
40,045–400,000	0.0266	0.805

Natural Convection, External Flow

Convection coefficient experimental data for heat exchange between a horizontal cylinder (diameter D, with outer surface temperature T_s) in a still fluid at temperature T_∞ (with properties β, k, ρ, c_p, μ and derived properties $\nu = \mu/\rho, \ \alpha = k/(\rho c_p))$ have been correlated by various authors (the "Notsurewho" correlation):

$$Nu_D = C[Gr_D Pr]^N$$

$$\frac{hD}{k} = C\left[\left(\frac{g\beta D^3 |(T_s - T_\infty)|}{\nu^2}\right)\left(\frac{\nu}{\alpha}\right)\right]^N$$

The constants C and N depend on the value of the Raleigh number $(Gr_D Pr)$.

Rayleigh $(Gr_D Pr)$ number range	C	N
$10^4 – 2.12(10)^7$	0.53	0.25
$2.12(10)^7 – 10^{12}$	0.13	0.3333

Note that convection coefficient depends on the absolute value of temperature difference. The coefficient of thermal expansion (β) for ideal gases is given by $\beta = 1/\left(T_{film} + 273\right)$, where the temperature is in Celsius.

Assume the exposed skin temperature is 32 °C (for natural convection cases).

For the diameters, use D = 0.02 m for a finger, D = 0.080 m for a forearm.

Use fluid properties evaluated at nominal room temperature (20 °C) from this table:

Property @ 20°C	*Units*	*Air*	*$H2O_{(L)}$*
k, thermal conductivity	W/m/K	0.0257	0.60
ρ, density	kg/m^3	1.205	1000
c_P, specific heat	J/kg/K	1009	4200
μ, dynamic viscosity	kg/sec/m	$1.82(10)^{-5}$	$1.0(10)^{-3}$
β, coef. of thermal expansion	1/K	$3.41(10)^{-3}$	$2.07(10)^{-4}$

RESULTS TABLE

Scenario	Fluid	Free Stream Fluid Velocity	T_∞	Convection Coefficient (W/m^2/K)
Exposed finger on a cold day with no wind	Air	0	0°C (= 32°F)	
Exposed finger on a *really* cold day with no wind	Air	0	-15°C (=4.4°F)	
Exposed finger on a cold day with a light wind (1 mph)	Air	0.447 m/s	NA	
Exposed finger on a cold windy day (10 mph)	Air	4.47 m/s	NA	
Exposed forearm on a cold day with no wind	Air	0	0°C	
Exposed forearm on a cold windy day (10 mph)	Air	4.47 m/s	NA	
Exposed finger dipped in ice/water	$H2O_{(L)}$	0	0°C	
Exposed finger under running water	$H2O_{(L)}$	1.0 m/s	NA	
Exposed forearm dipped in ice/water	$H2O_{(L)}$	0	0°C	
Exposed forearm under running water	$H2O_{(L)}$	1.0 m/s	NA	

Workshop 9.2. Forced Convection for Internal Flows

There are many applications that involve the exchange of heat between a fluid flowing in pipes (or ducts) and the inner walls of the pipe. For engineering calculations, the rate of heat transfer (q) is determined using Newton's Law of Cooling, that is, $q = h(\pi DL)(T_{wall} - T_{bulk})$, where D is the inner diameter, L is an appropriate length of pipe (which could be a differential length, Δx), T_{walls} is the temperature of the inner wall surface, and T_{bulk} is the average fluid temperature over the length in question. The value of the convection coefficient (h) depends on both the fluid type (in particular its properties such as thermal conductivity (k), density (ρ), specific heat (c_p), and viscosity (μ)) and on the character of the flow (the "wind chill" effect), and cannot be simply looked up in a Table. The goal of this workshop is to gain experience in using published Nusselt number correlations for internal, forced convection flow problems.

While many correlations have been published, the Dittus/Boelter equation for turbulent flow is arguably the most general, and would be a good starting place even if higher precision is warranted afterward. It is given by:

$$Nu_D = 0.023 R_{e,D}^{0.8} P_r^n$$

However, the Nusselt number cannot be less than a value of 3.66, the value obtained for laminar flow. The exponent "n" equals 0.4 for heating (walls hotter than fluid), and 0.3 for cooling (walls colder than fluid). Expressing the three dimensionless variables directly in terms of system parameters:

$$\frac{hD}{k} = 0.023 \left(\frac{\rho V D}{\mu}\right)^{0.8} \left(\frac{\nu}{\alpha}\right)^n$$

where V is the flow velocity and fluid properties (k, μ, $\nu = \mu/\rho$, $\alpha = k/\rho c_p$) are evaluated at the "film temperature," taken to be the average of the bulk fluid and wall temperatures. Also, this and even more sophisticated correlations yield ESTIMATES of the convection coefficient, and this uncertainty must be borne in mind in any application.

Pressure Drop

While thermodynamic analyses often neglect it, there is always a pressure drop in internal flows that can be estimated using the Fanning friction factor, f, a nonlinear function of Reynolds number and the relative roughness of the pipe. The pressure drop across a pipe of length L is given by: $\Delta P = f\left(\frac{L}{D}\right)\left(\frac{1}{2}\rho V^2\right)$ The pressure drop is not considered further here, except that it should be borne in mind that factors that give rise to high convection coefficients (high velocity, small diameter) tend to also give high pressure drops and therefore higher pumping costs.

Case Study

In a boiler application, liquid water enters a pipe of inner diameter (use D = 0.02 m as a default) whose inner walls are maintained at a higher temperature (a "heating" event). At the exit of a preheating section, the fluid is a saturated liquid (x = 0), and the value of the convection coefficient at that point will be calculated. The fluid continues to flow and phase change occurs in a boiler section until it is a saturated vapor (x = 1), and the value of the convection coefficient at that point will be calculated. The flow continues through a superheating section and exits as a superheated vapor. In a later chapter, the lengths of the heater sections will be calculated, for which appropriate convection coefficient values are needed.

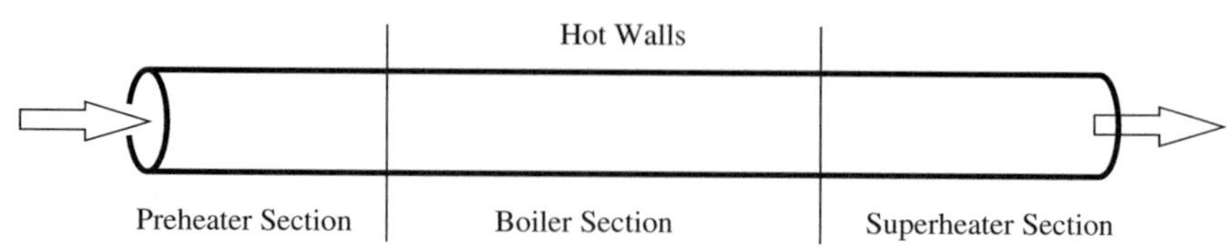

Properties of saturated H$_2$O @ 1 atm

	Liquid	Vapor	Unit	L/V ratio
k	0.646	0.028	W/m/K	23.1
ρ	989	0.518	kg/m^3	1,907.5
c_p	4,175	2,000	J/kg/K	2.1
μ	5.50E$-$04	1.40E$-$05	kg/m/s	39.3

In a condenser application, the above case is reversed. The fluid enters as a superheated vapor and exits as a compressed liquid. In this case, the inner walls are maintained at a lower temperature (a "cooling" event). This hypothetical case neglects pressure effects so that Nusselt number correlations can be used to compare convection coefficients of the same fluid type but in different phases.

Problem Statement

- Write a spreadsheet or computer code that takes as input parameters D (tube inner diameter), V (average flow velocity), and fluid properties (k, ρ, c_p, μ) and uses the Dittus/Boelter correlation to return the convection coefficient. Use an "if-then-else" statement to determine whether the flow is laminar or turbulent (do not use a transition Reynolds number, but rather ensure the Nusselt number exceeds the laminar flow value of 3.66).
- Write a spreadsheet or computer code that compares convection coefficients for the case study at the entrance and exit of the boiler section (properties given below) over a wide range of mass flow rates, spanning both laminar and highly turbulent flows. Prepare a plot of h vs. mass flow rate for saturated H$_2$O liquid and vapor at 1 atm with a diameter of 0.02 m (property values given). Put all four curves (heating and cooling for liquid and gas) on the same well-formatted log–log axes.

Workshop 9.3. Two-Node Mug of Coffee with Variable Convection Coefficients

Modify the 2-node mug of coffee analysis (resistance network repeated here) to include the effects of variable convection and radiation coefficients. As the mug and its inner and outer surface temperatures change, so do the convection coefficients. Use the Nusselt number correlations developed in this chapter (formulas of Fig. 9.9) to recalculate the convection coefficients at the mug wall at each step. For the coffee free surface, use the Nusselt number correlation for a heated horizontal plate of diameter D (from published or online source).

The simpler way to include these effects for this case study is to use the average mug temperature to determine the difference between the fluid and solid temperature, which makes the simulation an explicit one. That is, a closed-form expression for the future temperature is developed. This method is completely justified after the mug heating phase for this case because there is very little spatial variation across the mug wall (a low Biot number concept).

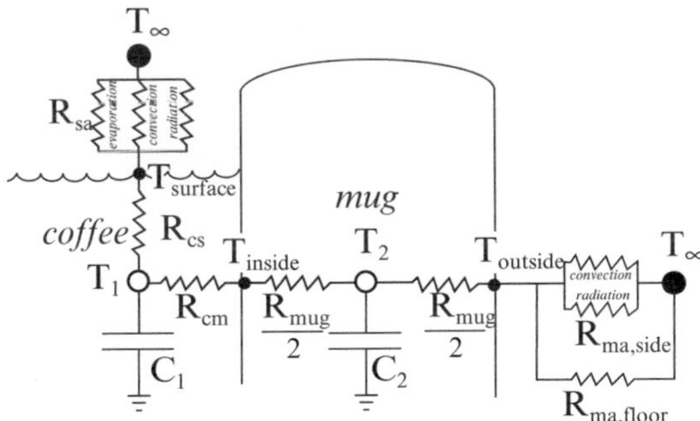

However, during the mug heating phase, the mug surface temperature is substantially different than the average mug temperature (particularly the inside temperature), and this effect can be accounted for by conducting an iterative scheme at each time step to determine the surface temperature and convection coefficients. This is a form of implicit formulation, which is not easily performed in a spreadsheet.

In addition to the coffee and mug temperatures, plot the convective and radiative coefficients vs. time, and consider the trends. Then, plot the dimensionless temperature vs. time on semi-log x axes to investigate the overall first-order behavior. A pure first-order behavior would result in a straight line. However, since the convection and radiative coefficients change, there will be some curvature. Define dimensionless temperature as:

$$\theta = \frac{T_{coffee} - T_\infty}{T_{LUMPED} - T_\infty}, \quad \text{where} \quad T_{LUMPED} = \frac{C_{coffee} T_{coffee,initial} + C_{mug} T_{mug,initial}}{C_{coffee} + C_{mug}} \quad \text{is the adiabatic}$$

temperature after the mug heating phase.

Convection Fundamentals

10

To this point, heat transfer modeling methods have been developed for closed thermodynamic systems, that is, for systems for which heat and work cross the system boundary, but mass does not. For applications where fluids (liquids, gases, slurries, etc.) flow into and/or out of the defined system, the mass that crosses the system boundary carries its energy (internal, kinetic, and potential) with it. In addition, a force (pressure times cross-sectional area) acting through a distance per unit time (velocity) is required to push fluid in or out, called "flow work." The term "flow power" is technically more accurate, but that would often be interpreted as the total head times the mass flow rate, a very different thing. The general mass balance for an open thermodynamic system with multiple inlets and outlets can be expressed as:

$$\frac{dM_{cv}}{dt} = \sum_{inlets} \dot{m} - \sum_{outlets} \dot{m}$$

where M_{cv} is the total mass inside the control volume (a triple integral), $\dot{m} = \rho V A$ is the mass flow rate of a stream which could be an inlet or an outlet (a double integral, with density and velocity being appropriately weighted averages). At steady-state, the mass in the control volume remains constant, and the net mass flow rate is balanced by the net mass flow rate out. That is:

$$\sum_{inlets} \dot{m} = \sum_{outlets} \dot{m} \quad \text{(at steady state)}$$

For a pipe or duct with a single inlet and single outlet, the mass flow rate of the inlet equals that of the outlet. The volumetric flow rate (VA) is only the same if the density is the same.

© Springer International Publishing Switzerland 2015
G. Sidebotham, *Heat Transfer Modeling: An Inductive Approach,*
DOI 10.1007/978-3-319-14514-3_10

The general energy balance for an open thermodynamic system with multiple inlets and outlets can be expressed as:

$$\frac{dE_{cv}}{dt} = \dot{Q}_{in} - \dot{W}_{shaft,out} + \sum_{inlets} \dot{m}\left(e + \frac{P}{\rho}\right) - \sum_{outlets} \dot{m}\left(e + \frac{P}{\rho}\right)$$

where

$$E_{cv} = \iiint_{CV} \rho e \, dV = \iiint_{CV} \rho\left(u + V^2/2 + gz\right)dV = M_{cv}\left(\bar{u} + \bar{V}^2/2 + g\bar{z}\right) \text{ is the total}$$

energy stored within the control volume,

u	is the internal energy per unit mass
$V^2/2$	is the kinetic energy of the fluid per unit mass
g	is the potential energy relative to a reference height per unit mass
$\dot{Q}_{in}$	is the rate of heat transfer crossing the boundary of the control volume
$\dot{W}_{shaft,out}$	is the net rate of shaft work (all non-flow work, including electric) crossing the boundary
$\dot{m}P/\rho$	is the rate of flow work associated with the inlet and exit streams

All cases considered in this text will be at steady-state (the rate of energy stored remains constant), there will be no mechanical devices exchanging work with the environment, kinetic and potential energy changes will be considered to be negligible. With these restrictions, the energy balance can be expressed in the form of Energy In = Energy Out:

$$\dot{Q}_{in} + \sum_{inlets} \dot{m}h = \sum_{outlets} \dot{m}h + \dot{W}_{shaft,out} \quad \text{(at steady state)}$$

where $h = u + P/\rho$ is the enthalpy of the streams as they cross the boundary. The enthalpy is a thermodynamic property that combines the internal energy carried in or out with the work required to push fluid in or out. A line is drawn through it to differentiate it from the convection coefficient, a fundamentally different quantity.

10.1 Boundary Layer Concepts

Convection coefficients are used as a boundary condition in transient and steady-state conduction problems. They are estimated in practice using Nusselt number correlations, a cookbook approach outlined in the previous chapter. An appreciation for the underlying physics of Newton's Law of Cooling requires a closer look at the fluid in the immediate vicinity of a solid object at a different temperature. In this chapter, a 1-node analysis of the flow across a thin flat plate is developed that reveals the underlying physics of convection.

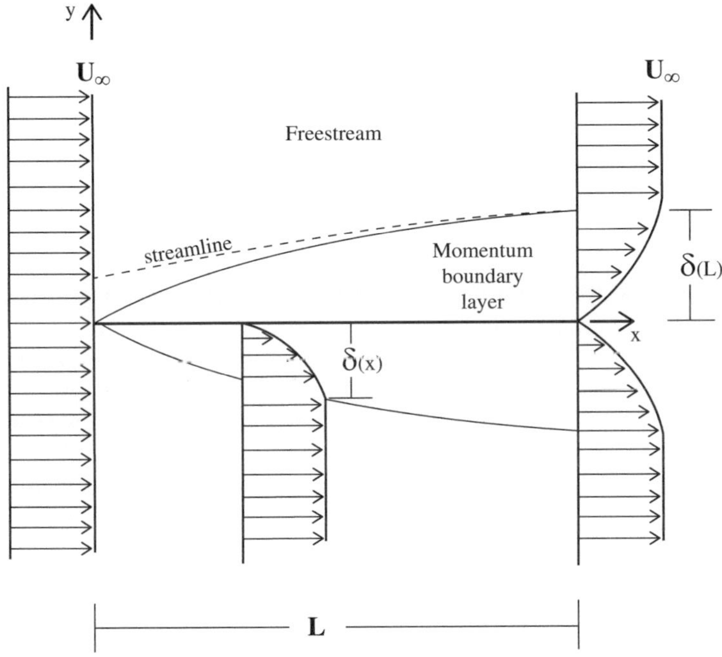

Fig. 10.1 Schematic of a momentum boundary layer on a flat plate placed in a uniform stream

The concept of a boundary layer is central to understanding convection coefficients. A boundary layer is a layer of fluid immediately adjacent to a solid that is generally thin compared to a characteristic dimension of the solid object. Outside the boundary layer, the fluid has the properties of the free stream (temperature and flow velocity). At the solid surface, the fluid properties are those of the solid. In between, there are large spatial gradients in velocity and temperature that occur across the boundary layer. For mass transfer applications, such as evaporation, there are gradients in chemical composition as well.

In heat transfer, there are two boundary layers that develop simultaneously: a momentum boundary layer and a thermal boundary layer. The momentum boundary layer is often called hydrodynamic for historical reasons in which liquid water was investigated, but momentum boundary layers occur for any fluid (gas or liquid). The retarding effects of viscosity occur within the momentum boundary layer. In fluid mechanics, the forces exerted by a fluid on an object are of primary interest and these are determined by the momentum boundary layer. Figure 10.1 shows a schematic of a momentum boundary layer on both sides of a thin plate placed in a velocity field of uniform flow velocity (U_∞). The plate has a streamwise length L and a width w into the page (not shown) and therefore a surface area Lw. The heavy solid line is the solid plate. At the inlet plane (at $x = 0$), a uniform velocity profile is shown. At the exit plane (at $x = L$), there is a velocity profile; the velocity is zero at the solid surface (the no slip condition at $y = 0$) and increases to the free stream velocity at the edge of the boundary layer (at $y = \delta_{(L)}$), with a zero gradient.

The lighter curved lines above and below the plate divide the free stream from the boundary layer. In the free stream, the flow velocity in the x-direction equals the free stream velocity. There is, however, a mild cross-stream flow velocity (in the y-direction) as a consequence of mass conservation.

The dotted line shown on the top surface is a streamline that is coincident with the edge of the boundary layer at the end of the plate. This streamline defines a streamtube that has a constant mass flow rate at every cross-section. Since the average velocity exiting the exit plane is less than the free stream velocity, the area of the streamtube is larger than at the entrance. On the other hand, the line that shows the boundary layer is not a streamline, and mass flows into the boundary layer.

A velocity profile near the midpoint is shown underneath the plate. Here, the boundary layer is thinner than at the exit, that is, $\delta_{(x)} < \delta_{(L)}$. Therefore, the boundary layer is a function of distance and generally increases with x. The velocity depends on both x and y. The velocity gradient at the wall is higher at the midpoint of the plate ($x = L/2$, say), and therefore the local shear stresses on the plate are highest near the leading edge and decrease with x.

In heat transfer, heat exchange between the solid object and the free stream occurs across a thermal boundary layer. However, the thermal boundary layer is influenced by the momentum boundary layer, so an understanding of both is needed. Figure 10.2 shows a schematic of a thermal boundary layer growing on one side of a flat plate fixed at a temperature T_w, with a uniform flow (not shown) of temperature T_∞ approaching from the left. In this case, the line separating the boundary layer from the free stream has a similar shape to the momentum boundary layer, but is not generally coincident with it. Outside the thermal boundary layer, the fluid temperature equals the free stream temperature. Temperature profiles are shown at the trailing edge ($x = L$) and at a midpoint. The temperature of the fluid gradually changes from its value at the wall (at $y = 0$), to the free stream value at the edge of the boundary layer (at δ_T), with zero slope. The shape of the temperature profile is similar, but the boundary layer is thicker at the exit than at the midpoint. The horizontal dashed line in this case represents the boundary of a control volume that will be used to analyze the thermal boundary layer. Notice that fluid crosses this control surface with a mass flow rate equal to the difference between the mass that flows into the control volume at the leading edge and the mass that exits at the trailing edge. Also the temperature of the fluid that exits the top surface equals the free stream value.

A powerful traditional method of analysis is the integral method, in which a polynomial function is used to approximate the temperature and velocity profiles across a boundary layer (y), and the x dependency is contained in the boundary layer thickness. In practice, these functions are formally derived assuming a polynomial fit to the velocity and temperature profiles that satisfy boundary conditions at the wall and edge of the boundary layer. For a second-order fit, three conditions are required to determine the coefficients of the function $f = ay^2 + by + c$. These conditions are the no slip condition and fixed wall

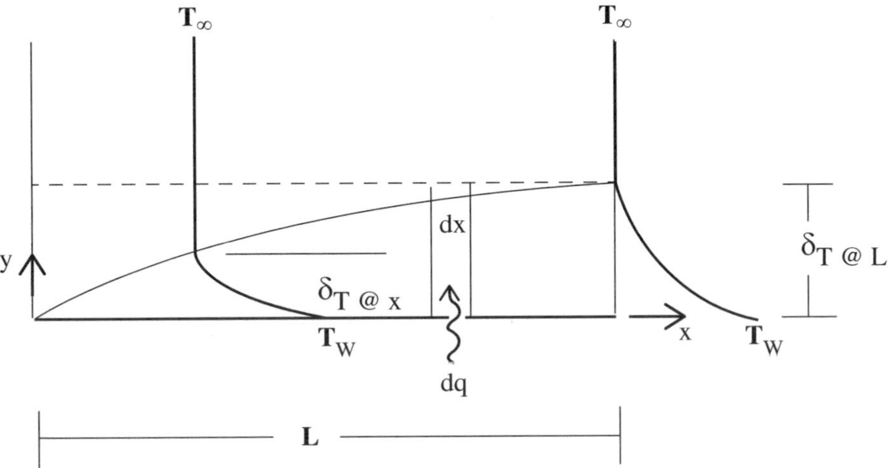

Fig. 10.2 Schematic of a thermal boundary layer on a flat plate in a uniform stream

temperature at the wall ($y = 0$), and a fixed velocity and temperature with zero slope at the edge of the boundary layer ($x = \delta$). The results are:

$$u_{(x,y)} = U_\infty \left[2\frac{y}{\delta_{(x)}} - \left(\frac{y}{\delta_{(x)}}\right)^2 \right]$$

$$T_{(x,y)} = T_{wall} + (T_{wall} - T_\infty)\left[2\frac{y}{\delta_{T(x)}} - \left(\frac{y}{\delta_{T(x)}}\right)^2 \right]$$

This velocity profile is reasonable for the case of forced convection only, where there is an imposed free stream velocity, and for laminar flow, a difficult statement to explain briefly. For natural convection, a third-order polynomial is the minimum order needed. However, a second-order polynomial fit to the temperature profile is reasonable for both natural and forced convection environments, with laminar flow.

In what follows, the 1-node boundary analysis is equivalent to the use of a linear velocity and temperature profiles. That is:

$$u_{(x,y)} = U_\infty \frac{y}{\delta_{(x)}}$$

$$T_{(x,y)} = T_{wall} + (T_{wall} - T_\infty)\frac{y}{\delta_{T(x)}}$$

Notice that the velocity and temperature within the boundary layer are functions of both x and y. The streamwise (x) dependency is contained in the boundary layer thickness (momentum and thermal), and the cross-stream (y) dependency is expressed as a polynomial approximation.

10.1.1 Newton's Law of Cooling

The rate of heat transfer from the solid into the fluid can be evaluated at the solid surface (on the fluid side of that discontinuity in material property, with solid below and fluid above) for the differential slice of fluid of thickness dx of Fig. 10.2 extending across the boundary layer. Then, Newton's Law of Cooling can be defined for the average heat transfer coefficient. That is, a general expression for the local heat transfer coefficient is:

$$dq = -k\frac{\partial T}{\partial y}\bigg|_{y=0} wdx \equiv h_{(x)}wdx(T_w - T_\infty)$$

Rearranging for the local convection coefficient:

$$h_{(x)} = \frac{-k\frac{\partial T}{\partial y}\big|_{y=0}}{(T_w - T_\infty)}$$

In first-order modeling, the temperature gradient at the wall (at an arbitrary distance from the leading edge, x) is considered to be proportional to the total change in temperature divided by the local thermal boundary layer thickness:

$$\frac{\partial T}{\partial y}\bigg|_{y=0} \sim \frac{(T_\infty - T_w)}{\delta_{T(x)}}$$

This slope is what would be obtained for a linear temperature profile. Therefore, the local convection coefficient is:

$$h_{(x)} \sim \frac{k}{\delta_{T(x)}}$$

The total heat transfer rate and average convection coefficient are determined by integrating the differential heat transfer rate across the length of the plate:

$$q = -\int_{x=0}^{L} k\frac{\partial T}{\partial y}\bigg|_{y=0} wdx \equiv \bar{h}Lw(T_w - T_\infty)$$

$$\bar{h} = \frac{-k}{L}\left[\frac{\int_{x=0}^{L} \frac{\partial T}{\partial y}\big|_{y=0} dx}{(T_w - T_\infty)}\right]$$

This integration cannot be formally conducted until the variation of boundary layer thickness with x is determined. However, in strict first-order modeling (linear temperature profile), the average temperature gradient is taken to be the overall temperature change divided by the geometric average of the boundary layer thickness at the leading edge ($\delta_{T(x=0)} = 0$) and at the trailing edge ($\delta_{T(x=L)} = \delta_T$). That is:

$$\int_{x=0}^{L} \frac{\partial T}{\partial y}\bigg|_{y=0} dx \sim \frac{(T_{wall} - T_\infty)L}{\frac{1}{2}(0 + \delta_{T(L)})} = \frac{2(T_{wall} - T_\infty)L}{\delta_{T(L)}}$$

That is, the average gradient over the length of the plate is twice that of the gradient at the trailing edge. This result will be shown to be true after the boundary layer thickness variation with x is approximated. The result, then, is that the average convection coefficient on the plate equals twice its value at the trailing edge of the plate:

$$\overline{h} = 2h_{(L)} \sim \frac{2k}{\delta_{T(@L)}}$$

The averaged Nusselt number is a dimensionless measure of the convection coefficient defined by:

$$Nu_L = \frac{\overline{h}L}{k} \sim 2\frac{L}{\delta_{T(L)}}$$

The local Nusselt number is:

$$Nu_x = \frac{hx}{k} \sim \frac{x}{\delta_{T(x)}}$$

The Nusselt number is therefore shown to be a measure of the ratio of plate length to boundary layer thickness. A thin boundary layer results in a large Nusselt number. The convective thermal resistance is defined as:

$$R_{convective} = \frac{1}{hA} \sim \frac{\delta_T}{kA}$$

Convection is, at its heart, a conduction process, with a thermal resistance that looks like the conductive resistance across a wall of thickness δ, and thermal conductivity k. The boundary layer acts as an insulating barrier to convection. The thicker the layer, the lower the heat transfer rate. As will be developed, the thermal boundary layer thickness also depends on thermal conductivity, so the temptation to say that the convection coefficient is proportional to the fluid thermal conductivity must be suppressed. The temperature gradient is steepest near the leading edge, and the local convection coefficient is high. The temperature gradient, and thus local convection coefficient, decreases with

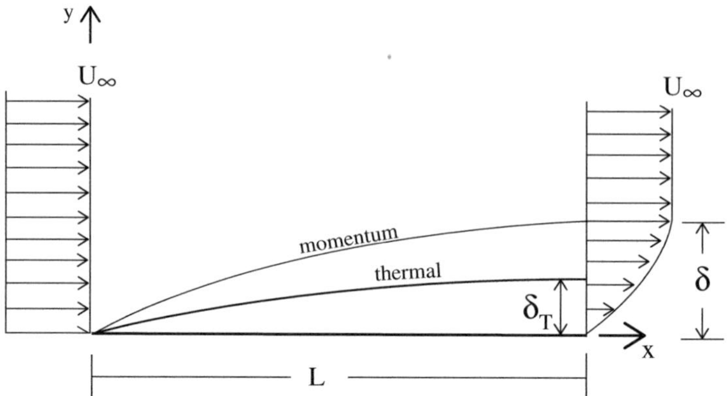

Fig. 10.3 Schematic of thermal and momentum boundary layers when the momentum boundary layer is thicker than the thermal boundary layer

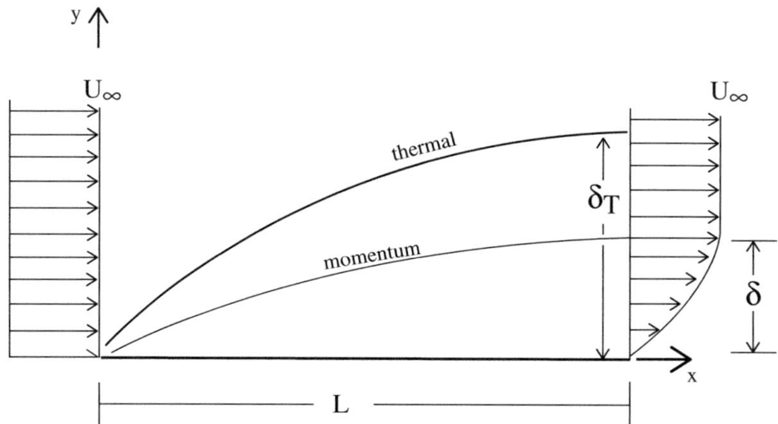

Fig. 10.4 Schematic of thermal and momentum boundary layers when the thermal boundary layer is thicker than the momentum boundary layer

distance from the leading edge. This functional dependence will be obtained in a simplified 1-node analysis of the thermal boundary layer.

The relative thickness of the two boundary layers will be shown to be related to the ratio between the fluid kinematic viscosity ($\nu = \mu/\rho$) and its thermal diffusivity ($\alpha = k/\rho/c$). When the kinematic viscosity is greater than the thermal diffusivity, the momentum boundary layer is thicker, as shown schematically in Fig. 10.3. In this case, the thermal boundary layer is contained entirely within the momentum boundary layer, and the velocity at the exit plane (at $x = L$) is everywhere below the free stream velocity. In the opposite case, the thermal boundary layer is thicker than the momentum boundary layer (Fig. 10.4). The exit velocity is uniform over a portion of the exit plane.

10.2 Forced Convection Across an Isothermal Flat Plate

The basic mechanism of convection is due to conduction across an insulated layer of fluid. All that needs to be done to obtain the heat transfer rate, and therefore define a convection coefficient, is to relate the thickness of that layer to physical properties. Traditionally, theoretical solutions (the Blasius exact solution to the boundary layer approximation equations, and integral approximate methods) for the case of laminar flow over a limited number of solid shapes (the flat plate being the simplest) are developed as the only cases for which theoretical solutions are possible. The approach taken here is to look at the boundary layer as a single node, and obtain a closed-form solution that predicts the exact same functional dependencies as these more detailed theoretical models. The final result differs only in the proportionality constant. All these methods fail to produce predictive models for turbulent flows, or for flows over more complicated shapes, but a deeper appreciation of the underlying physics associated with convection coefficients is gained.

An analysis of the momentum equation ($F = ma$) is first applied to obtain an expression for the momentum boundary layer thickness, and an expression for the drag coefficient, a dimensionless measure of the drag force. The thermal boundary layer is influenced by the momentum boundary layer, and these results are used as inputs to the thermal model.

10.2.1 Momentum Boundary Layer

Consider the control volume (CV) defined by the streamtube of Fig. 10.5. The top surface coincides with the streamline that divides fluid that enters the boundary layer, and the free stream. The free stream is gently displaced upward, but does not experience a retarding force. The left (east) face has an unknown height, y_{east}. All of the fluid that crosses this plane enters the boundary layer and then exits the CV through the right (west) face at the exit plane. The velocity varies from zero at the wall to the free stream velocity at the edge of the boundary layer (at the exit plane).

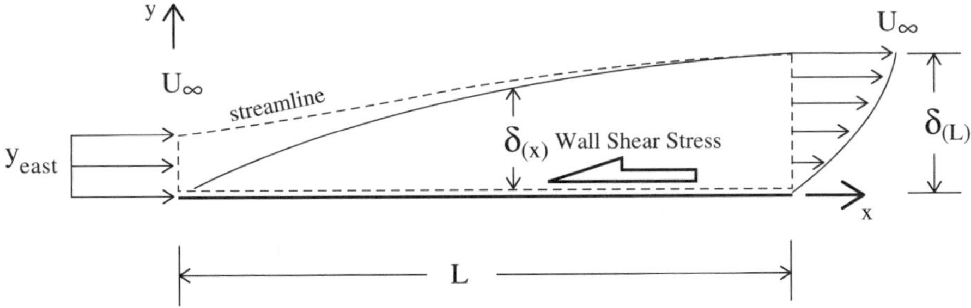

Fig. 10.5 Control volume (*dashed line*) used for analysis of momentum boundary layer

The bottom surface is coincident with the wall, where there is no flow velocity (no slip), but a shear stress is exerted by the wall on the fluid.

Mass balance. A mass balance will yield a relationship between the height of the control volume at the entrance (y_{east}) and the boundary layer thickness at the exit (δ). Taking a steady-state mass balance, the mass flux entering the left (east, at $x = 0$) plane equals the mass flux exiting the right (west, at $x = L$) face. Integrating across these planes:

$$\left[\int_{y=0}^{y_{east}} \rho u w\, dy\right]_{x=0} = \left[\int_{y=0}^{\delta} \rho u w\, dy\right]_{x=L}$$

The velocity is uniform at the entrance ($x = 0$), and the integral is easily expressed, but in terms of the unknown height of the streamtube, y_{east} :

$$\left[\int_{y=0}^{y_{east}} \rho u w\, dy\right]_{x=0} = \rho U_\infty w y_{east}$$

An average velocity at the exit plane is used at the exit plane:

$$\left[\int_{y=0}^{\delta_{(L)}} \rho u w\, dy\right]_{x=L} = \rho \overline{U} w \delta_{(L)} \sim \frac{1}{2}\rho w U_\infty \delta_{(L)}$$

Using the geometric average between the wall and free stream is equivalent to assuming a linear velocity profile with a discontinuous slope at the edge of the boundary layer. Putting the pieces together, the mass flux entering equals the mass flux leaving:

$$\rho U_\infty w y_{east} \sim \frac{1}{2}\rho w U_\infty \delta_{(L)}$$

The height of the control volume entering is therefore simply half of the boundary layer thickness at the exit, that is $y_{east} = \delta_{(L)}/2$. The result for a second-order velocity profile is $y_{east} = 2\delta_{(L)}/3$, that is, it has the same functional dependency, but a different coefficient.

x-Momentum balance. An integral force balance in the x direction applied to the control volume states that the rate of change of momentum stored (which is zero at steady-state) equals the momentum flux entering minus the momentum flux exiting plus the sum of the forces acting on the control volume. The only forces acting on this control volume is the retarding friction at the wall, expressed as an integral along the length. For flows across objects other than thin flat plates, pressure forces in the x-direction would also be present. In natural convection, gravity forces also come into play. The momentum balance is:

$$\frac{d(x - Mom)}{dt} = \left[\int_{y=0}^{y_{east}} \rho u^2 w\, dy\right]_{x=0} - \left[\int_{y=0}^{\delta} \rho u^2 w\, dy\right]_{x=L} - \int_{x=0}^{L} \tau_{wall} w\, dx = 0@SS$$

The first term, the momentum flux entering is:

$$\left[\int_{y=0}^{y_{east}} \rho u^2 w\, dy\right]_{x=0} = \rho U_{\infty}^2 w y_{east} \sim \rho U_{\infty}^2 w \frac{\delta_{(L)}}{2}$$

The momentum flux exiting is:

$$\left[\int_{y=0}^{\delta} \rho u^2 w\, dy\right]_{x=L} = \rho \overline{U}^2 w \delta_{(L)} \sim \rho \frac{U_{\infty}^2}{4} w \delta_{(L)}$$

The shear stress term is expressed by defining an average shear stress:

$$\int_{x=0}^{L} \tau_{wall} w\, dx \equiv \overline{\tau}_{wall} L w$$

Putting the pieces together:

$$0 = \frac{1}{2}\rho U_{\infty}^2 w \delta_{(L)} - \frac{1}{4}\rho U_{\infty}^2 w \delta_{(L)} - \overline{\tau}_{wall} L w$$

$$\overline{\tau}_{wall} = \frac{1}{4}\left(\rho U_{\infty}^2\right)\frac{\delta_{(L)}}{L}$$

This momentum balance relates the average shear stress to the boundary layer thickness.

For Newtonian fluids, the shear stress exerted by a fluid is proportional to the velocity gradient, with the proportionality being the dynamic viscosity, a fluid property. That is, the local shear stress at the wall ($y = 0$) is:

$$\tau_{wall} = -\mu \frac{\partial u}{\partial y}\bigg|_{y=0}$$

Therefore, an additional relationship between average shear stress and boundary layer thickness can be obtained by integrating the shear stress across the entire surface of the plate:

$$\overline{\tau}_{wall} = \frac{1}{L} \int_{x=0}^{L} \mu \frac{\partial u}{\partial y}\bigg|_{y=0} dx$$

The velocity gradient at the exit plane is:

$$\frac{\partial u}{\partial y}\bigg|_{x=L,y=0} \sim \frac{U_\infty}{\delta_{(L)}} \text{ for a linear velocity profile}$$

The average velocity gradient is expected to be higher than at the exit, but to exhibit the same functional dependency (i.e., the overall change in velocity across the boundary layer (U_∞) divided by the geometric average distance across the boundary layer $\delta_{(L)}$). Note that the shear stress at the entrance plane is infinite even in the exact Blasius solution. In reality, the Blasius solution is the exact solution to the boundary layer approximation equations, so that is not an exact model. The functional dependency of the average shear stress is therefore:

$$\overline{\tau}_{wall} = \frac{1}{4}(\rho U_\infty^2)\frac{\delta}{L} \sim \mu\frac{U_\infty}{\delta}$$

Solving for the boundary layer thickness:

$$\delta \sim 2\sqrt{\frac{\mu L}{\rho U_\infty}} = 2\sqrt{\frac{\nu L}{U_\infty}}$$

The dimensionless boundary layer thickness is:

$$\frac{\delta}{L} \sim 2\sqrt{\frac{\mu}{\rho U_\infty L}} = \frac{2}{\sqrt{\mathrm{Re}_L}}$$

This result has exactly the same functional dependence as more detailed models. The only difference is the constant, 2, compared to the Blasius solution of 5.0. This model underestimates the magnitude of the boundary layer thickness and is consistent with the use of the lower bound for the shear stress on the plate (i.e., the value at the edge of the plate). In addition, the shear stress using a linear velocity profile underestimates that of a more detailed model. The main goal here is to demonstrate the key functional dependencies associated with boundary layers.

10.2.1.1 Drag Coefficient

The forces acting on the plate are expressed conventionally in terms of a drag coefficient, which (for streamlined objects like a flat plate) is the force in the direction of flow normalized by a hypothetical force equal to the stagnation pressure times the planform area (shadow projected perpendicular to direction of flow). That is:

$$C_{Drag} = \frac{F_{drag}}{P_{stagnation}A_{planform}} = \frac{\int_{x=0}^{L} \tau_{wall} w \, dx}{\frac{1}{2}\rho U_\infty^2 L w} = \frac{\frac{1}{4}\rho U_\infty^2 \frac{\delta_{(L)}}{L} L w}{\frac{1}{2}\rho U_\infty^2 L w} = \frac{1}{2}\frac{\delta_{(L)}}{L}$$

$$C_{Drag} \sim \frac{1}{\sqrt{Re_L}}$$

10.2.1.2 Turbulence: When the Model Breaks Down

Fluid friction (shear stress) is fundamentally due to a transport of momentum at the molecular level, and the dynamic viscosity (μ) is the associated fluid property. Within the boundary layer, if the flow is laminar, then fluid particles travel smoothly along well-defined streamlines (coincident with pathlines) that are gently bumped away from the wall (a consequence of mass conservation).

If the flow within the boundary layer becomes turbulent, then another mechanism for the transport of momentum arises, namely, random fluctuations in the velocity field (in all three spatial dimensions). Streamlines and pathlines are no longer coincident nor well defined. The flow field is fundamentally transient and three dimensional, although a steady-state averaged velocity field is generally considered in analysis. The shear stress at the wall is no longer controlled by the viscous mechanism, and the model breaks down. So do the analytical models.

Largely from empirical data, for flows across a flat plate, there is a critical Reynolds number of $5(10)^5$ above which boundary layers become turbulent.

10.2.2 Thermal Boundary Layer

A different control volume is used to analyze the thermal boundary layer, shown in Fig. 10.6. The top (north) surface is horizontal (not a streamline) so mass crosses this surface, and it carries the enthalpy (internal energy plus flow work) associated with it. A 1-node model for the boundary layer is shown in which the node represents an average temperature within the control volume (taken to be the geometric average $\bar{T} = (T_\infty + T_{wall})/2$). Temperature-dependent current sources (rather than thermal resistances) are used to represent energy flows that cross the surface of the CV since these energy flows are not driven by a temperature difference. Fluid enters

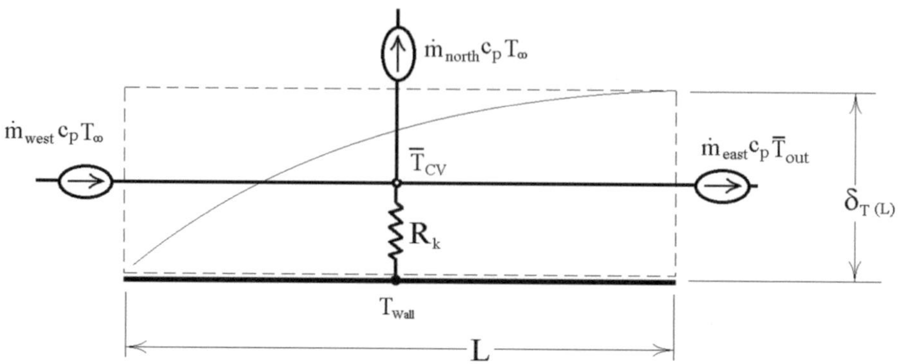

Fig. 10.6 Schematic of control volume and thermal resistance network for the thermal boundary layer on an isothermal plate

the left (east) face with a uniform velocity at the free stream temperature (T_∞). The fluid that exits the top surface is also at the free stream temperature (T_∞). The velocity and temperature of the fluid that exits the right (west) face vary from their value at the wall to their free stream values. The bottom surface, the wall, is maintained at a fixed temperature (by unspecified means), T_{wall}.

10.2.2.1 Mass Balance

A steady-state mass balance applied to the thermal boundary layer yields:

$$\frac{dM}{dt} = \dot{m}_{west} - \dot{m}_{north} - \dot{m}_{east} = 0@SS$$

Therefore, the mass flowing out the top (north) face is the difference between the inlet (east) and that exit (west) mass fluxes: $\dot{m}_{north} = \dot{m}_{west} - \dot{m}_{east}$

The velocity is uniform (at U_∞) as it enters the momentum boundary layer, and therefore the mass flux into the west plane is:

$$\dot{m}_{west} = \rho U_\infty w \delta_T$$

At the exit plane (east), the velocity changes from zero (no slip) to some non-zero value at the top edge of the thermal boundary layer. Defining an average velocity in the thermal boundary layer (whose value depends in part on the relative thickness of the momentum and thermal boundary layers, yet to be determined):

$$\dot{m}_{east} = \int_{y=0}^{\delta_T} \rho u w \, dy \equiv \rho \overline{U}_T w \delta_T$$

This average velocity is given a subscript "T" to distinguish it from the average velocity in the momentum boundary layer. An energy balance applied to the nodal network, considered to be in a steady-state is:

$$C\frac{d\overline{T}}{dt} = \dot{m}_{west}c_p\left(T_\infty - T_{ref}\right) - (\dot{m}_{west} - \dot{m}_{east})c_p\left(T_\infty - T_{ref}\right)$$

$$- \dot{m}_{east}c_p\left(\overline{T}_{east} - T_{ref}\right) + \frac{T_{wall} - \overline{T}}{R_k} = 0@SS$$

In this equation, each enthalpy stream is expressed relative to a reference temperature, and all the reference temperatures cancel. Mass conservation was used to eliminate the mass flow rate leaving the top surface. The heat transfer rate between the wall and the fluid in the thermal boundary layer is expressed with a conductive thermal resistance. The energy balance simplifies to:

$$\dot{m}_{west}c_p\left(\overline{T} - T_\infty\right) = \frac{T_{wall} - \overline{T}}{R_k}$$

This expression shows that the difference between the inlet and outlet enthalpy flows equals the rate at which heat flows into the control volume from the wall. Rearranging:

$$\dot{m}_{west}c_pR_k = \frac{T_{wall} - \overline{T}}{\overline{T} - T_\infty} = 1$$

The temperature ratio evaluates to 1 if the average temperature is taken to be the geometric average between the wall and free stream, that is, $\overline{T} = (T_{wall} + T_\infty)/2$.

The conductive resistance is defined as the average distance between the wall and the node representing the boundary layer divided by the thermal conductivity and area. That is, the functional dependency of the conductive resistance is given by:

$$R_k \sim \frac{\frac{1}{2}\left(0 + \frac{\delta_{T(L)}}{2}\right)}{kwL} = \frac{\delta_T}{4kwL}$$

The energy balance is:

$$\dot{m}_{west}c_pR_k \sim \left(\rho\overline{U}_Tw\delta_T\right)c_p\left(\frac{\delta_T}{4kLw}\right) = 1$$

Solving for the thermal boundary layer thickness:

$$\delta_T \sim 2\sqrt{\frac{kL}{\rho c_p\overline{U}_T}} = 2\sqrt{\left(\frac{\alpha L}{U_\infty}\right)\left(\frac{U_\infty}{\overline{U}_T}\right)}$$

The heat transfer rate from the wall to the fluid is driven by conduction, and can be expressed using Newton's Law of Cooling:

$$q \equiv hLw(T_{wall} - T_\infty) = \frac{T_{wall} - \overline{T}}{R_k} \sim \frac{T_{wall} - (T_\infty + T_{wall})/2}{\frac{\delta_T/4}{kLw}} = 2\frac{kLw}{\delta_T}(T_{wall} - T_\infty)$$

The convection coefficient is therefore:

$$h \sim 2\frac{k}{\delta_T}$$

For a given fluid, the thicker the thermal boundary layer, the lower the convection coefficient. Expressing the thermal boundary layer and rearranging:

$$h \sim \sqrt{k\rho c_p}\sqrt{\left(\frac{\overline{U_T}}{L}\right)}$$

The convection coefficient increases (to the ½ power) with the thermal conductivity, density, and specific heat of the fluid. This combination of fluid properties $\sqrt{k\rho c_p}$ is the thermal inertia, or thermal hardness of the fluid. The convection coefficient decreases with the length of the plate (because the boundary layer is thicker, giving a larger conductive distance). It increases with the average velocity at the boundary layer exit plane.

The Nusselt number (based on plate length, L) is given by:

$$Nu_L = \frac{hL}{k} \sim \sqrt{\left(\frac{U_\infty L}{\nu}\right)\left(\frac{\nu}{\alpha}\right)\left(\frac{\overline{U_T}}{U_\infty}\right)} \sim \sqrt{Re_L Pr\left(\frac{\overline{U_T}}{U_\infty}\right)}$$

The ratio of the momentum and thermal boundary layers is:

$$\frac{\delta}{\delta_T} \sim \sqrt{\frac{\nu\,\overline{U_T}}{\alpha U_\infty}} \sim \sqrt{Pr\frac{\overline{U_T}}{U_\infty}}$$

where the Prandtl number (ν/α), a combination fluid property, has been introduced. The average velocity at the exit (east) plane depends on the relative magnitude of the momentum and thermal boundary layers, but is always less than the free stream velocity. The Prandtl number is shown for select fluids as a function of temperature in Fig. 10.7. Air at atmospheric temperature has a Prandtl number near unity and it is effectively constant. The thermal and momentum boundary layer thicknesses are approximately the same. The Prandtl number for liquid water at nominal room temperature is on the order of 10 and decreases with temperature. The momentum boundary layer is somewhat thicker

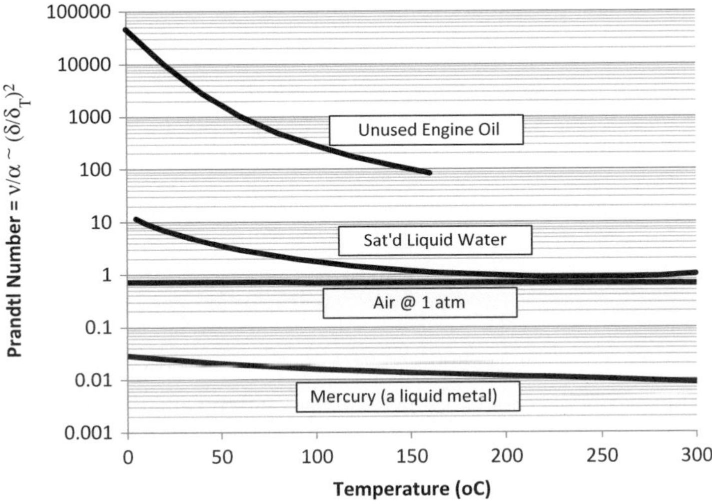

Fig. 10.7 Prandtl number of selected fluids as a function of temperature

than the thermal layer. Engine oil has a large Prandtl number and the thermal boundary layer is thin compared to the momentum boundary layer. These three cases are investigated in the next section. On the other hand, liquid mercury (and other liquid metals) has a very low Prandtl number, and the thermal boundary layer is much thicker than the momentum layer. These fluids are the simplest because the average velocity in the thermal boundary layer is approximately equal to the free stream velocity.

10.2.2.2 Limiting Case: Thermal Layer Thin Compared to Momentum Layer

The average velocity at the exit plane is now addressed for the case where the momentum boundary layer thickness is greater than the thermal boundary layer, that is, $\delta > \delta_T$. Given the linear velocity profile, the velocity at the edge of the thermal layer average velocity is:

$$u_{(y=\delta_T)} \approx \left.\frac{\partial u}{\partial y}\right|_{y=0} \delta_T = \left(\frac{U_\infty}{\delta}\right)\delta_T.$$

The average velocity at the exit plane of the thermal boundary layer is half that:

$$\overline{U}_T = \frac{U_\infty \delta_T}{2\delta}$$

Therefore, the ratio of boundary layers is:

$$\frac{\delta}{\delta_T} = \sqrt{\frac{1}{2}}\sqrt{\frac{\nu}{\alpha}\frac{\overline{U}}{U_\infty}} = \sqrt{\frac{1}{2}}\sqrt{\mathrm{Pr}\left(\frac{\delta_T}{\delta}\right)}$$

Rearranging and solving for the ratio of boundary layer thicknesses:

$$\frac{\delta}{\delta_T} = \left(\frac{1}{2}\right)^{\!1/3}\mathrm{Pr}^{1/3} = 0.793\ \mathrm{Pr}^{1/3}$$

The Prandtl number determines which boundary layer is thicker, momentum or thermal. The momentum boundary layer grows with kinematic viscosity ($\nu = \mu/\rho$), while the thermal boundary layer grows with thermal diffusivity ($\alpha = k/\rho/c_p$).

The Nusselt number is:

$$Nu_L = \frac{hL}{k} \sim \sqrt{\mathrm{Re}_L\mathrm{Pr}\left(\frac{\delta_T}{\delta}\right)} \sim \sqrt{\mathrm{Re}_L\mathrm{Pr}\left(\frac{2}{\mathrm{Pr}}\right)^{\!1/3}} \sim 1.12\sqrt{\mathrm{Re}_L}\mathrm{Pr}^{1/3}$$

The solution obtained from the integral method is:

$$Nu_L = \frac{hL}{k} = 0.664\sqrt{\mathrm{Re}_L}\mathrm{Pr}^{1/3} \quad \text{for} \quad \mathrm{Pr} < 1$$

The functional dependency is exactly the same.

10.2.2.3　Limiting Case: Thermal Layer Much Thicker Than the Momentum Layer

In this case, there is a retarded layer very close to the wall relative to the thermal boundary layer. The average velocity in the thermal boundary layer is therefore approximately equal to the free stream velocity. The Nusselt number (based on plate length, L) is given by:

$$Nu_L = \frac{hL}{k} = \sqrt{\left(\frac{U_\infty L}{\nu}\right)\left(\frac{\nu}{\alpha}\right)\left(\frac{\overline{U}_T}{U_\infty}\right)} = \sqrt{\mathrm{Re}_L\mathrm{Pr}}$$

The ratio of the momentum and thermal boundary layers is:

$$\frac{\delta}{\delta_T} = \sqrt{\frac{\nu}{\alpha}\frac{\overline{U}_T}{U_\infty}} = \sqrt{\mathrm{Pr}}$$

10.2.2.4 General Case: Thermal Layer Thicker Than Momentum Layer But of Comparable Magnitude

Recall the general expression for the thermal boundary layer:

$$\delta_T \sim 2\sqrt{\frac{\alpha L}{\overline{U_T}}}$$

The average velocity in the thermal boundary layer is determined by integrating the velocity profile (for a linear velocity in the momentum boundary layer):

$$\overline{U_T}\delta_T = \int_{y=0}^{\delta_T} u\,dy = \int_{y=0}^{\delta} \frac{U_\infty y}{\delta}\,dy + \int_{y=\delta}^{\delta_T} U_\infty dy = \frac{U_\infty}{\delta}\left(\frac{y^2}{2}\bigg|_0^{\delta}\right) + U_\infty(\delta_T - \delta)$$

$$= U_\infty\left(\delta_T - \frac{\delta}{2}\right)$$

Squaring the thermal boundary layer, rearranging and inserting the average velocity:

$$U_\infty\left(\delta_T - \frac{\delta}{2}\right)\delta_T \sim 4\alpha L$$

Rearranging in the form of a quadratic equation:

$$\frac{1}{2}\delta_T^2 - \frac{\delta}{4}\delta_T - \frac{2\alpha L}{U_\infty} = 0$$

Solving and recognizing that the thermal boundary layer must be positive:

$$\delta_T = \frac{\delta}{4} \pm \sqrt{\left(\frac{\delta}{4}\right)^2 + \frac{4\alpha L}{U_\infty}} = \frac{\delta}{4}\left[1 + \sqrt{1 + \frac{4^3\,\alpha L}{U_\infty \delta^2}}\right]$$

Recall the momentum boundary layer:

$$\delta \sim 2\sqrt{\frac{\nu L}{U_\infty}}$$

Inserting into the thermal boundary layer:

$$\frac{\delta_T}{\delta} = \frac{1}{4}\left[1 + \sqrt{1 + \frac{16\alpha}{\nu}}\right] = \frac{1}{4}\left[1 + \sqrt{1 + \frac{16}{Pr}}\right]$$

$$Nu_L = \frac{2L}{\delta_T} = 2\frac{L}{\delta}\frac{\delta}{\delta_T} = 2\left(\frac{\sqrt{\frac{U_\infty L}{\nu}}}{2}\right)\left(\frac{4}{1 + \sqrt{1 + 16/Pr}}\right)$$

$$Nu_L = \frac{4\sqrt{Re_L}}{1 + \sqrt{1 + 16/Pr}} \quad \text{for} \quad Pr > 1$$

10.3   Natural Convection Across an Isothermal Vertical Plate

When a solid object is surrounded by a stationary fluid at a different temperature, gravity will induce a flow as shown schematically in Fig. 10.8, which shows a rectangular plate (heavy line) of vertical length L, and width w (into the page). In the left schematic, the plate temperature (T_{wall}) is maintained at a higher temperature than the ambient fluid (T_∞). In the right schematic, the plate is colder than ambient. The coordinate system (x, y) is oriented with x vertical (in the direction of the main flow direction) and y horizontal (distance from the plate into the fluid). The origin is placed at the leading edge of the plate. Profile sketches of temperature (left side of plate) and velocity (right side) are shown at the trailing edge (at $x = L$) and at a midpoint. The light curved line is the outer edge of the boundary layer that grows from the leading edge. Unlike forced convection, there is only one boundary layer because the temperature field creates the density variation that results in a vertical buoyant force that drives the flow. That is, the momentum and energy equations are coupled.

For a hot plate in a cold fluid, the fluid in the boundary layer is usually at a lower density than that of the free stream, and the induced flow is in the upward direction (left schematic). The induced flow is downward if the plate is colder than the fluid. One important exception is the case of liquid water below 4 °C where the density is maximum. A plate fixed at the melting point (0 °C) surrounded by liquid water at 4 °C would experience an upward flow. Water is unusual that way.

Another difference between forced and natural convection, shown in Fig. 10.9, is that while the main direction of flow is along the direction of the plate (vertical), there is a horizontal component of flow toward the plate, not away from the plate as in forced convection. The streamline shown defines a streamtube whose cross-sectional area decreases in the direction of flow. The flow accelerates within the streamtube and coincides with the edge of the boundary layer at the end of the plate (at $x = L$). The mass flow that enters the streamtube does so with negligible momentum. All the momentum exiting the streamtube is acquired by a buoyant force (which is a pressure force), and there is a retarding frictional force exerted by the wall.

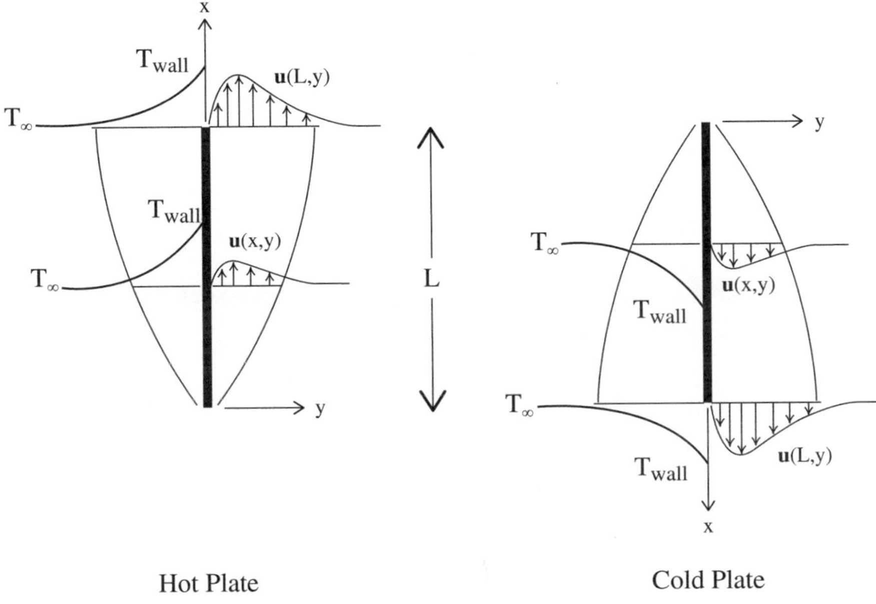

Fig. 10.8 Schematic of natural convection boundary layer on a vertical plate

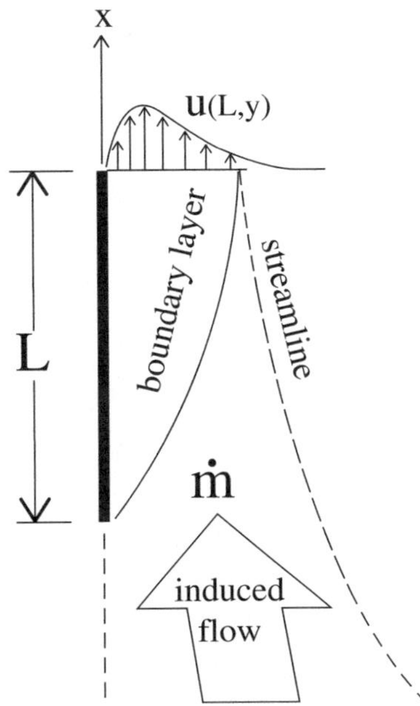

Fig. 10.9 Schematic of the
boundary layer and
streamline on a hot plate

10.3.1 Mass

The mass flow rate exiting the streamtube (which, at steady-state, equals the mass flow rate entering it in a much larger area) is:

$$\dot{m} = \int_{y=0}^{\delta} \rho u_{(L,y)} w \, dy \equiv \rho U^* w \delta$$

In this expression, a characteristic flow velocity, U^*, has been defined as the mass-average velocity at the exit plane of the boundary layer, and is to be determined.

10.3.2 x-Momentum

Taking a rectangular control volume that contains the boundary layer, the forces and momentum fluxes (in the x-direction) are shown in Fig. 10.10. The momentum flux entering the bottom and side surfaces are neglected in comparison to the momentum flux exiting, which is shown as a "virtual force" acting downward, equal to the mass flow rate times the average velocity. The pressure acting on the top and bottom surfaces are the hydrostatic forces and give rise to a net buoyant force. The weight of the fluid in the CV is a downward gravity force, and a shear stress on the wall acts to oppose the motion. This schematic is shown for a hot plate. For a cold plate, the shear stress would act in the opposite direction, and the exiting momentum flux would act upward on the bottom surface. The pressure and gravity forces would be the same as in this schematic.

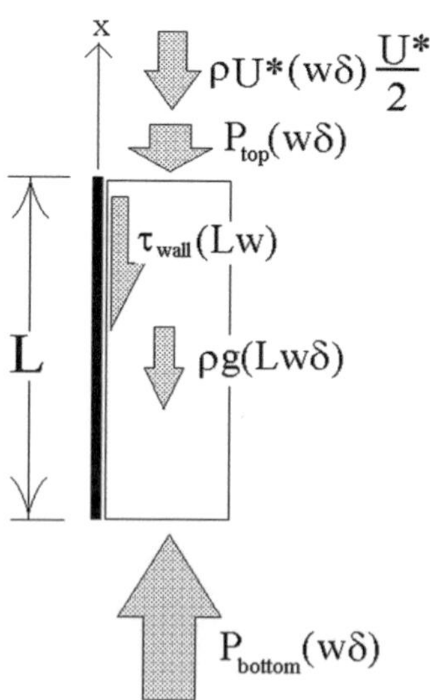

Fig. 10.10 Boundary layer control volume used for momentum balance on a hot plate with forces shown

At steady-state, the sum of the upward forces equals the sum of the downward forces:

$$\left(F_{pressure} - F_{gravity}\right) = F_{Net} = \pm \, F_{friction} \pm \text{"}F\text{"}_{momentum}$$

The pressure difference between top and bottom yields a net upward force, which is the buoyant force (equal to the weight of the displaced fluid, that is, Archimedes' principle). Gravity acts downward. The net upward force on the fluid in the boundary layer is the difference between the buoyant and gravity forces. The friction force and momentum virtual force are upward for a hot plate and downward for a cold plate. The analysis for a hot plate will be performed (+ signs), and the results are the same for a cold plate, with the direction of flow reversed.

Expressing each term further:

$$\left(P_{bottom} - P_{top}\right)w\delta - \rho g(Lw\delta) = \bar{\tau}_{wall}Lw + \frac{1}{2}\rho U^{*2}w\delta$$

The pressures are the hydrostatic pressures in the far field that are imposed across the boundary layer (which could be proven formally with a dimensional analysis of the y-momentum equation). That is, $\left(P_{bottom} - P_{top}\right) = \rho_{\infty}gL$. The momentum equation becomes:

$$\rho_{\infty}g(Lw\delta) - \rho g(Lw\delta) = \bar{\tau}_{wall}Lw + \frac{1}{2}\rho U^{*2}w\delta$$

Rearranging:

$$(\rho_{\infty} - \rho)g\delta = \frac{1}{2}\rho U^{*2}(\delta/L) + \bar{\tau}_{wall}$$

The average shear stress at the wall is taken to be proportional to the viscosity and the average velocity gradient taken to be the overall change in velocity over the half width of the boundary layer:

$$\bar{\tau}_{wall} = \frac{1}{L}\int_{x=0}^{L} \mu \frac{\partial u}{\partial y}\bigg|_{y=0} \sim \mu \frac{U^{*}}{\delta/2} = \frac{2\mu U^{*}}{\delta}$$

The x-momentum equation, in terms of two unknowns (δ and U*) is:

$$(\rho_{\infty} - \rho)g\delta = \frac{1}{2}\rho U^{*2}\frac{\delta}{L} + \frac{2\mu U^{*}}{\delta}$$

The underlying physics of this equation is that there is a net upward buoyant force (left hand side) that drives the flow. This force gives rise to a change in momentum of the heated fluid and is opposed by a frictional force on the wall.

10.3.2.1 Coefficient of Thermal Expansion, β

The difference in density between the fluid in the boundary layer and that in the free stream is what drives the flow. That difference is related to the difference in temperature through a fluid property, the coefficient of thermal expansion.

The difference in density between the boundary layer and free stream can be expressed in terms of temperature by taking a Taylor-type expansion of density, which, being a thermodynamic property, is in general a function of two other thermodynamic variables, namely:

$$\rho(T,P) = \rho_\infty + \frac{\partial \rho}{\partial T}\bigg|_P (T - T_\infty) + \frac{\partial \rho}{\partial P}\bigg|_T (P - P_\infty) + O\left((T - T_\infty)^2\right)$$
$$+ O\left((P - P_\infty)^2\right)$$

The coefficient of thermal expansion is a fluid property defined as:

$$\beta \equiv \frac{-1}{\rho}\frac{\partial \rho}{\partial T}\bigg|_P$$

Therefore, neglecting the pressure effect on density in comparison to the temperature effect (a justifiable approximation, from a more detailed dimensional analysis) and the second-order terms:

$$\rho_\infty - \rho \approx -\frac{\partial \rho}{\partial T}\bigg|_P (T - T_\infty) = \rho\beta(T - T_\infty)$$

The temperature for the control volume used is taken to be the geometric average:

$$\rho_\infty - \rho \approx \rho\beta\left(\frac{T_{wall} + T_\infty}{2} - T_\infty\right) = \frac{1}{2}\rho\beta(T_{wall} - T_\infty)$$

The momentum equation is therefore expressed in terms of the temperature, not density, difference:

$$\rho g\beta(T_{wall} - T_\infty)\delta = \rho U^{*2}\frac{\delta}{L} + \frac{4\mu U^*}{\delta}$$

There is, in general, a balance of three effects: net buoyancy, momentum change, and friction. Buoyancy drives the flow, adding momentum, which is opposed by friction.

10.3.3 Energy

A thermal resistance network for a 1-node steady-state model of the thermal boundary is shown in Fig. 10.11. The dashed line shows the control volume, with

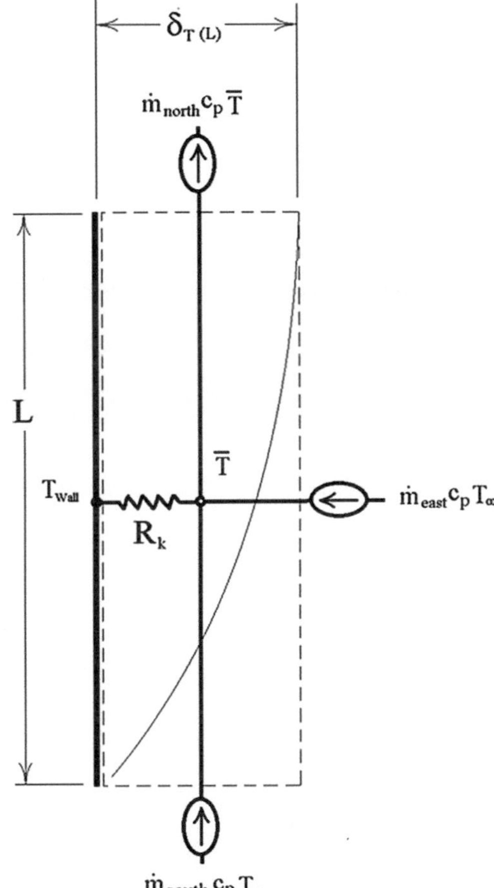

Fig. 10.11 1-Node energy model of natural convection boundary layer

ambient temperature fluid entering the bottom (south) and side (east) surfaces, modeled as current sources. The mass flow exiting the top (north) does so with an average temperature (taken to be the geometric average of the wall and ambient). The wall exchanges heat with the average temperature in the boundary layer through a conductive resistance.

The steady-state energy balance (energy in = energy out), recognizing the east and south mass flows both enter the control volume at ambient temperature, and that their sum equals the mass flowing exiting the top:

$$\dot{m}c_p\left(T_\infty - T_{ref}\right) + \frac{T_{wall} - \overline{T}}{R_k} = \dot{m}c_p\left(\overline{T} - T_{ref}\right)$$

Upon rearrangement, the reference temperature cancels:

$$\frac{T_{wall} - \overline{T}}{R_k} = \dot{m}c_p\left(\overline{T} - T_\infty\right)$$

This expression shows that heat conducted from the hot wall into the boundary layer equals the change in enthalpy flux between inlet and outlet flows. The temperature average is midway between the wall and ambient, and therefore the temperature terms are equal, and the underlying relationship for a natural convection boundary layer is:

$$\dot{m}c_p R_k = 1$$

Expressing the conductive resistance as an average difference to the midpoint of the boundary layer $(= \frac{1}{2}*(0+\delta/2)=\delta/4)$ divided by the thermal conductivity and surface area:

$$\frac{\dot{m}c_p(\delta/4)}{kLw} = \frac{\rho c_p U^* w\delta^2}{4kLw} = 1$$

The energy equation yields a simple relationship between the boundary layer thickness and the characteristic velocity, namely:

$$\frac{U^*\delta^2}{4\alpha L} = 1$$

10.3.3.1 Combined Momentum and Energy Equations

The momentum equation can be rearranged as:

$$\delta^2\left[g\beta(T_{wall} - T_\infty)L - U^{*2}\right] = 4\nu U^* L$$

Using the energy equation to eliminate the boundary layer thickness:

$$\frac{4\alpha L}{U^*}\left[g\beta(T_{wall} - T_\infty)L - U^{*2}\right] = 4\nu U^* L$$

Rearranging:

$$U^{*2} = \frac{g\beta(T_{wall} - T_\infty)L}{Pr + 1}$$

Back substituting the energy equation and rearranging:

$$\delta \sim 2\sqrt{\alpha}\left[\frac{(\text{Pr}+1)L}{g\beta(T_{wall}-T_\infty)}\right]^{1/4}$$

The boundary layer thickness normalized by the plate length (and multiplying and dividing by the square root of the kinematic viscosity):

$$\frac{\delta}{L} \sim 2\sqrt{\frac{\alpha}{\nu}}\left[\frac{(\text{Pr}+1)}{\frac{g\beta(T_{wall}-T_\infty)L^3}{\nu^2}}\right]^{1/4} = 2\frac{(\text{Pr}+1)^{1/4}}{\sqrt{\text{Pr}}(Gr)^{1/4}}$$

10.3.3.2 Newton's Law of Cooling
As before, Newton's Law of Cooling defines a convection coefficient:

$$qLw = \frac{T_w - \overline{T}}{R_k} \sim \frac{kwL}{\delta/4}\left(\frac{T_{wall}-T_\infty}{2}\right) = \overline{h}Lw(T_{wall}-T_\infty)$$

$$\overline{h} \sim \frac{2k}{\delta}$$

The average Nusselt number is:

$$Nu_L = \frac{\overline{h}L}{k} \sim 2\frac{L}{\delta} \sim \frac{Gr_L^{1/4}\sqrt{\text{Pr}}}{(1+\text{Pr})^{1/4}}$$

The ¼ power dependency on Grashof number is demonstrated. There is a more complicated dependency on Prandtl number than Grashof number. There are limiting cases, however:

$$Nu_L \sim Gr_L^{1/4}\sqrt{\text{Pr}} \qquad \text{for Pr} \ll 1 \text{ (i.e., liquid metals)}$$

$$Nu_L \sim (Gr_L \cdot \text{Pr})^{1/4} = Ra_L^{1/4} \qquad \text{for Pr} \gg 1 \text{ (i.e., cold liquid water, unused engine oil)}$$

As with forced convection, this model applies as long as the flow remains laminar. In reality, as the Raleigh number exceeds approximately 10^9, the flow within the boundary layer becomes turbulent, and an additional mechanism for the transport of momentum and energy becomes important. This turbulence mechanism is difficult to model simply.

Workshop 10.1. Boundary Layer Thickness for the Mug of Coffee

Calculate the thickness of the boundary layer and the corresponding characteristic flow velocity on the air and the coffee sides of the mug wall for the base coffee case at a time when the coffee is at 70 °C (results from Chap. 9, Figs. 10.9–10.11). Compare the boundary layer thickness to the diameter of the mug to determine whether the boundary layer is truly thin compared to the relevant physical dimension.

Internal Flows Models 11

It is common for engineering devices to be designed with pipes or ducts in which fluids (gas or liquid) flow. The fluid enters at an inlet temperature and the walls of the duct are at a different temperature. This chapter addresses internal flows in which a single ducted stream is involved. The next chapter considers heat exchangers, where one stream transfers heat to another. When mass flows into a control volume, it carries the energy associated with it. This mechanism of energy transfer is termed advection, and will be modeled as a fluid temperature-dependent current source. There are also momentum considerations, as the flows are driven by pressure (or gravity) and opposed by friction, so that pumps or compressors are generally needed to move the fluid.

In this chapter, straight sections of pipe are considered, but in practice, an internal flow frequently consists of straight sections followed by turns or coils. Conceptually, the flow path is treated here as if the actual flow path were straightened out into a single straight section of pipe with the same total length. The turns in practice can introduce turbulence, and thereby enhance the convection coefficients, as well as introduce additional pressure drops.

11.1 Thermal Length

The concept of a thermal length is useful for internal heat transfer situations and a tool for heat exchanger design. Consider Fig. 11.1 in which a fluid enters a pipe of inner diameter D_i with an average flow velocity V (or mass flow rate $\dot{m} = \rho V A$, where ρ is the fluid density and $A = \pi D_i^2/4$ is the cross-sectional area) at a uniform temperature T_{in}. Its environment (the fluid in the outer stream) is maintained at a temperature T_∞. If no phase change occurs inside, the fluid exits the pipe of length L at a different average temperature $T_{bulk,out}$. The mass flow rate is the same at any cross-section, but the flow velocity changes inversely with the fluid density. The bulk temperature is formally defined shortly.

© Springer International Publishing Switzerland 2015
G. Sidebotham, *Heat Transfer Modeling: An Inductive Approach*,
DOI 10.1007/978-3-319-14514-3_11

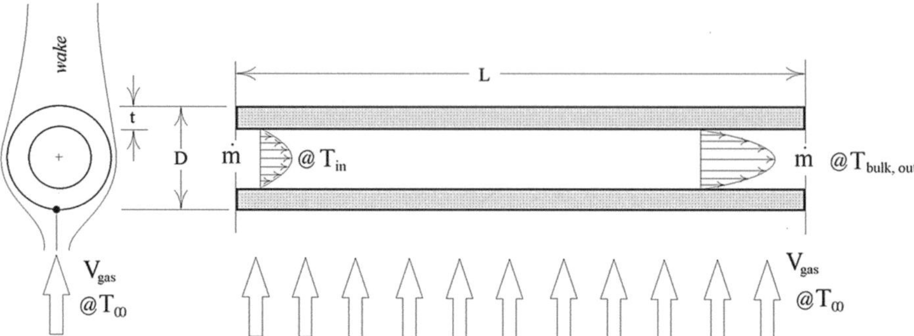

Fig. 11.1 Schematic of isolated tube of flowing fluid surrounded by a gas (or other fluid) at a different temperature. End view (*left*) and side view (*right*)

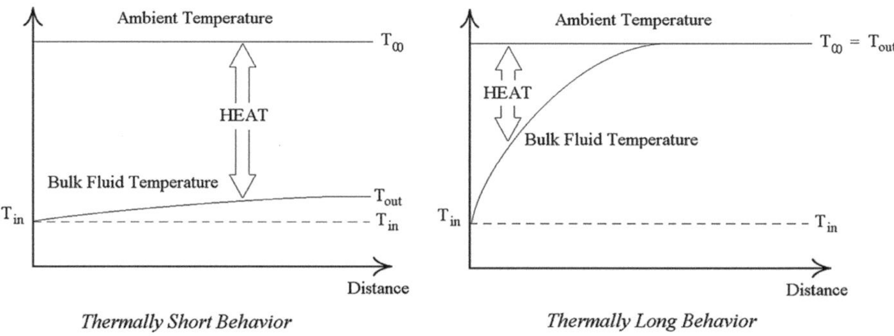

Fig. 11.2 Temperature profiles for thermally short (*left*) and thermally long behaviors (*right*)

There are two temperature differences that need to be carefully distinguished. On the one hand, there is the change of the inner fluid temperature as it flows in the pipe $(T_{bulk,out} - T_{in})$. On the other hand, there is the temperature difference that drives the heat transfer, which varies along the length of the duct (from $T_\infty - T_{in}$ at the inlet section to $T_\infty - T_{bulk,out}$ at the exit). This distinction is demonstrated by the sketches of temperature profiles for a heating event in Fig. 11.2, which shows axial profiles of the bulk average (defined shortly) fluid temperature plotted along the pipe. The temperature difference that drives the heat transfer is shown by the arrows. The effect it has on the fluid temperature is evidenced by the temperature as it approaches ambient. In the thermally short case, the temperature of the fluid does not change substantially, and the temperature difference that drives the heat changes very little with position. In contrast, for thermally long behavior, the fluid attains thermal equilibrium with ambient prior to exiting. The heat transfer rate is highest at the entrance and decreases to zero once equilibrium is attained.

Formally, a thermally short stream can be defined as one in which the exit temperature is much closer to the inlet temperature than it is to ambient temperature. That is:

$$|T_{out} - T_{in}| \ll |T_\infty - T_{in}| \quad \textit{Thermally Short Streams}$$

There is heat exchanged, but it does not substantially change the temperature of the fluid as it passes through the pipe. For example, in a car radiator, the air that is actively blown through (by a fan and/or the speed of the vehicle) is intended to behave as a thermally short stream. The inlet air temperature is typically 25 °C (T_{in}), while the inlet coolant flowing into the heat exchanger is typically 90 °C (which is considered T_∞ from the perspective of the air). The air exiting is not intended to exit at a substantially higher temperature and could be maybe 30 °C (T_{out}). So the change in air temperature as it flows through ($T_{out} - T_{in}$) is 5 °C, while the temperature difference driving the heat ($T_\infty - T_{in}$) is 65 °C at the inlet. The air is intended to act as close to an infinite reservoir as possible.

On the other hand, a thermally long stream is one in which the temperature exits much closer to the ambient temperature than to the inlet temperature:

$$|T_{out} - T_{in}| \gg |T_\infty - T_{out}| \quad \textit{Thermally Long Streams}$$

For thermally long streams, the fluid flowing in the pipe attains thermal equilibrium with the outer fluid.

The "thermal length" of a stream is defined when the change in temperature of the stream is comparable to the difference between the outlet temperature and the ambient temperature. That is:

$$|T_{out} - T_{in}| \approx |T_\infty - T_{out}| \quad \textit{Defines Thermal Length}$$

The functional dependency of thermal length can be obtained using a transient heat transfer analysis, by adopting a Lagrangian viewpoint, that is, one which follows a fluid element as it passes through. The thermal length (L_t) is defined when the time the fluid particle spends inside the pipe (its residence time) equals its thermal response time:

$$\textit{Residence Time} = \textit{Thermal Response Time}$$

$$\frac{L_t}{V} = RC$$

Applied to an element of fluid of length Δx, the thermal resistance can be expressed in terms of an overall heat transfer coefficient U_i (based on the inner diameter), which consists generally of three resistances in series: inner convection, across the walls of the pipe, and an outer convection. The capacitance of the element is the mass of the element times its specific heat:

$$\frac{L_t}{V} = \left(\frac{1}{U_i \pi D_i \Delta x}\right)\left(\rho c_p \frac{\pi D_i^2}{4} \Delta x\right)$$

Expressing the thermal length in terms of the average flow velocity:

$$L_t = \frac{\rho c_p V D_i}{4 U_i}$$

Expressing it in terms of the mass flow rate:

$$L_t = \frac{\dot{m} c_p}{\pi U_i D_i}$$

Notice that the physical length of the pipe (L) is not involved in the thermal length (L_t). Rather, a comparison of the actual length of pipe to the thermal length will determine whether a stream is thermally long, short, or intermediate. If the pipe is much longer than the thermal length ($L \gg L_t$), the pipe is thermally long. Making it any longer will not increase the heat transfer rate. Notice that a pipe of a given length can behave either as a thermally short pipe or a thermally long pipe merely by changing the flow rate. At low flows, the fluid spends a long time in the pipe and will reach thermal equilibrium (thermally long behavior) and vice versa at high flows.

The same relationship for thermal length will arise in the subsequent steady-state (Eulerian) analysis.

11.1.1 Basic Thermodynamic and Heat Transfer Relations

Internal flow and heat exchanger design theory involves a combination of basic open-system energy balances (thermodynamics) and rate equations (heat transfer). A thermodynamic specification would be, for example, that the fluid entering should be at a specific temperature, given an input mass flow rate, and should exit at another specific temperature, and the energy balance would yield the required heat transfer rate. A heat transfer calculation would be to determine the length of pipe required for a given pipe diameter, flow rate, and temperature of the ambient (outer). Alternatively, given the temperature of the ambient and a given pipe length, the exiting temperature could be sought. There are many different ways problems can be formulated, and it possible to inadvertently either over-specify or under-specify a problem.

An overall steady-state energy balance in the form enthalpy flow in = enthalpy flow out is:

$$\dot{m}\,\bar{h}_{in} + \dot{Q}_{in} = \dot{m}\,\bar{h}_{out} \quad (+KE + PE)$$

The unfortunate nomenclature conflict in which "h" is used to represent convection coefficients in heat transfer applications and enthalpy in thermodynamic applications is addressed here by drawing a line through h to represent enthalpy.

Kinetic and potential energy changes are generally considered to be negligible in heat transfer problems, but could be included when needed, such as in high speed or gravity-driven flows.

When there is no phase change, the enthalpy can be related to temperature through the specific heat (at constant pressure). For a uniform inlet stream (for which the temperature does not change with radial position), the enthalpy, relative to a reference temperature, is:

$$h_{in} = \int_{T_{ref}}^{T_{in}} c_p dT = \bar{c}_p \left(T_{in} - T_{ref} \right)$$

where the average specific heat is defined.

Internal flows are generally two dimensional. The temperature changes with distance from the entrance (x), and the temperature also varies with radial position (r). For a heating event (ambient hotter than inlet), the temperature of the fluid near the inner wall is higher than the fluid near the centerline. However, at a given distance downstream (x) there is an average temperature, called the bulk temperature, which yields the correct total enthalpy flux. The bulk temperature at any position x is therefore defined by:

$$\dot{m} c_p \left(T_{bulk} - T_{ref} \right) \equiv \int_{r=0}^{D_i/2} \rho u c_p \left(T - T_{ref} \right) 2\pi r dr$$

Assuming constant properties (ρ and c_p):

$$T_{bulk(x)} = \frac{2\pi\rho}{\dot{m}} \int_{r=0}^{D_i/2} u_{(r)} T_{(r)} r dr \quad \text{for a single phase process}$$

The energy balance equation, rearranged for the outlet temperature, becomes:

$$T_{bulk,out} = T_{bulk,in} + \frac{\dot{Q}_{in}}{\dot{m} c_p} \quad \text{for a single phase process}$$

For an internal flow, there is no such thing as an "ambient" temperature outside of a thermal boundary layer. The bulk temperature is an average temperature within the duct, intermediate to the wall temperature and fluid temperature at the centerline.

This thermodynamic equation places an overall energy balance constraint. It does not address how it is accomplished, for example, how long a pipe is needed. However, the rate of heat transfer between the inner and outer fluids is expressed in terms of system parameters, including flow rate, geometry, and thermal properties of both fluids and the walls separating them. All of these factors can be boiled down to a Newton's Law of Cooling type expression, with an overall heat transfer coefficient (U) substituted for a single convection coefficient (h):

$$\dot{Q}_{in} = \int_{x=0}^{L} UdA\left(T_{\infty} - T_{bulk(x)}\right) \equiv U_i(\pi D_i L)\left(T_{\infty} - \overline{T}_{bulk}\right)$$

The differential area ($dA = \pi D_i dx$) represents a specific surface area that separates the two fluids, for example, the inner surface of a differential element. The overall heat transfer coefficient, U, quantifies the effectiveness of the heat exchange and can be determined with thermal resistance methods developed in earlier chapters. In the integral, the bulk temperature, which varies with distance x along the duct, represents the mass flow weighted average temperature of the fluid at any cross-section.

The main objective of this chapter is to develop a modeling capability to estimate the rate of heat transfer between a fluid flowing in a duct and the walls of the duct. The principles will be introduced with a water-tube boiler example (inspired by Cooper Union student Julien Caubel, who designed and built a steam boiler for his senior project). A long, straight section of pipe of constant diameter is considered to be broken up into three sections (Fig. 11.3): a preheater (sometimes called an economizer) in which feedwater (as a subcooled liquid) is heated to its boiling point, a boiler section where the phase change occurs, and a superheater section where the steam is raised above its boiling point. Figure 11.4 shows, to scale, a plot of the temperature plotted against heat addition (per unit mass) for water at a pressure of 2.0 bar that enters at 100 °C, is heated to its boiling point (120.2 °C), then boiled completely, and finally heated to a target design temperature of 164 °C. This exit condition is that which, if delivered to a turbine or reciprocating piston, would exit at 1 atm as a saturated vapor, and is calculated assuming an isentropic expansion of a perfect gas from 2 to 1 bar, with cp = 2.02 kJ/kg/K and gas constant $R_{H2O} = 0.461$ kJ/kg/K. Note that the energy required to change the phase is much larger than that required to change the temperature of the liquid (in the preheater) and the vapor (in the superheater).

In practice, these three functions are often accomplished with very different design configurations due in part to the different fundamental heat transfer behavior and to the large change in fluid density associated with phase change. However, the response of water in a straight pipe from a compressed liquid all the way to a

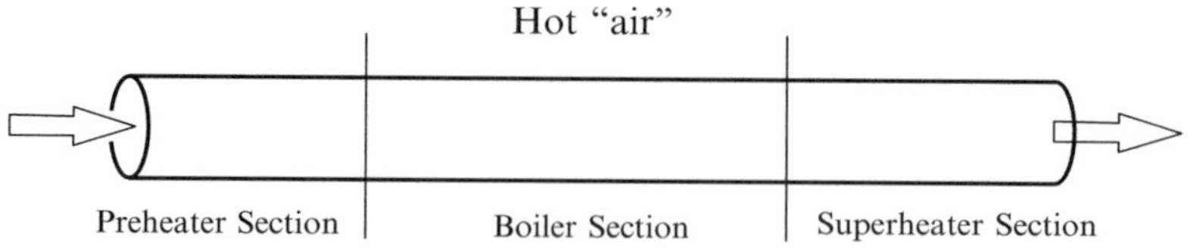

Fig. 11.3 Schematic of water-tube boiler

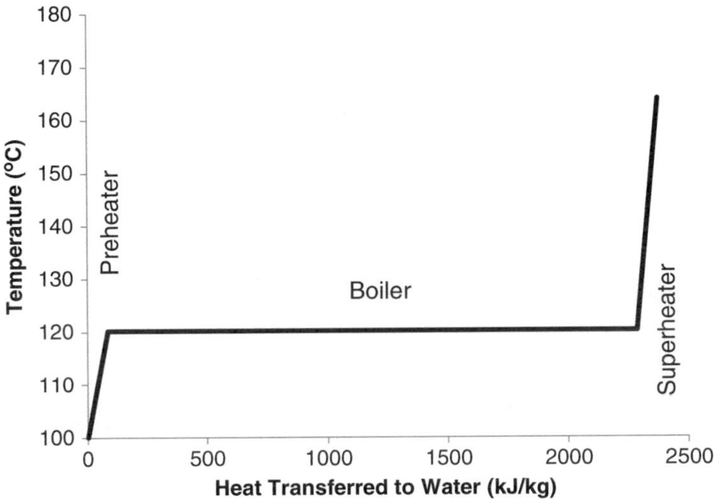

Fig. 11.4 Water temperature as a function of heat added to the water (entering at 100 °C) at a constant pressure of 2.0 bar (pressure drop neglected)

superheated vapor is a simple design configuration, and allows for a good example of the principle objectives of this chapter. The goal of the heat transfer analysis is to turn the x-axis from one of heat exchanged to physical length.

The first example considered is the boiler section, because the two fluid temperatures can be taken as the inlet temperatures, and there is no ambiguity about the average fluid temperature. On the water side, the fluid is changing phase along the entire length, and therefore does not change temperature as it is heated. On the air side, the flow is considered as an external flow, where the fluid is heated in a thermal boundary layer, but outside that layer, the temperature remains equal to the inlet temperature around the perimeter of the pipe. The cooling experienced by the air is considered to show up in the wake region. The second and third examples will treat the preheater and superheater, where the inner fluid temperature changes with position, while the external fluid (hot "air") is the same along the length. Given a boiler length in Example 1, the lengths of the preheater and superheater will be determined.

The approach to solving internal flow problems depends on how the problem statement is posed. In the following example, which has several parts, the geometry of the pipe (length, diameter, and wall thickness) in the boiler section is specified, as is the thermodynamic state of the entrance and exit. The internal mass flow rate (and therefore flow velocity) is to be determined. Since Nusselt number correlations will be used to estimate convection coefficients (rather than specifying them in the problem statement), an iterative procedure will be required to obtain an estimate of the mass flow rate. Then, with that mass flow rate, the required lengths of the preheater and superheater are determined.

11.2   Example: Water-Tube Boiler

In a water-tube boiler operating at steady-state (Fig. 11.1), saturated liquid water at an absolute pressure of 2.0 bar enters a 2.0 m long (L) K-type ¼″ copper pipe (with actual outer diameter $D_0 = 9.5$ mm and wall thickness t = 0.89 mm). The exterior of the pipe is subjected to a cross-flow of combustion projects at 1 atm, $T_\infty = 950$ K with an approach velocity of $V_{gas} = 1.0$ m/s. Determine the mass flow rate of water ($\dot{m}_w$) required to completely boil the water. Then determine the lengths of the preheat and superheat sections, assuming the inlet to the preheater is at 100 °C, and the exit of the superheater is at 164 °C.

11.2.1   1-Node Models with Advective Flow as a Current Source

A thermal resistance model for this problem is shown in Fig. 11.5. This model applies to each section (preheater, boiler, and superheater), provided the enthalpies and temperatures are suitably chosen. Space has been discretized by considering the entire inside section of the pipe as a control volume represented by a single node. The fluid entering the pipe is modeled as a current source, since enthalpy (internal

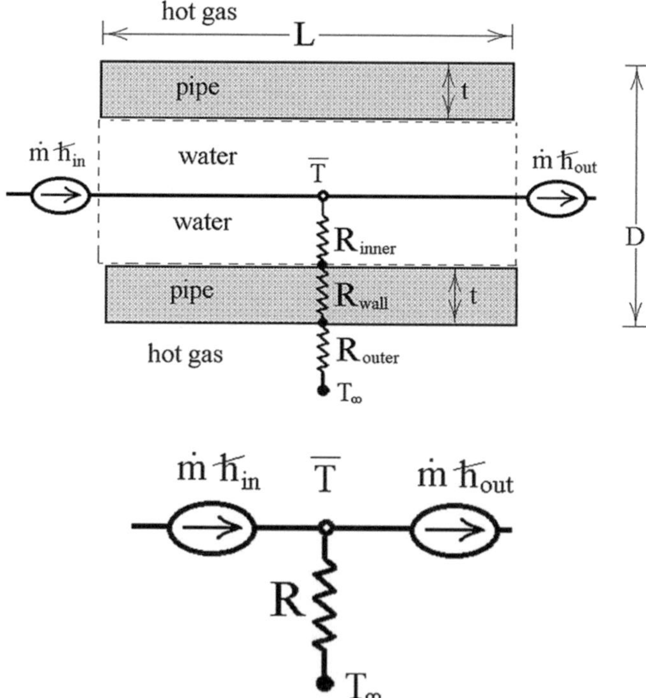

Fig. 11.5 Resistance network. *Top* image is superimposed on a side view pipe schematic

energy plus flow work) is carried into the control volume with the entering mass. It is not modeled as a resistance, since the energy crossing the boundary is not driven by a temperature difference. It is not heat transfer; it is mass transfer of a bulk flow (as opposed to a diffusional mass flow driven by concentration gradients). This mechanism of energy transport is termed advection. The fluid exiting is modeled as an advective current sink.

The main node represents an average temperature of the water inside the pipe $(\overline{T})$. It is in thermal contact with the hot combustion product gases through a series resistance channel which consists of a convective resistance on the inside (R_{inner}), a conductive resistance across the wall of the tube (R_{wall}), and a convective resistance on the outside surface (R_{outer}). The temperature of the gases will be lower in the wake region, but that has little influence in the thermal boundary layer region on the outside of the pipe. There are no capacitors shown in the model because the system is at steady-state. It would not be difficult to estimate a response time to a sudden change in inner or outer fluid temperature.

11.2.2 Energy Balance

Considering the general nodal equation for the main water node, with E representing the total energy stored in the control volume:

$$\frac{dE}{dt} = \sum \dot{Q}_{in} = 0 \ \ at \ Steady \ State$$

$$\dot{m}_w h_{in} - \dot{m}_w h_{out} + \frac{T_\infty - \overline{T}}{R} = 0$$

Solving for the desired mass flow rate of water:

$$\dot{m}_w = \frac{T_\infty - \overline{T}}{R h_{fg}}$$

In this expression, the conventional heat of vaporization nomenclature $h_{fg} = h_{out} - h_{in}$ has been used, recognizing that h_{out} is specified as saturated vapor and h_{in} is specified as saturated liquid for this particular problem. The water temperature, assuming thermodynamic equilibrium exists, is the saturation temperature corresponding to the internal pressure, which is assumed to be constant. For $P = 2.0$ bar $= 2.0(10)^5 \, N/m^2$, the corresponding saturation temperature is $120.4 \, ^\circ C$, obtained from steam tables. [1]If a significant pressure drop occurred, the change in corresponding saturation temperature would have to be considered.

---

[1] http://www.engineeringtoolbox.com/saturated-steam-properties-d_101.html.

11.2.3 Thermal Resistance

The thermal resistance, given an inner convection coefficient h_i and inner surface area A_i, pipe thickness t, thermal conductivity k_p, and average area (between inside and outside), outer convection heat transfer coefficient h_o, and outer surface area A_o, is expressed as:

$$R = \frac{1}{h_i A_i} + \frac{t}{k_p \bar{A}} + \frac{1}{h_o A_o}$$

The exact solution for the conduction across the pipe could be used, but the thin-wall approximation is valid in this case.

It is useful to express the resistance in terms of an overall heat transfer coefficient (U) based on a particular area. Using the pipe inner surface area as the basis, the resistance is expressed as:

$$R = \frac{1}{U_i A_i} = \frac{1}{A_i}\left(\frac{1}{h_i} + \frac{t A_i}{k_p \bar{A}} + \frac{A_i}{h_o A_o}\right)$$

For the cylindrical geometry:

$$\frac{A_i}{A_o} = \frac{\pi(D - 2t)L}{\pi D L} = 1 - \frac{2t}{D}$$

$$\frac{A_i}{\bar{A}} = \frac{\pi(D - 2t)L}{\frac{\pi D L + \pi(D - 2t)L}{2}} = \frac{1 - 2t/D}{1 - t/D}$$

For a thin-walled pipe ($t/D \ll 1$), the overall heat transfer coefficient is:

$$\frac{1}{U_i} = \frac{1}{h_i} + \frac{t}{k_p} + \frac{1}{h_o} \quad thin \ \ wall \ \ pipe$$

Here, the three terms represent the three resistances in series. If any one of them is significantly larger than the other two, it will control the heat transfer rate, and the other two can be neglected, in practice. The conductive resistance is easily calculated, and in this case is negligible in comparison to both convection coefficients (but still included in calculations). The outer convection environment is one of a mildly forced external convection of a gas, and a direct Nusselt number correlation can be used. The inner convection environment is one of forced internal convection. However, the inner flow velocity is not known from the problem statement. In addition, the fluid enters as a liquid, changes phase within the boiler section, and exits as a vapor (at the same mass flow rate, but very different flow velocity) with very different thermal flow rate. Therefore, an iterative solution method will be developed.

11.2.4 Nusselt Number Correlations

11.2.4.1 External Flow

The result of this particular problem is heavily dependent on the convection coefficient used on the outside of the pipe. A popular Nusselt number for forced convection across a cylinder of outer diameter D is the Knudsen/Katz correlation:

$$Nu_D = \frac{hD}{k} = CRe_D^n Pr^{1/3}$$

The values of the constant "C" and exponent "n" depend on the value of the Reynolds number of the cross-flow, and are presented in Table 11.1, with a modification of the published ranges to allow a smooth transition between ranges (as was done in the Chap. 9).

There are published correlations for a bank of tubes, which are more complicated once the flow patterns begin to interfere, but modeling the flow as an isolated cylinder is expected to provide a reasonable first approximation for the problem at hand.

The relevant properties for air at a film temperature of $(T_w + T_\infty)/2 \sim 400\ °C = 673\ K$ were obtained online[2] and input into the following equations:

$$Re_D = \frac{\rho V D_i}{\mu} = \frac{(0.524\,\text{kg/m}^3)(1\,\text{m/s})(0.00772\,\text{m})}{\left(3.28(10)^{-5}\text{kg/s/m}\right)} = 123.3$$

The constants from Table 11.1 are $C = 0.683$ and $n = 0.466$. The convection coefficient is therefore:

$$h_o = C\frac{k}{D}Re_D^n Pr^{1/3} = 0.683\frac{0.0515\,\text{W/m/K}}{0.00772\,\text{m}}(123.3)^{0.466}0.68^{1/3}$$

$$h_o = 26.7\frac{\text{W}}{\text{m}^2\text{°C}}$$

That number should seem reasonable by this time for forced convection of a gas.

Table 11.1 Correlation parameters for forced flow across a cylinder

Reynolds number range	C	n
0.4–4	0.989	0.330
4–35	0.911	0.385
35–4,083	0.683	0.466
4,083–40,045	0.193	0.618
40,045–400,000	0.0266	0.805

---

[2] http://www.engineeringtoolbox.com/air-properties-d_156.html.

11.2.4.2 Internal Flow

Estimating an appropriate convection coefficient value for the inside of the pipe requires use of Nusselt number-type correlations for the case when the fluid is boiling. Published approaches to this scenario model the convection coefficient as the sum of that obtained for a forced convection without boiling and a component attributed to the boiling. The boiling component depends on the viscosity, surface tension, heat of vaporization, liquid-specific heat, densities of saturated liquid and vapor, and Prandtl number of the liquid and can be difficult to interpret and implement. However, neglect of the boiling component results in an underestimate of the inner convection coefficient. In this case, it is anticipated that the convection coefficient on the outside will be much lower than even this underestimate, and therefore will be the controlling factor in the overall heat transfer coefficient.

The approach taken here will be to calculate the convection coefficient for forced convection without the boiling contribution at both the inlet (where it is liquid) and outlet (where it is steam). It might be expected that the former will be much larger than the latter (because it is a liquid vs. gas comparison), and that the latter will be comparable to the outer convection. However, the flow velocity at the outlet will be much larger than at the inlet (due to the large decrease in fluid density), making a strong forced convection, so judgment on that is deferred.

The flow inside the pipe is classified as forced convection for internal flows. The Dittus/Boelter correlation for fully developed turbulent flow in a pipe of inner diameter D is:

$$Nu_D = 0.023 Re_D^{0.8} Pr^n \quad for \;\; turbulent \;\; flow$$

The exponent on the Prandtl number (n) is 0.4 for heating and 0.3 for cooling.

For laminar flow, the Nusselt number is a constant:

$$Nu_D = 3.66 \quad\quad for \;\; laminar \;\; flow$$

Rather than defining a transition Reynolds number, it is recommended that the turbulent flow formula be evaluated (once the Reynolds and Prandtl numbers have been calculated). If that value is below 3.66, then set the Nusselt number to 3.66 (using an if statement, for example).

For the problem at hand, the mass flow rate is the same at the inlet and exit of the boiler section, but the flow velocity is different due to the change in density. Therefore, it is useful to express the Reynolds number in terms of the mass flow rate. Multiplying and dividing by the cross-sectional area $A = \pi D^2/4$:

$$Re_D = \frac{\rho V D}{\mu} = \frac{(\rho V A)D}{\mu(\pi D^2/4)} = \frac{4 \dot{m}_w}{\pi \mu D}$$

For the problem as posed, a guess for mass flow rate will be made, which allows the Reynolds number, hence Nusselt number, hence convection coefficient to be evaluated. It is intriguing that the Reynolds number depends only on the viscosity of the fluid for this particular problem (same mass flow and diameter at the entrance and exit). The entering liquid has a much higher viscosity and therefore a lower Reynolds number than the exiting vapor. Any difference in Reynolds number is not due to a different velocity.

It is instructive to express the convection coefficient (for turbulent flows) in terms of the fluid properties, and then determine the major contributing factors that distinguish liquids from vapor.

$$h_{inner} = 0.023 \frac{k}{D} \left(\frac{4\dot{m}}{\pi\mu D} \right)^{0.8} \left(\frac{\mu c_p}{k} \right)^n = 0.028 \left[\frac{k^{1-n} c_p^n}{\mu^{0.8-n}} \right] \left[\frac{\dot{m}^{0.8}}{D^{1.8}} \right]$$

For a fixed mass flow rate, there is no direct dependence of convection coefficient on density, while it increases with thermal conductivity and specific heat and decreases with viscosity.

Property values for water for saturated liquid and vapor at 2 bar pressure are shown in Table 11.2. The ratio of the liquid to vapor values is:

$$\frac{h_{Liquid}}{h_{Vapor}} = \left[\frac{\left(\frac{k_{liquid}}{k_{vapor}} \right)^{1-n} \left(\frac{c_{p,liquid}}{c_{p,vapor}} \right)^n}{\left(\frac{\mu_{liquid}}{\mu_{vapor}} \right)^{0.8-n}} \right] = \left[\frac{(42.8)^{1-n}(2.15)^n}{(23.5)^{0.8-n}} \right] \text{ for turbulent flow in both}$$

streams.

For heating (n = 0.4), the results are (showing effect of each fluid property):

$$\frac{h_{Liquid}}{h_{Vapor}} == \left[\frac{(9.52)(1.36)}{(3.53)} \right] = 3.66$$

That this ratio is precisely equal (to two significant figures) to the laminar Nusselt number is a complete coincidence. The thermal conductivity effect is almost a factor of 10, there is a mild specific heat effect, and a stronger viscosity effect (factor of 3.5). The higher viscosity of the liquid tends to suppress turbulent fluctuations and thereby reduce the effectiveness of the heat transfer between the tube inner wall and inner fluid.

Table 11.2 Fluid properties for saturated liquid and vapor water at 2 bar pressure (and their ratio)

H$_2$O properties			
Property	Sat'd liquid @ 2 bar	Sat'd vapor @ 2 bar	Ratio (liquid to vapor)
k (W/m/K)	0.685	0.016	42.8
ρ (kg/m^3)	959	1.23	779.8
c_p (J/kg)	4,248	1,970	2.2
μ (kg/s/m)	2.82E$-$04	1.20E$-$05	23.5
α (m^2/s)	1.68E$-$07	6.60E$-$06	0.025
ν (m^2/s)	2.94E$-$07	9.76E$-$06	0.030
Pr	1.75	1.48	1.2

FIXED PARAMETERS

ho	26.7	W/m2/K
D_i (m)	0.00772	m
t	0.00089	m
kpipe	400	W/m/K
L	2	m
hfg	2193000	J/kg
Tinf=	677	oC
Tw=	120.2	oC

ITERATION TABLE

	ENTRANCE REGION (Liquid)			EXIT REGION (Vapor)				RESISTANCES			
mdot (kg/s)	Reynolds #	Nusselt #	hinner (W/m2/K)	Reynolds #	Nusselt #	hinner (W/m2/K)	hinner AVERAGE	Rinner (oC/W)	Rwall (oC/W)	Router(oC/W)	Ui (W/m2/K)
0.00E+00	0	3.66	325	0	3.66	7.6	166.2	0.124	4.1E-05	0.627	27.4
3.38E-04	198	3.66	325	4643	23.07	47.8	186.3	0.111	4.1E-05	0.627	27.9
3.44E-04	201	3.66	325	4727	23.40	48.5	186.6	0.110	4.1E-05	0.627	27.9
3.44E-04	201	3.66	325	4729	23.41	48.5	186.6	0.110	4.1E-05	0.627	27.9
3.44E-04	201	3.66	325	4729	23.41	48.5	186.6	0.110	4.1E-05	0.627	27.9

Fig. 11.6 Spreadsheet used to calculate convection coefficients at the entrance and exit of the boiler tube, and then iteratively determine the mass flow rate

The spreadsheet used to iteratively determine the mass flow rate for the boiler is shown in Fig. 11.6. Fixed parameter values are compiled at the top. In the iteration table, the mass flow rate for the first iteration is set to zero. The Nusselt number correlation is calculated at the entrance and at the exit (vapor), the difference being in the fluid properties. An if-then-else statement is entered for the Nusselt number that using the Dittus/Boelter correlation with a lower limit of 3.66, the laminar flow value. The geometric average of the entrance and exit, and then the overall heat transfer coefficient (U_i, based on the inner diameter) is calculated. The mass flow is determined from the 1-node energy balance:

$$\dot{m}_w = \frac{T_\infty - T_w}{R\,h_{fg}} = \frac{U_i D_i (T_\infty - T_w)}{h_{fg}}$$

For the first iteration, where there is no inner flow, the corresponding convection coefficient values are 325 W/m^2/K for the inlet (liquid) and 7.6 W/m^2/K for the outlet (vapor). The average of these (166.2) is approximately six times as large as the outer coefficient so that the predicted overall heat transfer coefficient is controlled by the external convection. The mass flow rate is calculated using this overall coefficient for the next iteration. Now, the Reynolds number in the liquid region is 198, and the corresponding Nusselt number evaluates as the laminar flow limit (3.66). In the exit region, the Reynolds number is much larger (due to the lower absolute viscosity), and there is a much larger Nusselt number due to the strong forced convection. However, the convection coefficient (47.8) is still low

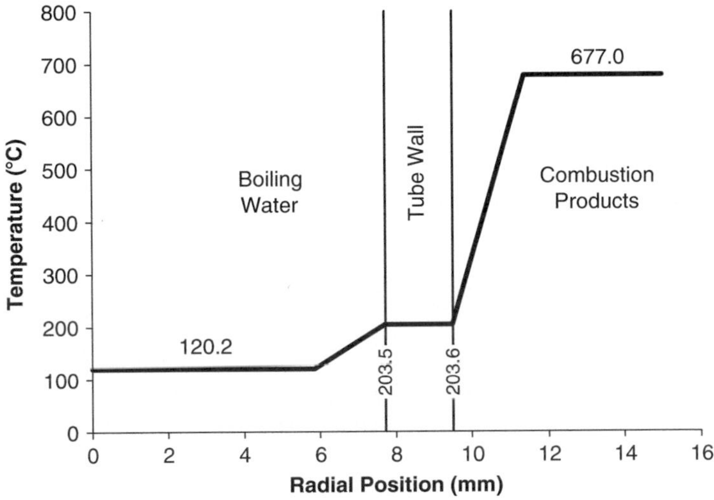

Fig. 11.7 Temperature vs. radial position in the boiler section

compared to the liquid side. The overall heat transfer coefficient for the second iteration differs by only a few percent from the first iteration because the outer convection dominates and is independent of flow rate inside the pipe. The iterative system converges very quickly to a self-consistent result with a final flow rate of:

$$\dot{m}_w = 3.44(10)^{-4}\text{kg/s}^* \left| \frac{3,600\,\text{s}}{\text{h}} \right| = 1.24\,\text{kg/h}$$

This seems like a low practical value (about 1.25 l/h of liquid water fed) and shows the challenge of designing practical boilers.

The radial temperature profile for the final converged solution is shown in Fig. 11.7. The thermal boundary layers are shown as straight lines (with a boundary layer thickness k/h). The conductive resistance across the tube wall is four orders of magnitude lower than the convective resistances, and therefore there is very little temperature change across the tube wall, which is maintained at 203.5 °C. The boiling action, conceptually, occurs in the inner boundary layer region, where the indicated temperature is above the boiling point and is a very complex phenomenon. The tube wall temperature is closer to the water temperature because the thermal resistance on the outside is approximately six times greater than the inside.

As a practical matter, the temperature of combustion product gases is frequently at a temperature well above the melting temperature of the containing vessel. This condition is only possible if the walls are cooled by the inner fluid, so that there is a tremendous jump in temperature across the thermal boundary layer between the hot gases and the containment wall. If, for some reason, the water flow was interrupted, a transient heating of the tube wall would occur, and could lead to rupture of the tube.

11.3   Momentum Considerations: Pressure Loss

An important secondary consideration for internal flows is the pressure drop due to fluid friction associated with the flow. Generally, the smaller the diameter (for a fixed mass flow), the higher the convection coefficient, hence heat transfer rates, but the higher the pressure drop, hence pumping requirements. The trade-off between pumping costs and heat transfer effectiveness can lead to an economic optimization problem. A brief outline is presented here of the pressure drop estimation, and fluid mechanics resources should be consulted if more detail on pressure effects is needed.

Pressure losses are categorized as major (for losses associated with a straight section of constant diameter) and minor (associated with changes in flow path area or direction). Considering the major losses, an overall force balance on a straight section of pipe (internal hydraulic diameter D and length L) with the inlet at state 1 (pressure P_1) and the exit at state 2 (P_2) is:

$$P_1\left(\pi D^2/4\right) = \tau_{wall}(\pi DL) + P_2\left(\pi D^2/4\right)$$

where τ_{wall} is the average shear stress that acts along the inner walls of the pipe (Fig. 11.8). Rearranging for the pressure drop:

$$P_1 - P_2 = \Delta P = 4\tau_{wall}\,\frac{L}{D}$$

The shear stress at the wall is expressed in terms of the Fanning friction factor (f) defined by:

$$\tau_{wall} = f\,\frac{1}{2}\,\rho V^2$$

where $\rho V^2/2$ is the dynamic pressure (the hypothetical pressure if the fluid were brought to rest without friction). The friction factor is a function of the Reynolds number of the flow, for incompressible flows and the relative roughness of the pipe (ε, the average roughness divided by the diameter). For laminar flows (Reynolds

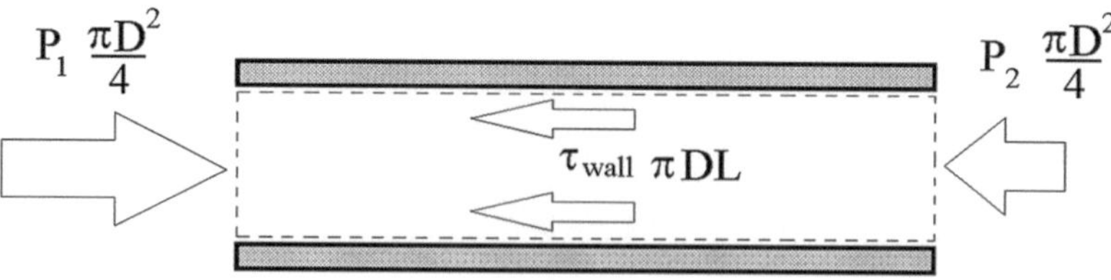

Fig. 11.8 Forces acting on a control volume in a straight section of pipe

Table 11.3 Pressure drop calculation for boiler section

mdot	3.44E−04 kg/s	
Di	0.00772 m	
ε, pipe roughness	2.00E−06 m	
ε/D, relative roughness	0.00026	
L	2 m	
	Entrance	Exit
ρ (kg/m^3)	959	1.23
Re	201	4,729
f (from Moody Chart)	0.080	0.010
dP/dx (Pa/m)	1.2	113.8
dP/dx*L (atm)	2.29E−05	2.25E−03

numbers below approximately 2,000), the friction factor is a closed-form expression:

$$f = 16/\mathrm{Re}_D \text{ for laminar flows, independent of roughness.}$$

For fully turbulent flows, the friction factor is a function of the relative roughness and is independent of Reynolds number. An implicit function, the Colebrook equation, can be used, or the friction factor can be obtained from a Moody chart. The pressure drop is therefore:

$$\Delta P = 2f\rho V^2 \frac{L}{D} = \frac{32f\dot{m}^2 L}{\pi^2 \rho D^5}$$

The second expression converted the velocity in terms of mass flow rate. The pressure drop is extremely sensitive to the inner diameter (fifth power).

In the boiler section, the density, velocity, and Reynolds number are very different at the exit than at the entrance. Therefore, it makes sense to consider the rate of change of pressure by setting the length L to a differential length (dx) and letting the finite change in pressure (ΔP) shrink to a differential change (dP). The rate of change of pressure with x is therefore:

$$\frac{dP}{dx} = \frac{2f\rho V^2}{D} = \frac{32f\dot{m}^2}{\pi^2 \rho D^5}$$

Results of a basic spreadsheet pressure drop calculation for the boiler section are shown in Table 11.3. The top block shows quantities that are invariant. The bottom block shows the relevant quantities at the entrance and exit sections. The friction factor was obtained from the laminar formula for the entrance case and from a Moody chart online (the Fanning-type, not the Darcy type) for the exit case. The pressure gradient (dP/dx) in units of Pascals per meter is shown to be higher at the exit than at the entrance, despite the lower friction factor, ultimately a density effect. However, when multiplied by the actual length, and converted to units of atmosphere, the pressure drop is of the order of 10^{-3} atm. The conclusion for this

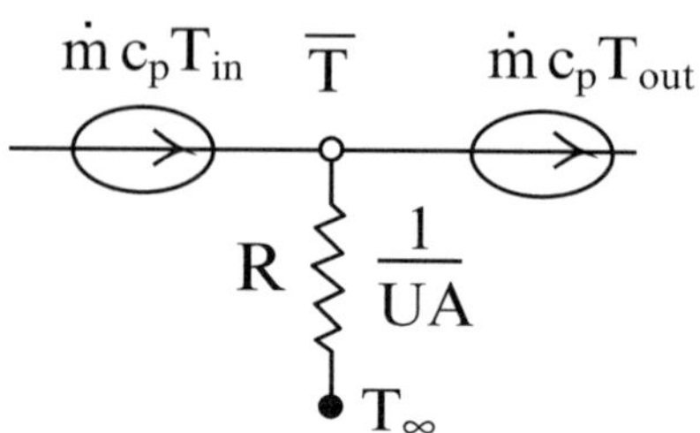

Fig. 11.9 Resistance network for preheater and superheater sections

case study is that the assumption of negligible pressure drop across the boiler is justified.

When there are changes in the flow path (turns or diameter changes), minor flow losses are introduced with a coefficient, K, multiplied by the dynamic pressure entering the element. Note that the magnitude of the minor losses can be larger than the major losses in an actual flow path. The total pressure drop is the sum of the pressure drops in all straight sections (major losses) and flow path elements (minor losses):

$$\Delta P = \sum_{\substack{major \\ section}} f\frac{L}{D}\frac{\rho V^2}{2} + \sum_{\substack{minor \\ element}} K\frac{\rho V^2}{2}$$

The minimum input power required to overcome this pressure drop, from an energy analysis, is:

$$W_{shaft,min} = \dot{m}\left(\frac{P_1}{\rho_1} - \frac{P_2}{\rho_2}\right)$$

11.3.1 Preheater and Superheater Lengths

Figure 11.9 shows a resistance network for the preheater and superheater sections considered as control volumes. Figure 11.5 also applies to this case. The central node, at temperature $\overline{T}$, represents a suitable average temperature of the water in the pipe in this section that is needed to calculate the rate of heat transfer between the outer fluid (hot gases) and inner fluid (liquid water in this section of pipe). Revisiting the problem statement for this application, water enters the preheater at 2 bar pressure as a compressed liquid (the inlet temperature, $T_{in} = 100\ °C$, is below the boiling point at that pressure of 120 °C). It carries flow energy into the pipe at the rate $\dot{m}_w(h_{in} + KE + PE)$, where the enthalpy consists of the internal energy of the inlet fluid (u) and the flow work (P/ρ) required to push it in. The fluid also has kinetic and potential energy. Since there is no phase change, the enthalpy (relative to a reference) can be related to the temperature through the specific heat. Neglecting kinetic and potential energy, the rate of energy flow in is $\dot{m}_w c_p (T_{in} - T_{ref})$. The reference temperature is not shown in Fig. 11.9 as it will be cancelled by the rate at which this reference enthalpy exits. The rate of energy flow out is expressed similarly. These energy flows refer to bulk temperatures at precise locations, where the fluid crosses the defined control volume.

Along the pipe length, heat is transferred from the hot gases outside (at T_∞) across the pipe wall through three thermal resistances in series (convection to outer wall, conduction across pipe, and convection to inner fluid). From a steady-state energy balance, the rate of energy flow out of the control volume is higher than the inlet energy flow rate by that amount. Unlike the boiler section, where the water temperature remains at the boiling temperature, the temperature varies with position, and so does the local rate of heat transfer. Treating the entire section as a single lumped element, the heat transfer rate can be approximated based on a suitably chosen average temperature.

Formally, the energy balance is:

$$C\frac{dE_{stored}}{dt} = \dot{m}_w c_p (T_{in} - T_{ref}) - \dot{m}_w c_p (T_{out} - T_{ref}) + \frac{T_\infty - \overline{T}_w}{R}$$

where C is the total heat capacitance of the system, and E_{stored} represents the energy stored within the control volume. For steady-state operation, the rate of change of energy stored is zero (or there is a balance between rates of energy crossing the boundary), and the energy balance can be rearranged in the following format, introducing an overall heat transfer coefficient (U):

$$UA = \frac{1}{R} = \dot{m}_w c_p \frac{(T_{out} - T_{in})}{(T_\infty - \overline{T}_w)}$$

The thermal resistance (this time based on the outer area, A_o, for variety) is the sum of the inner convective, pipe wall conductive (using the exact cylindrical form this time), and outer convective resistances:

$$R = \frac{1}{h_o \pi D_o L} + \frac{\ln(D_o/D_i)}{2\pi k_{pipe} L} + \frac{1}{h_i \pi D_i L} = \frac{1}{\pi D_o L}\left(\frac{1}{h_o} + \frac{D_o \ln(D_o/D_i)}{2 k_{pipe}} + \frac{D_o}{h_i D_i}\right)$$

The term in parentheses is the inverse of the overall heat transfer coefficient, based on the outer surface area.

$$\frac{1}{U_o} = \left(\frac{1}{h_o} + \frac{D_o \ln(D_o/D_i)}{2 k_{pipe}} + \frac{D_o}{h_i D_i}\right)$$

A closed-form expression for the pipe length is therefore:

$$L = \frac{\dot{m}_w c_p}{\pi D_o U_o} \frac{(T_{out} - T_{in})}{(T_\infty - \overline{T}_w)}$$

In this expression, for the defined problem, three of the four temperatures are well defined, while the average fluid temperature along the pipe is not. The fluid temperature is bound between the inlet and outlet temperatures, however. Figure 11.10 shows a calculation of the preheater and superheater lengths that determines a minimum, a maximum, and an average length, depending on what is chosen for the average water temperature. An underestimate, L_{min}, sets the average water temperature equal to the inlet water temperature. That model uses the largest possible temperature difference driving the heat all along the pipe length. An overestimate, L_{max}, sets the average temperature to the outlet temperature. An average length uses the geometric average of the inlet and outlet temperatures. What is apparent for this case study is that the predicted length changes by only a few percent from minimum to maximum. That is not always the case.

The only differences between the preheater and superheater sections are that the inner convection coefficient is substantially lower in the latter (making U lower, and the length longer), the heat capacity of vapor is approximately half that of liquid

INPUT PARAMETERS

Do	0.0095	m
t	0.00089	m
hinner	325	W/m2/K
houter	26.7	W/m2/K
Tin	100	oC
Tout	120	oC
Tinf	677	oC
mdot	0.000344	kg/sec
cp	4200	J/kg/K
kpipe	400	W/m/K

DERIVED PARAMS

Di	0.00772	m
U	24.2	W/m2/K
L_{min}	0.0692	m
L_{max}	0.0717	m
L_{ave}	0.0704	m

INPUT PARAMETERS

Do	0.0095	m
t	0.00089	m
hinner	48.5	W/m2/K
houter	26.7	W/m2/K
Tin	120	oC
Tout	164	oC
Tinf	677	oC
mdot	0.000344	kg/sec
cp	1970	J/kg/K
kpipe	400	W/m/K

DERIVED PARAMS

Di	0.00772	m
U	15.9	W/m2/K
L_{min}	0.1127	m
L_{max}	0.1224	m
L_{ave}	0.1173	m

Fig. 11.10 Spreadsheet calculation of preheater (left) and superheater (right) lengths

(making the length shorter) and the inlet and outlet temperatures are different (a larger temperature change, making the length longer). The average temperature difference driving the heat is also lower in this section, since the water temperature is closer to T_∞. The lengths of the preheater and superheater are both short compared to the length of the boiler section (2 m, an input to the problem) and that is primarily due to the much lower enthalpy change required in these sections (Fig. 11.4).

The way that open system problems are approached depends on the way a given problem is formulated. In this case, there were three sections of pipe. The length of the boiler section (2 m) was an input that was defined by a specific design objective. The analysis of that section results in a determination of the required mass flow rate. That flow rate became an input to the other two sections, whose inlet and outlet temperatures were also defined. Also, all the inlet and outlet temperatures were specified, and the required lengths determined. It is common for a given length of pipe and inlet temperature and flow rate to be specified and the resulting exit temperature determined. This example is intended to show several, but not all possible, variations.

11.4 Internal Heat Transfer Flow Models

In this section, a 1D steady-state heat transfer flow situation is defined in which fluid steadily enters a pipe of constant inner diameter D_i, wall thickness t, and length L at a mass flow rate $\dot{m}$ and inlet temperature T_{in} (Fig. 11.11). The pipe is exposed to an outer fluid at temperature T_∞ (constant along entire length). The fluid exit temperature is not specified in the problem statement (in contrast to the preheater and superheater of the previous section). The fluid does not undergo phase change, so any change in the fluid enthalpy changes its temperature. Posed with these inputs, there are three unknown quantities: the fluid exit temperature, the total heat transfer rate, and the average temperature of the fluid in the pipe.

Two of the three required relationships to solve for the three unknowns are an overall thermodynamic energy balance that relates the energy flows to the total heat transfer rate, and a heat transfer relationship that relates the total heat transfer rate to

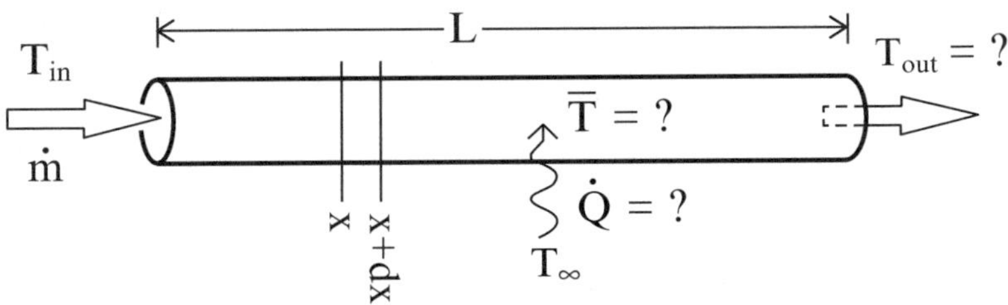

Fig. 11.11 Schematic diagram for internal flow heat transfer models

the average fluid temperature. The third relationship needed for closure requires an assumption for the average fluid temperature. Three 1-node models are developed, and their model predictions are then compared to a differential 1D flow model, which is considered to be an exact solution for comparison. That model is developed first.

11.4.1 Differential 1D Flow Model

A section of pipe of differential length dx is considered. A steady-state energy balance in the form energy in $=$ energy out is given by:

$$\dot{m}c_p\left(T_{(x)} - T_{ref}\right) + U\pi D_i dx\left(T_\infty - \overline{T}\right) = \dot{m}c_p\left(T_{(x+dx)} - T_{ref}\right)$$

In this expression, $T_{(x)}$ is the bulk temperature at location x (the inlet); $T_{(x+dx)}$ is the bulk temperature at the exit of the differential section; $\overline{T} \approx \left(T_{(x)} + T_{(x+dx)}\right)/2$ is the average temperature of the fluid in the section; and T_{ref} is a reference temperature, which cancels. The overall heat transfer coefficient is that based on the inner diameter (unlike the previous example, which based U on the outer diameter). Expressing the temperature at the exit as a Taylor expansion from the inlet:

$$T_{(x+dx)} = T_{(x)} + \frac{dT}{dx}\bigg|_x dx + \ldots$$

Expanding the energy balance:

$$\dot{m}c_p\left(T_{(x)}\right) + U\pi D_i dx\left(T_\infty - \frac{1}{2}\left(T_{(x)} + T_{(x)} + \frac{dT}{dx}dx\right)\right) = \dot{m}c_p\left(T_{(x)} + \frac{dT}{dx}dx\right)$$

Rearranging:

$$\frac{dT}{dx}\left(1 - \frac{1}{2}\left(\frac{U\pi D_i}{\dot{m}c_p}\right)dx\right) = \left(\frac{U\pi D_i}{\dot{m}c_p}\right)(T_\infty - T)$$

The second term on the left hand side (from the expansion of the average temperature in the heat transfer term) is a higher order term than the other two terms in the equation and becomes negligible in comparison to unity for a differentially sized element (i.e., as dx goes to zero). Another way of saying that is that the average temperature is approximately equal to the incoming fluid, which is going to be defined in the next section as a "thermally short" section. The preheater and superheater in the previous section behaved that way because the prescribed change in temperature through the pipe was small compared to the temperature difference between T_∞ and the entrance fluid temperature.

For the problem as stated, the outer fluid temperature is constant along the entire length of the duct (as opposed to heat exchangers where this temperature also varies with position), and therefore $dT_\infty/dx = 0$, and the energy balance can be expressed with $dT = d(T - T_\infty)$ as:

$$\frac{d(T - T_\infty)}{dx} = \left(\frac{U\pi D_i}{\dot{m}c_p}\right)(T_\infty - T)$$

The term in parentheses has units of inverse length and is defined as the thermal length, the same expression that was derived with a Lagrangian viewpoint:

$$L_t = \frac{\dot{m}c_p}{U\pi D_i}$$

Separating variables and applying limits from the entrance to an arbitrary point x within the pipe:

$$\int_{T_{in}}^{T} \frac{d(T - T_\infty)}{(T - T_\infty)} = -\int_0^x \frac{dx}{L_t}$$

Assuming the thermal length to be constant (the heat transfer coefficients could vary with position if entrance effects are important or fluid properties change substantially), the final result for temperature as a function of position (x) is:

$$\frac{T - T_\infty}{T_{in} - T_\infty} = e^{\frac{-x}{L_t}}$$

This solution is considered to be the exact solution of the 1D flow situation, but not a complete solution to the flow situation. The temperature T represents the average (bulk) temperature of the fluid at a particular position along the pipe. There must be 2D effects because this average fluid temperature is different from the temperature of the inner wall surface in order for heat to flow across the inner resistance. This expression has a very similar appearance and underlying physics to the lumped capacity transient temperature, but with time replaced with position, and time constant replaced with thermal length.

Consider the limits. The exit temperature is obtained by setting x to the pipe length (L):

$$\frac{T_{out} - T_\infty}{T_{in} - T_\infty} = e^{\frac{-L}{L_t}}$$

If the pipe length is much shorter than the thermal length ($L \ll L_t$), then the exit temperature is approximately equal to the inlet temperature. There is heat transfer, but it does not change the temperature of the inner fluid appreciably. That is "thermally short" behavior. If the pipe length is much longer than the thermal length, then the exit temperature approaches the outer fluid temperature (T_∞). The fluid has achieved thermal equilibrium with the outer fluid and is considered to be "thermally long" behavior. These concepts will be applied to the 1-node models in the next section.

The total heat transfer rate is determined by:

$$\dot{Q} = \dot{m}c_p(T_{out} - T_{in}) = \dot{m}c_p\left(T_\infty + (T_{in} - T_\infty)e^{\frac{-L}{L_t}} - T_{in}\right)$$

$$= \dot{m}c_p(T_\infty - T_{in})\left(1 - e^{\frac{-L}{L_t}}\right)$$

In order to compare models, it will prove useful to normalize the actual heat transfer rate by the maximum possible heat transfer rate, which is what would occur if the inner fluid were everywhere equal to the inlet temperature:

$$\hat{\dot{Q}} = \frac{\dot{Q}}{\dot{Q}_{max}} = \frac{\dot{Q}}{UA_i(T_\infty - T_{in})} = \frac{\dot{m}c_p}{UA_i}\left(1 - e^{\frac{-L}{L_t}}\right) = \frac{L_t}{L}\left(1 - e^{\frac{-L}{L_t}}\right)$$

Defining a dimensionless length as the pipe length normalized by the thermal length:

$$\hat{L} = \frac{U\pi D_i L}{\dot{m}c_p} = \frac{UA}{\dot{m}c_p} = \frac{L}{L_t}$$

An expression for the dimensionless heat transfer rate is therefore a function of the dimensionless length:

$$\hat{\dot{Q}} = \frac{\left(1 - e^{-\hat{L}}\right)}{\hat{L}}$$

This function, along with the results of the following 1-node models to follow, is shown in Fig. 11.12 (top figure). The dimensionless temperature is shown in the bottom figure.

11.4.2　1-Node Flow Models

The 1-node resistance model shown in Fig. 11.9 applies to this case. An overall energy balance is given by:

$$\dot{Q} = \dot{m}c_p(T_{out} - T_{in})$$

This thermodynamic relation has two unknowns ($\dot{Q}$ and T_{out}). It is useful, and common, in heat exchanger applications to define the maximum possible heat transfer rate that could be obtained, and that occurs simply when the fluid exits at

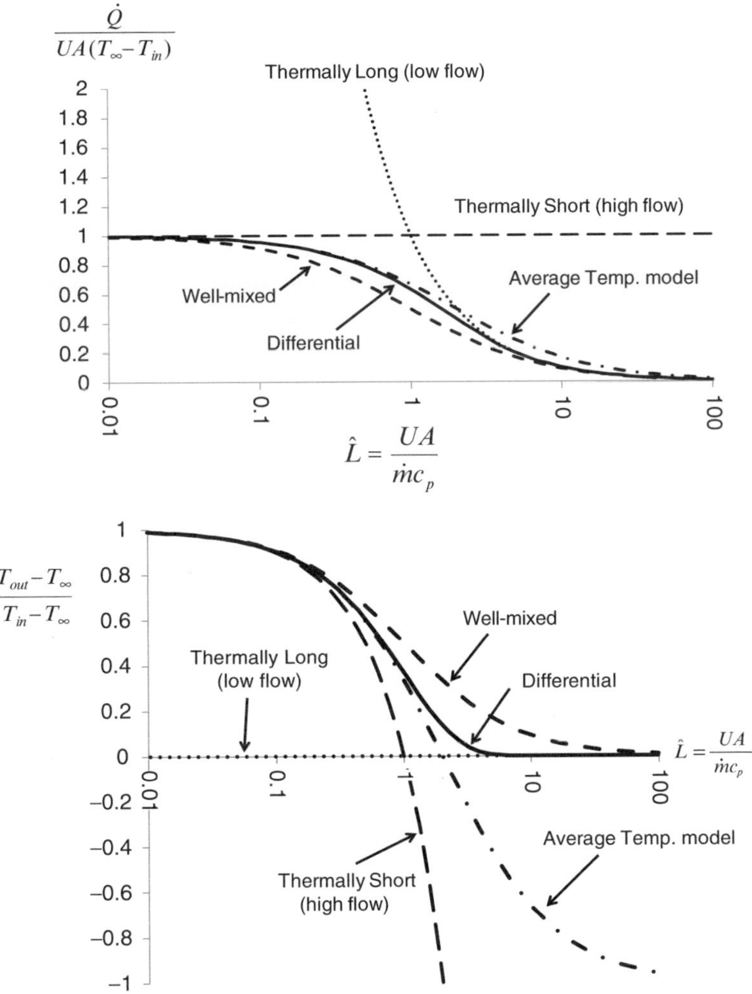

Fig. 11.12 Dimensionless heat transfer rate (*top*) and exit temperature (*bottom*) vs. dimensionless pipe length for several models

the ambient temperature, that is, for thermally long behavior. The maximum possible heat transfer rate is:

$$\dot{Q}_{\max} = \dot{m}c_p(T_\infty - T_{in})$$

This concept will apply to the "number of transfer units" (NTU) approach in heat exchanger applications in the next chapter.

The heat transfer rate between the outer and inner fluids is given by an overall heat transfer coefficient (U) across a defined surface area (A) as:

$$\dot{Q} = UA\left(T_\infty - \overline{T}\right)$$

This relation introduces the average temperature of the fluid. In the differential solution, this temperature was taken to be the geometric average between the inlet and outlet, but upon Taylor expansion, and shrinking the element size (dx → 0), the average temperature becomes approximately equal to the inlet temperature, for the purposes of calculating the heat transfer rate. In the 1-node model, the entire pipe length is taken as the element. In other words, the element is not differentially small, and the average temperature can be treated differently, as is shown in the following four cases:

11.4.2.1 Thermally Short Model

In the thermally short model, the temperature change of the fluid ($T_{out} - T_{in}$) is small compared to the temperature difference that drives heat $\left(T_\infty - \overline{T}\right)$. Heat is transferred to the fluid, but it does not appreciably change its temperature. In this case, a good approximation is that the average temperature equals the inlet temperature. The overall heat transfer rate, then, can be expressed directly as:

$$\dot{Q} = UA(T_\infty - T_{in})$$

Expressed in dimensionless form, normalized by the maximum possible heat transfer rate:

$$\frac{\dot{Q}}{\dot{Q}_{max}} = \frac{\dot{Q}}{UA_i(T_\infty - T_{in})} = 1$$

Or, simply

$$\hat{\dot{Q}} = 1$$

Note that there is in fact a change in fluid temperature predicted by this model. The average temperature was set equal to the inlet temperature for the purpose of estimating the heat transfer rate. The thermodynamic relationship yields the exit temperature:

$$T_{out} = T_{in} + \frac{\dot{Q}}{\dot{m}c_p} = T_{in} + \frac{UA}{\dot{m}c_p}(T_\infty - T_{in})$$

In terms of the dimensionless length:

$$T_{out} = T_{in} + \hat{L}\left(T_\infty - T_{in}\right)$$

$$\frac{T_{out} - T_\infty}{T_{in} - T_\infty} = 1 - \hat{L}$$

Consider the limits. For a pipe length much less than the thermal length ($\hat{L} \ll 1$), the exit temperature approaches the inlet temperature. For a pipe length equal to the thermal length, the exit temperature equals the outer fluid temperature, and the model is not expected to be very good. The predicted outlet temperature overshoots the outer fluid temperature for any pipe length greater than the thermal length. In short, the thermally short model is expected to be reasonable for situations where the pipe length is actually short compared to its thermal length, but violates the Second Law of Thermodynamics for pipe lengths that are longer than the thermal length.

11.4.2.2 Thermally Long Model

In the thermally long model, the fluid inside the pipe is considered to reach a thermal equilibrium with the outer fluid. This condition is attained by setting the exit temperature equal to the outer fluid temperature, and the thermodynamic balance is:

$$\dot{Q} = \dot{m}c_p(T_\infty - T_{in})$$

In dimensionless terms:

$$\frac{\dot{Q}}{\dot{Q}_{max}} = \frac{\dot{Q}}{UA_i(T_\infty - T_{in})} = \frac{\dot{m}c_p}{UA}$$

Or simply,

$$\hat{\dot{Q}} = \frac{1}{\hat{L}}$$

The dimensionless heat transfer exceeds the maximum possible for pipe lengths that are shorter than the thermal length, in violation of the Second Law. For pipe lengths greater than the thermal length, a reasonable trend is apparent. So the thermally long model gives reasonable trends for pipes that are long compared to the thermal length.

11.4.2.3 Average Temperature Model

It might seem that an improved model would be to set the average fluid temperature equal to the average of the inlet and exit temperatures, that is:

$$\dot{Q} = UA\left(T_\infty - \frac{T_{in} + T_{out}}{2}\right)$$

$$UA\left(T_\infty - \frac{T_{in} + T_{out}}{2}\right) = \dot{m}c_p(T_{out} - T_{in})$$

Solving for the exit temperature (and skipping details, a good exercise...):

$$T_{out} = \frac{\hat{L}\,T_\infty + T_{in}\left(1 - \hat{L}/2\right)}{1 + \hat{L}/2}$$

The dimensionless exit temperature is:

$$\frac{T_{out} - T_\infty}{T_{in} - T_\infty} = \frac{2 - \hat{L}}{2 + \hat{L}}$$

For thermally short pipes, the outlet temperature approaches inlet temperature. For thermally long pipes, the outlet temperature does not approach ambient. That is:

$$\lim_{\hat{L} \to \infty} T_{out} = 2T_\infty - T_{in}$$

The outlet temperature overshoots the outer fluid, but remains finite. The dimensionless heat transfer rate is:

$$\hat{Q} = \frac{2}{2 + \hat{L}}$$

This model gives good agreement with the exact model across thermally long and short cases, but does not give a good prediction for the exit temperature for thermally long cases. This situation is similar to the case in transient numerical analysis when a central difference approximation is used to express the first derivative of temperature with time. It is a more accurate numerical approximation than a backward difference when the numerical step size is sufficiently small, but it gives wrong limits for large step sizes.

11.4.2.4   Well-Mixed Thermal Model

In this model, the exit temperature is set equal to the average temperature. This model is called a well-mixed model (borrowing from stirred reactor chemical models) because it mimics the behavior of the type of case shown in Fig. 11.13 which is typical of, say, conditioned air entering a room through inlet vents, with the air in the room being well mixed (actively or passively), and exiting through open areas (i.e., underneath doors, through cracks in the wall) that are far removed from the inlet. There is a region of mixing near the inlet, but the fluid in the room is at a mixed average temperature. There is simultaneous heat exchange with the walls, but the temperature of the fluid as it exits equals the average well-mixed temperature.

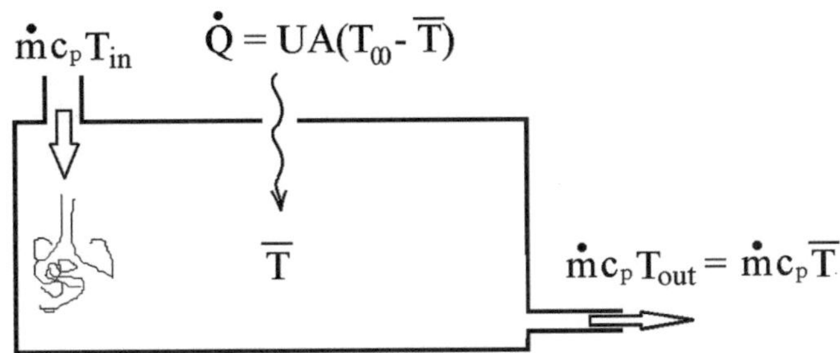

Fig. 11.13 Schematic of a well-mixed internal flow

The heat transfer rate is therefore:

$$\dot{Q} = UA\left(T_\infty - \overline{T}\right) = UA(T_\infty - T_{out})$$

Combining with the thermodynamic balance and rearranging for the outlet temperature:

$$T_{out} = \frac{UAT_\infty + \dot{m}c_pT_{in}}{UA + \dot{m}c_p} = \frac{\hat{L}T_\infty + T_{in}}{\hat{L} + 1}$$

Expressed as a dimensionless temperature:

$$\frac{T_{out} - T_\infty}{T_{in} - T_\infty} = \frac{1}{\hat{L} + 1}$$

The outlet temperature approaches the outer fluid for long pipes, and the inlet for short pipes, that is, the temperature exhibits the correct limits.

The dimensionless heat transfer rate is

$$\hat{\dot{Q}} = \frac{1}{1 + \hat{L}}$$

As shown in Fig. 11.12, this model underestimates the heat transfer rate and overestimates the dimensionless exit temperature, but gives reasonable trends over the entire range, and reasonable trends for the exit temperature. It is therefore considered to be the best general 1-node model to use for internal flow problems. It also can be advantageous in more detailed numerical simulations, as numerical stability is assured.

11.4.3 Pipe Flows with Constant Heat Flux

A common internal flow situation is to force heat into a fluid by, for example, wrapping an electrical heating tape around the outside of the pipe. Electrical energy is converted to thermal energy and heat is driven into the fluid. The same thermodynamic and heat transfer relations apply, but heat transfer rate is a known input quantity. The exit temperature and the temperature of the heating element (the role of ambient temperature) is unknown (and changes with position).

The total heat transfer rate forced into the fluid is given by $\dot{Q}$, and it is uniformly delivered across the entire length (L) of the pipe. For a differential element of length dx, a differential heat rate is added ($d\dot{Q}$), which is a constant along the pipe. It follows that a heating intensity (heat transferred per unit length) is given by:

$$\frac{d\dot{Q}}{dx} = \frac{\dot{Q}}{L}$$

The thermodynamic balance for this element is:

$$d\dot{Q} = \dot{m}c_p\left(T_{bulk(x+dx)} - T_{bulk(x)}\right)$$

Expanding in a Taylor expansion:

$$d\dot{Q} = \dot{m}c_p\left(T_{bulk(x)} + \frac{dT_{bulk}}{dx}dx + ,,, - T_{bulk(x)}\right) = \dot{m}c_p\frac{dT_{bulk}}{dx}dx$$

Rearranging:

$$\frac{dT_{bulk}}{dx} = \frac{d\dot{Q}}{dx}\frac{1}{\dot{m}c_p} = \frac{\dot{Q}}{L\dot{m}c_p}$$

The rate of change of bulk temperature with distance is a constant, and therefore, on integration, the bulk temperature increases linearly with x:

$$T_{bulk(x)} = T_{inlet} + \frac{\dot{Q}x}{L\dot{m}c_p}$$

The heat transfer balance is:

$$d\dot{Q} = U_i\pi D_i dx\left(T_{element,(x)} - T_{bulk,(x)}\right)$$

The overall heat transfer coefficient (U_i) based on the inner diameter (D_i) represents a series thermal resistance channel across any insulation around the heating tape, the wall of the pipe, and inner convection (assuming the temperature of the element itself is uniform, although conductive resistance of the heating element could be included). Solving for the temperature of the heating element:

$$T_{element,(x)} = T_{bulk,(x)} + \frac{\dot{Q}}{LU_i\pi D_i}$$

The element temperature is always hotter than the fluid. The more effective the heat transfer (U_i) and the larger the diameter, the closer the element temperature is to the fluid temperature.

Workshop 11.1. Finned Pipe

Water at temperature T_1, pressure P_1 enters an aluminum-finned steel pipe (dimensions defined in sketch) with a mass flow rate of $\dot{m}$ and flows with negligible pressure drop. The pipe is exposed to air at T_∞ with a cross-flow velocity (not shown in sketch) of V_{air}. For the parameter values given, determine the heat transfer rate ($\dot{Q}$), length of pipe (L), and number of fins (n_{fins}) if the exit is a saturated vapor. Treat the external flow as a forced convection over a flat plate of length w_{fin} and neglect radiation effects. Treat the internal flow as forced convection with fully developed flow.

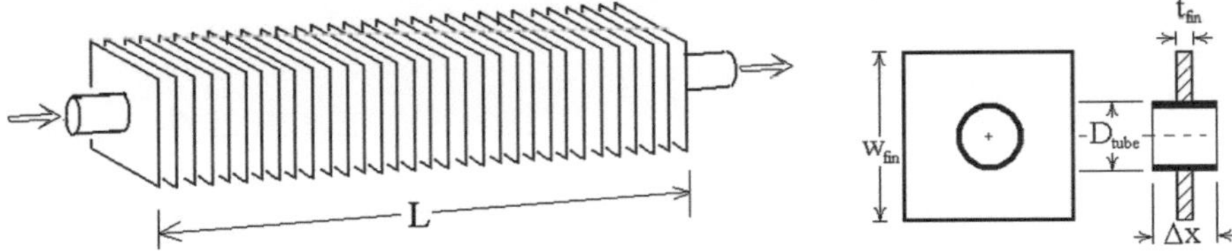

Default parameter values

Quantity	Symbol	Value	Unit
Mass flow rate	$\dot{m}$	0.02	kg/s
Ambient temperature	T_∞	25	°C
Inlet temperature	T_1	150	°C
Internal fluid pressure	P_1	1.0	Atm
Air cross-flow velocity	V_{air}	10	m/s
Pipe thermal conductivity	k_{pipe}	43	W/m/°C
Fin thermal conductivity	k_{fin}	212	W/m/°C
Pipe outer diameter	D_{pipe}	0.05	m
Pipe thickness	t_{pipe}	0.00655	m
Fin thickness	t_{fin}	0.00159	m
Fin width	w_{fin}	0.19	m
Fin spacing	Δx	0.01	m

Results

Quantity	Symbol	Value	Unit
Overall heat transfer coefficient (based on inner area)	U_i		W/m^2/K
Total heat transfer rate	$\dot{Q}$		Watts
Pipe length	L		m
Number of fins (round up)	n_{fins}		–

A detailed model development follows. It is a good idea to think about this problem and tackle it first before reading on.

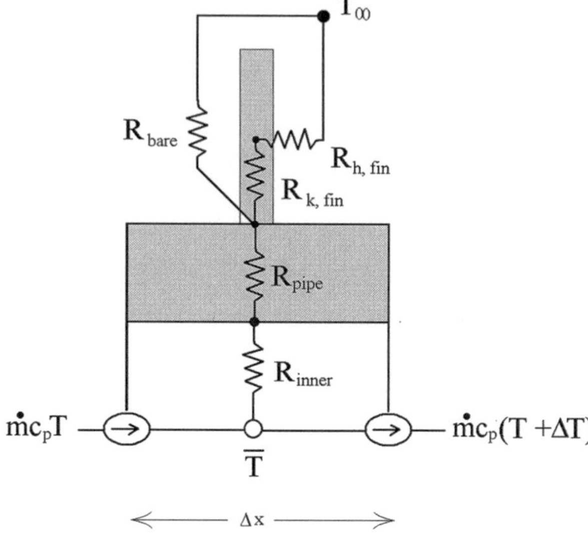

Fig. 11.14 Section detail. *Note*: the average fluid temperature is the fluid temperature at this section, which varies with distance along the pipe

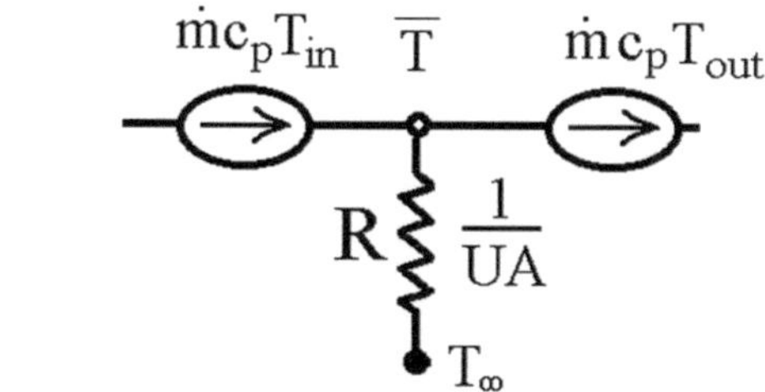

Fig. 11.15 Overall equivalent circuit. *Note*: the average fluid temperature represents the average between inlet and outlet of the pipe

Model Development

Use a 1-node fin resistance model (Fig. 11.14) and a 1-node overall flow model (Fig. 11.15). Since the inlet and exit temperature are specified (the exit is a saturated vapor), the average temperature can be set equal to the geometric average for the most accuracy. A single closed-form expression can be obtained that includes all the fins acting in parallel, or a more brute force computer simulation that steps through the pipe fin by fin until the final desired temperature is achieved.

Workshop 11.2. Solar Pipe in Continuous Mode: 2-Node Model

Water at temperature $T_1 = 15\ °C$ flows into a copper pipe ($k_{pipe} = 400$ W/m/K) of outer diameter $D = 0.0155$ m, thickness $t_{pipe} = 0.002$ m, length $L = 1.35$ m and is exposed to solar radiation (concentrated with a parabolic mirror) with an insolation value of $I_0 = 500$ W/m^2 and a solar collection width to rod diameter ratio of $n = 19.1$. An optical efficiency $\eta_{opt} = 0.9$ (defined as the ratio of the solar radiation that strikes the pipe to the collected radiation) accounts for collected solar energy not absorbed by the pipe. The angle between the incoming solar rays and the pipe is $\theta = 20°$. The absorption coefficient of incident radiation is $\alpha = 0.85$. The rod is exposed to air at a temperature $T_\infty = 25$ °C with a natural convection coefficient (from a Nusselt number correlation) that has a mild temperature dependency given by:

$$h_{c,outer} = 1.32 \left(\frac{T - T_\infty}{D} \right)^{1/4} \text{ W/m}^2/\text{K}$$

where T is the surface temperature of the pipe. The emissivity of the rod surface is $\varepsilon = 0.9$. The average temperature of the walls with which the rod exchanges thermal radiation is $T_{w\infty} = 30$ °C. The convection coefficient on the inside of the pipe can be approximated using the larger of Dittus/Boelter correlation or laminar flow lower limit. That is, the Nusselt number (SI units) is the higher of:

$$Nu_D = \frac{h_{inner}(D - 2t)}{k} = 0.023 \left(\frac{\rho V (D - 2t)}{\mu} \right)^{0.8} \left(\frac{\mu c_p}{k} \right)^{0.4} \qquad OR \qquad Nu_D = 3.66$$

For water properties, use $\rho = 1{,}000 \text{ kg/m}^3$, $c_p = 4{,}200 \text{ J/kg/K}$, $k = 0.6 \text{ W/m/K}$, and $\mu = 5.0(10)^{-4} \text{ kg/m/s}$. Assume the system is pressurized sufficiently to prevent water from changing phase.

Calculate, as a function of volumetric flow rate (AV):

- The steady-state average temperature of the pipe (T_{pipe}, in °C).
- The exit temperature of the water (T_2, in °C).
- The minimum pressure required to prevent phase change.
- The rate of heat transfer to the water (in Watts).
- The thermal efficiency, defined as the ratio of heat transfer rate to the water divided by the collected solar flux (solar radiation rate entering top plane).

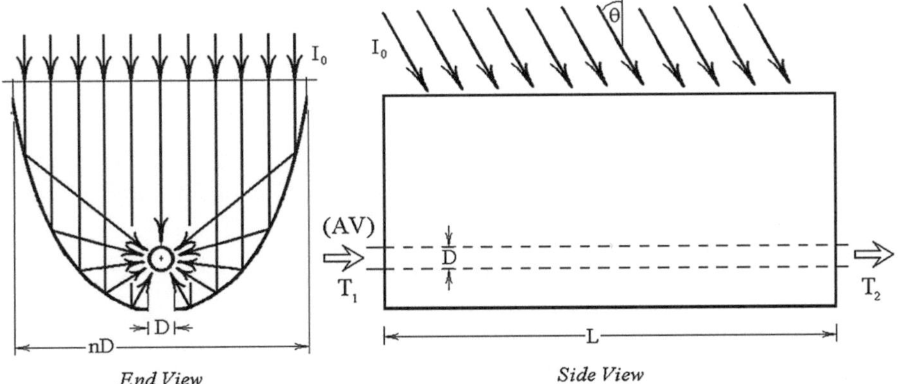

Think about and tackle yourself before reading on...

2-Node Well-Mixed Flow Model

The thermal resistance network shown for this system defines an absorbing surface node (the outer surface of the pipe, at an average temperature T_{pipe}), which receives solar radiation (modeled as a current source). This surface is in thermal contact with the ambient air surrounding it (at temperature T_∞) and the surrounding walls (at $T_{w\infty}$). Note that there are two very different radiation modes at play; the absorption of solar radiation (modeled as a current source), and thermal radiation of the pipe surrounded by the local environment (modeled as a resistance). The absorbing surface is in thermal contact with the fluid flowing through it (at an average temperature $\overline{T}_w$) through a conductive resistance across the pipe in series with an inner convective resistance. The energy flux associated with the water flow is modeled as current sources entering and leaving. A nodal energy balance on the absorbing surface node is:

$$0 = \dot{Q}_{abs} + \frac{T_\infty - T_{pipe}}{R_{h\infty}} + \frac{T_{w\infty} - T_{pipe}}{R_{r\infty}} + \frac{\overline{T}_w - T_{pipe}}{R_{pipe} + R_{h,inner}}$$

In the well-mixed flow model (which is numerically stable for all cases, and yields correct high and low flow limits), the average water temperature for the heat transfer term is set equal to the exit temperature ($\overline{T}_w = T_2$).

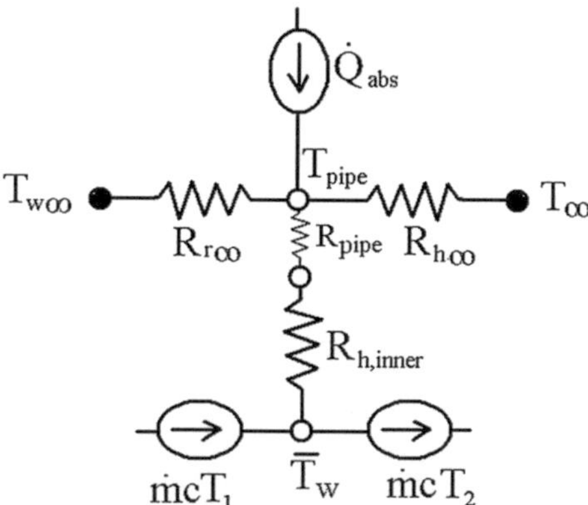

Using Newton's Law of Cooling and the Stefan–Boltzmann law for radiation (with $\sigma = 5.669(10)^{-8}$ W/m^2/K^4), the thermal resistances are defined as:

$$R_{h\infty} = \frac{1}{h_{c,outer}\pi DL}$$

$$R_{r\infty} = \frac{1}{h_r \pi D L}$$

$$= \frac{1}{\varepsilon\sigma\left[(T_{pipe} + 273)^2 + (T_{w\infty} + 273)^2\right](T_{pipe} + 273 + T_{w\infty} + 273)\pi D L}$$

$$R_{hinner} = \frac{1}{h_{c,inner}\pi(D - 2t_{pipe})L}$$

$R_{pipe} = \frac{t_{pipe}}{k_{pipe}\pi(D - t_{pipe})L}$ (thin-wall approximation, which is valid here)

The rate of heat absorption is defined by:

$$\dot{Q}_{abs} = \alpha\eta_{opt}nDLI_0\cos\theta$$

The nodal equation can be rearranged into a form suitable for the method of successive substitution, with a "k" superscript to indicate iteration number, noting that the thermal resistances between the outer pipe wall and ambient are functions of temperature:

$$T_{pipe}^{(k+1)} = \frac{\dot{Q}_{abs} + \frac{T_\infty}{R_{h\infty}^{(k)}} + \frac{T_{w\infty}}{R_{r\infty}^{(k)}} + \frac{T_2^{(k)}}{R_{pipe} + R_{h,inner}}}{\frac{1}{R_{h\infty}^{(k)}} + \frac{1}{R_{r\infty}^{(k)}} + \frac{1}{R_{pipe} + R_{h,inner}}}$$

An energy balance on the water node is:

$$\dot{m}c_p(T_2 - T_1) = \frac{T_{pipe} - \overline{T}_w}{R_{pipe} + R_{h,inner}}$$

where the mass flow is $\dot{m} = \rho A V$. Invoking the well-mixed model ($\overline{T}_w = T_2$), and rearranging for the exit water temperature in a form suitable for Method of Successive Substitution (note that the Jacobi method uses "k" and Gauss–Seidel uses "k + 1"):

$$T_2^{(k+1)} = \frac{\frac{T_1}{R_{flow}} + \frac{T_{pipe}^{(k) \ or \ (k+1)}}{R_{pipe} + R_{h,inner}}}{\frac{1}{R_{flow}} + \frac{1}{R_{pipe} + R_{h,inner}}}$$

where a new type of thermal resistance has been introduced: $R_{flow} = \frac{1}{\dot{m}c_p}$. This thermal resistance only works in this way for the well-mixed model.

The thermal efficiency of the panel is defined as the heat transferred to the water divided by the incident solar flux entering the top:

$$\eta_{thermal} = \frac{\text{Heat to Fluid}}{\text{Heat Collected}} = \frac{\dot{m}c_p(T_2 - T_1)}{nDLI_0\cos\theta}$$

Workshop

In teams of three or four, write program code with:

- A commented title section.
- A commented "INPUT PARAMETER" section in which all parameters are defined and enter values from the problem statement.
- A commented "DERIVED PARAMETERS" in which quantities that do not depend on the flow rate or unknown pipe temperature are calculated (i.e., R_{pipe}).
- A commented "ITERATION LOOP" section that uses an outer FOR loop to step through AV (flow rate), and an inner WHILE loop in which the method of successive substitution is implemented, and tested for convergence.
- A commented "RESULTS" section that plots pipe outer temperature, pipe inner temperature, and fluid exit temperature vs. flow rate (report in units of liters per hour), thermal efficiency vs. flow rate, and thermal efficiency vs. exit temperature.

Workshop 11.3. Solar Pipe in Continuous Mode: Multi-node Model

The problem statement is the same as Workshop 11.2. A multi-node mode is developed and applied.

Multi-node Well-Mixed Flow Model

A more complex model is required to account for spatial variations in the pipe surface temperature. On the one hand, such a variation would seem to be important for flow rates, where there is a substantial change in water temperature with position. On the other hand, conduction axially in the pipe would tend to maintain a constant wall temperature, especially for highly conductive pipes (like copper).

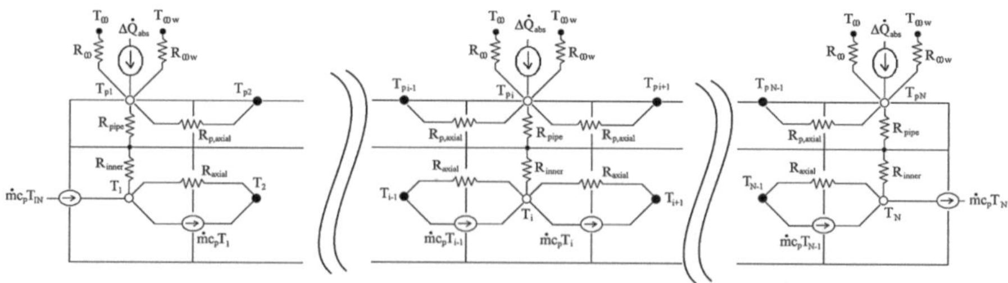

Consider the thermal resistance network shown (larger versions shown subsequently), in which the pipe and fluid spaces are broken into discrete elements (represented by nodes) and connected thermally by appropriate thermal resistances.

Each pipe surface node absorbs solar radiation, exchanges heat with the environment (by convection and radiation acting in parallel), with the adjacent inner fluid (by a conductive resistance across the pipe in series with a convective resistance inside the pipe), and with the neighboring pipe nodes (by a conductive resistance through the pipe wall). The end segments (T_{p1}, T_{pN}, T_1, and T_N) differ from the interior nodes (T_{p2}...T_{pN} and T_2...T_N) in that they communicate with only one neighbor. Each fluid node is in thermal contact with the pipe outer surface through a series thermal resistance across the pipe wall and convection on the inner surface. There is an inlet flow and outlet flow to each fluid node and the well-mixed model sets the average temperature equal to the outlet temperature. Finally, heat is conducted forward and backward through an axial conductive resistance (in parallel with the flow). This latter mechanism can be important for highly conductive pipes like copper. The forward and backward conduction in the fluid is in parallel with the flow and is a mechanism whose relative importance will be shown to depend on the size of the spatial grid.

For a numerical choice of N segments (2N nodes), the grid spacing is given by:

$$\Delta x = \frac{L}{N}$$

All the resistance values and solar absorption term are the same as that in the 2-Node model, except that the dimension L is replaced with the grid spacing, Δx. The axial pipe resistance and axial resistance for the inner fluid are new to this model and are given by:

$$R_{p,axial} = \frac{\Delta x}{k_{pipe}(\pi/4)\left(D^2 - (D - 2t_{pipe})^2\right)} = \frac{\Delta x}{k_{pipe}\pi D t_{pipe}\left(1 - t_{pipe}/D\right)}$$

$$R_{axial} = \frac{\Delta x}{k_f \pi (D - 2t)^2/4}$$

A nodal energy balance on each of the 2N unknown temperatures, written in the form suitable for a Gauss–Seidel iteration, results in the following system of algebraic equations (nonlinear when the outer convection and radiation resistances are functions of temperature):

For the first pipe node:

$$T_{p1} = \frac{\Delta \dot{Q}_{abs} + \frac{T_\infty}{R_\infty} + \frac{T_{w\infty}}{R_{w\infty}} + \frac{T_{p2}}{R_{p,axial}} + \frac{T_1}{R_{pipe}+R_{h,i}}}{\frac{1}{R_\infty} + \frac{1}{R_{w\infty}} + \frac{1}{R_{p,axial}} + \frac{1}{R_{pipe}+R_{h,i}}}$$

For the interior pipe nodes ($i = 2$ to $N - 1$):

$$T_{pi} = \frac{\Delta \dot{Q}_{abs} + \frac{T_\infty}{R_\infty} + \frac{T_{w\infty}}{R_{w\infty}} + \frac{T_{p(i-1)}}{R_{p,axial}} + \frac{T_{p(i+1)}}{R_{p,axial}} + \frac{T_i}{R_{pipe}+R_{h,i}}}{\frac{1}{R_\infty} + \frac{1}{R_{w\infty}} + \frac{2}{R_{p,axial}} + \frac{1}{R_{pipe}+R_{h,i}}}$$

For the last pipe node:

$$T_{pN} = \frac{\Delta\dot{Q}_{abs} + \frac{T_\infty}{R_\infty} + \frac{T_{w\infty}}{R_{w\infty}} + \frac{T_{p(N-1)}}{R_{p,axial}} + \frac{T_N}{R_{pipe}+R_{h,i}}}{\frac{1}{R_\infty} + \frac{1}{R_{w\infty}} + \frac{1}{R_{p,axial}} + \frac{1}{R_{pipe}+R_{h,i}}}$$

An energy balance on the first fluid node, rearranging into a form suitable for MOSS, and invoking the well-mixed flow model, with $R_{flow} = 1/\dot{m}c_p$ and:

$$T_1 = \frac{\frac{T_{IN}}{R_{flow}} + \frac{T_{p1}}{R_{pipe}+R_{h,i}} + \frac{T_2}{R_{axial}}}{\frac{1}{R_{flow}} + \frac{1}{R_{pipe}+R_{h,i}} + \frac{1}{R_{axial}}}$$

For interior fluid nodes ($i = 2$ to $N - 1$):

$$T_i = \frac{\frac{T_{i-1}}{R_{flow}} + \frac{T_{pi}}{R_{pipe}+R_{h,i}} + \frac{T_{i-1}}{R_{axial}} + \frac{T_{i+1}}{R_{axial}}}{\frac{1}{R_{flow}} + \frac{1}{R_{pipe}+R_{h,i}} + \frac{2}{R_{axial}}}$$

For the last (exit) fluid node:

$$T_N = \frac{\frac{T_{N-1}}{R_{flow}} + \frac{T_{pN}}{R_{pipe}+R_{h,i}} + \frac{T_{N-1}}{R_{axial}}}{\frac{1}{R_{flow}} + \frac{1}{R_{pipe}+R_{h,i}} + \frac{1}{R_{axial}}}$$

This system of equations can be programmed into an iterative loop to solve for all nodal temperatures, and then any desired quantity.

Consider the relative importance of the axial conduction and the axial advection:

$$\frac{R_{axial}}{R_{Flow}} = \frac{\frac{\Delta x}{k_f \pi (D-2t)^2/4}}{1/\dot{m}c_p} = \frac{4\dot{m}c_p \Delta x}{\pi k_f D_i^2}$$

The axial conductive mechanism becomes more important than the advective flow resistance when the spacing between nodes becomes sufficiently small.

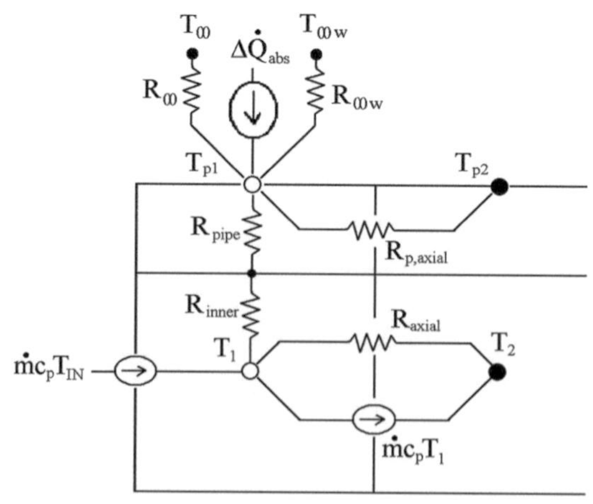

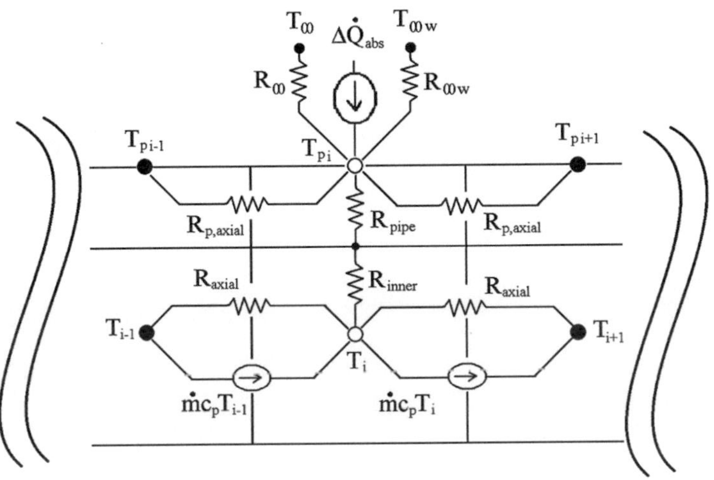

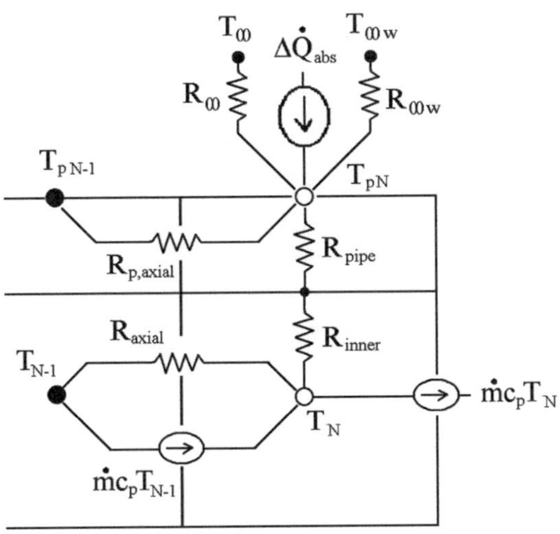

Heat Exchanger Fundamentals   12

The previous chapter focused on the flow of a single stream that exchanges heat with its environment. This chapter considers the exchange of heat between two streams at different temperatures separated by a solid barrier. The aim is to develop a simplified modeling approach that can be used as a preliminary design tool. A closed-form 2-node model is developed that captures the key relationships between fluid types, their flow rates and temperatures, and the heat transfer surface between them. A more detailed model of a specific design configuration (a tube in a tube, or double-pipe) is developed as a case study for a more detailed investigation.

There are many different configurations in engineering practice, and there are numerous resources available to obtain more comprehensive treatments of heat exchangers. In the end, practitioners and designers must learn how to interpret manufacturer specifications, which can be presented in various forms. A good fundamental understanding of the key principles involved facilitates that practice.

12.1   Functional Requirements of a Heat Exchanger

A heat exchanger can be defined as an engineering device whose primary function is to transfer heat from one fluid to another. Heat exchangers are hidden everywhere. For example, nearly every household refrigerator has two heat exchangers. An evaporator (usually hidden from view) is located inside the refrigerated space and is designed so that heat from the cold space will transfer across a barrier to an even colder refrigerant that flows through it. A condenser (usually placed in back of or under the refrigerator) is designed to transfer heat from a hot refrigerant being pumped through a cycle to ambient air. The heat rejected by the condenser consists of the sum of the heat picked up by the evaporator, and the work added to the refrigerant as it is compressed from a low pressure vapor to a high pressure vapor.

As another example, in nearly every automobile, the engine block is cooled by a liquid coolant (termed "water" cooled, although it is not water) that flows into spaces in the engine block to keep it from overheating. The engine block itself is

© Springer International Publishing Switzerland 2015

G. Sidebotham, *Heat Transfer Modeling: An Inductive Approach*,

DOI 10.1007/978-3-319-14514-3_12

technically not a heat exchanger because its primary function is to convert chemical energy into mechanical energy, not to transfer heat from one fluid to another. On the other hand, the heat picked up by the coolant must be "dissipated" somehow, and therefore the radiator is a true heat exchanger that transfers the heat picked up in the engine block to ambient air. The hot coolant that leaves the engine block enters the radiator and passes through a long pipe (not straight but doubled back and forth) that has heat transfer fins placed on it (to increase the effective overall heat transfer coefficient). Air is blown across the fins and heat is transferred from the hot coolant, through the fins and to ambient. The coolant exits the radiator colder than its inlet (but hotter than ambient) and returns to perform its main function of keeping the engine block cool.

Other examples include heat exchangers designed to create high pressure steam from a high temperature source fluid (i.e., from hot combustion products). These are widely used for the generation of mechanical power (which can be used to drive an electric generator) or merely as a means of distributing heat to an industrial process. After passing through a turbine, the low pressure steam must be condensed in another heat exchanger that uses ambient air, or river water as the second fluid. Cooling towers involve evaporation of some condensing water.

A schematic of a typical "water in tube" boiler design is shown in Fig. 12.1. In contrast, a "fire tube" boiler is one in which hot combustion products flow inside tubes that are surrounded by liquid water which cause boiling on the outside surface. The left sketch shows an end view, with hot "air" (i.e., combustion products are hot, and have thermal properties nearly identical to those of air) entering from below and encountering a bank of tubes that contain water at a lower temperature. Heat is transferred from the gas to the tubes and the gases exit the top at a lower temperature. How much lower depends on many factors such as thermal properties of both fluids, flow rates of both streams, geometry and thermal properties of the walls that separate the two fluid streams. The right sketch shows a side view, with water entering each tube row as a liquid and emerging from the right as a vapor. In practice, each tube could be delivered fluid with the same inlet conditions (and the tubes are connected in parallel, as shown in Fig. 12.1), or the tubes could be connected in series into a single long pipe with multiple passes (with connections outside the field of view).

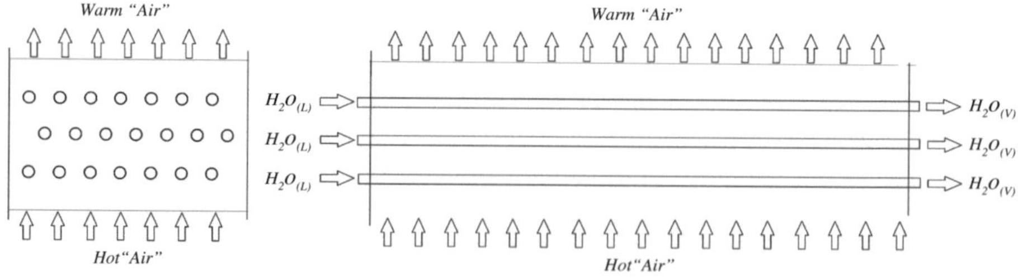

Fig. 12.1 Schematic of a boiler section consisting of a bank of tubes in a cross-flow of hot gases

In the boiler example considered in the previous chapter, the outer fluid was considered to be at a single temperature (T_∞) along the entire length of the pipe. The inner fluid enthalpy changed with position. This case is considered a heat exchanger, in the sense that heat is exchanged from one fluid to another across a wall. However, the temperature of only one stream changed with position (for the preheater and superheater), and the outer fluid temperature was constant along the entire length of the duct. In contrast, both fluids in a heat exchanger often experience a significant change in temperature with position, each with an inlet temperature, and a different outlet temperature. This chapter considers means to estimate the performance of such engineering devices. There are many designs, but most of them can be analyzed by the methods developed here, at least approximately.

Broadly, heat exchanger problems are often formulated by specifying all four temperatures (two inlets, two outlets) and the desired heat transfer rate, and then it becomes a matter of determining required flow rates, or the size of the heat exchanger. Alternatively, the flow rates, inlet temperatures, and the size of the heat exchanger can be specified, and the heat transfer rate and exit temperatures need to be determined. The key parameter that defines the size of the heat exchanger is the surface area that separates the two fluids. A large surface area will bring one or both of the fluids to near thermal equilibrium (thermally long behavior). A small surface area will exchange heat, but not substantially change the temperature (thermally short behavior).

12.2   Double-Pipe Heat Exchangers

A schematic representation of the so-called double-pipe heat exchangers (a common configuration) is shown in Fig. 12.2. In both the co-flow and counter-flow arrangements, one fluid (A) enters a pipe at an inlet temperature (T_{Ain}), receives (or rejects) heat, and exits at a different temperature (T_{Aout}). This inner fluid is like those considered in the previous chapter, except that the temperature of the fluid outside the inner pipe varies with position. Fluid B enters through a larger, concentric pipe and flows through an annular space. The Second Law of Thermodynamics (heat flows from hot to cold) requires that the hotter fluid must be at a higher temperature than the colder fluid at every location within the pipe and that the temperature change along the path must be monotonic (increasing or decreasing, or remaining constant for phase changing systems). These constraints yield very different temperature profiles for the co-flow and counter-flow arrangements.

In the co-flow arrangement, fluid B enters from the same side as A and flows in the same direction. Assuming that A is the hotter fluid, its temperature must decrease monotonically in the direction of flow, as B increases along the same direction. Therefore, the exiting temperature of A must be greater than the exiting temperature of B. The sketch of temperature vs. position shows that the lowest temperature of fluid A (at its exit) is greater than the highest temperature of fluid B (at its exit).

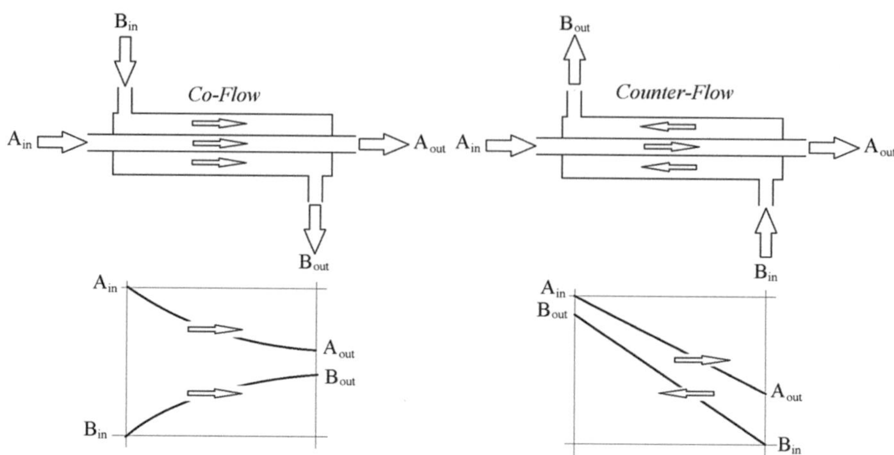

Fig. 12.2 Schematic representation of co-flow (*left*) and counter-flow (*right*) double-pipe heat exchangers. Representative temperature profiles (temperature vs. position) are shown below each schematic for non-phase changing operation

In the counter-flow arrangement, fluid B enters where fluid A exits, and vice versa, and flows in the opposite direction. Assuming A is the hotter fluid, its temperature must decrease monotonically from entrance to exit (left to right), as B increases from its entrance to exit (right to left). At every location along the pipe, A must be at a higher temperature. The Second Law restriction on A is that it must exit at a temperature above the inlet of B (not the exit, as in the co-flow case). Similarly, the exit of B must be below the inlet of A. That restriction yields a temperature profile that is very different from the co-flow arrangement.

In the co-flow arrangement, the local temperature difference between A and B is a maximum near the entrance, and decreases along the length of the pipe. In the counter-flow arrangement, the temperature difference changes less, and in fact can be a constant throughout the entire length of the pipe.

12.2.1 Averaged Temperature Differences

The heat transfer rate (q, in Watts) between the two fluid streams can be expressed in a Newton's Law of Cooling form, namely:

$$q = U_i A_i \left(\overline{T}_A - \overline{T}_B \right)$$

In this expression, the overall heat transfer coefficient U_i is based on the inner surface area (A_i), but could be based on the outer (or any other surface area). The temperature difference between the fluids is given as averages, and there are different possible approaches. The most obvious and easily derived is the average mean temperature difference (AMTD) in which the geometric average of the inlet and outlet fluid temperatures are used. That is:

$$q = U_i A_i AMTD = U_i A_i \left[\left(\frac{T_{Ain} + T_{Aout}}{2} \right) - \left(\frac{T_{Bin} + T_{Bout}}{2} \right) \right]$$

This result is the same for the co-flow and counter-flow arrangement. In the next section, 2-node models will be developed that yield closed-form solutions for the exit temperatures and heat transfer rates. A differential analysis, not developed here, yields a similar expression for the heat transfer rate, but with a less intuitive expression for the averaged temperature difference, called the log mean temperature difference (LMTD). The LMTD is different for the co-flow and counter-flow arrangements:

$$q = U_i A_i LMTD$$

where

$$\left(\overline{T}_A - \overline{T}_B \right) = LMTD_{CO-FLOW}$$

$$= \frac{\left(T_{Ain} - T_{Bin} \right) - \left(T_{Aout} - T_{Bout} \right)}{\ln \left[\frac{\left(T_{Ain} - T_{Bin} \right)}{\left(T_{Aout} - T_{Bout} \right)} \right]} \quad \text{for co-flow heat exchangers}$$

$$\left(\overline{T}_A - \overline{T}_B \right) = LMTD_{COUNTER-FLOW}$$

$$= \frac{\left(T_{Ain} - T_{Bout} \right) - \left(T_{Aout} - T_{Bin} \right)}{\ln \left[\frac{\left(T_{Ain} - T_{Bout} \right)}{\left(T_{Aout} - T_{Bin} \right)} \right]} \quad \text{for counter-flow heat exchangers}$$

The three different types of averages are shown for a specific case in Fig. 12.3. The inlet temperature of A is set at 100 °C and the inlet of B is set at 0 °C. The exit of A is set at 50 °C, and the average temperature from the three forms is shown as a function of all possible exit temperatures for B. It is apparent that the AWTD and $LMTD_{COUNTER-FLOW}$ give similar results, with significant deviations for the $LMTD_{CO-FLOW}$ case.

For problems in which all four temperatures are specified, the averaged temperature difference can be calculated upfront. Figure 12.3 shows that the AMTD, which is easily remembered, can be used for counter-flow arrangements, and that it is generally preferable to use the LMTD for co-flow arrangements.

The following general problem statement is written for those cases, where the inlet temperatures and flow rates of both streams are specified, as is the effective heat transfer coefficient U_i. The exit temperatures and heat transfer rates are to be determined as a function of total effective length (surface area/perimeter) of the pipe.

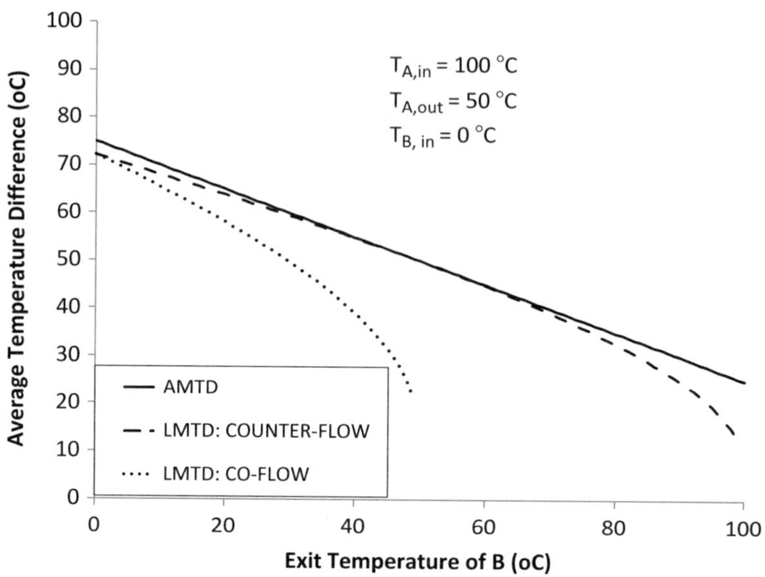

Fig. 12.3 Comparison of averaged temperature differences for the case, where fluid A enters at 100 °C and exits at 50 °C, and fluid B enters at 0 °C. The average mean temperature difference (AMTD) and two log mean temperature difference (LMTD) expressions are shown for all possible exit temperatures of fluid B. Note that for the co-flow arrangement fluid B cannot exit at a higher temperature than the exit of A (50 °C in this case)

12.2.1.1 Problem Statement

In a double-pipe heat exchanger (Fig. 12.2) of total length L, fluid A (with density and specific heat) enters a pipe (inner diameter D_{inner}, wall thickness t, thermal conductivity k) with a volumetric flow rate $(AV)_A$ at a temperature $T_{A,in}$. In the annular region of a concentric pipe with inner diameter D, fluid B (with density and specific heat) enters with a volumetric flow rate $(AV)_B$ at a temperature $T_{B,in}$. Neglect stray heat loss between the outer pipe and its ambient and assume no phase change in either stream. Calculate, as a function of pipe length, the exit temperatures of the two streams, the internal heat transfer rate between the two streams, and the heat exchanger effectiveness (defined as the actual heat transfer rate normalized by the maximum possible heat transfer rate, where one fluid exits at the inlet temperature of the other, and the other fluid does not overshoot the inlet). Consider both co-flow and counter-flow arrangements.

12.2.2 2-Node Analysis: Well-Mixed Model

Invoking a 1-node model for each fluid stream, and using a well-mixed model, a closed-form solution for the exit temperatures can be obtained. A thermal resistance model (Fig. 12.4) indicates that heat exchange between the two fluids takes place

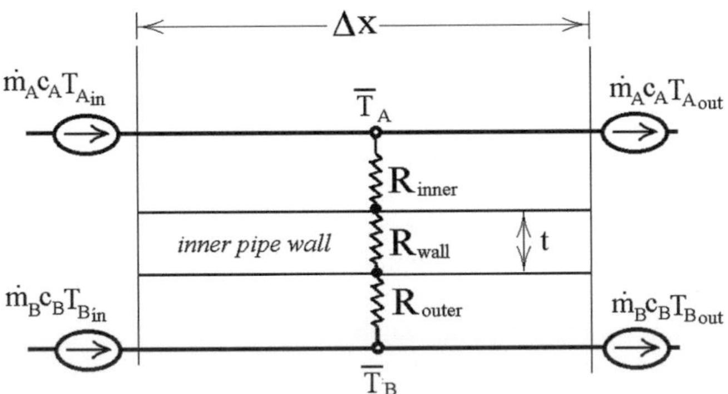

Fig. 12.4 Thermal resistance network for the 2-node heat exchanger model

with an overall heat transfer coefficient (U) based on a defined area of the wall separating the streams. The overall heat transfer coefficient for the double-pipe arrangement, referenced to the inner surface area of the pipe is defined by the three thermal resistances in series between the two fluids (convection at the inner wall, conduction across the wall of thickness t, and convection at the outer wall):

$$U_i \pi D_i L = \frac{1}{R_{inner} + R_{wall} + R_{outer}} = \frac{1}{\left(\frac{1}{h_{inner} \pi D_i L} \right) + \left(\frac{t}{k \pi (D_i + t) L} \right) + \left(\frac{1}{h_{outer} \pi (D_i + 2t) L} \right)}$$

where the thin-wall approximation has been used (the exact formula for conduction across a cylindrical shell could be used instead). Solving for the overall heat transfer coefficient:

$$U_i = \frac{1}{\frac{1}{h_{inner}} + \frac{t D_i}{k (D_i + t)} + \frac{D_i}{h_{outer} (D_i + 2t)}}$$

Nusselt number correlations can be used for the convection coefficients, recognizing that the outer flow region is an annular region and the Reynolds number is based on an equivalent hydraulic diameter (that gives the same cross-sectional area as the annular region):

$$\frac{\pi D_H^2}{4} = \frac{\pi \left(D^2 - (D_i + 2t)^2 \right)}{4}$$

Solving for the hydraulic diameter:

$$D_H = \sqrt{D^2 - (D_i + 2t)^2}$$

Nodal equations applied to nodes A and B are:

$$\dot{m}_A c_{pA}\left(T_{Ain} - T_{Aout}\right) + U_i A_i\left(\overline{T}_B - \overline{T}_A\right) = 0$$

$$\dot{m}_B c_{pB}\left(T_{Bin} - T_{Aout}\right) + U_i A_i\left(\overline{T}_A - \overline{T}_B\right) = 0$$

The temperatures in the heat transfer terms represent the average temperatures in each section. Treating the two inlet temperatures, mass flow rates and specific heats as inputs, there are four unknown temperatures for these two equations (two exit temperatures and two average temperatures).

Invoking the well-mixed model, the average temperature of the fluid in the pipe is set equal to the exit temperature, adding two more equations to close the set. This model will underestimate the average temperature difference between the two streams, and therefore the model underestimates the total heat transfer rate. Dividing each equation by the mass flow rate and specific heat, and rearranging:

$$T_{Aout}\left(1 + \frac{U_i A_i}{\dot{m}_A c_{pA}}\right) = T_{Ain} + \left(\frac{U_i A_i}{\dot{m}_A c_{pA}}\right)T_{Bout}$$

$$T_{Bout}\left(1 + \frac{U_i A_i}{\dot{m}_B c_{pB}}\right) = T_{Bin} + \left(\frac{U_i A_i}{\dot{m}_B c_{pB}}\right)T_{Aout}$$

It is convenient and insightful to define two dimensionless lengths as the physical length (L) normalized by thermal length for each stream:

$$\hat{L}_A = \frac{U_i A_i}{\dot{m}_A c_{pA}} = \frac{U_i \pi D_i L}{\dot{m}_A c_{pA}} = \frac{L}{L_{t,A}}$$

$$\hat{L}_B = \frac{U_i A_i}{\dot{m}_B c_{pB}} = \frac{U_i \pi D_i L}{\dot{m}_B c_{pB}} = \frac{L}{L_{t,B}}$$

Notice that there are two thermal lengths, one for each stream. The longer of the two (with the lower thermal flow rate) is considered the controlling stream because it is the closest to achieving thermal equilibrium with its partner, and its value is referred in practice as number of transfer units (NTU). Combining the two nodal equations:

$$T_{Aout}\left(1 + \hat{L}_A\right) = T_{Ain} + \hat{L}_A\left(\frac{T_{Bin} + \hat{L}_B T_{Aout}}{1 + \hat{L}_B}\right)$$

Rearranging:

$$T_{Aout}\left(1 + \hat{L}_A - \frac{\hat{L}_B \hat{L}_B}{1 + \hat{L}_B}\right) = T_{Ain} + \left(\frac{\hat{L}_A}{1 + \hat{L}_B}\right)T_{Bin}$$

Finally, the exit temperature of fluid A is expressed in terms of input parameters:

$$T_{Aout} = \frac{T_{Ain} + \left(\frac{\hat{L}_A}{1+\hat{L}_B}\right)T_{Bin}}{\left(1 + \hat{L}_A - \frac{\hat{L}_A\hat{L}_B}{1+\hat{L}_B}\right)}. \text{ Similarly for fluid B:} T_{Bout} = \frac{T_{Bin} + \left(\frac{\hat{L}_B}{1+\hat{L}_A}\right)T_{Ain}}{\left(1 + \hat{L}_B - \frac{\hat{L}_A\hat{L}_B}{1+\hat{L}_A}\right)}.$$

The heat transfer rate between the two streams is obtained from one of three forms:

$$q_{A\rightarrow B} = \dot{m}_A c_{pA}(T_{Ain} - T_{Aout}) = \dot{m}_B c_{pB}(T_{Bout} - T_{Bin}) = U_i A_i(T_{Aout} - T_{Bout})$$

The heat exchanger effectiveness is defined as the actual heat transfer rate normalized by the maximum possible heat transfer rate. The maximum possible heat transfer rate of either stream is that which would occur if the stream exited at the inlet temperature of the other stream. That is:

$$q_{max,A} = \dot{m}_A c_{pA}(T_{Ain} - T_{Bin})$$

$$q_{max,B} = \dot{m}_B c_{pB}(T_{Ain} - T_{Bin})$$

The quantity $(\dot{m}c_p)$ is considered to be a thermal mass flow rate. The smaller of these defines the maximum possible heat transfer rate. A higher heat transfer rate would cause the other stream to overshoot the inlet of the other, in violation of the second Law. That is:

$$q_{max} = \left(\dot{m}c_p\right)_{min}(T_{Ain} - T_{Bin})$$

The heat exchanger effectiveness is therefore:

$$\varepsilon = \frac{q}{q_{max}}$$

This 2-node model can give surprisingly good results, and it can be applied to any heat exchanger type (when properly interpreted). However, it cannot distinguish between co-flow and counter-flow behavior. The well-mixed model was used because it works well across all ranges of flow. If, as is common, the inlet and exit temperatures and mass flow rates of one stream are specified, then the geometric average temperature can be used instead, and then the required mass flow of the other stream and the effective UA required can be determined more directly.

The NTU method is commonly used in design calculations, where NTU is a dimensionless measure of the effective length of a double-pipe heat exchanger, defined as the longer of the two thermal lengths:

$$NTU = \frac{UA}{\left(\dot{m}c_p\right)_{min}} = \hat{L}_{max}$$

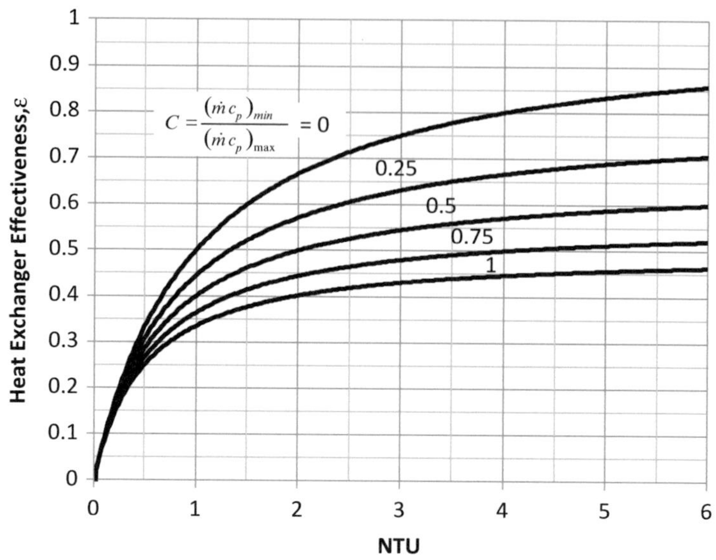

Fig. 12.5 Heat exchanger effectiveness vs. number of transfer units (NTU) (longer of the two thermal lengths) for 2-node well-mixed heat exchanger for selected values of the ratio of the thermal mass flow rate

The heat exchanger effectiveness is plotted against pipe length (as NTU) in Fig. 12.5 for fixed ratios of the minimum to maximum thermal flow rates (C) for the 2-node well-mixed heat exchanger model: Use of this plot will give good initial estimates of heat exchanger performance, and reveal key trends that can guide design decisions. For example, for a situation where the thermal mass flow rates of the two streams are comparable ($C = 1$), then there is no need to select a heat exchanger with an NTU of greater than approximately 2. The uncertainty in estimating UA is generally considerable in any case.

The results and temperature profiles for three cases are considered in which the temperature of the hotter fluid enters at 100 °C and the colder fluid enters at 0 °C. The calculated average temperatures for the appropriate model are shown below the plot of temperature vs. position. In the first case (Fig. 12.6 with NTU = 1, C = 0), the effectiveness is 0.5, and the co-flow and counter-flow results are identical. The cold exhibits high flow, or thermally short, behavior. Its temperature does not change in the heat exchanger, and the direction of flow does not influence the response of the hotter fluid. For longer heat exchangers (higher NTU) but with $C = 0$, the exit temperature of the hotter fluid will be closer to the cold fluid, but the shapes will be similar to this case.

Figure 12.7 shows a case with a longer heat exchanger (NTU = 3) and with the thermal mass flow rate of the hotter fluid half that of the colder fluid (C = 0.5). In this case, both fluid temperatures change, but the hotter fluid changes more due to its lower thermal flow rate. The counter-flow temperature profiles are not parallel,

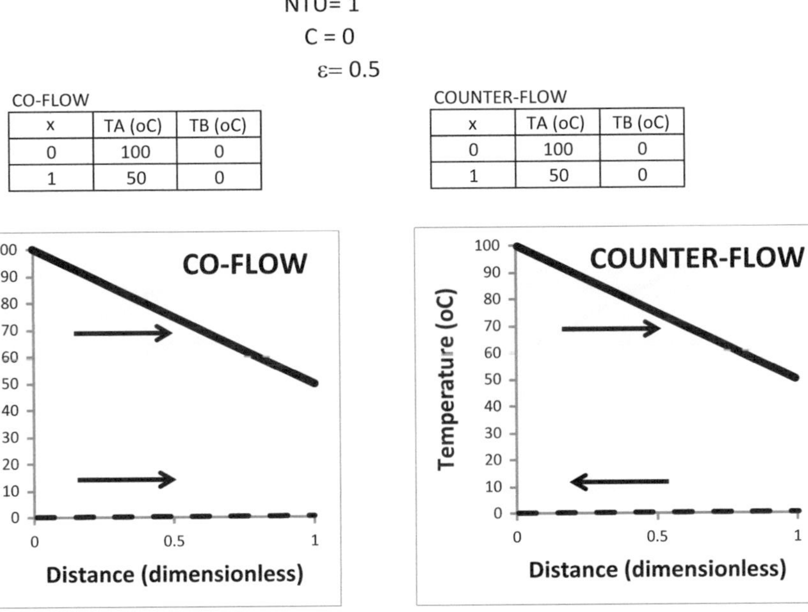

CO-FLOW		
x	TA (oC)	TB (oC)
0	100	0
1	50	0

COUNTER-FLOW		
x	TA (oC)	TB (oC)
0	100	0
1	50	0

AMTD	75	oC
LMTD	72.13	oC

AMTD	75	oC
LMTD	72.13	oC

Fig. 12.6 Temperature profiles from the well-mixed 2-node model for the case of NTU = 1 (thermal length of hotter fluid), and C = 0 (infinite thermal flow rate of colder fluid, or thermal length of zero)

and the restriction of the well-mixed model does not allow the exit of the cold fluid to exceed the exit of the hotter.

Figure 12.8 shows a case with a very long heat exchanger (NTU = 6) and with equal thermal flow rates of the two streams (C = 1). In this case, the exit streams of both streams are close together. In the counter-flow case, the temperature profiles are parallel (which will be true for all cases with C = 1), and the temperature difference is the same along the entire length of the pipe. The response of either stream is that of a constant heat flux.

The well-mixed model forces the outlet temperature of each stream to be equal to the average temperature. Therefore, the outlet temperature of the hotter stream must be greater than the outlet temperature of the colder stream. For this reason, for very long heat exchangers, the well-mixed model is better suited to the co-flow arrangement than the counter-flow arrangement, in which case the exit of the colder fluid can exceed the exit of the hotter fluid.

The model considered next is a better model for the counter-flow case, where the thermal mass flow rates are comparable, but the length is sufficiently long that the exit of the cold fluid is significantly higher than the inlet of the hotter fluid.

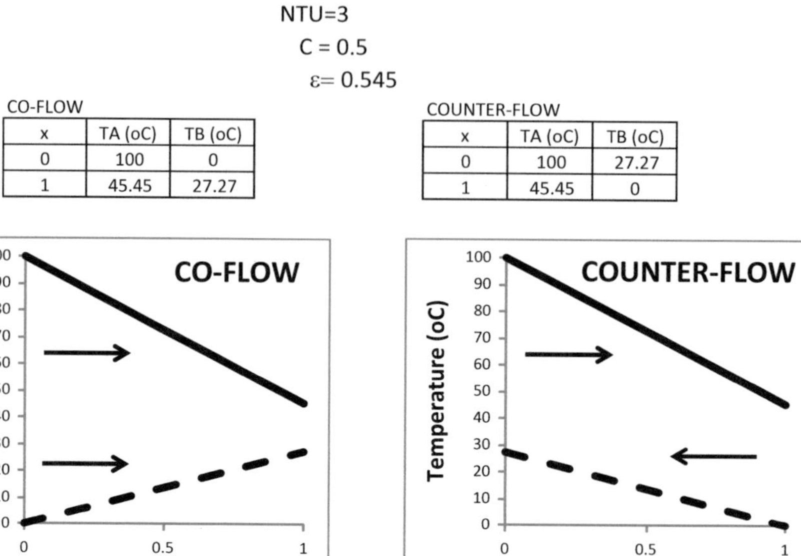

CO-FLOW

x	TA (oC)	TB (oC)
0	100	0
1	45.45	27.27

COUNTER-FLOW

x	TA (oC)	TB (oC)
0	100	27.27
1	45.45	0

AMTD	59.09	oC
LMTD	47.99	oC

AMTD	59.09	oC
LMTD	58.03	oC

Fig. 12.7 Temperature profiles from the well-mixed 2-node model for the case of NTU = 3 (thermal length of hotter fluid), and C = 0.5 (thermal flow rate of hotter fluid half that of colder fluid)

12.2.3 2-Node Analysis: Average Temperature Model

The well-mixed model underestimates the heat transfer rate, but has the advantage of not violating the Second Law of Thermodynamics. It is expected to work better for co-flow arrangements than counter-flow arrangements. As with the well-mixed model, the heat transfer rate can be expressed three different ways, based on the direct heat transfer, or as the thermodynamic change:

$$q = UA\left(\overline{T}_A - \overline{T}_B\right) = \left(\dot{m}c_p\right)_A (T_{Ain} - T_{Aout}) = \left(\dot{m}c_p\right)_B (T_{Bout} - T_{Bin})$$

For the average temperature model, the temperature difference is expressed as the geometric average of the temperature difference at the inlet and that at the outlet. Expressing that difference in the form for the counter-flow arrangement:

$$\overline{T}_A - \overline{T}_B = \frac{1}{2}[(T_{Ain} - T_{Bout}) + (T_{Aout} - T_{Bin})]$$

Expressing this difference for the co-flow arrangement yields the same result. Algebraic manipulation (a good exercise) results in the following expression for

NTU=6

C = 1

ε= 0.461

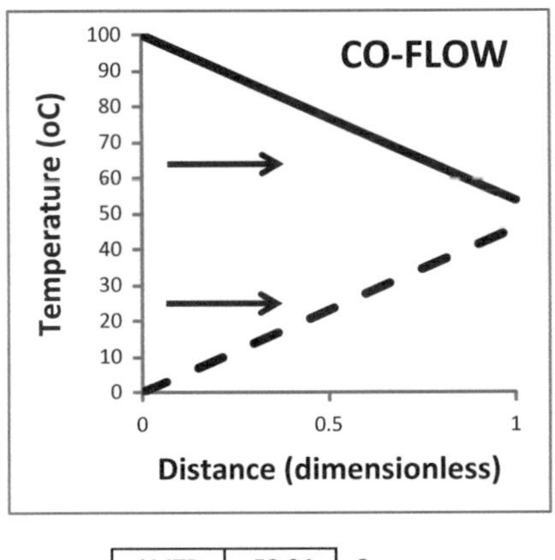

CO-FLOW

x	TA (oC)	TB (oC)
0	100	0
1	53.85	46.15

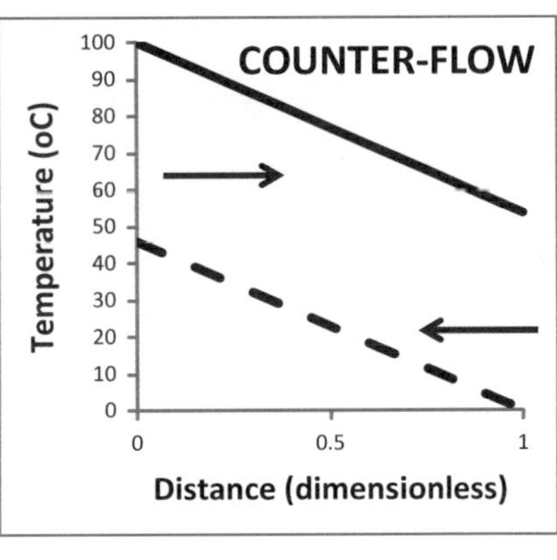

COUNTER-FLOW

x	TA (oC)	TB (oC)
0	100	46.15
1	53.85	0

AMTD	53.84	oC
LMTD	35.99	oC

AMTD	53.84	oC
LMTD	53.86	oC

Fig. 12.8 Temperature profiles from the well-mixed 2-node model for the case of NTU $= 6$ (thermal length of hotter fluid) and C $= 1$ (thermal flow rate of hotter fluid equal to that of colder fluid)

the outlet temperature of A in terms of the inlet temperatures and the two thermal lengths, defined as with the well-mixed model.

$$T_{Aout} = \frac{\left(2 + \hat{L}_B - \hat{L}_A\right)T_{Ain} + 2\hat{L}_A T_{Bin}}{2 + \hat{L}_A + \hat{L}_B + \hat{L}_A\hat{L}_B}$$

$$T_{Bout} = \frac{\left(2 - \hat{L}_B\right)T_{Bin} + \hat{L}_B\left(T_{Ain} - T_{Aout}\right)}{2 + \hat{L}_B}$$

The outlet of B is expressed in terms of the outlet for A, which must therefore be calculated first. The effectiveness of the heat exchanger as a function of the NTU (longer of the two thermal lengths) for various ratios of the thermal flow rates (C) is shown in Fig. 12.9 which exhibits the same general shape as the well-mixed model. However, for lower values of C, the heat exchanger effectiveness exceeds unity, in violation of the Second Law. This result should not be surprising, based on the limiting case for a single stream model from the previous chapter. For that reason, this model should be used with caution.

One temperature profile case is shown in Fig. 12.10, with NTU $= 3$ and C $= 1$. The exit temperature of the hotter fluid is lower than the exit of the colder fluid. This behavior violates the second Law for the co-flow arrangement but is acceptable for the counter-flow arrangement.

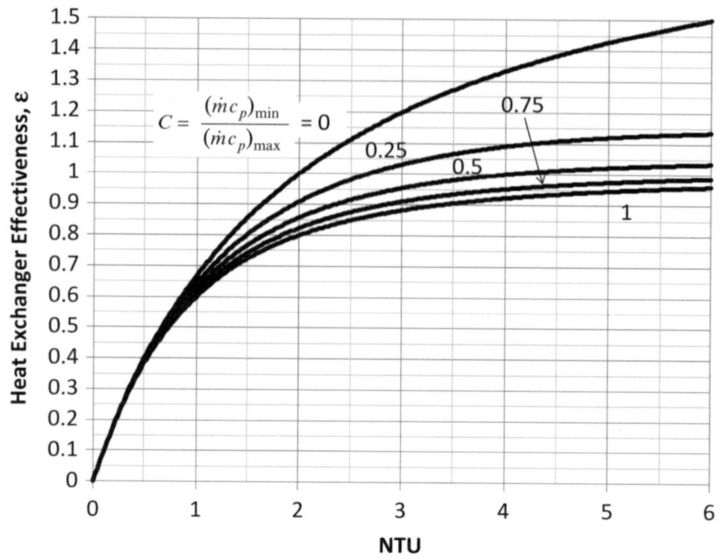

Fig. 12.9 Heat exchanger effectiveness vs. NTU (longer of the two thermal lengths) for 2-node average temperature heat exchanger model for selected values of the ratio of the thermal mass flow rate. Cases that exhibit an effectiveness greater than unity violate the Second Law of Thermodynamics

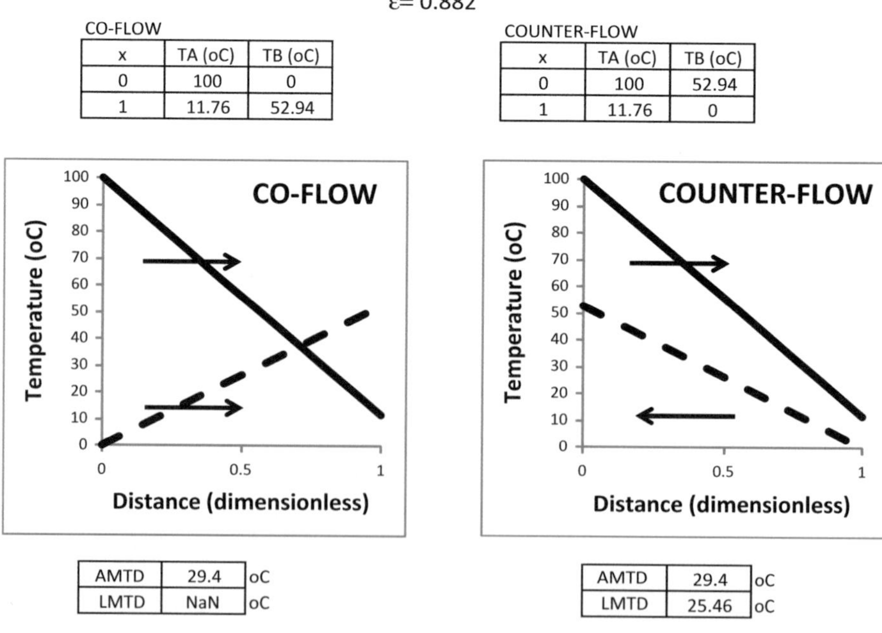

CO-FLOW

x	TA (oC)	TB (oC)
0	100	0
1	11.76	52.94

COUNTER-FLOW

x	TA (oC)	TB (oC)
0	100	52.94
1	11.76	0

AMTD	29.4	oC
LMTD	NaN	oC

AMTD	29.4	oC
LMTD	25.46	oC

Fig. 12.10 Temperature profiles from the average temperature 2-node model for the case of NTU = 3 (thermal length of hotter fluid = 3) and C = 1 (thermal flow rate of hotter fluid equal to that of colder fluid). The co-flow case violates the Second Law of Thermodynamics

A general conclusion or recommendation is that the well-mixed model can be used for all cases to establish the basic performance of most, if not all, heat exchangers. The average temperature model gives improved accuracy in certain areas (short heat exchangers), but can give results that violate the Second Law. An important practical case where the average temperature model is advantageous is for counter-flow arrangements with thermally long pipes with comparable thermal flow rates of the two streams. This arrangement is common in practice.

12.2.4 2n-Node Model

More quantitative accuracy and a distinction between co-flow and counter flow arrangements is obtained by discretizing space, as expected by now. The pipe is broken into n equally spaced sections of length $\Delta x = L/n$. A resistance model for an interior cell "i" is shown in Fig. 12.11 for both the co-flow and counter-flow arrangements. There is a node for each fluid (A and B). An energy balance applied to node A(i) using the well-mixed model (in which the exit temperature from a cell

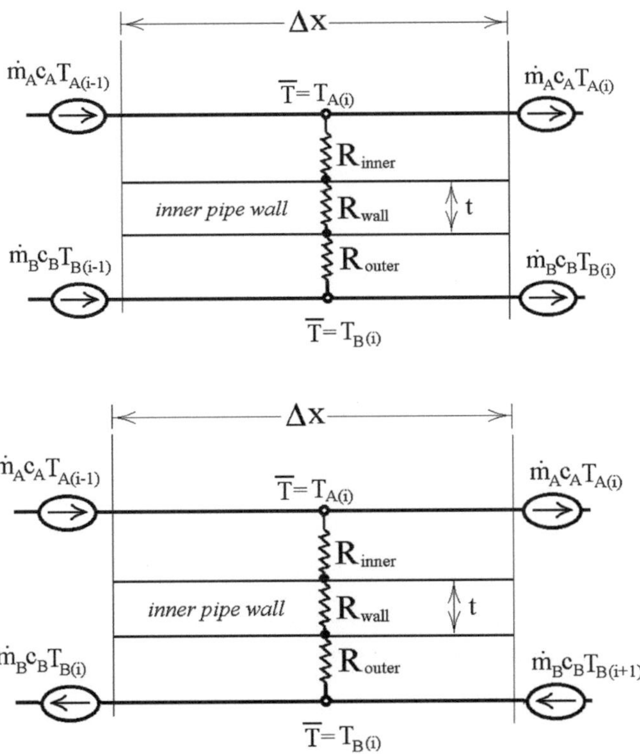

Fig. 12.11 Thermal resistance model for node "i" in the co-flow (*top*) and counter-flow (*bottom*) arrangements. The well-mixed model is used (in which the average temperature of each fluid is set equal to the exit temperature from that cell)

equals the average temperature used to calculate the heat exchange between the corresponding fluid B cell) is:

$$\dot{m}_A c_{pA}(T_{A,i-1} - T_{Ai}) + U_i \pi D_i \Delta x(T_{Bi} - T_{Ai}) = 0$$

where U_i is an overall heat transfer coefficient based on the inner surface area. Rearranging for T_{Ai} (a form suitable for Method of Successive Substitution, with superscript "k" to indicate iteration number):

$$T_{Ai}^{(k+1)} = \frac{\dot{m}_A c_{pA} T_{A,i-1}^{(k)} + U_i \pi D_i \Delta x T_{Bi}^{(k)}}{\dot{m}_A c_{pA} + U_i \pi D_i \Delta x}$$

Similarly, an energy balance applied to cell B(i) yields:

$$T_{Bi}^{(k+1)} = \frac{\dot{m}_B c_{pB} T_{B,i-1}^{(k)} + U_i \pi D_i \Delta x T_{Ai}^{(k)}}{\dot{m}_B c_{pB} + U_i \pi D_i \Delta x} \quad \text{(co-flow arrangement)}$$

For the counter-flow arrangement, with fluid A flowing left to right and B flowing right to left, the nodal equation for fluid B must recognize that the flow term entering the cell comes from the right (T_{B+1}) and the nodal equation is:

$$T_{Bi}^{(k+1)} = \frac{\dot{m}_B c_{pB} T_{B,i+1}^{(k)} + U_{inner} \pi D_{inner} \Delta x T_{Ai}^{(k)}}{\dot{m}_B c_{pB} + U_{inner} \pi D_{inner} \Delta x} \text{(counter-flow arrangement)}$$

Workshop 12.1. Double-Pipe Heat Exchangers

The following instructions for an in-class workshop were followed to create an excel file. The base case input parameter values are for a 1 m long water to water heat exchanger. The flow rates were adjusted (using "goal seek") so that the thermal lengths of both streams were set to unity. What follows are results for several cases, varying pipe length, flow rates of one stream and substituting air for water. These results are not explicitly discussed. Rather, the reader is encouraged to practice technical writing skills by describing the trends observed for each case shown and to run different case studies.

Suggested Spreadsheet Solution

- In a worksheet entitled "params," write a separate block for input parameters (from problem statement) that a user can change, and a block for relevant derived parameters. Add a block to display final results (that confirms convergence). Consider inner and outer convection coefficients to be input parameters (but separate worksheets could be used to calculate them based on Nusselt number correlations).

- In a worksheet[1] entitled "Iterations_co-flow," for the heat exchanger broken into ten equal sized sections, define a row with ten columns for fluid A and ten columns for fluid B. Enter the inlet fluid temperatures as initial "guessed" values (reference to "params"). In the next row, enter the recursion formulas for the fluid temperatures that reference the row above it. Add two columns to check for convergence of the exit streams. Copy this iterative row as far as needed to obtain a reasonable convergence.
- Copy the "Iterations_co-flow" worksheet, rename it "Iterations_counter-flow" and modify it for the counter-flow arrangement.
- In a worksheet entitled "Results" lay out a block for the heat exchanger with four rows of temperatures (fluid A, inner wall, outer wall, fluid B) for co-flow. Reference the fluid temperatures to the final converged values from "iterations" and enter voltage divider formulas for the wall temperatures. Repeat for counter-flow. Make well-formatted plots of all four temperatures vs. position for both arrangements (Figs. 12.12, 12.13, 12.14, 12.15, 12.16, 12.17, 12.18, 12.19, 12.20, 12.21, 12.22, 12.23, 12.24, and 12.25).
- Use the worksheet as a design tool...

CO-FLOW ARRANGEMENT
Max Change = 0.00E+00
dx=0.1 m

node	1	2	3	4	5	6	7	8	9	10	
x	0	0.05	0.15	0.25	0.35	0.45	0.55	0.65	0.75	0.85	0.95	1
TA	100.00	91.67	84.72	78.93	74.11	70.09	66.74	63.95	61.63	59.69	58.07	
Tinner	40.44	42.03	43.36	44.47	45.39	46.16	46.80	47.33	47.78	48.15	48.46	
Touter	39.71	41.42	42.85	44.05	45.04	45.87	46.55	47.13	47.61	48.01	48.34	
TB	0.00	8.33	15.28	21.07	25.89	29.91	33.26	36.05	38.37	40.31	41.93	

COUNTER-FLOW ARRANGEMENT
Max Change = 0.00E+00
dx=0.1 m

node	1	2	3	4	5	6	7	8	9	10	
x	0	0.05	0.15	0.25	0.35	0.45	0.55	0.65	0.75	0.85	0.95	1
TA	100.00	95.24	90.48	85.71	80.95	76.19	71.43	66.66	61.90	57.14	52.38	52.38
Tinner	68.80	66.88	62.11	57.35	52.59	47.83	43.06	38.30	33.54	28.78	24.02	24.02
Touter	68.42	66.53	61.77	57.01	52.24	47.48	42.72	37.96	33.19	28.43	23.67	23.67
TB	47.62	47.62	42.86	38.10	33.34	28.57	23.81	19.05	14.29	9.52	4.76	0.00

Fig. 12.12 Worksheet entitled "Results" which checks for convergence, reports the converged fluid temperatures and calculates the surface temperatures of the inner pipe

check A	check B	A 1	A 2	A 3	A 4	A 5	A 6	A 7	A 8	A 9	A 10	B 1	B 2	B 3	B 4	B 5	B 6	B 7	B 8	B 9	B 10
		100	100	100	100	100	100	100	100	100	100	0	0	0	0	0	0	0	0	0	0
9.092581	9.092581	90.9	90.9	90.9	90.9	90.9	90.9	90.9	90.9	90.9	90.9	9.1	9.1	9.1	9.1	9.1	9.1	9.1	9.1	9.1	9.1
7.43908	7.43908	91.7	83.5	83.5	83.5	83.5	83.5	83.5	83.5	83.5	83.5	8.3	16.5	16.5	16.5	16.5	16.5	16.5	16.5	16.5	16.5
6.086271	6.086271	91.7	84.9	77.4	77.4	77.4	77.4	77.4	77.4	77.4	77.4	8.3	15.1	22.6	22.6	22.6	22.6	22.6	22.6	22.6	22.6
4.979473	4.979473	91.7	84.7	79.2	72.4	72.4	72.4	72.4	72.4	72.4	72.4	8.3	15.3	20.8	27.6	27.6	27.6	27.6	27.6	27.6	27.6
4.073948	4.073948	91.7	84.7	78.9	74.5	68.3	68.3	68.3	68.3	68.3	68.3	8.3	15.3	21.1	25.5	31.7	31.7	31.7	31.7	31.7	31.7
3.333094	3.333094	91.7	84.7	78.9	74.0	70.6	65.0	65.0	65.0	65.0	65.0	8.3	15.3	21.1	26.0	29.4	35.0	35.0	35.0	35.0	35.0
2.726965	2.726965	91.7	84.7	78.9	74.1	70.0	67.4	62.3	62.3	62.3	62.3	8.3	15.3	21.1	25.9	30.0	32.6	37.7	37.7	37.7	37.7
2.231062	2.231062	91.7	84.7	78.9	74.1	70.1	66.6	64.7	60.0	60.0	60.0	8.3	15.3	21.1	25.9	29.9	33.4	35.3	40.0	40.0	40.0
1.82534	1.82534	91.7	84.7	78.9	74.1	70.1	66.8	63.7	62.5	58.2	58.2	8.3	15.3	21.1	25.9	29.9	33.2	36.3	37.5	41.8	41.8

Fig. 12.13 Worksheet entitled "iterations_co-flow" which shows the first few rows of the iteration. A similar worksheet "iterations_counter-flow" is not shown

---

[1] It is advised to set up this iteration table in the "params" worksheet, and once it is working and debugged, cut and paste it to its own sheet. Alternatively, name key cells in "params" and use the named cells in the formulas.

INPUT PARAMETERS

L=	1	m	pipe length
rhoA=	1000	kg/m^3	density of A
cpA =	4200	J/kg/K	specific heat of A
rhoB=	1000	kg/m^3	density of B
cpB=	4200	J/kg/K	specific heat of B
Dinner =	0.02	m	inner diameter of inner pipe
t =	0.005	m	pipe thickness
D =	0.2	m	inner diameter of outer pipe
k =	40	W/m/K	thermal conductivity of pipe
AV,A=	8.91E-06	m^3/s	volumetric flow rate of A
AV,B=	8.91E-06	m^3/s	volumetric flow rate of B
Tain=	100	oC	inlet temperature of A
Tbin=	0	oC	inlet temperature of B
hinner =	1000	W/m2/K	inner convection coefficient
houter =	1000	W/m2/K	outer convection coefficient

DERIVED PARAMETERS

dx=	0.1	m	length of numerical segment (for 10 section grid)
Rinner =	0.159155	oC/W	thermal resistance between A and inner wall
Rwall=	0.001941	oC/W	thermal resistance across pipe wall
Router=	0.106103	oC/W	thermal resistance between B and outer wall
Uinner=	595.6416	W/m2/K	overall heat transfer coefficient
mdotA=	8.91E-03	kg/s	mass flow rate of A
mdotB =	8.91E-03	kg/s	mass flow rate of B
(mdot*cp)A=	3.74E+01	oC/W	Flow capacitance of A
(mdot*cp)B=	3.74E+01	W/K	Flow capacitance of B
LthermalA=	0.999798	m	thermal length of A
LthermalB=	0.999798	m	thermal length of B
A =	1.000202		L/LthermalA, dimensionless length of A
B =	1.000202		L/LthermalB, dimensionless length of B
qmaxA =	3741.77	W	maximum possible heat transfer rate of A
qmaxB =	3741.77	W	maximum possible heat transfer rate of B
qmax	3741.77	W	maximum possible heat transfer rate

RESULTS   (change = 0.00E+00   0.00E+00)

	2-Node	co-flow	counter-flow		
TAout=	66.7	58.1	52.4	oC	exit temperature of A
TBout=	33.3	41.9	47.6	oC	exit temperature of B
qA=	-1247.3	-1568.8	-1782.0	W	heat transfer rate to A
qB=	1247.3	1568.8	1782.0	W	heat transfer rate to B
effectiveness=	0.333	0.419	0.476		heat exchanger effectiveness

NTU= UA/Cmin	1.00	"Number of Transfer Units" = Longer of the 2 thermal lengths
Cmin/Cmax=	1.00	ratio of thermal lengths
AV,A=	5.35E-01	LPM	volumetric flow rate of A
AV,B=	5.35E-01	LPM	volumetric flow rate of B
Lt,A=	0.999798	m	Thermal length of A
Lt,B=	0.999798	m	Thermal length of B
AREA=	0.062832	m^2	Inner Surface Area

Fig. 12.14 Parameters for the base case: 1 m long water to water heat exchanger with thermal lengths equal unity

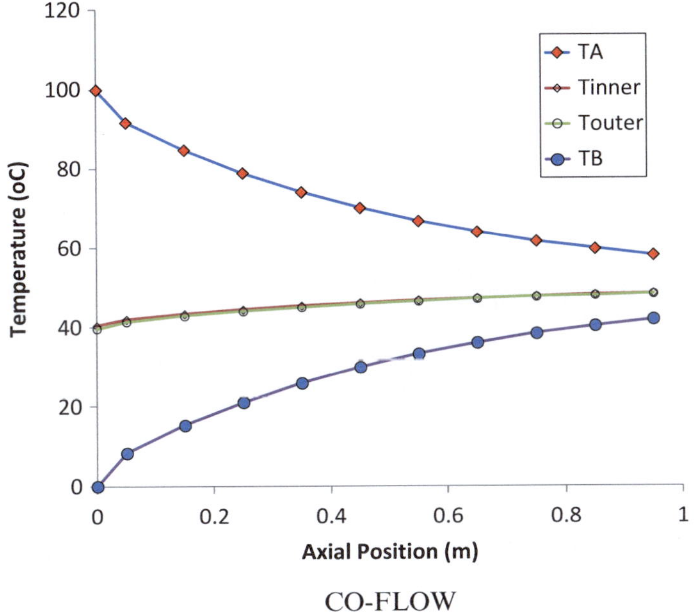

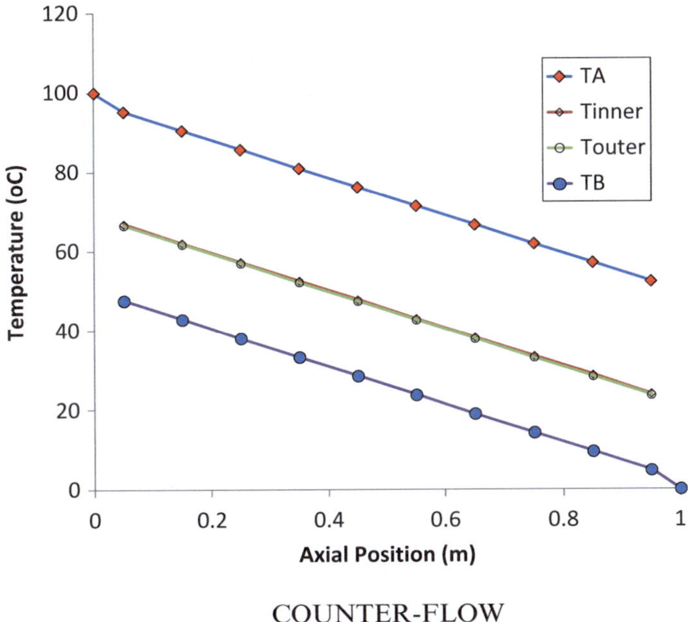

Fig. 12.15 Temperature profiles for base case water to water heat exchanger with unity thermal lengths

INPUT PARAMETERS

L=	2	m	pipe length
rhoA=	1000	kg/m^3	density of A
cpA =	4200	J/kg/K	specific heat of A
rhoB=	1000	kg/m^3	density of B
cpB=	4200	J/kg/K	specific heat of B
Dinner =	0.02	m	inner diameter of inner pipe
t =	0.005	m	pipe thickness
D =	0.2	m	inner diameter of outer pipe
k =	40	W/m/K	thermal conductivity of pipe
AV,A=	8.91E-06	m^3/s	volumetric flow rate of A
AV,B=	8.91E-06	m^3/s	volumetric flow rate of B
Tain=	100	oC	inlet temperature of A
Tbin=	0	oC	inlet temperature of B
hinner =	1000	W/m2/K	inner convection coefficient
houter =	1000	W/m2/K	outer convection coefficient

DERIVED PARAMETERS

dx=	0.2	m	length of numerical segment (for 10 section grid)
Rinner =	0.079577	oC/W	thermal resistance between A and inner wall
Rwall=	0.00097	oC/W	thermal resistance across pipe wall
Router=	0.053052	oC/W	thermal resistance between B and outer wall
Uinner=	595.6416	W/m2/K	overall heat transfer coefficient
mdotA=	8.91E-03	kg/s	mass flow rate of A
mdotB =	8.91E-03	kg/s	mass flow rate of B
(mdot*cp)A=	3.74E+01	oC/W	Flow capacitance of A
(mdot*cp)B=	3.74E+01	W/K	Flow capacitance of B
LthermalA=	0.999798	m	thermal length of A
LthermalB=	0.999798	m	thermal length of B
A =	2.000404		L/LthermalA, dimensionless length of A
B =	2.000404		L/LthermalB, dimensionless length of B
qmaxA =	3741.77	W	maximum possible heat transfer rate of A
qmaxB =	3741.77	W	maximum possible heat transfer rate of B
qmax	3741.77	W	maximum possible heat transfer rate

RESULTS   (change = 0.00E+00   0.00E+00)

	2-Node	co-flow	counter-flow		
TAout=	60.0	51.7	37.5	oC	exit temperature of A
TBout=	40.0	48.3	62.5	oC	exit temperature of B
qA=	-1496.8	-1806.2	-2338.8	W	heat transfer rate to A
qB=	1496.8	1806.2	2338.8	W	heat transfer rate to B
effectiveness=	0.400	0.483	0.625		heat exchanger effectiveness

NTU= UA/Cmin	2.00	"Number of Transfer Units" = Longer of the 2 thermal lengths
Cmin/Cmax=	1.00	ratio of thermal lengths
AV,A=	5.35E-01	LPM	volumetric flow rate of A
AV,B=	5.35E-01	LPM	volumetric flow rate of B
Lt,A=	0.999798	m	Thermal length of A
Lt,B=	0.999798	m	Thermal length of B
AREA=	0.125664	m^2	Inner Surface Area

Fig. 12.16 Parameters and results for a pipe length twice that of the base case

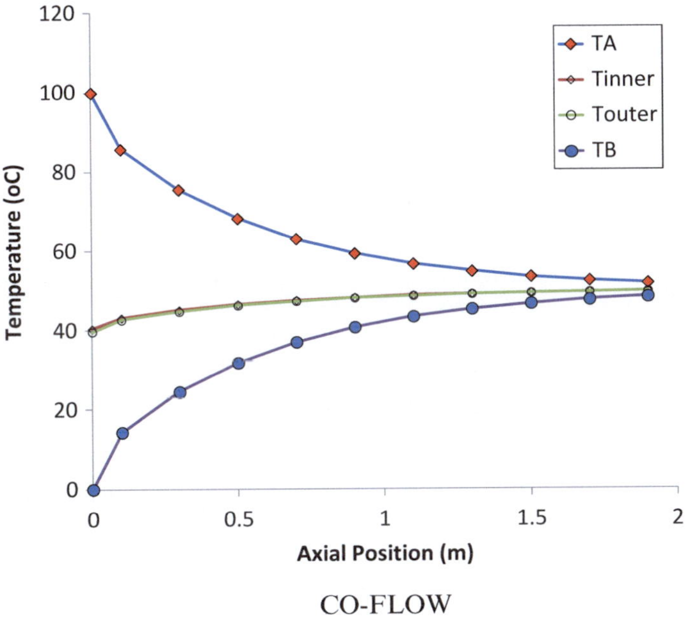

CO-FLOW

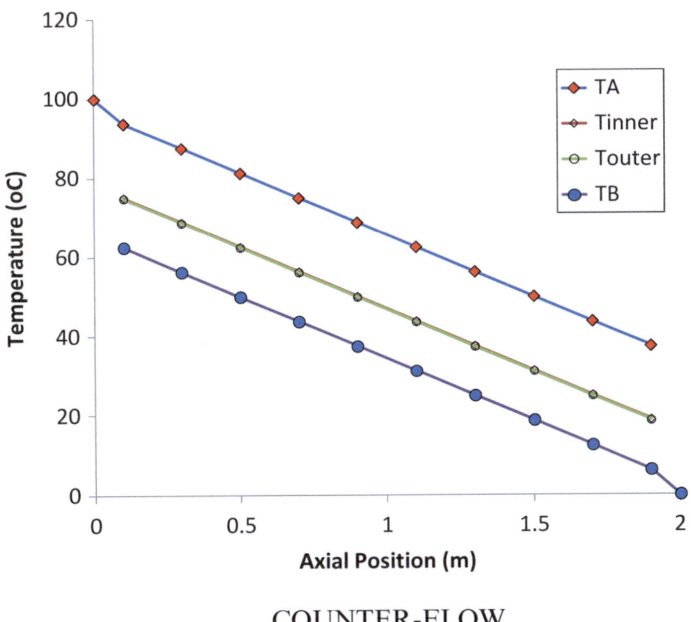

COUNTER-FLOW

Fig. 12.17 Temperature profiles for a pipe length twice as long as the base case

INPUT PARAMETERS

L=	0.5	m	pipe length
rhoA=	1000	kg/m^3	density of A
cpA =	4200	J/kg/K	specific heat of A
rhoB=	1000	kg/m^3	density of B
cpB=	4200	J/kg/K	specific heat of B
Dinner =	0.02	m	inner diameter of inner pipe
t =	0.005	m	pipe thickness
D =	0.2	m	inner diameter of outer pipe
k =	40	W/m/K	thermal conductivity of pipe
AV,A=	8.91E-06	m^3/s	volumetric flow rate of A
AV,B=	8.91E-06	m^3/s	volumetric flow rate of B
Tain=	100	oC	inlet temperature of A
Tbin=	0	oC	inlet temperature of B
hinner =	1000	W/m2/K	inner convection coefficient
houter =	1000	W/m2/K	outer convection coefficient

DERIVED PARAMETERS

dx=	0.05	m	length of numerical segment (for 10 section grid)
Rinner =	0.31831	oC/W	thermal resistance between A and inner wall
Rwall=	0.003882	oC/W	thermal resistance across pipe wall
Router=	0.212207	oC/W	thermal resistance between B and outer wall
Uinner=	595.6416	W/m2/K	overall heat transfer coefficient
mdotA=	8.91E-03	kg/s	mass flow rate of A
mdotB =	8.91E-03	kg/s	mass flow rate of B
(mdot*cp)A=	3.74E+01	oC/W	Flow capacitance of A
(mdot*cp)B=	3.74E+01	W/K	Flow capacitance of B
LthermalA=	0.999798	m	thermal length of A
LthermalB=	0.999798	m	thermal length of B
A =	0.500101		L/LthermalA, dimensionless length of A
B =	0.500101		L/LthermalB, dimensionless length of B
qmaxA =	3741.77	W	maximum possible heat transfer rate of A
qmaxB =	3741.77	W	maximum possible heat transfer rate of B
qmax	3741.77	W	maximum possible heat transfer rate

RESULTS   (change = 0.00E+00   0.00E+00)

	2-Node	co-flow	counter-flow		
TAout=	75.0	69.3	67.7	oC	exit temperature of A
TBout=	25.0	30.7	32.3	oC	exit temperature of B
qA=	-935.5	-1149.7	-1207.2	W	heat transfer rate to A
qB=	935.5	1149.7	1207.2	W	heat transfer rate to B
effectiveness=	0.250	0.307	0.323		heat exchanger effectiveness

NTU= UA/Cmin	0.50		"Number of Transfer Units" = Longer of the 2 thermal lengths
Cmin/Cmax=	1.00		ratio of thermal lengths
AV,A=	5.35E-01	LPM	volumetric flow rate of A
AV,B=	5.35E-01	LPM	volumetric flow rate of B
Lt,A=	0.999798	m	Thermal length of A
Lt,B=	0.999798	m	Thermal length of B
AREA=	0.031416	m^2	Inner Surface Area

Fig. 12.18 Parameter values for a pipe length half that of the base case

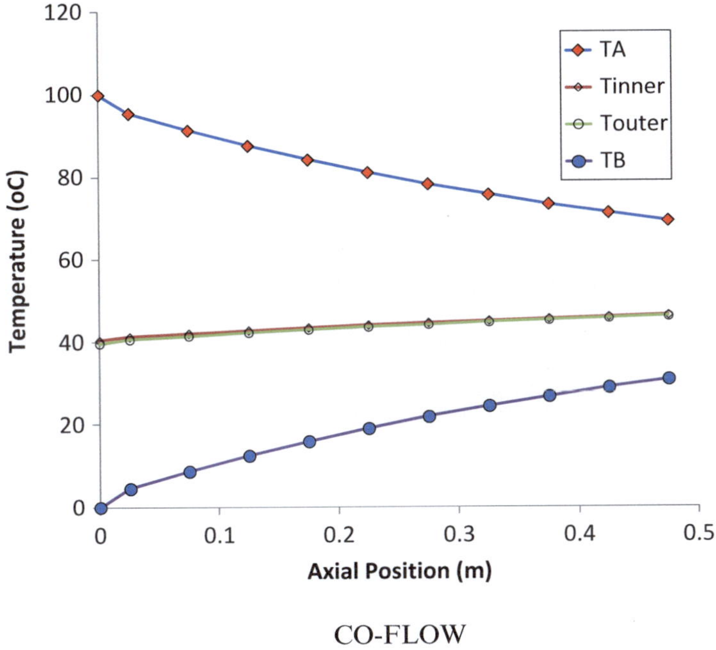

CO-FLOW

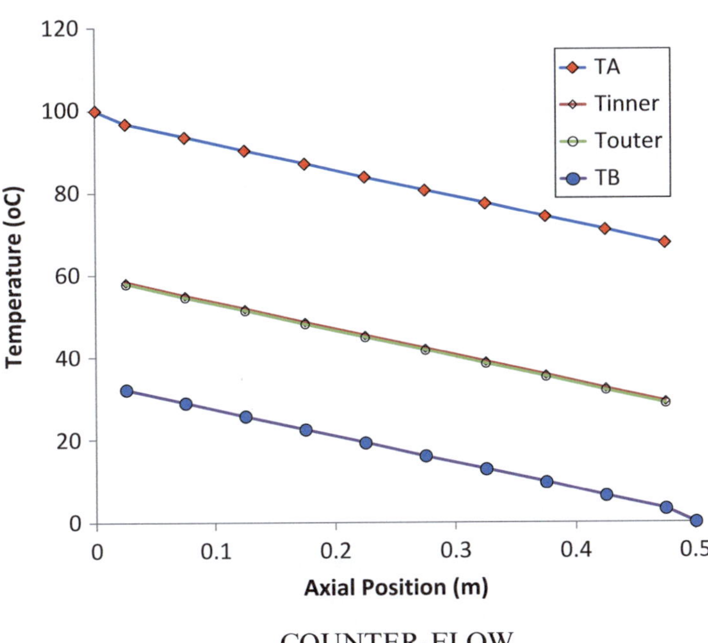

COUNTER-FLOW

Fig. 12.19 Temperature profiles for a pipe length half that of the base case

INPUT PARAMETERS

L=	1	m	pipe length
rhoA=	1000	kg/m^3	density of A
cpA =	4200	J/kg/K	specific heat of A
rhoB=	1000	kg/m^3	density of B
cpB=	4200	J/kg/K	specific heat of B
Dinner =	0.02	m	inner diameter of inner pipe
t =	0.005	m	pipe thickness
D =	0.2	m	inner diameter of outer pipe
k =	40	W/m/K	thermal conductivity of pipe
AV,A=	8.91E-06	m^3/s	volumetric flow rate of A
AV,B=	8.91E-05	m^3/s	volumetric flow rate of B
Tain=	100	oC	inlet temperature of A
Tbin=	0	oC	inlet temperature of B
hinner =	1000	W/m2/K	inner convection coefficient
houter =	1000	W/m2/K	outer convection coefficient

DERIVED PARAMETERS

dx=	0.1	m	length of numerical segment (for 10 section grid)
Rinner =	0.159155	oC/W	thermal resistance between A and inner wall
Rwall=	0.001941	oC/W	thermal resistance across pipe wall
Router=	0.106103	oC/W	thermal resistance between B and outer wall
Uinner=	595.6416	W/m2/K	overall heat transfer coefficient
mdotA=	8.91E-03	kg/s	mass flow rate of A
mdotB =	8.91E-02	kg/s	mass flow rate of B
(mdot*cp)A=	3.74E+01	oC/W	Flow capacitance of A
(mdot*cp)B=	3.74E+02	W/K	Flow capacitance of B
LthermalA=	0.999798	m	thermal length of A
LthermalB=	9.997978	m	thermal length of B
A =	1.000202		L/LthermalA, dimensionless length of A
B =	0.10002		L/LthermalB, dimensionless length of B
qmaxA =	3741.77	W	maximum possible heat transfer rate of A
qmaxB =	37417.7	W	maximum possible heat transfer rate of B
qmax	3741.77	W	maximum possible heat transfer rate

RESULTS   (change = 0.00E+00   0.00E+00)

	2-Node	co-flow	counter-flow		
TAout=	52.4	41.1	40.0	oC	exit temperature of A
TBout=	4.8	5.9	6.0	oC	exit temperature of B
qA=	-1782.0	-2203.9	-2244.0	W	heat transfer rate to A
qB=	1782.0	2203.9	2244.0	W	heat transfer rate to B
effectiveness=	0.476	0.589	0.600		heat exchanger effectiveness

NTU= UA/Cmin	1.00		"Number of Transfer Units" = Longer of the 2 thermal lengths
Cmin/Cmax=	0.10		ratio of thermal lengths
AV,A=	5.35E-01	LPM	volumetric flow rate of A
AV,B=	5.35E+00	LPM	volumetric flow rate of B
Lt,A=	0.999798	m	Thermal length of A
Lt,B=	9.997978	m	Thermal length of B
AREA=	0.062832	m^2	Inner Surface Area

Fig. 12.20 Parameter values with fluid B having ten times the flow rate of the base case

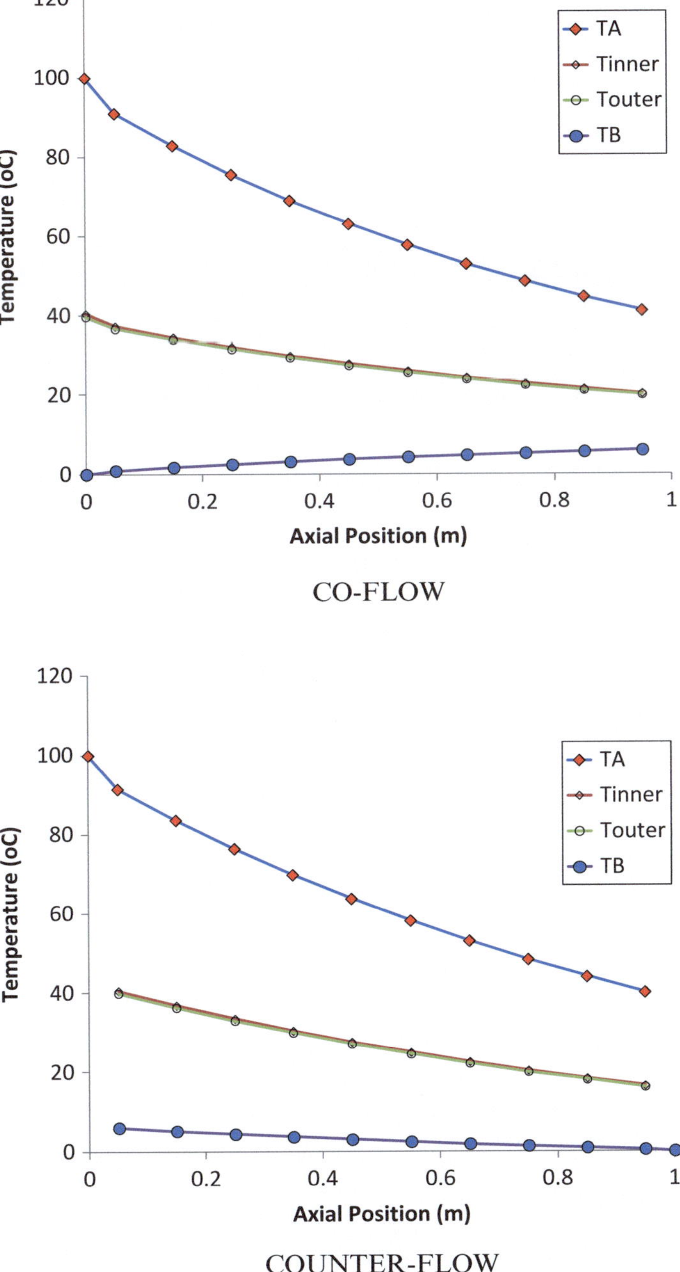

Fig. 12.21 Temperature profiles with fluid B having ten times the flow rate as the base case

INPUT PARAMETERS

L=	1	m	pipe length
rhoA=	1000	kg/m^3	density of A
cpA =	4200	J/kg/K	specific heat of A
rhoB=	1000	kg/m^3	density of B
cpB=	4200	J/kg/K	specific heat of B
Dinner =	0.02	m	inner diameter of inner pipe
t =	0.005	m	pipe thickness
D =	0.2	m	inner diameter of outer pipe
k =	40	W/m/K	thermal conductivity of pipe
AV,A=	8.91E-06	m^3/s	volumetric flow rate of A
AV,B=	8.91E-07	m^3/s	volumetric flow rate of B
Tain=	100	oC	inlet temperature of A
Tbin=	0	oC	inlet temperature of B
hinner =	1000	W/m2/K	inner convection coefficient
houter =	1000	W/m2/K	outer convection coefficient

DERIVED PARAMETERS

dx=	0.1	m	length of numerical segment (for 10 section grid)
Rinner =	0.159155	oC/W	thermal resistance between A and inner wall
Rwall=	0.001941	oC/W	thermal resistance across pipe wall
Router=	0.106103	oC/W	thermal resistance between B and outer wall
Uinner=	595.6416	W/m2/K	overall heat transfer coefficient
mdotA=	8.91E-03	kg/s	mass flow rate of A
mdotB =	8.91E-04	kg/s	mass flow rate of B
(mdot*cp)A=	3.74E+01	oC/W	Flow capacitance of A
(mdot*cp)B=	3.74E+00	W/K	Flow capacitance of B
LthermalA=	0.999798	m	thermal length of A
LthermalB=	0.09998	m	thermal length of B
A =	1.000202		L/LthermalA, dimensionless length of A
B =	10.00202		L/LthermalB, dimensionless length of B
qmaxA =	3741.77	W	maximum possible heat transfer rate of A
qmaxB =	374.177	W	maximum possible heat transfer rate of B
qmax	374.177	W	maximum possible heat transfer rate

RESULTS   (change = 1.42E-14   0.00E+00)

	2-Node	co-flow	counter-flow		
TAout=	91.7	90.9	90.0	oC	exit temperature of A
TBout=	83.3	90.9	99.8	oC	exit temperature of B
qA=	-311.8	-340.0	-373.3	W	heat transfer rate to A
qB=	311.8	340.0	373.3	W	heat transfer rate to B
effectiveness=	0.833	0.909	0.998		heat exchanger effectiveness

NTU= UA/Cmin	10.00		"Number of Transfer Units" = Longer of the 2 thermal lengths
Cmin/Cmax=	0.10		ratio of thermal lengths
AV,A=	5.35E-01	LPM	volumetric flow rate of A
AV,B=	5.35E-02	LPM	volumetric flow rate of B
Lt,A=	0.999798	m	Thermal length of A
Lt,B=	0.09998	m	Thermal length of B
AREA=	0.062832	m^2	Inner Surface Area

Fig. 12.22 Parameter values with the flow rate of B ten times lower than the base case

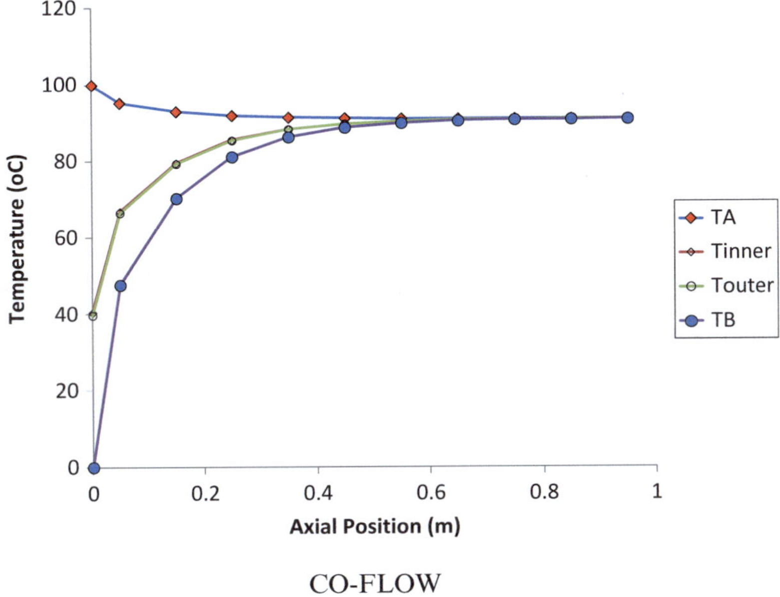

CO-FLOW

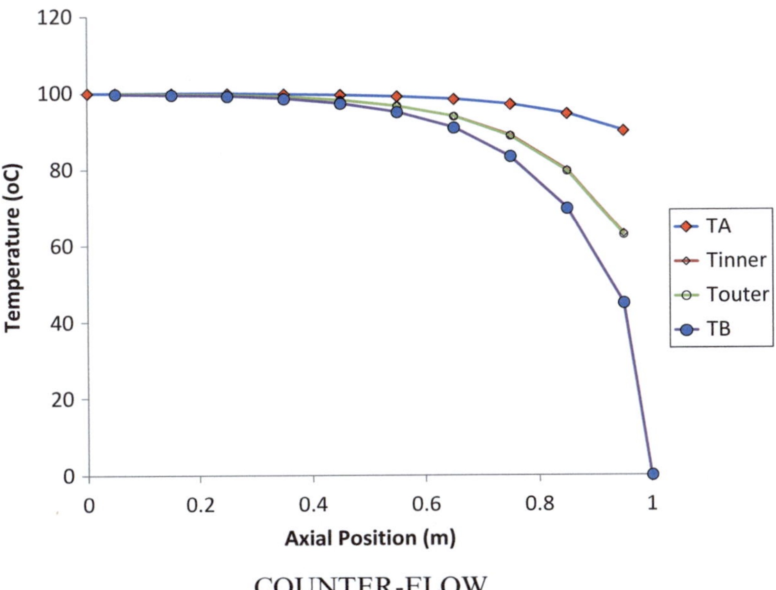

COUNTER-FLOW

Fig. 12.23 Temperature profiles with the flow rate of B ten times lower than the base case

INPUT PARAMETERS

L=	1	m	pipe length
rhoA=	0.946278	kg/m^3	density of A
cpA =	1005	J/kg/K	specific heat of A
rhoB=	1000	kg/m^3	density of B
cpB=	4200	J/kg/K	specific heat of B
Dinner =	0.02	m	inner diameter of inner pipe
t =	0.005	m	pipe thickness
D =	0.2	m	inner diameter of outer pipe
k =	40	W/m/K	thermal conductivity of pipe
AV,A=	3.19E-03	m^3/s	volumetric flow rate of A
AV,B=	7.24E-07	m^3/s	volumetric flow rate of B
Tain=	100	oC	inlet temperature of A
Tbin=	0	oC	inlet temperature of B
hinner =	50	W/m2/K	inner convection coefficient
houter =	1000	W/m2/K	outer convection coefficient

DERIVED PARAMETERS

dx=	0.1	m	length of numerical segment (for 10 section grid)
Rinner =	3.183099	oC/W	thermal resistance between A and inner wall
Rwall=	0.001941	oC/W	thermal resistance across pipe wall
Router=	0.106103	oC/W	thermal resistance between B and outer wall
Uinner=	48.35856	W/m2/K	overall heat transfer coefficient
mdotA=	3.02E-03	kg/s	mass flow rate of A
mdotB =	7.24E-04	kg/s	mass flow rate of B
(mdot*cp)A=	3.04E+00	oC/W	Flow capacitance of A
(mdot*cp)B=	3.04E+00	W/K	Flow capacitance of B
LthermalA=	0.999798	m	thermal length of A
LthermalB=	1.000492	m	thermal length of B
A =	1.000202		L/LthermalA, dimensionless length of A
B =	0.999508		L/LthermalB, dimensionless length of B
qmaxA =	303.7844	W	maximum possible heat transfer rate of A
qmaxB =	303.9953	W	maximum possible heat transfer rate of B
qmax	303.7844	W	maximum possible heat transfer rate

RESULTS   (change = 0.00E+00   0.00E+00)

	2-Node	co-flow	counter-flow		
TAout=	66.7	58.1	52.4	oC	exit temperature of A
TBout=	33.3	41.9	47.6	oC	exit temperature of B
qA=	-101.3	-127.4	-144.7	W	heat transfer rate to A
qB=	101.3	127.4	144.7	W	heat transfer rate to B
effectiveness=	0.333	0.419	0.476		heat exchanger effectiveness

NTU= UA/Cmin	1.00	"Number of Transfer Units" = Longer of the 2 thermal lengths
Cmin/Cmax=	1.00	ratio of thermal lengths
AV,A=	1.92E+02	LPM	volumetric flow rate of A
AV,B=	4.34E-02	LPM	volumetric flow rate of B
Lt,A=	0.999798	m	Thermal length of A
Lt,B=	1.000492	m	Thermal length of B
AREA=	0.062832	m^2	Inner Surface Area

Fig. 12.24 Parameters with air as fluid A, and the flow rate adjusted to achieve a thermal length of unity in both streams

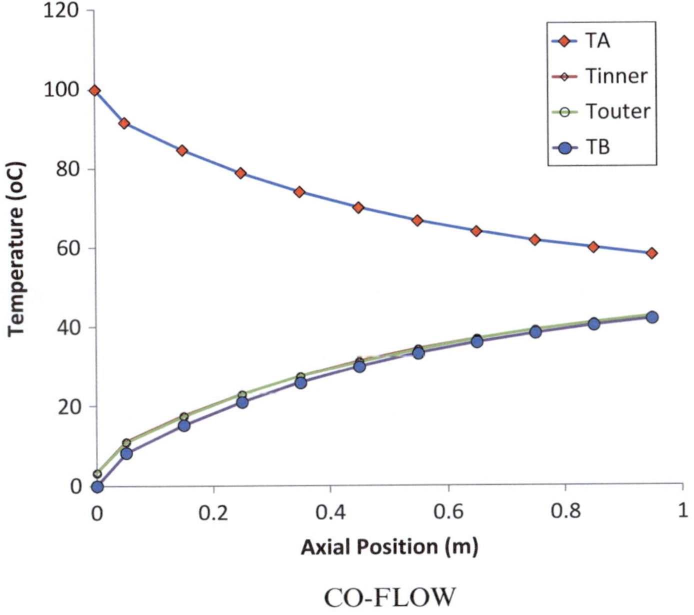

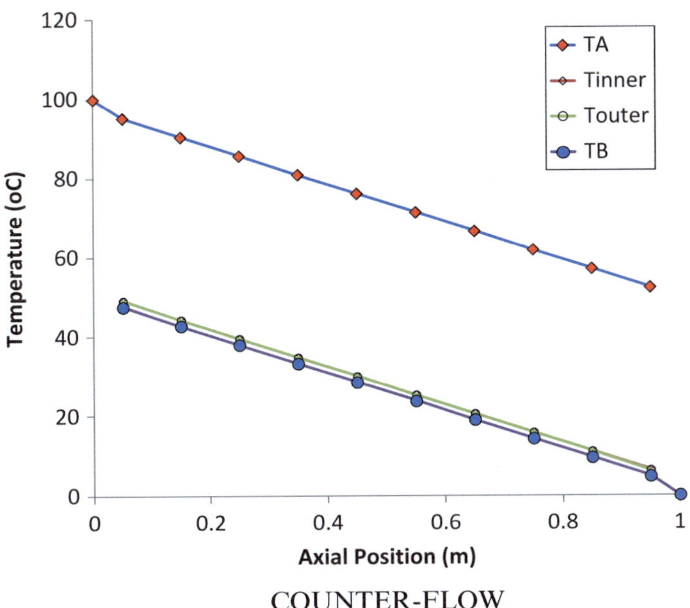

Fig. 12.25 Temperature profiles with air as fluid A and both flow rates adjusted to achieve a thermal length of unity

Workshop 12.2. Serving Coffee at the Right Temperature

Motivation. A real revolution in the coffee bar experience would be if a customer could specify the serving temperature of coffee. Coffee must be brewed at close to the boiling point of water to obtain the desired chemical extraction, but that temperature is generally too hot to drink (except for my mom). So, as a capstone workshop to this textbook, a preliminary heat exchanger design is outlined whose function is to cool coffee from its brewing temperature (defined as 95 °C) to a consumer-selected outlet temperature (ranging from, say 60–80 °C). However, rather than dissipating the excess heat to the environment, heat is transferred from the coffee to feedwater which offsets the heat required in the brewing process. The problem is set up to be a continuous process, which would be difficult to execute in practice, as coffee flowing through a tap is generally intermittent. The spirit of the exercise is to practice heat exchanger calculations while creating food (rather, drink) for thought.

Problem statement. A double-pipe heat exchanger is used to cool coffee to a desired serving temperature while preheating feedwater. Coffee at $T_{Ain} = 95$ °C steadily enters the center pipe (3/8″ K-type copper tubing, with inner diameter of 0.402 in. and outer diameter 0.500 in.). Feedwater at $T_{Bin} = 15$ °C enters the annular region (1/2″ K-type copper tubing with inner diameter of 0.527 in. and outer diameter 0.625 in.) at a 20 % higher flow rate than the coffee (to account for evaporation and water retention in the coffee beans). The outer pipe is insulated to minimize heat loss. A valve with discrete settings controls the flow rate and thereby exit temperature of the coffee. The system is designed to deliver a 16 fluid ounce coffee at $T_{Aout} = 80$ °C in 2 seconds considered to be the base case. Coffee at lower temperatures (75, 70, 65, 60 °C) is delivered by a valve setting that lowers the flow rate to a specific corresponding value. The flow rate of the feedwater is 20 % higher than the coffee flow rate (to account for evaporation and water absorbed in coffee beans). Determine:

- The inner and outer convection coefficients for the base case.
- The outlet temperature of the feedwater for the base case.
- The length of the heat exchanger using the well-mixed model and the average temperature model. Use the latter going forward.
- The filling times (for a 16 oz. coffee) and feedwater exit temperatures required for the lower coffee serving temperatures.
- Be an entrepreneur and bring this system to market!

Bonus. To tie in concepts from earlier in the course, analyze what happens when the flow rate is stopped, and the heat exchanger is full of coffee and feedwater. What is the final equilibrium temperature, and how long does it take to achieve it. Note that the volume of the annular region (feedwater) was deliberately chosen to be small compared to the inner region (coffee) so that the coffee would not be TOO cold. This bonus question should raise a lot of issues with the reliability of the practical application of this idea.

Evaporation and Mass Transfer Fundamentals 13

This chapter introduces fundamental mass transfer concepts through the modeling of the evaporation process that takes place at the free surface of a mug of coffee. An evaporative heat transfer coefficient, h_{evap}, is defined that can be included in any of the transient numerical simulations from earlier chapters (i.e., lumped 1-node, few node, multi-node). Evaporation is driven by concentration gradients, not a temperature difference, and defining an evaporation coefficient this way allows direct comparison of the relative effects of evaporation, convection, and radiation on the cooling of the free surface. As will become apparent, this approach makes sense only for cases when the liquid surface temperature differs significantly from ambient. The case where the coffee reaches thermal equilibrium with the environment and the slower evaporation continues is treated differently, and it will be seen that the coffee attains a temperature below ambient, a form of evaporative cooling.

Specifically, coffee (treated as water and in a quasi-steady state) is at an average temperature T_{coffee} in a cylindrical container of inner diameter D, with a free surface a distance L below the top rim. Ambient air is at a temperature T_∞ with a relative humidity r_h. The goal is to estimate the rate of energy transport by the evaporation process in a Newton's Law of Cooling form, that is:

$$q_{evap} = h_{evap}S(T_s - T_\infty)$$

where S is the surface area, and T_s is the surface temperature of the water. Compare, as a function of coffee temperature, the magnitude of this mechanism to the convection and radiation processes that act in parallel.

13.1 Background

The free surface of the coffee is the only fluid/fluid interface in the cooling mug of coffee scenario. Something different happens there than at fluid/solid interfaces, namely, evaporation. Technically, evaporation is not heat transfer; it is mass transfer.

© Springer International Publishing Switzerland 2015
G. Sidebotham, *Heat Transfer Modeling: An Inductive Approach*,
DOI 10.1007/978-3-319-14514-3_13

Heat transfer is energy that crosses a boundary due to a temperature difference. Mass transfer is mass that crosses a boundary due to molecular diffusion driven by a concentration gradient. That mass has energy associated with it, and therefore, thermal consequences. It is a fundamentally different mechanism than conduction or radiation. It is also a fundamentally different mode of energy transfer than advection, in which case energy flows into a control volume with the bulk mass.

Empirically, the rate of evaporation appears to be much slower than the rate of cooling. The coffee reaches room temperature without any noticeable change in coffee volume (barring sipping, of course). However, there is phase change involved, and a small amount of mass in the liquid phase that exits at a much higher energy level as vapor is more important from an energy perspective than from a mass perspective.

The evaporation process is influenced by humidity, the familiar weather term. The absolute humidity, not reported in weather reports, is generally defined as the mass of water vapor in a given volume of humid air. In weather reports, humidity is given as a percentage, called the relative humidity. Figure 13.1 shows a schematic snapshot of a sample of humid air, a term which means that there is air (small dots) with a few water molecules dispersed in it (big dots). If a snapshot were taken a short time after this one, the picture would look the same, but every molecule would be in a different place. All the molecules migrate, performing a random walk, but there is no net drift. At any one time, if a molecule moves up, another is moving down. In reality, the water molecules are a little smaller than air molecules, but they are shown as big dots because they are the main focus of the present discussion.

Air, with all water vapor removed, is a mixture of fixed component species consisting of 78.0840 mol% N_2, 20.9460 % O_2, 0.9340 % Argon, 0.0390 % CO_2

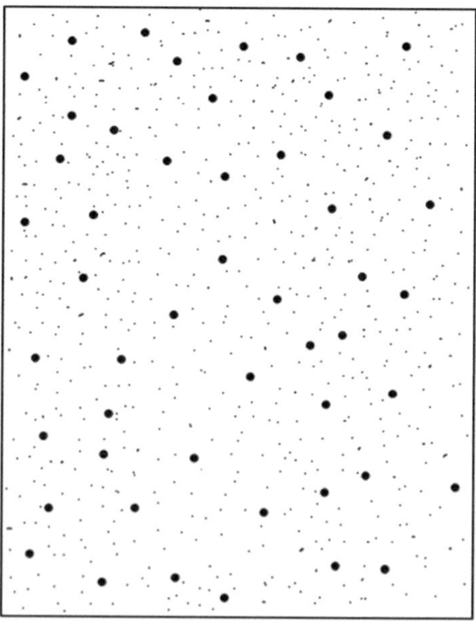

Fig. 13.1 Schematic representation of a snapshot of humid air, with air molecules represented as *small dots*, and water molecules represented as *large dots*

(and rising due to fossil fuel combustion) and many other components in decreasing concentration.[1] These components are considered to be noncondensable, except under sufficiently low temperatures and/or high pressures, and can be treated together (in the absence of chemical reactions), with an equivalent molecular weight of 28.8 kg/kmol.

On the other hand, water vapor is condensable. If humid air is sufficiently cooled (isobarically), it will condense (at its dew point temperature). The concentration (expressed as a number per unit volume, mole fraction, or mass fraction) of water vapor in humid air is variable and limited by the liquid/vapor phase change properties of water. Dalton's Law defines partial pressures for gas mixtures, in which each component of the mixture exerts a pressure on the walls of the vessel as if the other components did not exist. The total pressure is the sum of the partial pressures. The partial pressure exerted by water vapor in humid air is shown as a function of temperature in Fig. 13.2 for several values of the relative humidity. The partial pressure exerted by water vapor for 100 % relative humidity air corresponds to the saturation pressure, that is, the pressure at which pure water would boil for a given temperature. Relative humidity curves are simple fractions of that curve.

For nominal room temperature air (say 25 °C at 1 atm total pressure), the partial pressure exerted by water vapor in 100 % humid air is only 0.025 atm. If a drop of

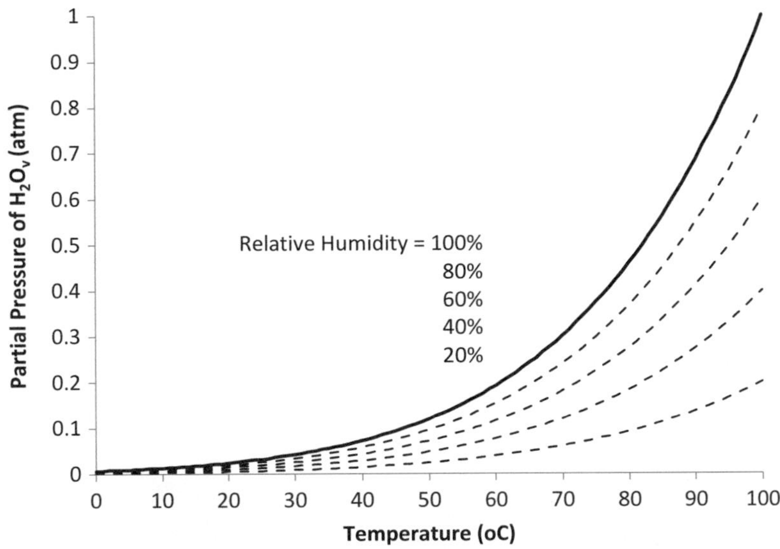

Fig. 13.2 Saturation (vapor) pressure of water as a function of temperature between the normal freezing and boiling points (0 and 100 °C) for various values of relative humidity. The saturation curve (100 % relative humidity) is formed from a fit of the form $P = Ae^{B/T}$ (consistent with the Clausius–Clapeyron equation) using tabulated points at 25 and 100 °C

---

[1] http://en.wikipedia.org/wiki/Atmosphere_of_Earth, accessed 19 November 2010.

liquid water were added, that drop will simply fall to the floor of the container, and none of it would evaporate. However, if that initial 100 % relative humidity air at 25 °C were heated at constant total pressure to, say, 70 °C, it would exert the same partial pressure, but would now correspond to a relative humidity of less than 20 %. At the higher temperature, the water vapor is capable of exerting a much higher partial pressure, but the amount of water vapor is fixed by what was initially present. If a drop of liquid water were now added, that liquid would evaporate, raising the relative humidity. More liquid water could be added, and it would continue to evaporate until the partial pressure of water was 0.3 atm, at which point the air would be saturated.

Consider the scenario in Fig. 13.3 in which hot coffee is poured into a container and a lid is suddenly placed on it, trapping room air. The sketch on the left shows the situation immediately after placing the lid. The sketch on the right shows the situation after some period of time long enough that the air becomes saturated (right), but short enough that the coffee is still hot. There is a vent hole that allows the total pressure to remain atmospheric and therefore room air can flow in, or the vapor/air mixture inside can flow out, depending on the circumstances. The trapped air starts with the same humidity as the ambient air, which is less than or equal to 100 %. Inside the mug, there are two regions in different phases, namely, a liquid (coffee, shown as solid black), and a gas above it, which is a mixture of water molecules (big dots) and air molecules (small dots). The air will naturally be warmed to a value between the coffee temperature and ambient. A good exercise would be to estimate its temperature using a resistance network. If the container and

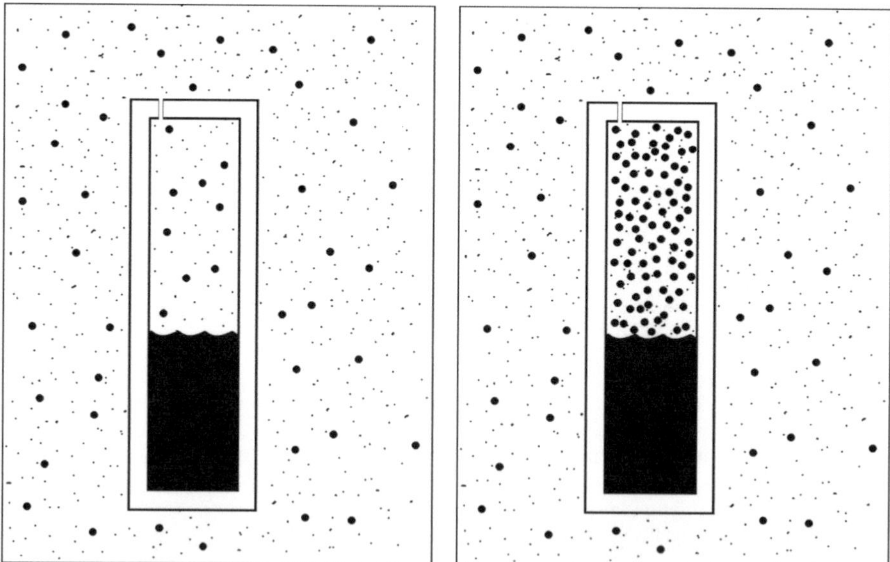

Fig. 13.3 Schematic representation of a container of liquid water with trapped room air immediately after a lid is placed (*left*) and after enough time for equilibrium to be established (*right*)

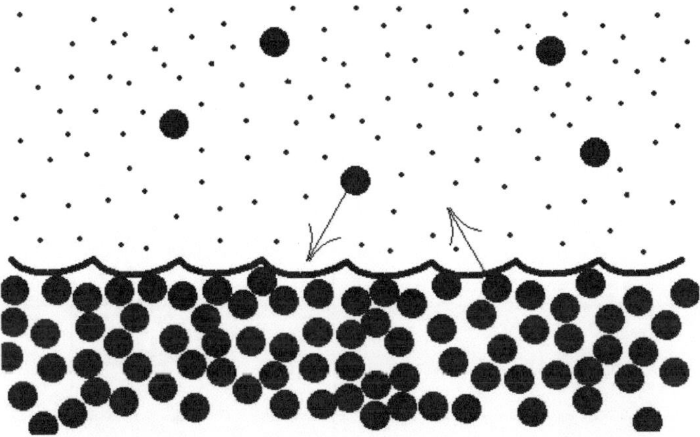

Fig. 13.4 Schematic representation of the dynamic quasi-equilibrium at a water/humid air interface. Gas phase water molecules that strike the surface become absorbed in the surface as some liquid molecules near the surface can break free

lid are well insulated, the air temperature will be close to the coffee temperature. Given time, presumably short compared to the cooling time, enough liquid will evaporate until the now warm humid air will become saturated.

Figure 13.4 is a close-up snapshot of the dynamic equilibrium associated with the coffee-free surface, with two water molecules shown with arrows to indicate their motion. The molecules in the liquid phase are all vibrating, but are held together by intermolecular forces. Liquid molecules near the free surface, however, can occasionally vibrate so hard that they break free from their neighbors, and enter into the gas phase. The higher the temperature of the liquid, the harder they vibrate, and the more leave in a given time frame. These molecules are evaporating (going from liquid to gas phase), and the number of them evaporating in a given period of time increases strongly with liquid temperature.

The water molecules in the gas phase, by contrast, are not heavily influenced by intermolecular forces, until they experience a collision. They travel in straight lines until they encounter either an air molecule, another water molecule, the free surface or the walls of the container. During collisions between molecules, the intermolecular forces are very high, short duration, and repulsive. The collision serves to redirect both collision partners. Collisions with the free surface of the liquid are different, however. A water molecule that collides with the free surface can be absorbed into the liquid phase. Those molecules condense (going from gas to liquid phase). The number of molecules that condense in a given period of time, then, depends directly on the number of collisions with the free surface that take place in that time. It is generally assumed (and justifiable) that a quasi-equilibrium is quickly established in the near vicinity of the liquid surface in which the rate of molecules evaporating equals the rate of molecules condensing. Therefore, the partial pressure of the water vapor molecules is generally assumed to be equal to the saturation pressure just on the gas side of the free surface.

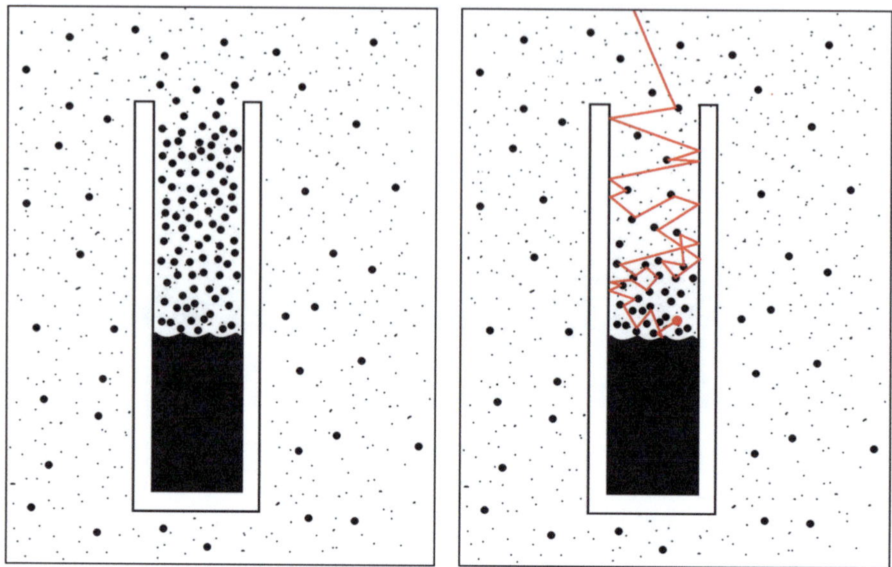

Fig. 13.5 Schematic representation after the lid is first removed (*left*) and the subsequent quasi-steady scenario, with the migratory path of a target molecule shown in *red* (*right*)

Figure 13.5 shows the mug just after the lid is removed. Those vapor molecules that would otherwise have hit the lid and bounced back inside now escape to the room. The concentration (number density) therefore decreases near the open top of the container. Those escaping molecules are replaced by molecules from below, some of which then escape, and a form of "concentration wave" gradually works its way down toward the free surface of the coffee. At the free surface, molecules from the liquid continue to evaporate into the gas phase, while gas phase water molecules that hit the liquid are reabsorbed. The concentration of vapor molecules just above the coffee surface is therefore maintained at some level near the saturation condition associated with the coffee surface.

A quasi-steady state situation is achieved (in short order, in practice, compared to the cooling rate) in which vapor molecules escape at a steady rate. The right sketch shows the path of an individual vapor molecule, painted red. It starts out near the free surface and is intermittently redirected at each collision. The average length of each straight line, called the "mean free path," is the average distance of travel between collisions. Eventually, the molecule works its way up and out. The speed of the molecules is close to the speed of sound (i.e., fast), but the speed that it moves in the vertical direction is much slower than that. This migration from regions of high concentration to low concentration is called diffusion and is the fundamental mechanism required for evaporation to occur.

With this backdrop of the basic physics of vapor pressures and diffusion, some simplified models aimed at quantifying the evaporation rate, and its thermal implications, are developed.

13.2 Model Development

A series of models to quantify fundamental diffusion effects is developed. First, the energy balance on the lumped capacity model is modified to include the effects of evaporation. Then a model of pure diffusion, called Model 1 across the air cylinder above the coffee surface is derived heuristically. Model 2 recognizes that along with the diffusion of water vapor through the air cylinder, there is a net drift, or bulk flow, that was ignored in Model 1. Model 3 considers a control volume outside the mug and yields a working model more suitable when the mug is nearly full. Model 4 combines Models 2 and 3. These models are intended to predict the effect of evaporation during the cooling phase of the coffee, and the change in mass of the coffee/mug system is neglected. A separate final analysis focuses on the change in mass for a mug of coffee left indefinitely on a table. How long does it take to completely evaporate?

13.2.1 Lumped Model Revisited

Consider the original system used to analyze the cooling mug of coffee. The boundary was placed on the free surface of the mug, and the mass that crossed it, hence the evaporation, was neglected. That is, it was treated as a closed system, to which the conservation laws of mass and energy apply. To address the phenomenon of evaporation, the system must be treated as an open system, that is, one in which mass crosses the boundary.

The First Law of Thermodynamics for this particular open system is not derived in detail here. The result, which adds a simple term to account for evaporation to the lumped capacity model, is:

$$(Mc)_{Coffee+Mug} \frac{dT_{coffee}}{dt} = \dot{Q}_{FreeSurface} + \dot{Q}_{sides} + \dot{Q}_{floor} - \dot{m}_{evap} h_{fg}$$

The heat transfer rate into the system from the free surface, sides, and floor are appropriate convection and radiation modes of heat transfer. The free surface term, however, represents the heat transfer between the free surface and ambient, which acts in parallel to the evaporation process, and can be considered a convective/radiative mode. The last term, the evaporative cooling term, is the focus of this chapter. The rate of evaporation, $\dot{m}_{evap}$, is the rate of evaporation, in kg/s, and h_{fg} is the enthalpy of evaporation of water. Water that evaporates enters the free surface as a liquid, and exits as a vapor, with a higher specific energy content. The energy required to change the phase has the effect of reducing the energy stored in the

CV. There is nothing to stop a definition of an evaporative "convection-like" coefficient:

$$\dot{m}_{evap}\, h_{fg} \equiv h_{evap} S (T_\infty - T_S)$$

where $S = \pi D^2/4$ is the area of the free surface of the coffee. Rearranging to define the evaporative heat transfer coefficient:

$$h_{evap} = \frac{\dot{m}_{evap}\, h_{fg}}{S(T_\infty - T_S)}$$

This evaporative heat transfer coefficient model really makes the most sense when a hot liquid is cooling, and the surface temperature of the liquid is substantially different than ambient. When the coffee temperature equals ambient, evaporation is not zero, and the coefficient is undefined (divided by zero).

A traditional way to express the mass transfer process, analogous to Newton's Law of Cooling, is to define an evaporative coefficient, γ_{evap}, that is the proportionality factor (not necessarily constant) for evaporation driven by a difference in concentration of the evaporating species, expressible as the mass fraction, Y:

$$\dot{m}_{evap} = \gamma_{evap} S (Y_S - Y_\infty)$$

Here, Y_S is the mass fraction of the evaporating component (water vapor) just above the surface, and Y_∞ is that in the free stream (which is where humidity of the air comes into play). This latter model is more appropriate for a pure evaporation process, not one in which cooling of the bulk liquid is taking place simultaneously with the evaporation. That is, the evaporation process after the coffee cools to ambient (or very close to it). In practice, correlations similar to Nusselt number correlations are used to determine mass transfer coefficients.

Something surprising happens at steady-state, namely, the coffee temperature does not equal ambient temperature:

$$\dot{Q}_{FreeSurface} + \dot{Q}_{sides} + \dot{Q}_{floor} = \dot{m}_{evap}\, h_{fg} \quad \text{(at steady state)}$$

The energy required for evaporation is supplied by heat transfer from the environment to the coffee. Using a lumped capacity resistance network, this heat transfer rate can be expressed in terms of an overall heat transfer coefficient, U, based on the surface area of the mug of coffee:

$$\dot{Q}_{FreeSurface} + \dot{Q}_{sides} + \dot{Q}_{floor} = UA \left(T_\infty - T_{coffee,SS} \right) = \dot{m}_{evap}\, h_{fg}$$

Solving for the steady-state coffee temperature:

$$T_{coffee,SS} = T_\infty - \frac{\dot{m}_{evap}\, h_{fg}}{UA}$$

The only way heat can be transferred to the coffee is if the coffee is at a lower temperature than ambient. This phenomenon is difficult to measure, but is real, and is why we sweat.

13.2.2 Model 1: Pure Diffusion Model and Fick's Law

To model the rate of evaporation, consider a control volume that consists of the humid air column between the free surface and the exit plane below the rim, shown in Fig. 13.6. On the bottom of the control volume, at $y = 0$, the air is assumed to be at the temperature of the liquid at the surface (not necessarily equal to the bulk coffee temperature), and with a mass fraction Y_{v0}. An upper limit to Y_{v0} would be to assume the air is saturated there. At the top of the control volume, at $y = L$, the air has a mass fraction Y_{vL}, which has a lower limit defined by the humidity of ambient room air. Even if the liquid is at ambient temperature, the concentration of water vapor at the surface is higher than that in the ambient, and a concentration gradient exists, unless the room is at 100 % relative humidity and the same temperature as the liquid. The actual concentration of water vapor at the exit plane is generally higher than that of the ambient air, but air currents from the room tend to make it equal, in the limit, especially for large values of L. For liquids colder than ambient, the concentration is lower and water vapor diffuses in and a net condensation occurs.

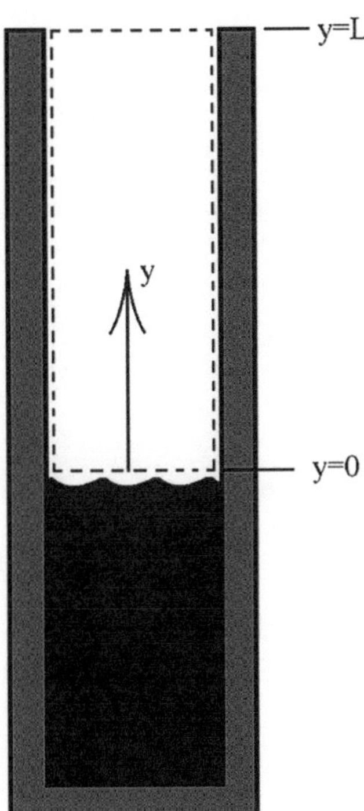

Fig. 13.6 Control volume of the humid air column between the coffee surface (at $y = 0$) and the exit plane (at $y = L$), shown as *dashed line*

Since evaporation is driven by a difference in concentration, it is postulated that the rate of evaporation, expressed as a mass flow rate of water vapor through the CV, is proportional to the difference in concentration, expressed as a mass fraction, Y, between the top and bottom of the control volume:

$$\dot{m}_{evap} \sim (Y_{v0} - Y_{vL})$$

It is further postulated that the evaporation rate varies inversely with the distance between the two surfaces, L:

$$\dot{m}_{evap} \sim \frac{1}{L}$$

These two effects together make up a gradient, a change in mass fraction over a distance.

Furthermore, if the surface area $(S = \pi D^2/4)$ is doubled, the evaporation rate is expected to double, meaning the evaporation rate is proportional to S:

$$\dot{m}_{evap} \sim S$$

Now, the more dense the humid air, the more molecules are packed in a given volume, all of which are diffusing out:

$$\dot{m}_{evap} \sim \rho$$

Putting these four effects together:

$$\dot{m}_{evap} \sim \frac{\rho S(Y_{v0} - Y_{vL})}{L}$$

The proportionality factor includes all other effects, most importantly, how easy it is for molecules to migrate from the concentrated area to the less concentrated area. In keeping with convention, a diffusion coefficient D_v is introduced:

$$\dot{m}_{evap,MODEL\,1} = \frac{\rho D_v S}{L}(Y_{v0} - Y_{vL})$$

The SI units of D_v are m^2/s. A literature value for the diffusion coefficient of water vapor into air (at nominal room conditions) is:

$$D_v = 25.6(10)^{-6}\,\text{m}^2/\text{s}$$

Fick's Law of Diffusion has just been derived heuristically. It is analogous to Fourier's Law of Conduction. This model neglects the fact that there is a bulk flow leaving the mug, which complicates matters and is addressed in the next model. A more general expression in terms of a local concentration gradient is:

$$\dot{m}_{diffusion} = -\rho D_v \frac{\partial Y_v}{\partial y} S$$

The negative sign forces the direction of diffusion to be from high to low concentration.

In Model 1, the gradient of mass fraction is taken as being the overall change in mass fraction divided by the overall length. Consider the limits. The mass fraction of vapor just above the free surface cannot exceed the saturation condition:

$$Y_{v0} \leq Y_{SAT,surface}$$

At the exit plane, the mass fraction cannot be lower than ambient:

$$Y_{vL} \geq Y_{\infty}$$

Both of these limits work in the same direction, namely, to reduce the value of the concentration difference from the free surface to the exit plane, and therefore:

$$Y_{v0} - Y_{vL} \leq Y_{SAT,surface} - Y_{\infty}$$

For Model 1, the limiting case is considered, and the inequality is treated as an equality.

The Appendix details the calculation of mass fractions for this case. The mass fractions are directly related to the partial pressure exerted by the vapor, and therefore are very sensitive to temperature.

Model 1 Results
Results of model predictions are presented for a base case, with parameter values shown in Fig. 13.7. The mug has an 8.93 cm inner diameter (D) and it is filled with coffee at various temperatures in a room with air at 25 °C and a 70 % relative humidity. The amount of coffee added is also varied, and therefore the distance from the free surface to the exit plane (the air cylinder height, L, which is the distance water must diffuse to exit the mug).

The semi-logy plot shows the evaporation rate as a function of instantaneous coffee temperature for several values of the air cylinder length, L. At a fixed value of L, the evaporation rate is zero at a temperature somewhat below ambient (approximately 19 °C). This temperature is the dew point temperature and corresponds to the condition in which the mass fraction of vapor on the free surface is equal to the mass fraction in the ambient. The evaporation rate increases sharply as the temperature moves away from the dew point on both sides. At lower temperatures, however, the evaporation rate is negative (which cannot be plotted on semi-log axes so the absolute value is shown), indicating that water vapor is diffusing into the mug and condensing. At ambient temperature, the evaporation rate is not zero because the room air has a relative humidity less than 100 %.

INPUT PARAMETERS

D	0.0893	m	mug inner diameter
L	VARY	m	distance from free surface to exit plane
Pinf	1.013E+05	N/m2	atmospheric pressure
Tinf	25	oC	ambient temperature
rh	0.7		relative humidity of ambient

PHYSICAL CONSTANTS

Dv	2.56E-05	m2/s	Diffusion coefficient (water vapor into air)
DA	2.06E-05		Diffusion coefficient (air into itself)
hfg	2.33E+06	J/kg	Heat of Vaporization of water
A	9.1496E+10	N/m2	CLAUSIUS-CLAPYRON CONSTANTS FOR WATER
B	-5.1152E+03	K	$$P_{sat}(T) = Ae^{\frac{B}{(T+273)}}$$

DERIVED PARAMETERS

S	0.00626315	m^2	Free Surface Area
Xinf	0.0222		mol fraction of ambient
Yinf	0.0138		mass fraction of ambient
rho	1.1857	kg/m3	dry air density
X0,sat	0.0317		saturated surface mol fraction at ambient
Y0,sat	0.0199		saturated surface mass fraction at ambient
Dew Point	18.93	oC	Dew Point temperature

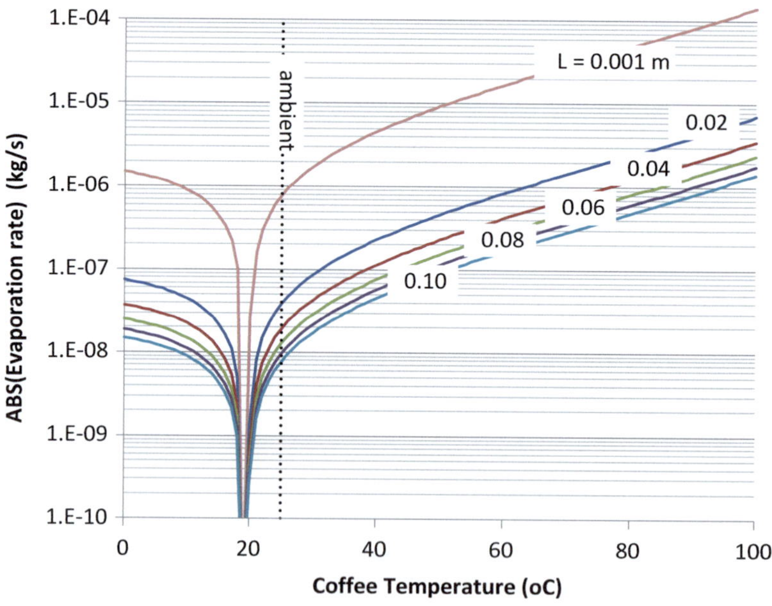

Fig. 13.7 Evaporation rate for Model 1 (pure diffusion) as a function of coffee temperature for various lengths of the air column above the free surface. Below 18.93 °C (the *dew point*), the evaporation rate is negative

As temperature increases above ambient at a fixed L, the evaporation rate increases steadily to the boiling point, changing by over two orders of magnitude. This exponential growth is a direct result of the vapor pressure above the liquid surface. At a fixed temperature, the evaporation rate decreases with L, the distance across which vapor must diffuse. The case for a very nearly full mug (L = 0.001 m) is shown to demonstrate that the evaporation becomes very sensitive to L as L approaches zero, at least as predicted by Model 1. This behavior will be shown to represent a breakdown of the model, which imposed an unrealistic constraint that the vapor mass fraction at the exit plane is equal to the ambient value. The lower the value of L, the worse that assumption is expected to be. This limitation of the pure diffusion model will be addressed in Model 3.

A more subtle limitation to the pure diffusion model is that with evaporation, there is a net mass flow upward. This effect is one of advection, and is addressed in Model 2.

13.2.3 Model 2: Advection/Diffusion Model

There are two parallel means by which water vapor is transported through the air column. There is diffusion across the concentration gradient, and simultaneously, there is a bulk flow that carries it. Model 1 does not account for the bulk flow effect. To do so, it is insightful (and convenient) to view things from the perspective of the air. In this two-component gas system (air and water vapor), the local mass fractions of air and water must everywhere account for all species:

$$Y_v + Y_{air} = 1$$

Taking the derivative:

$$\frac{dY_v}{dy} + \frac{dY_{air}}{dy} = 0$$

This relation demonstrates that if there is a concentration gradient in the water vapor, there must be a gradient in the air of equal magnitude and opposite sign. In the humid air column, air diffuses in the opposite direction as the vapor, toward the coffee surface. However, air does not experience a net flow downward, as it is blocked by the free surface of the coffee (which is considered to be fixed, although it is receding slowly). So as air diffuses downward, it is equally swept upward by the bulk flow. To express these two effects mathematically, with D_A being a diffusion coefficient for air into humid air, the mass flow rate of air crossing any plane inside the humid air column is:

$$\dot{m}_{air} = 0 = -\rho D_A \frac{\partial Y_{air}}{\partial y} S + \dot{m}_{evap} Y_{air}$$

The mole fractions are assumed to be uniform in the horizontal direction, and the partial derivative can be replaced by a total derivative. Separating variables:

$$\frac{dY_{air}}{Y_{air}} = \left(\frac{\dot{m}_{evap}}{\rho D_A S}\right) dy$$

Integrating across the entire control volume (from the free surface to the exit plane), recognizing that the mass flow rate through the column is the evaporation rate and is independent of position:

$$\int_{Y_{A0}}^{Y_{AL}} \frac{dY_{air}}{Y_{air}} = \int_{y=0}^{L} \left(\frac{\dot{m}_{evap}}{\rho D_A S}\right) dy$$

$$\ln\left(\frac{Y_{AL}}{Y_{A0}}\right) = \left(\frac{\dot{m}_{evap}}{\rho D_A S}\right) L$$

Expressing the air mass fraction in terms of the vapor mass fractions (using $Y_A + Y_V = 1$), and rearranging for the evaporation rate:

$$\dot{m}_{evap} = \frac{\rho D_A S}{L} \ln\left(\frac{1 - Y_{vL}}{1 - Y_{v0}}\right)$$

In keeping with tradition, it is useful to define a mass transfer number, B_{0L}, not to be confused with a Biot number, as follows:

$$\left(\frac{1 - Y_{vL}}{1 - Y_{v0}}\right) \equiv 1 + B_{0L}$$

The two subscripts refer to the two boundaries (coffee surface and exit plane). Performing some algebra:

$$B_{0L} = \left(\frac{1 - Y_{vL}}{1 - Y_{v0}}\right) - \left(\frac{1 - Y_{v0}}{1 - Y_{v0}}\right) = \left(\frac{Y_{v0} - Y_{vL}}{1 - Y_{v0}}\right)$$

The final result for the evaporation rate is:

$$\dot{m}_{evap} = \frac{\rho D_A S}{L} \ln(1 + B_{0L})$$

where

$$B_{0L} = \frac{Y_{v0} - Y_{vL}}{1 - Y_{v0}}$$

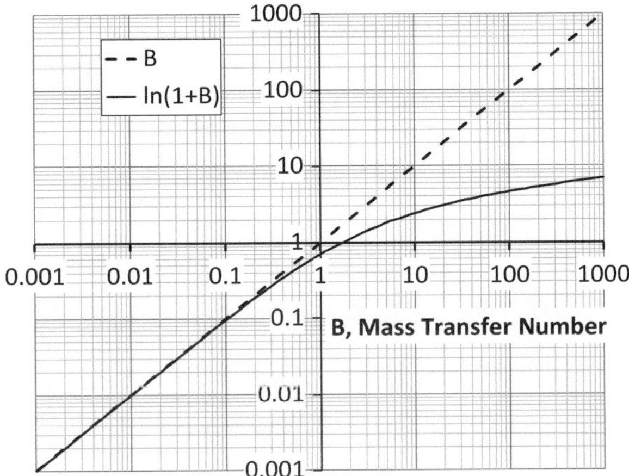

Fig. 13.8 Demonstration of low Biot number limit. The functions B and ln(1+B) are plotted versus B

The mass transfer number is useful since $\ln(1) = 0$, and the evaporation rate is zero for $B = 0$, is positive for $B > 0$, corresponding to evaporation, and negative for $B < 0$, corresponding to condensation. Furthermore, there is a useful limit for low values of B, namely:

$$\lim_{B \to 0} \ln(1 + B) = B$$

As shown in Fig. 13.8, this limit is valid for $B < 0.1$, approximately.

To make quantitative predictions, the mass fractions must be estimated. Considering the limits, the mass fraction of vapor just above the free surface cannot exceed the saturation condition:

$$Y_{v0} \leq Y_{SAT, surface}$$

At the exit plane, the mass fraction cannot be lower than ambient:

$$Y_{vL} \geq Y_{\infty}$$

Both of these limits work in the same direction, namely, to reduce the value of the transfer number, so it can be stated that:

$$B_{0L} \leq \frac{Y_{SAT, surface} - Y_{\infty}}{1 - Y_{SAT, surface}}$$

A diffusion coefficient for air into humid air is different from the diffusion coefficient of water vapor into humid air, namely:

$$D_A = 20.6(10)^{-6} \, \text{m}^2/\text{s}$$

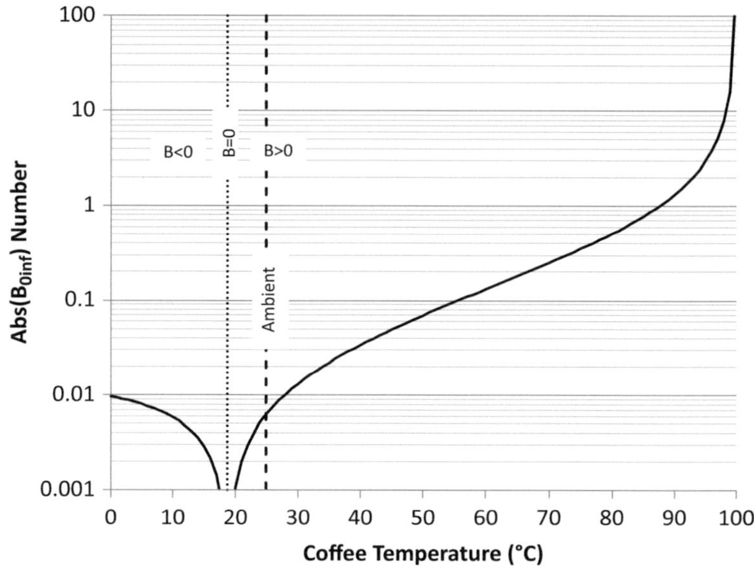

Fig. 13.9 Mass transfer number (B) as a function of coffee temperature, assuming saturation conditions at the coffee surface, and ambient conditions at the exit plane

This difference is due fundamentally to the different size of the molecule. Model 1 was derived from the point of view of water diffusing out, while Model 2 is based on air diffusing in. These approaches are not inconsistent, just different points of view.

Model 2 Results

The results of Model 2 presented assume the mass fraction limits hold in evaluating the B number. The accuracy of that assumption is expected to improve as the distance between the free surface and exit plane (L) increases. Figure 13.9 shows the absolute value of the mass transfer number (B) as a function of coffee temperature on a semi-logy scale. At approximately 18.5 °C, the B number is zero (as is $\ln(1+B)$), corresponding to the mass fraction on the free surface taking on the ambient value. This temperature corresponds to the dew point temperature of the ambient air, that is, the temperature at which water would begin to condense on isobaric cooling. The B number (and therefore $\ln(1+B)$) is negative below the dew point, and it is positive above. Negative evaporation rates correspond to condensation.

The B number increases by over two orders of magnitude between 30 and 90 °C. Above 90 °C, it turns sharply upward as the mass fraction at the surface ($Y_{SAT,0}$) approaches unity, and the denominator goes to zero. In this calculation, the surface temperature is set at a particular value. The evaporation process has a cooling effect on the surface of the coffee, and it becomes more difficult to maintain the surface temperature the closer to the boiling point. To demonstrate that effect, consider the boundary condition at the coffee surface, where the heat transfer to the surface from the adjacent fluid is modeled with heat transfer coefficients:

$$h_{s,coffee}S\left(T_{coffee} - T_S\right) + \dot{m}_{evap}\,h_f = h_{s,air}S(T_S - T_{ambient})$$
$$+ \dot{m}_{evap}\,h_g \quad \text{(surface energy balance)}$$

Solving for the surface temperature:

$$T_S = \frac{h_{s,sir}ST_{ambient} + h_{s,coffee}ST_{coffee} - \dot{m}_{evap}\,h_{fg}}{h_{s,air}S + h_{s,coffee}S}$$

Assuming the heat transfer coefficient between the surface and coffee is much greater than that between the surface and ambient air ($h_{s,coffee} \gg h_{s,air}$):

$$T_S = T_{coffee} - \frac{\dot{m}_{evap}\,h_{fg}}{h_{s,coffee}S}$$

The source of the energy required for the evaporation process is heat transfer from the adjacent fluids, and therefore the surface temperature is below what it would be in the absence of evaporation. As the boiling point is approached, the evaporation rate increases to very large values (as the mass fraction of vapor approaches unity).

Figure 13.10 compares the evaporation rates for the pure diffusion model with the advection/diffusion model for the base case (with L = 0.05 m), with the evaporation

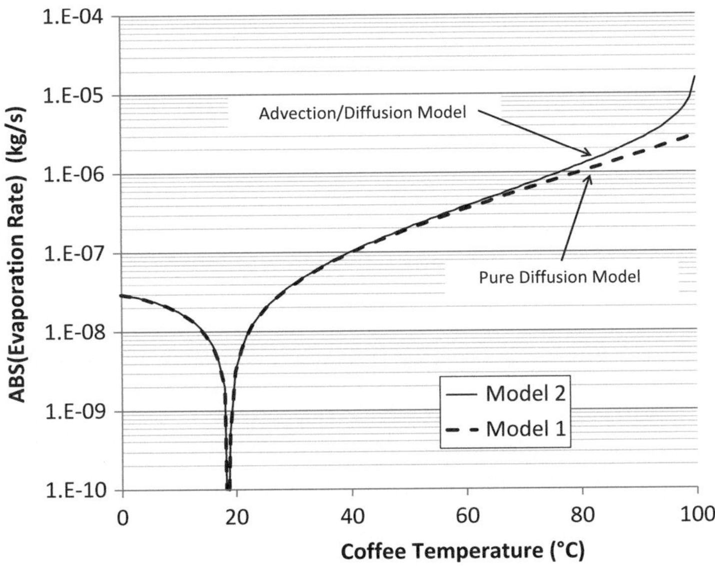

Fig. 13.10 Comparison of the evaporation rate from the pure diffusion and advection/diffusion models

rate plotted against coffee temperature. The comparison is qualitatively similar for all non-zero values of L (and are indeterminate for L = 0, where the evaporation rate is infinite). The models agree almost exactly for temperatures below approximately 70 °C. Above that, the advection/diffusion model predicts an increasingly higher evaporation rate than the pure diffusion model. Physically, there are two parallel mechanisms for the transport of vapor across the air cylinder in Model 2.

Both Models 1 and 2 assume that the condition at the exit plane is that of the ambient air. This artificial constraint becomes less realistic as L decreases.

It is apparent that the effects of advection are only important as the temperature approaches the boiling point of the liquid. Returning to the original problem of the cooling coffee cup, where the coffee starts at a temperature close to its boiling point and gradually cools, the advection/diffusion model represents a significant improvement over the pure evaporation model. It has not been demonstrated (yet) whether the evaporation process has a significant impact on the cooling time and that will be addressed after the final model has been developed.

13.2.4 Model 3: Diffusion from a Point Source

The models considered thus far have focused on a control volume about the air column inside the mug. Evaluation of the evaporation rate requires an assumption about the value of the water mass fraction at both the free surface (with an upper limit at the saturation condition) and at the exit plane of the mug (with a lower limit of the ambient absolute humidity). For the condition at the free surface, it can be argued that setting the mass fraction (Y_{v0}) to the saturation limit ($Y_{sat,0}$) is a good approximation, except perhaps as the boiling point is approached, and the evaporation rate is highest.

On the other hand, to say that the mass fraction of water vapor at the exit plane is equal to the ambient value seems reasonable for a large value of L, but not when the mug is nearly full. Visual inspection of the vapors leaving a hot cup of coffee suggests that there is a significantly higher vapor concentration at the exit plane than exists in the free stream. Furthermore, for full mugs ($L \rightarrow 0$), the predicted evaporation rate increases dramatically, making a difference between the exit plane and ambient even more likely.

In order for water vapor to leave the free surface, it must pass through the air column to the exit plane, and then pass from the exit plane to the far field away from the mug. To get a handle on that outer world, a simplified quasi-steady one-dimensional analysis can be developed. Figure 13.11 shows a schematic that models the exit plane, at a fixed mass fraction, to be a hemispherical shell of radius R_0. The area of the shell is the same as the area of the exit plane:

$$\frac{\pi D^2}{4} = \frac{1}{2}\left(4\pi R_0^2\right)$$

Fig. 13.11 Radial coordinate system outside the mug, with the surface as a hemisphere whose surface area equals that of the coffee-free surface area

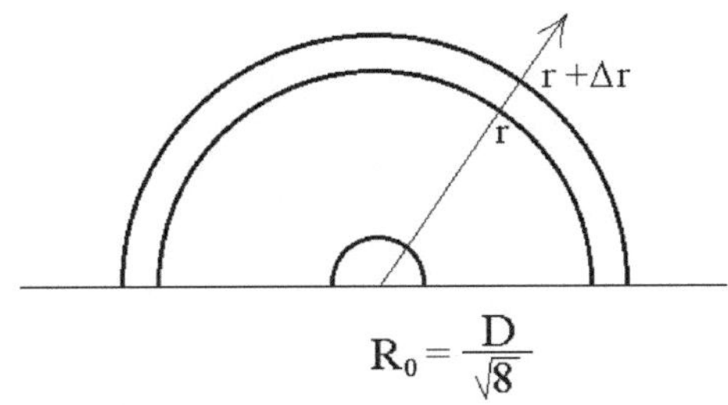

$$R_0 = \frac{D}{\sqrt{8}}$$

The use of the hemisphere is conceptually accurate if the exit plane is at the same level as the floor, and not a height H above it. This estimate therefore is an underestimate of the ability for water vapor to diffuse away. The effects of the global natural convection have a much larger effect in practice. Therefore, the result of this model provides a lower limit on the ability for the environment to maintain the exit plane at ambient conditions.

At an arbitrary radial position from the center line, there is no net mass flux of air:

$$\dot{m}_{air} = 0 = \dot{m}_{evap} Y_a + \left(-\rho D_A \frac{dY_A}{dr} \frac{4\pi r^2}{2} \right)$$

Separating variables:

$$\frac{dY_A}{Y_A} = \left(\frac{\dot{m}_{evap}}{2\pi \rho D_A} \right) \frac{dr}{r^2}$$

Integrating from the exit plane (at R_0, where $Y_A = Y_{AL}$) to "infinity" (where $Y_A = Y_\infty$):

$$\int_{Y_A = Y_{aL}}^{Y_\infty} \frac{dY_A}{Y_A} = \left(\frac{\dot{m}_{evap}}{2\pi \rho D_A} \right) \int_{r=R_0}^{\infty} \frac{dr}{r^2}$$

$$\ln\left(\frac{Y_\infty}{Y_{AL}} \right) = \left(\frac{\dot{m}_{evap}}{2\pi \rho D_A} \right) \left(\frac{-1}{r} \right) \Big|_{r=R_0}^{\infty}$$

Evaluating the limits, expressing mass fractions in terms of water $(1 = Y_v + Y_A)$ and rearranging for the evaporation rate, with the point source radius (R_0) expressed in terms of the mug diameter (D, not to be confused with the diffusion coefficient D_A):

$$\dot{m}_{evap} = 2\pi \frac{D}{\sqrt{8}} \; \rho D_A \ln\left(\frac{1 - Y_{v\infty}}{1 - Y_{vL}}\right)$$

$$\dot{m}_{evap} = \frac{\pi}{\sqrt{2}} \; D\rho D_A \ln(1 + B_{L\infty})$$

where the transfer number applies between the exit plane (L) and ambient (∞):

$$B_{L\infty} = \frac{Y_{vL} - Y_{v\infty}}{1 - Y_{vL}}$$

The major assumption of the point source model is that the only air movement outside the coffee cup is that due to the evaporation process. In reality, especially when the coffee is hot, natural convection currents will create a vertical flow that will bring ambient air closer to the exit plane, and ease the ability for vapor to diffuse from the exit plane to ambient. Therefore, the point source represents a conservative prediction for the evaporation rate.

Model 3 Results

Figure 13.12 shows the evaporation rate as a function of coffee temperature on semi-logy axes for the point source model (Model 3, dotted line) and the advection/diffusion model (Model 2) for several values of air column lengths (L). The point source model prediction is independent of the length of the air column (L), so only

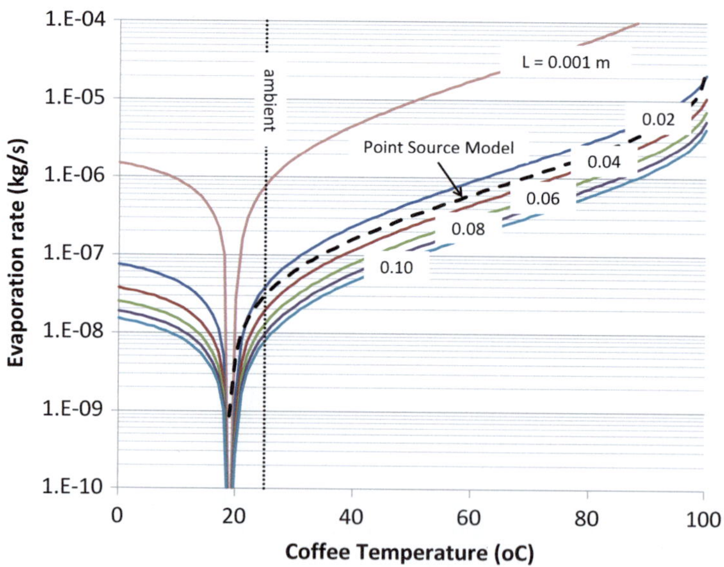

Fig. 13.12 Model 2 (air cylinder with advection/diffusion) and Model 3 (point source) evaporation rates

one curve is shown. The shape of the curve is similar for both models because both are being driven by the same transfer number, B_{0L}.

The evaporation rate predicted by the air column model (Model 2) is greater than that predicted by the point source model for air column lengths less than approximately 3 cm. Conceptually, the exit plane is shared by both models, but it corresponds to the exit plane for the air column and is the inlet plane for the point source. The point source curve (dotted) line represents the maximum rate at which vapor can diffuse away from the exit plane into the room (subject to the major assumptions of the model). For $L < 0.03$, then, the mass delivered to exit plane by the air column exceeds the ability for the point source to remove it.

One way around this contradiction in implementing the model would be to calculate the evaporation rate from both models, and choose the lower of the two as being controlling. Another, more sophisticated, way around it is to relax the imposition of the mass fraction at the exit plane to be at one extreme or the other (ambient in the air column model and surface saturation in the point source model). That leads to Model 4.

13.2.5 Model 4: Matching the Air Column with the Point Source

This model matches Model 3 (point source) with Model 2. In order to escape from the free surface to the room, water vapor must first diffuse, with the help of a convective push, across the air cylinder to the exit plane. From there, it must diffuse away to ambient. Models 2 and 3 result in simple expressions for the evaporation rate, but the transfer number depends on the mass fractions at the boundaries of each model. The two mass transfer regions (inside the air cylinder and the spherical shell outside) are in series. Expressing the evaporation rate in terms of mass thermal resistances, first across the air column, and then for the point source outside the mug:

$$q_{evap} = \dot{m}_{evap}\, h_{fg} = \frac{Y_{v0} - Y_{vL}}{R_{m,0L}} = \frac{Y_{vL} - Y_{v\infty}}{R_{m,L\infty}}$$

Rearranging for the intermediate mass fraction (Y_{vL}) and dropping the "m" from the subscripts:

$$Y_{vL} = Y_{v\infty} + \frac{R_{L\infty}}{R_{0L} + R_{L\infty}}(Y_{v0} - Y_{v\infty})$$

The resistances cannot be evaluated completely from input variables. They depend on Y_{vL}, making the equation nonlinear, and difficult, if not impossible, to solve analytically. An iterative numerical method is viable, however. Forming a resistance ratio:

$$Y_{vL} = Y_{\infty} + \frac{1}{R_{0L}/R_{L\infty} + 1}(Y_{v0} - Y_{\infty})$$

Entering the appropriate Models 2 and 3 results and expressing S in terms of D:

$$\frac{R_{0L}}{R_{L\infty}} = \frac{\frac{L(Y_{v0}-Y_{vL})}{\rho D_a h_{fg} S \ln(1+B_{0L})}}{\frac{\sqrt{2}(Y_{vL}-Y_{v\infty})}{\pi D \rho D_a h_{fg} \ln(1+B_{L\infty})}} = \sqrt{8}\left(\frac{L}{D}\right)\left(\frac{\ln(1+B_{L\infty})}{\ln(1+B_{0L})}\right)\left(\frac{Y_{v0}-Y_{vL}}{Y_{vL}-Y_{v\infty}}\right)$$

An implicit expression for the intermediate mass fraction (i.e., at the exit plane) is:

$$Y_{vL} = Y_{\infty} + \frac{(Y_{v0}-Y_{\infty})}{\sqrt{8}\left(\frac{L}{D}\right)\left(\frac{\ln(1+B_{L\infty})}{\ln(1+B_{0L})}\right)\left(\frac{Y_{v0}-Y_{vL}}{Y_{vL}-Y_{v\infty}}\right)+1}$$

The transfer numbers both depend on Y_{vL}, making this equation nonlinear. The equation in this form is a suitable recursion relation for application of the method of successive substitution, however, in which case a value for Y_{vL} is guessed, followed by evaluation of the right hand side, whose result is used as the next guess. An iterative loop is run until the method either converges, or fails. If it fails, another method can be attempted.

Model 4 Results

Implementation of Model 4, the matched model, requires an iterative solution. It is cumbersome, though possible, to solve the transient coffee problem using Excel (with macros and/or user-defined functions) because an iterative loop is required for each time step. It is easier (but still requiring a macro) to write an excel file that merely calculates everything as a function of instantaneous coffee temperature. A macro-enabled excel file was written to do so. A macro was written that performs the copy and "paste as values" required to conduct an iteration. A test for convergence is conducted. However, the user is required to check to see if the solution is converged, and run the macro repeatedly until it does. Every time a change is made to an input variable (i.e., L), the iteration macro must be run.

Figure 13.13 shows the evaporation rate as a function of coffee temperature for several values of air column length (L) for the matched model (Model 4) and the point source model (Model 3, which is independent of L). This plot shows the same cases as for Fig. 13.12, where the air column model was compared with the point source model. In that latter case, the air column model predicted higher evaporation rates for nearly full mugs (L < 0.03 m). In the matched model, the point source model places an upper limit on the air cylinder. The problem from the straight air column model, where the evaporation increases dramatically L approaches zero is eliminated.

Model 4 is considered to be the most comprehensive model, yet it must be emphasized that it is based on a one-dimensional model, and the effects of natural

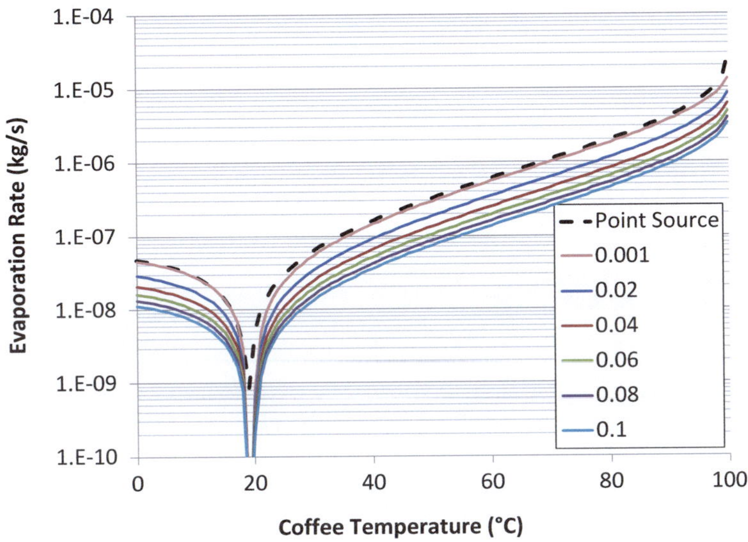

Fig. 13.13   Model 4 evaporation rate predictions for different air column lengths (in m)

convection currents would render that approximation questionable at best. Model 4 is therefore considered to somewhat underestimate the actual evaporation rates.

13.2.6 Comparison of Heat Transfer Coefficients

To this point, only the evaporation rate has been investigated from a purely mass transfer point of view. It is clear that the evaporation depends strongly on the instantaneous temperature of the liquid, driven by the highly nonlinear saturation pressure. What is not yet clear is whether the evaporation has a significant effect on the original problem of the cooling of a cup of coffee. From before, an evaporative heat transfer coefficient can be defined as:

$$h_{evap} = \frac{-\dot{m}_{evap}\, h_{fg}}{S(T_\infty - T_S)}$$

Inserting the evaporation rate expression:

$$h_{evap} = \left(\frac{\rho D_v\, h_{fg}}{L}\right)\frac{\ln(1 + B_{0L})}{(T_S - T_\infty)}$$

where, again,

$$B_{0L} = \frac{Y_{v0} - Y_{vL}}{1 - Y_{v0}}$$

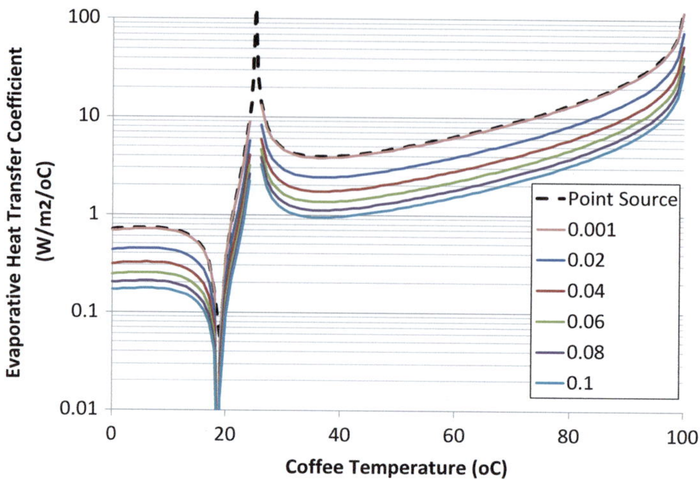

Fig. 13.14 Evaporative heat transfer coefficients predicted by Model 3 (point source) and 4 (matched) for various values of L (distance between coffee surface and exit plane)

Figure 13.14 shows the evaporative heat transfer coefficients plotted on semi-logy axes against coffee temperature for several values of the air column length (L) using the matched model (Model 4) and the point source model (Model 3). The evaporative heat transfer coefficient is very sensitive to temperature and exhibits some dramatic behavior when the coffee is close to ambient, to be discussed shortly.

Three parallel mechanisms for energy transfer from the free surface to ambient (radiation, convection, and evaporation) are compared in Fig. 13.15 for the base case (L = 0.05 m, a half-mug of coffee). The radiation model used assumes the free surface is completely surrounded visually by ambient. In reality, the inner walls of the mug will be somewhat higher, and the effective radiative heat loss will be lower than predicted by this model, so the radiation curve is an upper estimate. The convection model shown is based on published Nusselt number correlation for a heated horizontal plate. That is the model that would seem most applicable if the free surface were treated as a non-evaporating surface, and the mug were filled to the brim. The actual heat transfer rate by conduction to the surface would be lessened somewhat by the evaporative flow (especially near the boiling point), and also by the free surface being below the rim. So the convection curve is also an upper limit. On the other hand, the evaporation curve is an underestimate because use of the point source model as controlling the diffusion away from the mug is the slowest possible case, particularly when the coffee is hot and generating a strong vertical natural convective flow.

When the coffee is first poured, the evaporative heat loss is very high, and dominates the heat loss from the free surface. However, not modeled here, this effect would tend to lower the free surface temperature below the average coffee temperature, reducing the effects of evaporation. As the coffee continues to cool, evaporation effects drop dramatically and becomes a small contributor. The natural

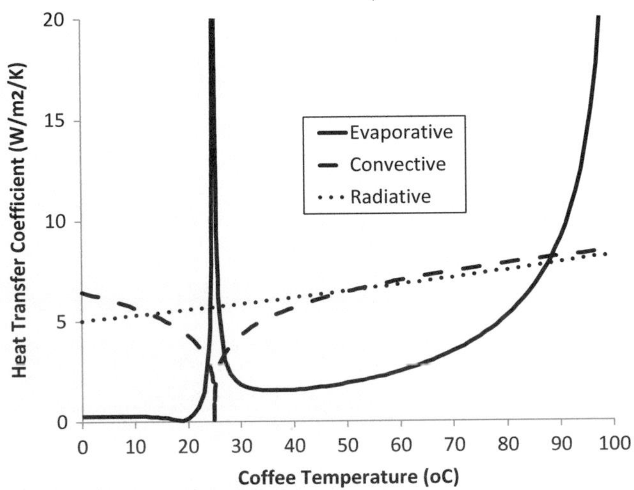

Fig. 13.15 Comparison of the evaporative, natural convective and radiative heat transfer coefficients associated with the coffee-free surface as a function of coffee temperature

convection and radiation models are equally important, except as the coffee approaches ambient, where the natural convection drops below to zero if the coffee temperature equals ambient.

Therefore, for the overall cooling of the coffee, evaporation will contribute to a high heat loss rate at the very beginning, at the same time the cold mug is being heated, and subsequently is noticeable, but not dominant until the coffee is nearly at room temperature.

Mathematically, the evaporative heat transfer coefficient is infinite when the coffee temperature equals ambient, but the heat flux is finite. Use of an evaporative heat transfer coefficient is reasonable when the coffee is cooling (or heating), and when the coffee is near ambient. The energy transfer (not heat) associated with evaporation is driven by diffusion, not by a temperature difference.

When the coffee is at the dew point temperature, the evaporation rate is identically zero (no concentration difference) and the evaporative heat transfer coefficient is zero. On the other hand, when the coffee temperature is precisely equal to ambient, the evaporation rate is finite, but the temperature difference is zero.

13.3 Evaporation of a Forgotten Cup of Coffee

After the coffee cools to room temperature, in several hours, the event is over, right? Wrong. The cooling event is over, but the coffee continues to evaporate. Unless the room is at 100 % relative humidity, the concentration of vapor just above the free surface will be higher than that in the room, and water vapor will continue

Fig. 13.16 Photo of
evaporation experiment.
A ceramic cup is initially
filled to the brim with water
and placed on a weighing
scale and left in a room,
maintained at nominal room
temperature. The mass and
distance from the free surface
are measured as functions
of time

to diffuse. Left undisturbed, the liquid will all eventually evaporate into the room.
The goal of this section is to predict the evaporation time and compare it to an
experiment.

13.3.1 Experiment

Figure 13.16 shows a photo of the evaporation experiment. A ceramic mug with an
inner diameter of approximately 6.2 cm and inner height of 10 cm is placed on a
food scale. Empty, the mug weighs 0.210 kg. Two hundred and fifty milliliters of
nominal room temperature water is added, filling the mug to the brim. The mug was
placed in a corner of a room (maintained nominally at 22 °C, during winter, when
the inside relative humidity is expected to be low) and forgotten about, except for
occasionally, when a data point (mass and distance) was collected. The ambient
conditions were not well controlled.

The experimental results are shown in Fig. 13.17. The length was measured
periodically for some 63 days. The data is plotted and a power law fit applied. The
fit is extrapolated until the length equals the depth of the cup (0.1 m), which shows
105 days to complete evaporation. Unfortunately, the experimentalist forgot to take
data after day 63. . . .

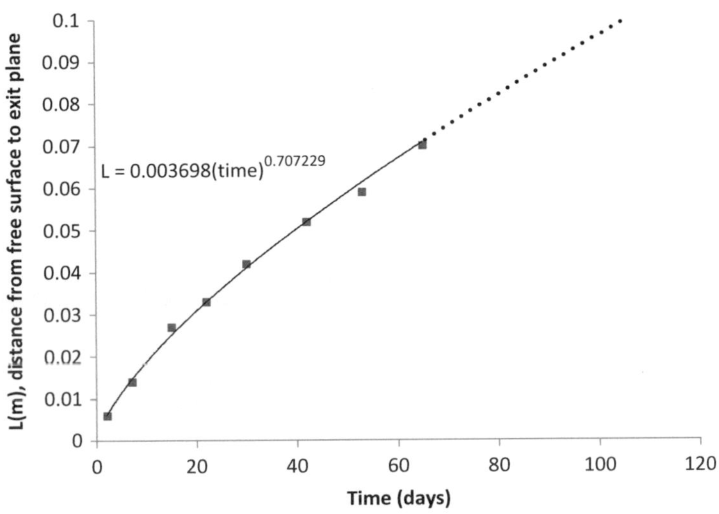

Fig. 13.17 Measured distance vs. time. A power law fit is applied, and the *dotted line* extrapolates until complete evaporation (L = H = 0.1 m). A total evaporation time of 105 days is predicted

INPUT PARAMETERS

Pinf	1.01E+05	N/m2	atmospheric pressure
Tinf	22	oC	ambient temperature
H	0.1	m	inner height of mug
D	0.056	m	inner diameter of mug
rh	0.3	--	relative humidity

PHYSICAL CONSTANTS

rholiq	1000	kg/m^3	liquid density
Dv	2.56E-05	m^2/sec	mass diffusivity of vapor
hfg	2.33E+06	J/kg	heat of vaporization
A	9.1496E+10	N/m2	CLAUSIUS-CLAPYRON CONSTANTS FOR WATER
B	-5.1152E+03	K	$P_{sat}(T) = A e^{\frac{B}{(T+273)}}$

DERIVED PARAMETERS

rhogas	1.20	kg/m^3	gas density
Xvinf	0.0080	molV/mol	mol fraction of vapor in ambient
Yvinf	0.0051	massV/mass	mass fraction of vapor in ambient
X0	0.0266	molV/mol	mol fraction of vapor on free surface
Y0	0.0167	massV/mass	mass fraction of vapor on free surface
B0inf	1.17E-02		Transfer number, surface to ambient

Fig. 13.18 Input and derived parameter values for evaporation experiment

13.3.2 Theory

Three different models will be developed and a case study run based on the input parameters shown in Fig. 13.18, which correspond loosely to the experimental conditions. For derived parameters, the gas density is obtained from the ideal gas law. The mole fraction of vapor in the ambient room is given by the relative humidity times the saturation pressure (from the Clausius–Clapeyron fit) divided by the total pressure. The mass fraction is obtained by the conversion derived in the Appendix. The free surface is assumed to be at ambient temperature for this case. A calculation will show that the free surface is theoretically colder than ambient (because of evaporative cooling), but the heat transfer response time is so much faster than the evaporation time, that an isothermal situation is an excellent approximation. This approximation would have to be revisited if the ambient temperature were close to the boiling point. The mole and mass fractions of vapor at the free surface, then, represent the saturation state at ambient temperature. The mass fraction at the free surface is 0.0167, showing that the partial pressure of water is only about 1.7 % of the total, the balance being air. This behavior is referred to as the "dilute gas approximation," where the partial pressure exerted by the vapor is much less than the total pressure. This behavior holds when the ambient temperature is sufficiently far from the boiling point, as in this case.

The overall transfer number, $B_{0\infty}$, is defined by:

$$B_{0\infty} = \frac{Y_0 - Y_\infty}{1 - Y_0}$$

This transfer number represents the maximum it could possibly be and has a value of 0.0117, which shows that the low B limit model applies, that is, for the transfer number defined between generic surfaces A and B:

$$\ln(1 + B_{0\infty}) \approx B_{0\infty} = \frac{Y_0 - Y_\infty}{1 - Y_0}$$

In addition, the air is always dilute in the vapor $(Y \ll 1)$, and therefore the denominator is approximately equal to unity. Therefore, the low B number, dilute system approximation is:

$$B_{0\infty} \approx Y_0 - Y_\infty$$

Different evaporation models are developed shortly that all result in a B number, but with different assumptions about the mass fraction values.

13.3.3 System Mass Balance

The goal of this study is to model the receding surface level, and therefore the distance between the free surface and the exit plane is not a constant. A quasi-equilibrium approximation is used, however, in which the rate of evaporation is determined at any time as if that length were constant.

Application of the conservation of mass principle to the system defined (as before) to include the mug and the coffee, as a word statement, is:

$$\begin{pmatrix} \text{Rate of Change} \\ \text{of Mass Stored} \end{pmatrix} = \begin{pmatrix} \text{Mass Flow} \\ \text{Rate In} \end{pmatrix} - \begin{pmatrix} \text{Mass Flow} \\ \text{Rate Out} \end{pmatrix}$$

Introducing appropriate symbols, recognizing that there are two subcomponents, mug and coffee, no mass inflow, and the exiting mass flow is the evaporation rate:

$$\frac{d\left(M_{coffee} + M_{mug}\right)}{dt} = 0 - \dot{m}_{evap}$$

Since the mug is not evaporating, the rate of change of the mass of the mug is zero, and the mass balance becomes:

$$\frac{dM_{coffee}}{dt} = -\dot{m}_{evap}$$

Expressing the coffee mass as the liquid density (ρ_{Liquid}) times the instantaneous coffee volume (in terms of the length of the air column, L):

$$\frac{d\left(\rho_{Liquid}S(H - L)\right)}{dt} = -\dot{m}_{evap}$$

Simplifying and rearranging:

$$\frac{dL}{dt} = \frac{\dot{m}_{evap}}{\rho_{Liquid}S}$$

The evaporation rate depends on the model used. The rate of change of the coffee surface is related to the rate of evaporation and depends inversely on the density of the liquid and the free surface area (S). This expression will be revisited after developing models for the evaporation rate. From experience, this height should not change very much during the initial cooling process. It does change if left unattended a few days.

Early Time Model: L ≪ D

Starting with a completely full mug, the free surface "sees" a nominal hemispherical air environment around it, with little movement. Modeling the evaporation as a diffusion process from a hemispherical source, with a surface area equal to that of the coffee-free surface, the differential equation for L is:

$$\frac{dL}{dt} = \frac{\dot{m}_{evap}}{\rho_{Liquid}S} = \frac{\frac{\pi}{\sqrt{2}}D\rho_{gas}D_v \ln(1 + B_{L\infty})}{\rho_{Liquid}\,{\pi D^2}/{4}}$$

Rearranging:

$$\frac{dL}{dt} = \frac{\sqrt{8}\rho_{gas}D_v\ln(1 + B_{L\infty})}{\rho_{Liquid}D}$$

The transfer number for this case is:

$$B_{L\infty} = \frac{Y_L - Y_\infty}{1 - Y_L}$$

Y_L represents the mass fraction of vapor at the exit plane. When the mug is nearly full, the free surface is near the exit plane, and the mass fraction at the exit plane is assumed equal to the mass fraction on the free surface in this model limit:

$$Y_L \approx Y_0$$

The transfer number becomes equal to the transfer number applied to the system, from the free surface to ambient:

$$B_{L\infty} \approx B_{0\infty} = \frac{Y_0 - Y_\infty}{1 - Y_0}$$

Furthermore, the low B number limit for dilute gas systems is applied and the differential equation for air column length becomes:

$$\frac{dL}{dt} = \frac{\sqrt{8}\rho_{gas}D_v(Y_0 - Y_\infty)}{\rho_{Liquid}D}$$

The velocity of the receding free surface (dL/dt) is a constant in this model, and the equation can be integrated directly:

$$L = \left[\frac{\sqrt{8}\rho_{gas}D_v(Y_0 - Y_\infty)}{\rho_{Liquid}D}\right]t + L_0$$

L_0 is the length of the air column at the start, which will be taken to be zero (a full mug). The total evaporation time (t_{evap}) can be calculated by setting the length of the air column equal to the inner height of the mug (H):

$$t_{evap,LowL} = \frac{\rho_{Liquid}D(H - L_0)}{\sqrt{8}\rho_{gas}D_v(Y_0 - Y_\infty)}$$

Inserting values for the case study considered:

$$t_{evap,LowL} = \frac{\left(1,000\frac{kg}{m^3}\right)(0.563\,m)(0.1-0)m}{\sqrt{8}\left(1.22\frac{kg}{m^3}\right)\left(2.56(10)^{-5}\,\frac{m^2}{s}\right)(0.0266-0.0167)}$$

$$t_{evap,LowL} = 5.46(10)^6 s * \left[\frac{h}{3,600\,s}\right] * \left[\frac{day}{24\,h}\right]$$

$$t_{evap,LowL} = 63\,days$$

This model predicts the complete evaporation of water from the mug will be about 2 months. It is about 40 days shorter than measured.

Late Time Model: $L \gg D$
Once the free surface recedes below the exit plane of the mug, the diffusion of water vapor through the air column between the free surface and the exit plane is assumed to become the basic means of evaporation. Inserting the evaporation model for this diffusion process into the derivative for L:

$$\frac{dL}{dt} = \frac{\dot{m}_{evap}}{\rho_{Liquid}S} = \frac{\frac{\rho_{gas}D_v S\ln(1+B_{0L})}{L}}{\rho_{Liquid}S}$$

Rearranging:

$$\frac{dL}{dt} = \left[\frac{\rho_{gas}D_v\ln(1+B_{0L})}{\rho_{Liquid}}\right]\frac{1}{L}$$

The transfer number for this case is:

$$B_{0L} = \frac{Y_0 - Y_L}{1 - Y_0}$$

Y_L represents the mass fraction of vapor at the exit plane. When the mug is nearly empty, the free surface is far from the exit plane, and the mass fraction at the exit plane is assumed equal to the mass fraction of ambient air in this model limit:

$$Y_L \approx Y_\infty$$

The transfer number becomes equal to the transfer number applied to the system, from the free surface to ambient:

$$B_{0L} \approx B_{0\infty} = \frac{Y_0 - Y_\infty}{1 - Y_0}$$

Furthermore, the low B number limit for dilute gas systems is applied, and the differential equation for air column length becomes:

$$\frac{dL}{dt} = \left[\frac{\rho_{gas} D_v (Y_0 - Y_\infty)}{\rho_{Liquid}} \right] \frac{1}{L}$$

In this model, the speed of the receding free surface (dL/dt) depends inversely on the instantaneous value of the length of the air column (L). As the surface slowly recedes, the rate of diffusion from the free surface to the exit plane decreases due to the reduced concentration gradient that drives the mass transfer.

Separating variables (with the factor 2 applied for later convenience) and applying limits of integration:

$$\int_{L=L_0}^{L} 2L \, dL = \int_{t=0}^{t} \left[\frac{2\rho_{gas} D_v (Y_0 - Y_\infty)}{\rho_{Liquid}} \right] dt$$

$$L^2 - L_0^2 = \left[\frac{2\rho_{gas} D_v (Y_0 - Y_\infty)}{\rho_{Liquid}} \right] t$$

This model results in the so-called L^2 Law, which has an important counterpart in the evaporation of liquid droplets. L_0 is the length of the air column at the start, which will be taken to be zero (a full mug). The total evaporation time (t_{evap}) can be calculated by setting the length of the air column equal to the inner height of the mug (H):

$$t_{evap, \, Large \, L} = \frac{\rho_{Liquid} \left(H^2 - L_0^2 \right)}{2\rho_{gas} D_v (Y_0 - Y_\infty)}$$

Inserting values for the case study considered:

$$t_{evap, \, Large \, L} = \frac{\left(1{,}000 \frac{kg}{m^3} \right) \left(0.1^2 - 0 \right) m^2}{2 \left(1.22 \frac{kg}{m^3} \right) \left(2.56(10)^{-5} \frac{m^2}{s} \right) (0.0266 - 0.0167)}$$

$$t_{evap, \, Large \, L} = 1.37(10)^7 s * \left[\frac{h}{3{,}600 \, s} \right] * \left[\frac{day}{24 \, h} \right]$$

$$t_{evap, \, Large} = 158 \, days$$

This model predicts the complete evaporation of water from the mug will be about 5 months, more than twice that predicted from the point source model.

Matching Model
In both the early and late evaporation models, the mass fraction of water vapor at the exit plane was assumed to be at one extreme (free surface) or the other

(ambient). In this final model, these assumptions will be relaxed, and an expression for the evaporation rate that applies for the entire evaporation event is developed. To do so, the evaporation rate is expressed in both forms:

$$\dot{m}_{evap} = \frac{\pi}{\sqrt{2}}\, D\rho_{gas}D_v \ln(1 + B_{L\infty}) = \frac{\rho_{gas}D_v S \ln(1 + B_{0L})}{L}$$

Simplifying and rearranging:

$$\sqrt{8}\left(\frac{L}{D}\right)\ln(1 + B_{L\infty}) = \ln(1 + B_{0L})$$

Applying the low B number limit for dilute gas systems:

$$\sqrt{8}\left(\frac{L}{D}\right)(Y_L - Y_\infty) = (Y_0 + Y_L)$$

Solving for Y_L:

$$Y_L = \frac{Y_0 + \sqrt{8}\left(\frac{L}{D}\right)Y_\infty}{1 + \sqrt{8}\left(\frac{L}{D}\right)}$$

Notice that in the limit $L/D \to 0$, the mass fraction at the exit plane is the free surface value (Y_0), and in the $L/D \to \infty$, the mass fraction at the exit plane is the ambient value (Y_∞). The evaporation rate, using the air cylinder model with the low B, dilute gas approximations is:

$$\dot{m}_{evap} = \frac{\rho_{gas}D_v S \ln(1 + B_{0L})}{L} \approx \frac{\rho_{gas}D_v S(Y_0 - Y_L)}{L}$$

Substituting for Y_L:

$$\dot{m}_{evap} \approx \frac{\rho_{gas}D_v S\left(Y_0 - \left(\frac{Y_0 + \sqrt{8}\left(\frac{L}{D}\right)Y_\infty}{1 + \sqrt{8}\left(\frac{L}{D}\right)}\right)\right)}{L}$$

$$\dot{m}_{evap} = \frac{\sqrt{8}\rho_{gas}D_v S(Y_0 - Y_\infty)}{D\left(1 + \sqrt{8}\frac{L}{D}\right)} = \frac{\rho_{gas}D_v S(Y_0 - Y_\infty)}{\left(D/\sqrt{8} + L\right)}$$

Now, to enter this expression into the mass balance equation:

$$\frac{dL}{dt} = \frac{\dot{m}_{evap}}{\rho_{Liquid}S} = \frac{\dfrac{\rho_{gas}D_v S(Y_0 - Y_\infty)}{\left(D/\sqrt{8} + L\right)}}{\rho_{Liquid}S}$$

Separating variables, and applying integration limits, setting L_0 to zero (to simplify subsequent algebra):

$$\int_{L=0}^{L} \left(\frac{D}{\sqrt{8}} + L\right) dL = \int_{t=0}^{t} \left[\frac{\rho_{gas} D_v (Y_0 - Y_\infty)}{\rho_{Liquid}}\right] dt$$

Performing the integrations results in a quadratic equation for the air cylinder length (L) as a function of time:

$$\frac{1}{2}L^2 + \frac{D}{\sqrt{8}}L - \left[\frac{\rho_{gas} D_v (Y_0 - Y_\infty)}{\rho_{Liquid}}\right] t = 0$$

Solving for L:

$$L = -\frac{D}{\sqrt{8}} + \sqrt{\frac{D^2}{8} + \left[\frac{2\rho_{gas} D_v (Y_0 - Y_\infty)}{\rho_{Liquid}}\right] t}$$

The total evaporation time is obtained by setting L to H:

$$t_{evap,Matched} = \frac{\rho_{Liquid} \left(\frac{H^2}{2} + \frac{DH}{\sqrt{8}}\right)}{\rho_{gas} D_v (Y_0 - Y_\infty)}$$

Rearranging:

$$t_{evap,Matched} = \frac{\rho_{Liquid} H^2 \left(1 + \frac{D}{H\sqrt{2}}\right)}{2\rho_{gas} D_v (Y_0 - Y_\infty)}$$

Notice that this result reduces to the point source model ($L \ll D$) when $D \gg H$ and to the air cylinder model ($L \gg D$) when $H \gg D$. Evaluating for the case study:

$$t_{evap,Matched} = \frac{\left(1{,}000\frac{kg}{m^3}\right)(0.1\,m)(0.1\,m + 0.056\,m)}{2\left(1.22\frac{kg}{m^3}\right)\left(2.56(10)^{-5}\,\frac{m^2}{s}\right)(0.0266 - 0.0167)}$$

$$t_{evap,Matched} = 1.95(10)^7 s * \left[\frac{h}{3{,}600\,s}\right] * \left[\frac{day}{24\,h}\right]$$

$$t_{evap,Matched} = 225\,days$$

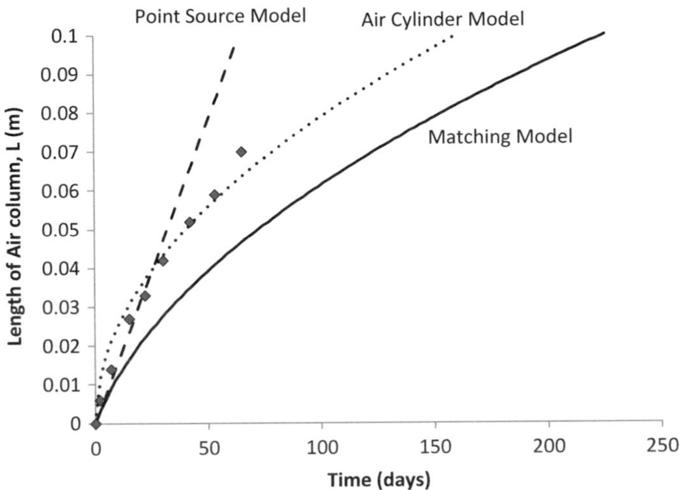

Fig. 13.19 Air column length vs. time. Model predictions and experimental data

13.3.4 Results and Discussion

Figure 13.19 shows the length L as a function of time for the three models, and with the experimental data. For the point source model, the air cylinder length increases linearly, while the mass (not shown) decreases linearly. When all of the water is evaporated (L = H), the event is over. The air diffusion cylinder model predicts an initial vertical slope and subsequently the slope decreases with time as the concentration gradient weakens. If the mug in question were twice as deep as the one used, the difference between the point source and air diffusion cylinder would be much more dramatic. The final model matches the air cylinder to the point source. The simulation follows the point source model initially, and as time proceeds, gradually takes the shape of the air cylinder model, shifted in time to account for the initial period where the model breaks down.

The experimental points appear to follow the air cylinder model most closely. However, the functional dependence of length with time is different. Figure 13.20 shows the three theoretical curves plotted, and then a power law fit to them. The point source and air cylinder models exhibit the linear and square root dependencies as expected. However, the matched model, whose solution does not conform to a simple power law fit, exhibits a power law exponent of 0.69, which agrees well with the experimental value of 0.707 (from Fig. 13.17). Therefore, it is concluded that the more involved matched model exhibits a similar functional dependency as the experiment, but an overall slower evaporation rate.

Figure 13.21 shows the mass fraction at the exit plane (Y_L) as a function of time. The point source model forces the mass fraction to be the saturated surface value. The air cylinder forces it to be the ambient value. For the matching model, the exit

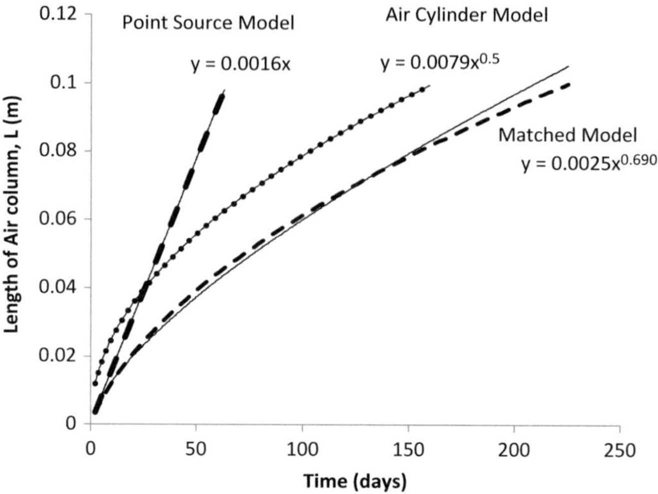

Fig. 13.20 Model predictions, with power law fits. For each model, the *dashed* or *dotted line* is the model prediction, the *light line* is a power law fit, with the equation shown

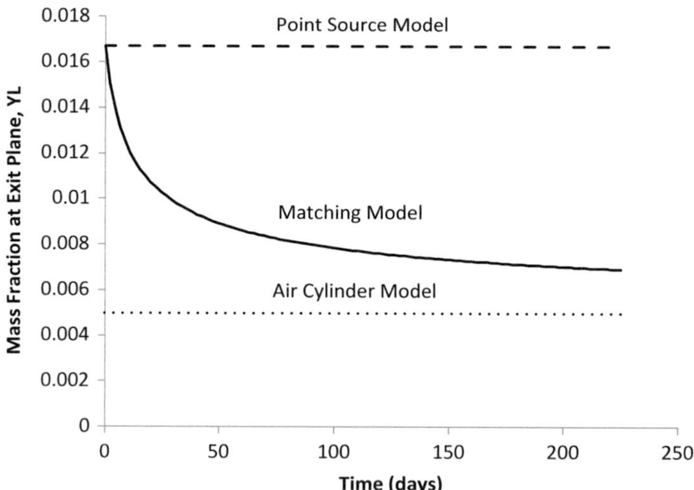

Fig. 13.21 Predicted mass fraction at the exit plane for the three evaporation models as a function of time

plane is intermediate, and the mass fraction there gradually varies as the instantaneous length to diameter ratio changes. The mass fraction drops rapidly and very gradually approaches the ambient value. Note that this particular mug has a modest L/D ratio of 1.8.

The experiment shows a more rapid evaporation process than predicted by the matching model for the assumed experimental conditions. However, the theoretical

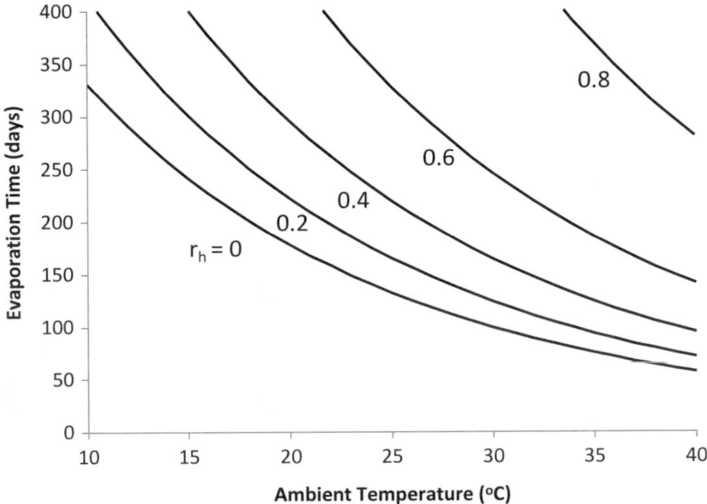

Fig. 13.22 Evaporation time as a function of ambient temperature and relative humidity for the matched model

prediction is very sensitive to the assumed ambient temperature and humidity, as shown in Fig. 13.22, which shows the predicted evaporation time as a function of ambient temperature for various relative humidity values.

Perhaps a bigger deviation from the model would be due to air currents in the room, both induced by the natural uncontrolled movements in the room, as well as those created by the mug of water itself. Figure 13.23 shows a suggested mechanism. Since the molecular weight of water is less than that of air, the density of the humid air inside the mug is less than that of the room. The effect is small, from the point of view of generating an updraft, but any updraft would have a profound effect on the concentration field, and thus evaporation rates. The point source model is one which would be appropriate for a completely still room, in the absence of gravity. So it is reasonable to expect somewhat faster evaporation than predicted by these models, whose main goal is to demonstrate the fundamental processes associated with mass transfer.

13.3.5 Surface Temperature

The final resistance network of this textbook is one that shows that evaporation causes the surface temperature of the evaporating mug of coffee, left unattended in a room, to be below the ambient temperature, in apparent violation of the Second Law. Figure 13.24 shows a resistance network to demonstrate the basic heat and mass transfer channels. There are no capacitances shown, as this situation is after

Fig. 13.23 Suggested mechanism for enhanced evaporation rates compared to the one-dimensional models

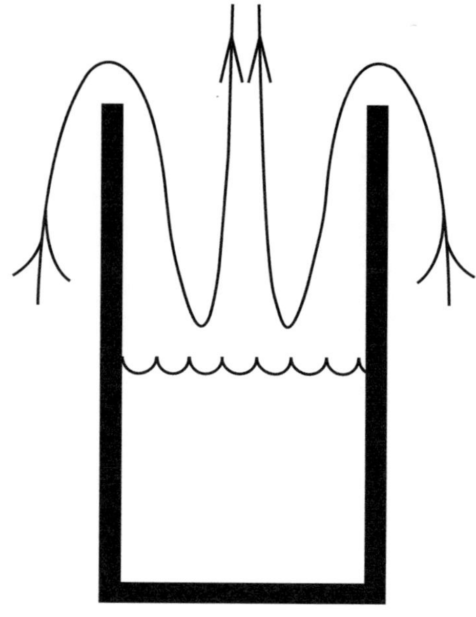

Fig. 13.24 Resistance network for estimating the coffee and surface temperature of an evaporating mug of coffee

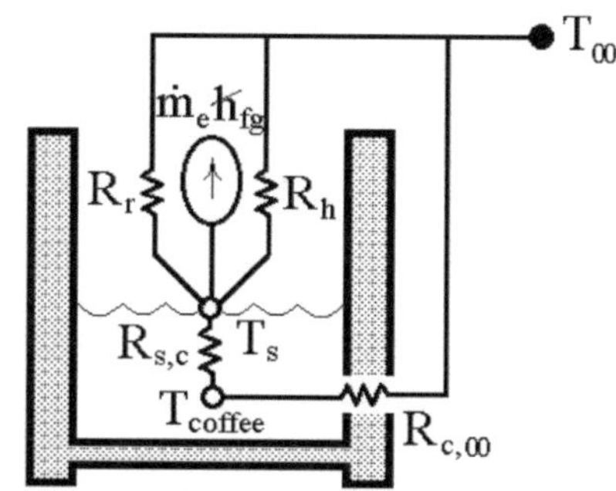

the cooling period is over, and any transient effects are considered to be unimportant, an assertion that could be justified with a full transient analysis.

The free surface of the coffee experiences three heat transfer channels: convective (R_h) and radiative (R_r) parallel channels directly to ambient and a convective channel to the bulk of the coffee ($R_{c,s}$). The coffee in turn is in thermal contact with ambient ($R_{c,\infty}$ which is a complex channel through the wall and to ambient). The evaporation process is shown as a sink, since it is mass transfer driven (not temperature). It is this evaporation process that acts as a cooling mechanism for the surface, which must be below ambient in order for heat to flow into it.

A nodal balance on the free surface yields the temperature of the free surface, noting that the series channel from ambient through the coffee is used:

$$T_s = \frac{T_\infty \left(\frac{1}{R_r} + \frac{1}{R_h} + \frac{1}{R_{c,\infty}+R_{s,c}} \right) - \dot{m}_e \, h_{fg}}{\left(\frac{1}{R_r} + \frac{1}{R_h} + \frac{1}{R_{c,\infty}+R_{s,c}} \right)}$$

This equation is implicit, since the evaporation rate is a function of the surface temperature. Once thermal resistances are approximated, this equation is in a form suitable for the Method of Successive Substitutions.

The coffee temperature, being a voltage divider between the surface and ambient can be obtained next, as:

$$T_{coffee} = \frac{\frac{T_S}{R_{S,C}} + \frac{T_\infty}{R_{C,\infty}}}{\frac{1}{R_{S,C}} + \frac{1}{R_{C,\infty}}}$$

The effect can be enhanced by using an insulated mug, in which case $R_{c,\infty}$ can be made large compared to all other thermal resistances. The limit as $R_{c,\infty}$ goes to infinity is:

$$T_s = T_\infty - R_{eq,s\infty} \dot{m}_e \, h_{fg}$$

$$T_{coffee} = T_s$$

where the equivalent resistance between the surface and ambient is:

$$\frac{1}{R_{eq,s\infty}} = \frac{1}{R_r} + \frac{1}{R_h}$$

Numerical determination of this effect is expected to show a small, perhaps immeasurable temperature drop, and is reserved for the curious reader to tie together in a Workshop.

Workshop 13.1. Evaporation Experiments

During the first week of class, procure samples of clear tubing. Choose samples with a much larger length to diameter ratio than the value of 1.8 investigated in this chapter. Mount them vertically with an open end up top, and sealed at the bottom. Mark them with a scale. Find locations for them where they will not be disturbed, fill them to the top with water (eliminate all air bubbles trapped), and start recording L vs. t. Consider using different liquids, and note that if liquid mixtures are used with components of different volatility (i.e., isopropyl alcohol/water mixtures), that the composition will change with time. Modeling this effect requires some basic understanding of distillation-type processes. Compare results to theoretical predictions by the end of the semester.

Workshop 13.2. Surface Temperatures During Evaporation

Calculate the surface and coffee temperatures for an evaporating mug of coffee as a function of ambient temperature (assume zero relative humidity). Demonstrate that the effect becomes very large as the temperature approaches the boiling point of the liquid. Use the point source model as giving the largest effect modeled here. Alternatively, use published correlations for the mass transfer coefficients.

Workshop 13.3. A Cooling Mug of Coffee with Evaporation Effects

Include a lumped capacitance to the coffee node of Fig. 13.24 and include the effects of evaporation onto the basic cooling response of the coffee. Use a well-insulated mug and treat the thermal resistance between coffee and ambient ($R_{c,\infty}$) as a constant. Use either an evaporative heat transfer coefficient (h_{evap}) or directly include the evaporation as a sink. Implement an implicit Euler method in which the surface temperature, evaporation rates, radiation, and natural convection are all calculated (iteratively) at each time step. Start with the coffee at its boiling point, and notice how much lower the surface temperature is than the average coffee temperature early on. Plot the dimensionless temperature vs. time on semi-logy axes to observe deviations from first-order behavior.

Appendix. Evaluating Dew Points and Mass Fractions

An important detail in evaluating evaporation rates is to relate the mass fractions in terms of input variables, which requires a little more background into modeling gas mixtures. The gas phase is modeled as being a mixture of two ideal gases, air and water vapor.

For humid air at temperature T, the partial pressure of water is given by:

$$p_V = r_h p_{sat}(T)$$

where r_h is the relative humidity.

The dew point temperature ($T_{DewPoint}$) of air with a relative humidity (r_H) at an ambient temperature (T_∞) is defined by the temperature at which the partial pressure of saturated vapor equals the partial pressure of vapor in the humid (but unsaturated) air. That is:

$$p_{sat}(T_{DewPoint}) = r_h p_{sat}(T_\infty)$$

The relationship between the saturation pressure and the temperature can be modeled using a curve fit of the form given by the Clausius–Clapeyron relation, namely:

$$p_{sat}(T) = A e^{\frac{B}{T}}$$

where A and B are constants for a given species, and T is the absolute temperature. The values of A and B can be calculated by fitting two points taken from saturation tables in the vicinity of the problem in question. This curve for water was plotted in Fig. 13.2.

The partial pressure of water vapor is related to the concentration of water vapor. For n_v moles of water vapor at absolute temperature T occupying a volume V, the ideal gas law is:

$$p_v V = n_v \hat{R} T$$

$$\hat{R} = 8,314 \, \frac{\text{J}}{\text{kmol K}}$$

is the universal gas constant. Multiplying and dividing by the total number of moles and rearranging:

$$p_v = \left(\frac{n_v}{n}\right)\left(\frac{n\hat{R}T}{V}\right) = X_v P$$

where P is the total pressure (the sum of all partial pressures). The mole fraction of vapor, X_v is the number of moles of vapor divided by the total number of moles in the gas mixture. The mass fraction and mole fractions are related to each other through the molecular weights of the mixture and vapor by the following word string:

$$\left(\frac{\text{Mass of Vapor}}{\text{Mass of Mixture}}\right) = \left(\frac{\text{Moles of Vapor}}{\text{Moles of Mixture}}\right)\left(\frac{\text{Moles of Mixture}}{\text{Mass of Mixture}}\right)\left(\frac{\text{Mass of Vapor}}{\text{Moles of Vapor}}\right)$$

$$Y_v = (X_v)\left(\frac{1}{\text{MW}_{mix}}\right)(\text{MW}_{vapor})$$

The molecular weight of the mixture is obtained from the following word string:

$$\left(\frac{\text{Mass of Mixture}}{\text{Moles of Mixture}}\right) = \left(\frac{\text{Mass of Vapor} + \text{Mass of Air}}{\text{Moles of Mixture}}\right)$$

$$\left(\frac{\text{Mass of Mixture}}{\text{Moles of Mixture}}\right) = \left(\frac{\text{Moles of Vapor}}{\text{Moles of Mixture}}\right)\left(\frac{\text{Mass of Vapor}}{\text{Moles of Vapor}}\right)$$

$$+ \left(\frac{\text{Moles of Air}}{\text{Moles of Mixture}}\right)\left(\frac{\text{Mass of Air}}{\text{Moles of Air}}\right)$$

$$MW_{mix} = X_{vapor}MW_{vapor} + X_{air}MW_{air}$$

$$Y_v = (X_v)\left(\frac{MW_{vapor}}{X_v MW_{vapor} + (1 - X_v)MW_{air}}\right)$$

$$Y_v = \frac{1}{1 + \left(\frac{1-X_v}{X_v}\right)\left(\frac{MW_{air}}{MW_{vapor}}\right)}$$